Urbanisation, Land Use, Land Degradation and Environment

Urbanisation, Land Use, Land Degradation and Environment

— Editors —
Munir Ozturk
Ahmet Ruhi Mermut
Ali Celik

DENIZLI MUNICIPILITY,
DENIZLI, TURKEY

PAMUKKALE UNIVERSITY,
DENIZLI, TURKEY

SCIENTIFIC AND TECHNOLOGICAL RESEARCH
COUNCIL OF TURKEY (TÜBİTAK),
ANKARA, TURKEY

CENTRE FOR SCIENCE & TECHNOLOGY OF THE
NON-ALIGNED AND OTHER DEVELOPING COUNTRIES
(NAM S&T CENTRE)

2011
DAYA PUBLISHING HOUSE®
Delhi - 110 002

ISBN 9789351242222

Centre for Science and Technology of the Non-Aligned and Other Developing Countries (NAM S&T Centre)
Core-6A, 2nd Floor, India Habitat Centre, Lodhi Road,
New Delhi-110 003 (India)
Phone: +91-11-24644974, 24645134, Fax: +91-11-24644973
E-mail: *namstct@gmail.com*
Website: *www.namstct.org*

Published by : **Daya Publishing House**
A Division of
Astral International Pvt. Ltd.
– ISO 9001:2008 Certified Company –
4760-61/23, Ansari Road, Darya Ganj,
New Delhi - 110 002
Phone: 23245578, 23244987
Fax: (011) 23260116
e-mail : *dayabooks@vsnl.com*
website : *www.dayabooks.com*

Laser Typesetting : **Classic Computer Services**
Delhi - 110 035

Printed at : **Chawla Offset Printers**
Delhi - 110 052

PRINTED IN INDIA

Message

In the last days of September 2009, more than 200 scientists assembled at Pamukkale University for an International Workshop on "Urbanization, Land Use, Land Degradation and Environment". The University put lots of effort in collaboration with TÜBİTAK and NAM S &T CENTRE for a successful realization of this event. The theme of this workshop was highly topical aiming at promoting scientist-to-scientist contacts, interactions, exchanges and sharing of information on various aspects of the land and its use.

Land as a part of our culture provides distinctive networking in the community. Land use and land management practices have a major impact on natural resources including water, soil, nutrients and plants, and the living beings. Land degradation affects the earth's productivity. Main reasons for this are natural ones but, it also occurs where there is a deficiency in land use planning or in its orderly execution, or the existence of financial or legal incentives leading to the wrong land use decisions, or one-sided central planning leading to over-utilization of the land resources. This creates a bad situation for a major part of the local population and destroys the valuable ecosystems. To overcome this problem we need to replace this situation by using integrated and holistic techniques for the planning and management of land resources, which will ensure long-term quality of our land, prevention of social conflicts related to land use, and the conservation of ecosystems of high biodiversity. Moreover, planning of a viable land use programme requires an appreciation of the historical past, as well as an analysis of the environment in which it is supposed to operate.

Urbanization is the physical growth of urban areas as a result of population migration from rural to the urban areas. The outcome is a change in population density and pressure on administration services. The expansion of urban areas and infrastructure is encroaching on productive land and natural habitats throughout the world. This encroachment at the same time results in problems like pollution, especially in urban areas, which is one of the biggest challenges the world faces today affecting us economically, physically and in our daily lives. Water pollution causes contamination of rivers, streams, lakes, underground water and larger water masses thus killing plants and animals as well as severe health problems in humans. Soil pollution contaminates our lands with hazardous wastes and other substances which do not belong to it. Thermal and heat pollution is a cause of greenhouse effect and global warming. All these pollution types affect the non-living and living components of life, which disturbs the ecological balance and thereby leads to a variety of detrimental conditions.

The overall programme of the workshop was organized in three parallel "Technical Sessions" and a joint "Concluding Session". The two technical sessions were primarily for the Turkish speakers and audience. One session was mainly for the foreign delegates sponsored by the NAM S&T Centre and a large number of Turkish and other foreign speakers.

After full 4 days of presentations and fruitful discussions, Pamukkale declaration was prepared for onward submission to the governing bodies in NAM member countries as well as associate members in other countries.

The presentations in the workshop collected, selected and published in this book include discussions which will go beyond the boundaries of Pamukkale University. The chapters will pave the way for new discussions, research projects for adopting a sustainable attitude towards our nature.

It was a great honor for us to host the conference and meet leading scientists from different counties. I am sure that, we will have new opportunities to see all of them and many others in our university again. I want to thank our organization committee and coordinators for their great effort. They did their best for the success of this workshop. Our special thanks go to Prof. Dr. Nüket Yetiþ, President of "TÜBİTAK" for her support in the organization of this international event. Also I won't forget Prof Arun Kulshreshtha. He supported us continuously with his kind and experienced guidance.

Nihat ZEYBEKCİ

Grand Mayor of Denizli

Foreword

Urbanization concept, which is a consequence of rapid industrialization and population growth, has caused many problems in our country as well as in the world. In addition to socio- economic reasons such as stress, noise, hard economical conditions; we all know that urbanization has negative effects on the environment such as air pollution, water pollution and soil pollution.

The close relationship between unplanned rapid urbanization and environmental problems is holding the first place in the list of problems to be solved while planning for our future. We can indicate that poor management of population growth and urbanization process cause our daily life troubles and environmental health problems.

In urbanization, for both infrastructure and superstructure, the most common term is "Social Density". It includes measuring water passing through a pipe length per unit and calculation of structures depending on it or built in a unit area which is worked out according to the expected population of the area. However, one of the main problems in our cities is the density problem. In other words, if more people live in unit area than expected to be and infrastructure services, green spaces, air, water, soil, etc. become insufficient, we can say that the city is poorly managed.

Urbanization is an incoming and ongoing process. At the end of this process, it a literal urbanization is desired to be mentioned, physical, cultural and social changes have to be realized at the same time. In the lack of any one of these elements, a literal urbanization cannot be mentioned for a city.

In order to prevent negative effects of urbanization for environment and socio-economic life, necessary measures should be taken. This topic assigns major responsibilities to central and especially local administrations which work for the future of the country and the city.

In view the gravity of the hazards of rapid urbanisation and intense land use leading to immense damage to humans, more efforts and investments are needed to learn from land degradation. There is thus an urgent need for proper planning in urbanisation, land use planning practices, dissemination of knowledge and public awareness. A proper planning and projection configures urbanization process in a suitable way. If we keep on working with environmental consciousness, as a legacy to future generations, we can leave a well planned, organized, safe and livable city which has clean and pure water sources.

The Mayorship of Denizli would like to convey their thanks to the Non-Aligned and Other Developing Countries Science and Technology Centre (NAM) for organising the International Workshop under question on "Urbanisation, Landuse, Land Degradation and Environment" in our city. Our special thanks go to Pamukkale University, Turkish Scientific and Technical Research Organisation (TUBITAK) and Ministry of Environment and Forests for their collaboration in making this important event a success.

We hope that very fruitful developments will take place in Denizli and other Provinces of Turkey as well as in the contries whose participants joind thier hands in the realisation of this workshop.

Fazil Necdet Ardiç

Rector, Pamukkale University

Preface

Land transformations, especially in the developing world, are becoming a major threat to agricultural lands and food production causing massive loss of biodiversity and productivity. Many of the NAM and other countries are hit by recurring natural disasters. Rapid urbanisation and intense land use pattern have lead to an immense damage to human habitats. There is an urgent need for proper land use planning practices, dissemination of knowledge and public awareness with respect to environmental degradation. The culture of learning from disasters is growing at a very slow pace, but efforts can be increased to ensure that no more lives will be lost on recurrence of such events in the future. This book discusses the current challenges of land use and proposes recommendations to the concerned managers, policy makers, planners, builders, researchers, educators, and NGOs. It consists of selected articles presented during the International Workshop on "Urbanisation, Land use, Land degradation, and Environment" jointly organised by the Pamukkale University, Turkish Scientific and Technical Research Organization (TUBITAK) and the Center for Science and Technology of the Non-Aligned and Other Developing Countries (NAM S and T Center) during 28th of September-1st October,2009 in Denizli, Turkey. Scientists, technologists, professionals and administrators from Argentina, Bulgaria, Canada, Northern Cyprus, Egypt, Indonesia, Iran, Italy, Kenya, Mauritius, Myanmar, Pakistan, South Africa, Sri Lanka, Syria, Tanzania, Turkey, U.K., and Vietnam participated actively in the workshop.

The workshop activity included 4 sections.

In land degradation section papers from Sri Lanka, Turkey, Kenya, Mauritius, Myanmar, Egypt, Vietnam, Syria, and Iran (two papers) were presented. Land degradation includes heavy soil losses, high sediment yields, soil fertility decline and reduction of crop yields, marginalization of agricultural lands, land slides, salinization, alkalinization, acidity. Major causes for this are natural i.e, geomorphological conditions and rainfall characteristics as well as human related.

It was stressed that there is a great necessity for a proper monitoring program coupled with related database.

In urbanization section following topics were discussed:1) urban ecology, organization, and urban roads in Turkey; 2) sustainable utilization of natural resources in urban and peri-urban areas of the eastern Cape of South Africa, and 3) computation of urban heat islands in Twin Cities of Pakistan. Last century witnessed a rapid urbanization due to population outburst leading to a great loss of very productive agricultural lands. This was especially observed in the industrialising world where food production plays a very significant role.

In the environment section variety of subjects were discussed namely; oil degrading microbes for land and beach reclamation in Indonesia, green house gases emissions from cement manufacturing industry and mitigation measures in Tanzania, role of communication technologies within the perspective of European Union in Turkey, land use planning through the strategic environmental assessment, a case study in Turkey, effects of continuous thermophilic and mesophilic treatments on maturity of corncob compost and reducing of composting time in Iran, drinking water system in Denizli -Turkey, vegetable and fruit pollution by intestinal parasites and their larval stages in Mosul City in Iraq, resources use optimization in main food legume crop production in River Nile State of Sudan, and erosion, pollution, and urbanization in agricultural lands in Turkey.

Five papers were presented in salinity section. These were: 1) 'halophytes of the Arabian Peninsula', 2) salinity tolerance in the halophyte Prosopis strombulifera, 3) 'halophytes as biofuel', 4) 'halophytes as medicinal plants', and 5) 'an overview of halophytic plant diversity from Salt Lake area, Central Turkey'. The papers in this section show that halophytes are an untapped source of bio-energy and medicinal plants.

We are thankful to Pamukkale University, Turkish Scientific and Technical Research Organization (TUBITAK) and the Center for Science and Technology of the Non-Aligned and Other developing Countries (NAM S and T Center) for the encouragement and support given to us to materialize this important workshop. We hope that the articles presented in this book will encourage further discussion, research and development on land degradation and mitigation to ensure safe food and sufficient production for human survival, taking into account environmental sustainability.

Munir Ozturk

Ahmet Ruhi Mermut

Ali Celik

Introduction

Land primarily refers to geographic locations and it has been the cause of many conflicts and disputes in the past. Land reform programmes, which are designed to redistribute possession and use of geographic land, are often the cause of great controversies since times immemorial. Land also refers to natural resources, river and water masses and mineral deposits, and conflicts over their use too have contributed to numerous civil wars and continually remains a topic of concern between many nations and localities. Further, the land use and land management practices have a major impact on natural resources including water, soil, nutrients and plants and also the living beings. For instance, water bodies in a region that has been deforested or has eroded will have different water quality than those in areas that are forested. Land use information can be used to develop solutions for natural resource management issues such as salinity and water quality.

Environmental pollution is defined as the addition of any substance or form of energy to the environment at a rate faster than the environment can accommodate it by dispersion, decomposition, recycling or storage in some harmless form. Such pollution, especially in urban areas, is one of the biggest challenges the world faces today affecting us economically, physically and each day of our lives. Water pollution is the contamination of streams, lakes, underground water and larger water masses by substances that are harmful to living being. Impure water kills plants and animals. It also causes humans to fall sick. Land pollution is similar to that of water. It is the contamination of land with hazardous waste like garbage and other waste materials that do not belong to the land, pesticides, herbicides, oil spills, petroleum products, etc. Many of these items are consumed by plants and animals and then when the next consumer feeds on either the plant or the animal, it piles up and contaminates the body. Thermal and heat pollution is a cause of greenhouse effect and global warming. Thus any type of pollution affects the abiotic (non-living) and biotic (living) components of life, which disturbs the ecological balance and thereby leads to a variety of detrimental conditions.

Urbanization is the physical growth of urban areas as a result of population migration from rural areas to an existing urban area. The outcome is a change in population density and pressure on administration services. In the developing and developed countries, the expansion of urban areas and infrastructure is encroaching on productive land and natural habitats.

Land degradation affects the earth's arability. Besides natural causes, land degradation occurs where there is a deficiency in land use planning or in its orderly execution, or the existence of financial or legal incentives leading to the wrong land use decisions, or one-sided central planning leading to over-utilisation of the land resources, for instance for immediate production at all costs. This often results in misery for large segments of the local population and destruction of valuable ecosystem. It is highly desirable to replace such narrow approaches by using integrated and holistic techniques for the planning and management of land resources in order to ensure the long-term quality of the land for human use, prevention or resolution of social conflicts related to land use, and conservation of ecosystems of high biodiversity.

Land is a part of our culture and society providing a distinctive networking in the community. Planning a viable land use programme requires an appreciation of the historical past, as well as an analysis of the environment in which it is supposed to operate, including its public and its stakeholders. With this in view the Centre for Science and Technology of the Non-Aligned and Other Developing Countries (NAM S&T Centre) organised a 4 days International Workshop at Denizli, Turkey on 'Urbanisation, Land use, Land Degradation and Environment', jointly with the Pamukkale University and TÜBİTAK, the Turkish Scientific and Technological Research Council.

There were 25 overseas participants from Argentina [Prof. Maria Virginia Luna, Universidad Nacional de Rio Cuarto, Argentina]; Bulgaria [Dr. Anely M. Nedelcheva, Assistant Professor, Department of Botany, Sofia University]; Canada [Prof. Ahmet R. Mermut, Department of Soil Science, Saskatchewan University]; Cyprus [Dr. Salih Gücel, Assistant Professor, Environmental Sciences Institute, Near East University, Northern Cyprus]; Egypt [Dr. Wafik Noseir, Environmental Engineering Consultant, AUC, Cairo]; Indonesia [Mrs. Dwi Susilaningsih, Senior Researcher, Biotechnology Research Center, Indonesian Institute of Sciences. Bagor]; Iran [Prof. Gholamreza Zehtabian, General Director of Academic Affairs, University of Tehran; Mr. Mehran Kiani Rad, Faculty member, Biotechnology Department, IROST, Tehran; and Prof. Mohammad Kafi, Head, Agronomy Department, Faculty of Agriculture, Ferdowsi University of Mashhad]; Ireland [Dr. Patrick S.M. Dunlop, Research Fellow, Nanotechology and Integrated Bio-Engineering Centre–NIBEC–University of Ulster]; Italy [Prof. Vincenzo Belgiorno, Salerno University and Dr. Süreyya Meriç-Pagano from the Universities of Napoli and Salerno]; Kenya [Ms. Joyce Wanjiru Ngure, Chief Research Officer, Ministry of Higher Education, Science and Technology, Directorate of Research Management and Development]; Mauritius [Mrs. Cecily Cyparsade, Environment Officer, Ministry of Environment and National Development Unit, Port Louis]; Myanmar [Dr. May Thae Su Moe, Associate Professor, Department of Technical and Vocational Education, Ministry of Science and Technology]; Pakistan

[Prof. M. Ajmal Khan, Department of Botany, Karachi University, Karachi and Prof. M. Anwar Baig, Institute of Geographic Information System (IGIS), National University of Sciences and Technology (NUST), Islamabad]; South Africa [Mr. Lulama L. Sombalo, Researcher, Agricultural Research Council, ARC-ISCW, Pretoria]; Sri Lanka [Mr. M.A.K. Munasinghe, Research Officer, Natural Resources Management Centre, Department of Agriculture, Peradeniya]; Syria [Er. Chadi Badra, Head of Division of Combat Desertification, Ministry of State for External Affairs, Damascus]; Tanzania [Dr. Dugushilu Mafunda, Principal Research Officer, Commission for Science and Technology of Tanzania, Dar-es-salaam]; U.K. [Ms. Shahina A. Ghazanfar, Head, Temperate Regional Team, Herbarium, Library, Archives and Art, Royal Botanic Gardens, Kew, Richmond, Surrey and Dr. Cathy McCullagh, The Robert Gordon University, Aberdeen, Scotland]; and Vietnam [Ms. Ngo Thi Loan, Vice Head of Project Department, Centre for Regional Research and Development, Ministry of Science and Technology (MOST), Hanoi; and Mrs. Luu Nam Phuong, Officer, Department of International Cooperation, Ministry of Science and Technology (MOST), Hanoi].

The overall programme of the workshop was held in three parallel Technical Sessions and a joint Concluding Session. In these technical sessions, two were primarily for the Turkish speakers and audience. One session however was exclusively for the English speaking audience comprising the foreign delegates sponsored by the NAM S&T Centre and some other Turkish and foreign speakers. The papers presented in this session were on 'Paradox of Halophytes Utilization as Biofuel Resources and Land sustainability' by Prof. Mohammad Kafi; 'Country Status Report for Mauritius' by Mrs. Cecily Cyparsade; 'Computation of Urban Heat Island created due to Changes in Land Cover and Land Use–Case Study of Twin Cities of Pakistan' by Dr. M. Anwar Baig; 'Urbanisation, Land use, Land Degradation and Environment in Myanmar' by Dr. May Thae Su Moe; 'Green House Gases Emission from Cement Manufacturing Industry and Mitigation Measures' by Dr. Dugushilu Mafunda; 'Evaluation of Sustainable Urban Agriculture in India for Applicability in Zwilitsha Township, Eastern Cape, South Africa' by Mr. Lulama Ludumo Sombalo; 'Land Degradation in Sri Lanka and Its Mitigation' by Mr. M. A. K. Munasinghe; 'Soil Protection Techniques from Erosion in the Sloping Areas of Vietnam' by Ms. Ngo Thi Loan; 'Global Environmental Information System' by Dr. Wafik Noseir; 'Effects of Continuous Thermophilic and Mesophilic Treatments on Maturity of Corncob Compost and Reducing of Composting Time' by Mr. Mehran Kiani Rad; 'Land Use in Kenya' by Ms. Joyce Wanjiru Ngure; 'Characterizing and Screening Oil Degrading Microbes for Land and Beach Reclamation in Indonesia' by Dr. Dwi Susilaningsih; 'The Case of Desertification in Syria' by Mr. Er.Chadi Badra; and 'Urban Ecology of Edirne-Turkey' by Dr. Salih Gücel. A special invited talk on 'Halophytes–Potential of Cash Crops using Saline Resources' was given by Prof. Ajmal Khan and Prof. Arun P. Kulshreshtha made a presentation on 'the NAM S&T Centre and Its Role in South-South Collaboration in Science and Technology'.

The papers presented by Turkish participants in this session were on 'Watershed Rehabilitation Projects' by Dr. Serpil Erkmen of the Turkish Ministry of Environment and Forestry; on 'Integrating Environment into Land-Use Planning Through Strategic Environmental Assessment: A Case Study on Environment Plans in Turkey' by

Dr. S. Sinem Ozyurt; on 'Extractable Boron in Naturally-Occurring High Boron Soils Around Bigadic Mining Area' by Dr. M. Serdal Sakçali, Assistant Professor and Chairperson, Biology Department, Fatih University, Istanbul; on 'Long-term Planning and Governance Concerns in Disaster Risk Mitigation and Sustainable Land Use Management: Lessons for City of Denizli' by Dr. Pınar Savaş Yavuzçehre of Pamukkale University; on 'Socio-Economic Effects of Irregular Settlement and Industrialization on Land Use: The Research of Kahramanmaras Urban Area' by Mehmet Pinar Yilmaz of Pamukkale University; and on 'Water Network Leak Detection and Pressure Managment in Denizli' by Mr. Cantûrk Ümit of Denizli Municipality.

In addition to the above, there were poster presentations by the Turkish and overseas scientists. The concluding part of the session with foreign participants was devoted mainly to the discussion on Pamukkale Recommendations on Land Use, Land Degradation and Environment.

This publication is an edited compilation of the presentation made during the Pamukkale workshop and I sincerely trust that it would be a valuable reference material for the entire scientific community especially those working in the fields of Urbanisation, Land use, Land Degradation and Environment. This would also help the developing countries in executing their plans and strategies in this area.

I gratefully acknowledge the dynamic involvement and untiring effort of Mr. M. Bandyopadhyay, Senior Expert and Administrative Officer (formerly Adviser, International Cooperation, Department of Science and Technology, Government of India) and Dr. V.P. Kharbanda, Publications Adviser in the NAM S&T Centre for extensive support in technical editing of this publication. The valuable services provided by the entire team of the NAM S&T Centre, particularly, Ms. Debanjana Dey and Ms. Isha Parmar in compiling the presented papers and giving a shape to this volume are also deeply appreciated.

Finally, I thankfully acknowledge the partial financial support given by the Grand Mayor of Denizli, Turkey for publication of this book.

Arun P. Kulshreshtha

Director, NAM S&T Centre

Contents

Chapter 1

The Role of Communication Technologies for Access to Environmental Information and Environmental Policies: Perspective of European Union

Mustafa Ökmen[1] *and Serhat Baştan*[2]

[1]*Celal Bayar University, School of Higher Education, Salihli-Manisa, Turkey*

[2]*Celal Bayar University, School of Applied Sciences, Manisa, Turkey*

ABSTRACT

Whoever is responsible is and whatever the reason may be, a serious deterioration of ecosystems and the beginning of its hazardous effects against the life on the earth are turning into one of the main crisis of the humanity today. This situation menaces the whole world along with every aspects from developed to underdeveloped, the Easterner to Westerner, the Southerner to Northerner or socialists to capitalists. As the nature has increased the frequency of warning signals pertaining to the serious results of this fact which has been continuously being expressed by scientists, it has become a major national issue on which various expedients and attempts of solutions have been intensified. Today it has been understood that environmental problems with their cumulative characteristics are closely interrelated with many economic, social, cultural, political and administrative parameters. Ecological deterioration is one of the most important problems such as fast growth in population, consumption of natural resources, starvation, terrorism and regional conflicts being caused by occurrences which are more or less related to industrial and urban phenomenon which has been accelerating since the beginning of second half of the 20th Century. Moreover it is possible to claim that this is the most important problem among others threatening common future of the whole world. Consequently, environmental questions identically require global solutions in accordance with their global characteristics. There is a common thought in scientific community and public opinion in the world that Western policies, economic development models and lifestyles depending on consumption and prosperity are primary reasons of environmental exhaustion. Though the responsibility is imputed to the West at the great extent by this way, environmental problems stand as a common matter of the whole humanity with their supra-system characteristics. Moreover, the necessity to keep the ecological changes with their dynamic attributes under control is gaining more importance and constantly getting harder yet. Consequently, these facts require some common preferences and objectives to be determined and effective policies to be generated beyond the national level.

European Union, which as a reconstruction has emerged in the context of global and regional constitutions. It is a concrete, influential and important model bearing a potential to generate policies, norms, values and standards on the subject of environment as it does in many other issues. Many technological, economic and social instruments have been put forward concerning environmental policies along with the progress and the experience of EU acquis and integration. Information and communication technologies have been one of these main instruments in carrying out environmental policies of EU. The computerized information and communication has taken on a dominant role on flexibility in manufacturing, productivity in the usage of production factors, activity in decision support and problem solving processes and advantage in access to knowledge for citizens of democratic societies. Thus environmental matters have become one of the primary titles of European information society objectives. In this context, applications relevant to use of information-technology infrastructure on solving environmental problems in member or candidate countries are gradually gaining priority in the scope of e-Europe 2002 and e-Europe+ 2003 action plans and their successor initiatives. These technologies have multi-dimensional effects from technical and operational viewpoint such as monitoring/measuring environmental happenings and empiric data exchange between scientific and expert organizations to political and social aspects of the subject such as public access to information and formation of public opinion concerning environmental agenda about environmental matters. Consequently, it is possible to propound that information and communication technologies have become the one of the main parameters of environmental issues.

INTRODUCTION

The current serious deterioration of ecosystems is becoming dangerous for the living beings no matter whatever the reason and whoever responsible is. It is gradually becoming one of the basic questions for the whole humanity. This situation is threatening the whole world in every aspect equally; the developed and developing nations, the Easterner to Westerner, the Southerner to Northerner or socialist to capitalist. Lately, as the frequency of nature's warning signals related to these serious problems is increasing it is expressed by the scientists too. In fact it has become a major international issue on which intensification of attempts to put forth different solutions is also increasing. It is now well understood that environmental problems with their cumulative characteristics are closely interrelated with many economic, social, cultural, political and administrative parameters. Ecological deterioration is one of the most important problems related to such developments as rapid population growth, consumption of natural resources, starvation, terrorism and regional conflicts emanating from industrialization and urbanization due to their acceleration since the beginning of second half of the 20th Century. In fact, it can be regarded as the most important problem among all others threatening the common future of our world on global basis. Consequently, environmental problems demand global solutions due to their global nature. The European Union; which has emerged as one of the important structures from the global and regional developmental perspective; is a concrete, influential and important model with a potential to generate policies, norms, values and standards on mutual values, standards and politics on the subject of environment as is the case with many other issues. All through the formation and experience of EU acquis and integration, technological, economic, social and similar other instruments and solutions on the topic of environment have been put forward concerning environmental policies within the context of standards of European origin. Along with this progress the elasticity, productivity concept in the use of resources, effectiveness during the period of decision making and problem solving, and priority in the procurement of information by the citizens of the democratic communities the information and communication technology too has become one of the main and basic instruments in carrying out environmental policies in EU. In this way environmental problem has become one of the precedent topics in the aims of European information community. In this context e-Europe 2002 and e-Europe+ 2003 action plans and the interferences following these infrastructure of the information technology is gaining importance towards their use in solving the environmental problems in all the member countries of the Union and standing candidates. The technologies under question have multidimensional impacts by allowing public to reach the technical and functional information, in creating public awareness in the field of environmental agenda including their social aspects through tracking and measuring of environmental developments as well as environmental data transfer among the scientific and expert foundations. As such, it is possible to say that information and communication technologies have taken a position among the main parameters of environmental problems.

1. ENVIRONMENTAL PROBLEMS AND ENVIRONMENTAL POLITICS BEYOND THE NATIONAL LEVEL

Humans are face to face with an important problem originating from the destruction of our environs and generally termed as environmental problems. From the very start due to close interactions with the environs humans have been trying to dominate and change these in accordance with their needs and wishes. But, while doing so they have created a chain of very complex environmental problems resulting in a destruction of the surroundings by upsetting the ecological balance due to its overuse and by applying pressures beyond the limits of nature. The events like increase in the population, fast and unbalanced urbanisation, and industrialisation and over consumption together with the economical aspects are leading towards an expansion of the environmental problems. The negative impacts originating in this connection due to an advantageous position in the production quantity by the pollution creating monopoles has lead environmental problems towards a focus of discussion at local, regional and international level. Today the environmental topics are bringing different countries on a very complex reciprocal dependence. The danger of extinction in the environment is compelling nations to come to a mutual understanding on global basis. The concept of trans-boundary pollution too is making environment as an international issue rather than national, thus pressing for regional and universal cooperation[1]. In general the first global environmental political decisions were taken by UNEP (United Nations Environmental Programme) established by the United Nations soon after its creation in 1972 after the meeting in Stockholm. The report on "Common Future"; published in 1987 by the United Nations Environment and Development Commission (Brundtland Commission); brought the model of Sustainable Development to the forefront and 20 years later, in 1992 Rio Conference the present situation of the world was scrutinized. The most important point of common understanding among 179 nations was; how economy, society and environment can be integrated in to the sustainable future development of our world. The topics dealt with in the Rio Conference declaration were arranged under global environmental politics and principals that will lead in the politics of environment-development were assembled under Agenda 21. According to this, the agenda feels that sustainable development is actually under the responsibility of respective governments which needs national strategy, plan and politics. The national efforts should be in collaboration with international bodies like UN. An active position of non-governmental associations and other groups together with a wide public participation should be encouraged. The aim of Agenda 21 necessitates that developing nations will get substantial new financial aid. These nations need this support as a compensation for dealing with the global environmental problems and speed up sustainable development programme. There is a need for a support from international bodies as well in order to fulfill these aims[2]. The sustainable development theme from the Rio Conference and Agenda 21 in Rio declaration stresses a need for a local, regional and global consensus on the prevention of environmental problems and politics related to this[3]. The headings of global environmental politics and framework of the international environmental politics accepted within this context are enough to play a functional role in outlining the national environmental politics as well. The concrete efforts related to this above the national level are outlined in the World Summit held in 2000 in Johannesburg on Sustainable Development, which was very important as different actors dealt with economy and environmental problems in a balanced way, and because this summit was the latest joint standing of humans against environmental problems[4].

Since industrialisation is above the level of nations, it has given the environmental problems a status beyond the national level. As such, environmental politics is gradually gaining weight in the international and national environmental politics due to its regulatory principals. The national politics has great importance due to its responsibilities and authority, because international problems of ecology and economy are first faced by the nation. This government at the same time is often helpless against world economy and its problems. National government although accepted as a single great actor against global environmental problems actually is in a week position[5]. In this context, Bell's "for big problems very small, for small problems very big" well known assumption confirms that effectiveness of national government gets reduced in the face of problems and politics upcoming in parallelism with globalisation of environment. In other words bindings of national government politics run against each other *de facto* to globalisation of environmental problems[6]. In this sense, from national governments it moves towards web communities[7]. Today, on one side globalisation is occurring, but at the same time settlement attitudes are strengthening. In this two ways developmental process, globalisation attitudes are changing the traditional structure of national governments, as against this settlement period reproduces and rejoins the historical, cultural and physical identities, due to its one-sided centralistic structure within the globalisation itself, in order to support a world more humanitarian and worth living. The changes in the economic and political functioning of the national governments, due to globalisation, regionalisation and settlement attitudes coming to the fore front and experienced by the communities during post industrialisation, in particular lead towards depressions because of suppressions and integralism, and this has been overcome in the West through developments in democracy. But this is not certainly the solution and in the coming decades, localisation and regionalisation attitudes will increase. Environment is getting major share from the problems created by these attitudes due to its national and international characteristics. Parallel to the new aspects of the environment, efforts to develop global solution has attained new dimensions within these changes. In this period, in addition to the global and local aspects the problems are considered from the economic, political and cultural point of view directly related to the regionalisation.

International organisations are considered under two headings. These are cooperative activities and integrations like European Community. Former are respecting the sovereignty of member countries because decisions not acceptable to them cannot be taken, as the decisions are taken on full vote basis. As against this, EU uses some transferred sovereignty rights, member countries have given up some powers in favour of member states. This double sided development is seen in the functioning and cooperation on environmental matters. [8]

In this connection during the approaches towards the international environmental politics, international rules and regulations due to pressures of rising organisations from sovereign nations and local-national actors affiliated to the international structures as spreading organisations become apparent. [9] The present day topic of curiosity and discussion is up to what extent the force of globalisation or global environmental organisations will be effective. More important is, in this age of globalisation how much effective national environmental politics will be? [10] The strong discussions taking place while outlining the politics during the transfer period of European Economic Community to European Union are observed on the topic of environment too between the Union and member nations.

2. ENVIRONMENTAL POLITICS, ENVIRONMENTAL INFORMATION AND COMMUNICATION TECHNOLOGIES

All efforts to protect and solve the environmental problems are a part of environmental politics. However, measures taken by the governments for the future development and protection of natural and created environments can be termed as environmental politics. Environmental politics in the sense of searching for solutions of concrete negative exposures beyond the observational trends, reactions and ideological speeches, takes different shapes in each country according to its characteristics but at almost all places understanding on some characteristics of environmental politics are met with. A creation of healthy environments for humans, protection and development of community environmental values, and sharing of the necessary load of environmental politics applications mean an adaptation to the principals of public justice. [11]

Generally different solutions are preferred in the case of environmental politics in different countries on our earth. A similar discrimination is observed among the environmental specialists, scientists, writers, philosophers and political organisations working for the protection of environment. The economic system in a country could be capitalistic, socialistic or capitalistic dominated mixed system, these effect environmental politics as well like all other political decisions, depending upon the situation in the political regime of a country, whether run by a multi-parliamentarian democratic system or single part rule. [12] However, beyond all these differences, in this connection politics generated in general can be dealt with under two headings namely; protective and repairing. Although it is not possible to isolate environmental problems from political ideology, one of the important themes emerging from this politics is environmental management which is directed by the economical and technological feasibilities beyond this ideology. Not only public but also in private sector a suitable communication, planning and control is needed and for this purpose a working group for this system should be created, wherein environmental management will form a basis for environmental politics, so that all living beings live in a healthy and balanced environment, as well as for a protection of natural resources, their evaluation and development. This period embodies like execution, planning and supervision of all functions related to the environmental planning and development within the framework of central and local governing bodies. [13] One of the most important instruments in the environmental politics is environmental information embodying efforts related either directly to the constitution or related laws or other texts in jurisprudence, from government programmes to developmental planning arrangements within a wide perspective. The most important side of the environmental information during the protective political measures applied before the environmental problems appear and those developed after the environmental problems come into being is information technology. Information and transportation technology is an important instrument of environmental information in the unsolved environmental problems in addition to being an important parameter in the changes involving transformation from industrial society to the one beyond industrial level (scientific society). The environmental information produced after an effective application of this instrument can play an important role in the environmental politics at local, regional and global levels against environmental problems. The present day situation dictating international politics is environmental information and within this context the scientific technology is gaining more weight in particular due to the boundary less nature and cumulative characteristics of environmental problems.

This topic is of the same importance in the environmental politics in Turkey as well as in Europe which is accepted as a popular example in the regional organisational matters in this context. The topic is carrying great weight due to interesting developments in environmental problems, in particular due to latest developments between EU and Turkey.

3. ENVIRONMENTAL INFORMATION AND ENVIRONMENTAL POLITICS IN THE EUROPEAN UNION

3.1 Environmental Politics of European Union from the Past to Present: Qualities and Problems

The developments of environmental politics during the formation of European Union have followed a long path after passing through a complex framework of consultations, scientific approaches in socio-economic aspects, legal instruments and technical facilities. [14] The joint ventures in environmental field and collective politics have become necessary for obtaining a free competition and free movement, accepted as one of the basic requirements in the unity of Europe. An application of different politics in the member countries, in particular determination of different environmental measurements, results in a difference in the costs of products. Similarly, quality standards in some member countries restrict the entrance of some products produced by other member nations. In addition, in some member countries investments needed in order to protect the air and water pollution increases the costs of products to a great extent. Due to this and similar other reasons the question of free competition and free movement of products among the member nations come into being. In order to overcome this situation need for a joint environmental politics has been stressed. For a betterment of the living standard attained by the member countries, development and continuation of natural living conditions in a healthy manner is the other development necessitating the need for a common environmental politics. It would not be understandable to remain as a stranger towards the most important factor like protection of environment and natural resources for a continuation and upgrading of the quality towards a healthier living of humans by the European Union which is a project created for a better quality and comfortable living of public on the whole in the member states. Other reason for EU to produce environmental politics related to its own spheres is the political one. The differences in the environmental politics due to differing living conditions and under different political systems are regarded as a position not acceptable by the countries affiliated to each other within the same economic circle. [15]

In the agreements of the EU foundation the topic of environment is not mentioned openly up to 1987. The situation was dealt with the help of operation planning and through guiding of national laws, till one Europe document was signed in 1986. The reason toping the list in the development of oneness politics through practical application of Rome Agreement; regarded as EU constitution; is quite apparent from how deep the legal limitations in the union by law are. The most important point here is instead of limiting the wishes of masses via economic agents; whose responsibility lies on its shoulders; its expansion to the public factor. As will be clear from the later realities the subject was approached absolutely economically, but the masses addressed included citizens on the whole within this Union. Due to a harmony in the economies of the member states and realization of balanced developmental target environmental pollution was defined as a limitation and the subject of environment was included as three subheadings under a single subject in 1986 in the highest document of EU laws. [16] According to the reason given in the joint environmental politics of the union in Rome Agreement and One Europe Instrument, for a harmonious development joint studies are required for continuous and balanced development as well as economic activities in the member states.

According to the statement on mutual environmental politics in Rome Agreement and One Europe Accord, joint actions are needed for a balanced and continuous development, as well as for harmonious economic activities in the members of the Union. One Europe Accord put into force in 1987 brought about basic change in the European Community founding agreement, which basically aimed to remove the physical, technical and financial obstacles up to 1993, lying in front of the common market for the realisation of a single market. One Accord on one side brought changes in the departmental structure of the European Community; on the other hand it widened the power of members within this framework. One of the areas covered by One Accord openly included the powers of community members regarding the protection of environment and environmental politics. The article 25 of European One Accord implies to include number VII Environmental Heading in the Rome Agreement. In this way, environmental problems are included in the founding agreements for the first time since the foundation of European Community. The heading added to the founding agreement includes articles 130 R, 130 S and 130 T. Here 130 R is the basic article, defined with an aim to protect and improve the quality of environment for the sake of protection of human health and rational use of natural resources. [17]

In order to stress the urgency of this topic, in addition to the changes in this article revising the Rome Agreement, European Community publicised 21 March, 1987–20 March, 1988 as European Environmental Year. Parallel to this, the 4th Operational Programme covering the years 1987-1992 was ratified and put in to force. The form and contents of the programme under question, although embodying most of the characteristics of earlier programmes, also considers new problems and rules. Here, comparatively strong protective measures have been considered and determined, both as a result of industrialisation as well as social living. In addition, community legislations applied theoretically aimed to get these activated practically as well. It was accepted that application of an effective environmental politics and environmental education of individuals will provide practical operation of the legislation. [18] In spite of both European One Accord as well as important contents of the Operational Programmes an increase in the criticism in the inadequacy of the European community environmental politics and occupation of environmental problems on the whole world agenda due to their cumulative characteristics, community was forced to make more realistic and concrete efforts. The determining date on this point is 7 February, 1992 when Maastricht Agreement transforming European Community into European Union was signed. This date, in addition to being a concrete step towards a political union of the European Communities, is important also for a realisation of politics of several important topics including the environment. [19]

Environmental politics of European Union has been initiated within the scope of some targets and principals both before and after Maastricht Agreement. In particular after the agreement under question, the V. Environmental Action Plan which came into force in 1993 and legal arrangements made thereafter these targets and principals have been defined more openly. The basic targets in the legal documents on the environmental politics of the Union can be summarised as follows; protection of the environment, environmental look after, upgrading of environmental quality standards, to be cautious about the damages on ecological balance arising from the use of nature-natural resources, provide means for their intelligent management, protection of human health, follow development in a balanced way with the quality requirements, in particular pave way for development of working conditions and living environs, arrange for more consideration of the environmental impacts in the use of soils and urban planning, and search for joint solutions on global environmental problems together with the countries outside the Union particularly international organisations. Some principals given in the basic documents of the environmental politics provide important steps in the application and formulation of an integral politics in the Union.

The basic applications directed by the environmental politics of the European Union are becoming more varietal and effective. Protection of air tops these applications. Protection of air quality is made possible by prevention of contact between solid pollutants with air, preventive measures adopted in energy, transportation, tourism and planning of land. For this purpose environmental politics applied both at the level of Union as well as national and local levels requires important changes. The problems of ozone layer depletion, control of acidification and ground level ozone and other pollutants, climatic changes and global warming are the major areas considered for the protection of the air in the Union. Each problem is affected from different public and industrial sectors and need solutions peculiar to every one. The Union is operating in the case of ozone depletion with the permission and support of UNEP within the framework of the programme accepted by more than 120 countries. The target for gradual stop over of the production of chlorofluorocarbons and halogens damaging ozone layer has been reached in 1996. Union is also working on the formulation of politics on low cost emission limits and fuel quality standards in the industries related to the petrol and motor vehicles production by pushing them to decrease the consumption of fuel and improve air quality in spite of an increase in the traffic level. [20]

Climate change sector too is dealt with among the joint political topics. In Europe important temperature increases have been observed during the last 30 years and developments related to a decrease in the seasonal changes recorded. [21] The Union is taking front line in the struggle for climate change. From 19th Century onwards world ground temperature has increased by a mean value of 0.3-0.6 °C. The research investigations too prove that negative impacts due to human activities like use of fossil fuel and destruction of trees cause emission of carbon dioxide and other gases resulting in greenhouse effect. Union has outlined the framework of the strategy for struggle against climate change in the beginning of 1990 and supplemented the UN framework agreement on Climate Change accepted in 1992 during Rio Summit. European Union is encouraging the production of electricity from renewable energy and improvement in fuel economy, in addition to several initiatives related to climate. The Kyoto protocol regarding the climate change accepted by the member States of UN in 1997 too has been signed by the Union. [22] Kyoto protocol has determined a target of reducing the greenhouse gases 5 per cent below the 1990 level until 2008-2012 from the industrialised nations. A study undertaken recently confirms earlier investigations that for a balance in climate change on long term basis there is a need for greater reduction of global emissions. Some member states of European Union have outlined the targets for reducing the emissions to a great extent. [23] In addition, it is a fact that as compared to the period before 1998 there has been an important development in the reduction of emissions in European Union. In this connection, the importance of Kyoto Protocol cannot be neglected in its effectiveness in the politics related to the climate change. [24]

The steps taken after Maastricht Summit and in particular after 5th Environmental Action Programme need a special mention in the concretisation, diffusion and effectiveness of all these principals and applications towards the environmental protection. Moreover, European Environmental Agency; providing an institutional structure in the politics; has contributed a lot in the functioning of this activity. Maastricht agreement put into force in 1993 is accepted as a turning point in the unity of Europe. It allowed changes in the institutional structures on one side, but on the other hand it provided measures for a provision of economic and monetary union. It also provided a calendar for application of precautionary measures in the political union towards a unity in defence, security, law and internal affairs. Moreover, together with addition of new politics within the unity context some existing political fields were strengthened. Environmental politics too has been strengthened by the Maastricht Agreement and for the first time principle of environmental protection has been included in the targets of European Community (Union) within this framework. It has been stressed that economic development should take place in a balanced way with the environment; as such Union has been asked to develop environmental politics. [25]

The criticism towards the insufficiency of environmental politics in the European communities due to an increased awareness towards the environmental problems has stressed a need for taking more effective steps on the topic of environment without affecting the sustainable development of member states. [26] The first stages of changes in the founding agreement in order to make preparations for a wave of new and greater expansion of European unity in 21st Century, has been agreed upon during the Amsterdam Summit on 16-17 June, 1997. In this connection, in addition to modest expansion in the political aspects left within the authority of European Union, some smaller changes in the institutional structures have also been realised. Moreover, a consolidation in the agreement has been done by summing up and itemising the changes made at different times in the founding agreement. In the new changes brought in the field of environmental politics within European unity with the functioning of Amsterdam agreement which become effective in 1999, sustainable development principal was added to the 7th paragraph in the preface of European Union agreement and in order to arrange an application of the same principal some changes were made in the articles related to the environment in the European Community Agreement. [27] In this connection; in addition to the article 2 in the rearranged community agreement depicting protection of environment at the highest level and improvement of its quality, article 6 embodying rule of integration of community activities and politics too has been revised in a balanced way with article 3 related to the formation of politics in the environmental field. Looking at the contents we can say that no important changes have occurred in the structural aspects as a result of Maastricht Agreement.

An important development which needs a mention regarding the environmental politics of the community is the summit of Heads of the States of European Union held during 7-8 December, 2000. It can be regarded as one of the turning points in the European Union. The intergovernmental conference continuing after signing of the Amsterdam agreement in 1997, this Summit has concluded with Nice agreement. The agreement under question contains changes in the institutional structure of Europe, in particular effective after expansion and aiming at effective functioning. In the current institutional structure and in particular interruptions faced by the European Union even during the decisions, in the case of expansion from 15 to 28 members, the dangers in its non-functioning on the whole has lead to such an introduction before the completion of expansion of member states. Nice Agreement (2000) has not brought any important changes in the European Community Agreement regarding the rules in the environmental politics. [28]

One of the topics in the environmental politics of the Union is the annual Management Plan prepared by the European Commission. As one of the 36 General Directorates within the Commission, the Administrative Plan prepared in 2005 by the Directorate General of Environment gives important notes on the latest situation regarding the environmental politics of the Union. The "sustainability of the sustainable development" has been fully stressed in the protection of environment for the present and future generations in addition to its sustainability and development. The topics like climatic change, protection of nature's diversity, sustainable production and consumption are highlighted specially within the 6th Environmental Action programme for the long term strategies regarding the environment and attention has been drawn towards internal (EU members) and external (Global) actors. Moreover during this period, activities and environmental planning of international actors, environmental civil defence and information exchange on this topic too has been stressed. For a better planning and application of the environmental politics, attention has been drawn towards a creation of all sorts of scientific information webs and their effective exchange on economic, social and technical aspects in order to fill in the gaps in the knowledge [29]. "European Union Environmental Politics-2004" draft prepared by the Commission and discussed in the Council and Parliament, is an important document [30] enlightening us on the latest situation in the environmental politics of EU. This draft has been published on 27th January, 2005 and embodies information on the 2004 developments and important events on environmental politics together with environmental projections for 2005. The projections mention about the contradictions expected to rise from the new 10 members of the Union in particular Romania, Bulgaria due to a fast economic development. 2004 has been a highly productive year for environmental protection in Europe and the World and it has been stressed that this action has been supplemented by the Sixth Environmental Action Programme implemented by the Union. The climatic change, conservation of biodiversity and similar other topics brought to the forefront by this programme have been subjects of priority in 2005 as well. [31]

The common level of environmental standards and their applications in the member nations in the EU has a 30 years solid experience of application, most of the time depending on the most advanced level reached by a nation during the acceptance of bylaws. EU laws have lead the world on such topics as testing of the dangerous chemicals and their labelling, control of biotechnological investigations and products, and control of major industrial accidents. During 2000- 2010 basic aims of the EU environmental politics can be summarised as; improvement of an application of the bylaws, arrange for their adaptation by the members, arrange for a punishment from the Council of Justice for those who do not adapt these, integration of environmental politics with other political aspects, change consumption attitudes, arrange public participation in the solution of problems. [32] In addition, the developments including the topics regarding the active participation of citizens within the concept of "good environmental management" too can be regarded as important in the environmental politics. The "good environmental management" in the White Paper of 2001 European Constitution includes five main principles like freedom, participation, responsibility, effectiveness and consistency, for a formation of the justified political environment related to the environment in Europe. This principle and understanding has been ratified in the fifth Action plan as well which brings to the frontline the rights of citizens based on basic environmental standards and in the meantime gives important responsibility to the non-governmental organizations. [33] The NGO's dealing with environment are gaining weight particularly in creating public awareness through environmental education in Western Europe, by reinforcing administrative duration and hold an important position in the structural and functional development of civilians. [34] The importance of NGO's in the EU environmental politics cannot be neglected particularly due to their efforts related to public participation.

3.2 Position of Environmental Information in the Environmental Politics of European Union and European Environmental Agency

One of the most important components both during the formation as well as enforcement of the EU environmental politics under question has been Environmental Information. A fast development in the information technology for procuring data and analysis techniques has proved of great help to ecology and environmental science. [35] Knowledge and information technologies are capable of producing potential positive or negative effects on every step of sustainability. [36] Because, a need for functioning of the environmental politics parallel to the globalisation of environmental problems is to a great extent dependent on the establishment and management of a good information infrastructure. In this sense, success of the application of environmental politics by the Union is closely related to the effective use of scientific technology and diffusion of information webs that will be created within this framework. One of the foundations occupying the frontline in the case of sustainable development in the Union, particularly during the period following Maastricht, is "European Environment Agency". In order to provide healthy and comparable information to the Union and the member states on the position of environment, the EEA;established in 1990 and put into force in 1993; has undertaken the function of providing centrally oriented sustainable development strategy during the periods of developing politics towards environmental protection. [37] The aim of the agency was to help EU and member states in the improvement of environment, integration of problems related to the environment into economical politics and decision making on the progress towards sustainable development. [38] Among the working fields of agency are procurement and analysis of environmental information, coordination of EIONET and an information system, provide services to the European Commission, member countries of EEA, international organisations, conventions and agreements, provide suggestions related to the politics, and other topics related to this on a broader scale. The strategy of EEA for 2004–2008 is the third perennial working programme prepared according to the outline mentioned in the establishment format of the Agency. It has been accepted in the meeting held on 25 November, 2003 by the EEA governing body and is on the same lines as sixth Action Programme. The activities of the Agency cover 4 basic themes: follow climate change, protect biodiversity loss, follow habitat changes, protection of human health and living standard, use and management of natural resources. During the coming 5 years EEA will produce a series of evaluations on these topics and sectoral interactions in the environment. Some basic limited indicators will be used for reporting the effects of sectoral environmental activities and their impact on the ecosystems. [39]

Member countries and EU will have the, transfer of knowledge and information within EEA through EIONET. One of the most difficult issues waiting for the standing member states is the enforcement and following of EU laws on environment. The difference between EU and the standing member states at the level of environmental protection and restrictedness of the old responsibilities regarding the environment will require long term strategies with known aims. On the other hand, member candidates possessing wide untouched virgin areas will make a great contribution to the biodiversity of Europe. A protection of these resources and at the same time develop and administer an economically and environmentally sustainable framework, member states will require an effective cooperation with EU on some topics. Presently, EU and the standing member states are signing and approving agreements related to the EEA and EIONET (European Environment Information and Observatory Network). The membership in EEA will prove beneficial for both EU and standing members. In this way, standing member states will have direct access to the Agency's information web and provide effective participation covering comparative analysis. [40] There is a need for more discussions on the importance of activities of EIONET in the functioning of environmental politics, which is one of the basic activities of the EEA. EIONET aims at providing knowledge and expertise through dependable and good quality data in time regarding the pressures and position of environment in Europe. This situation, allows politicians to make decisions for proper protective measures of environment at national and European level and observe the effectiveness of the politics and protective measures followed. EIONET is a joint web of the member and standing members of EEA. It itself embodies many centres related to Europe, over 300 national environmental agencies from 37 countries, and a circle of more than 900 experts from other foundations dealing with environment information. These are national focal points and national reference centres. The partnership of EIONET is important for the supports of EEA, in order to collect, arrange, develop and distribute this information.

Infrastructure of the information technology (sometimes called e-EIONET) supports web regulations and individuals. EIONET included 15 EU member countries out of 18 EEA member states in 1994. In 1996, 10 standing member countries from Central and East Europe too started contributing in the activities of EEA and developed their national networks. Others too joined EEA together with Cyprus, Malta and Turkey in January, 2002. In 2003 all became members of EEA and joined EIONET, raising the number of countries to 31. EEA is the first organisation which opened the doors to all expanding and standing members of EU. At present Switzerland, Albania, Bosnia Herzegovina, Croatia, Macedonia and Serbia Montenegro are participating in advance in e-EIONET and EEA programmes till they become member states. With the help of e-EIONET EEA is compiling on time nationally approved environmental data obtained from each country to provide high quality information. This serves as the basis for the information on integrated environmental assessment publicised and opened for communication by EEA through its website. This information allows formation and evaluation of environmental politics at the European, global and national levels and serves to support the public participation.

European Topic Centre (ETC) is the union of companies with expertise in special environmental fields formed by the organisations of the EEA member states under the agreement signed by EEA for supporting working plan. ETC's are expert centres in the field under the agreement signed by EEA in order to fulfil certain duties mentioned in the annual management plans and in the EEA strategy(5 year working plan), and appointed by the governing body of EEA as an extension of EEA on certain topics after competitive selection period and activity. Each ETC is composed of a leading organisation from environmental research and scientific group, with expert partnership organisations, which bring together sources on their special expertise fields. ETC work together with the members and participating countries, making easy the transfer of information coming from member countries and submission of reports as well as other services to the EIONET. At present there are five ETC's; air and climate changes, water, terrestrial environs, flow of waste and materials, nature protection and biodiversity.

National Focal Point (NFP) is a group of experts found in the national environmental organisations approved by EEA as the main contact point financially supported and appointed by the countries. In order to support the applications of EEA working programme, NFP coordinates national web composed of an infinite number of national reference centres (NRC). NFP's are the main contact points for

EEA in the member countries responsible for the cooperation with EEA and ETC's as well as arranging national coordination related to the EEA strategy. The working procedures differ between the countries. This reflects the different natural structures of national environmental systems. For example, some NFP's are affiliated to the environmental agencies, others are a part of the Ministry of Environment, a few in the national central administrations, some not administered centrally or working within the federal administrative systems. NFP's protect and develop the national web, coordinate contacts, requirements and deliveries at national and EU levels. Moreover, it serves as an advisor to the members of EEA governing body, arranges contacts with other related webs like Eurostat. Many NFP's organise the publication of news bulletins for publicising the important EEA publications.

National Reference Centres (NRC) are experts or group of experts in the national environmental organisations appointed and financed by the national governments in order to work with the EEA and other European centres in definite environmental topics related to the EEA working programme. NRC's are determined by the member states. These centres are experts are groups of experts appointed by using national sources, with suitable information on different environmental topics and present in the organisations compiling or providing environmental data at the national level, with suitable knowledge on different environmental topics and their tracking or modelling. NRC's have been established in the air pollution, climate change, quality of streams, water pollution, waste production, biodiversity, energy and many other environmental fields, playing a role in the technical cooperation of these fields and working together with EEA and related ETC's.

EEA has determined the flow of some annual data's in accordance with the information coming from EIONET partners. At present the data collected on air quality, air emission, terrestrial waters, marine and inland waters, polluted soils, protection of nature and land cover is used to update core environmental indicator sets which serve as the basis of EEA reports and evaluations. While presenting this data an annual progress report on the country performance is prepared by the EEA. This work is a useful tool for determining the country performance and development of responsibility in all countries. The data and information provided by the countries is used as much as possible for EIONET as well, within EU and international obligations. This means that the data collected once at the national level can be used for many purposes at national and international as well as EU levels. An accommodative European approach is needed for solving the environmental problems. EIONET has played a foremost role in the procurement of this unity covering a major part of Europe with an experience of 10 years development and sustainability. This web brings together the best national experts on environmental topics and politics makers at national, international and European levels. The countries while developing their national efficiencies have made use of sharing the suggestions, expertise and experience in the production and publication of information, particularly in the field of administration and data compilation that properly fits into their politics. The ETC's established embody basic environmental topics and by inclusion in the EIONET strengthen the contacts between the countries and EEA. A dense exchange between NRC and ETC's has developed the data quality at the national and European levels and has helped many countries in strengthening the capacity of present observation systems. An early EEA and EIONET membership has helped in upgrading the profile of environmental topics and affected the characteristics and contents of national observation systems in the new EU member states. The responsibility to report the country data at the European level has the aim to provide table of the position of environment in Europe and pressures on it which is openly compiled and solved by the EEA and ETC's. In this way, a determination of environmental performance of the countries also becomes possible. In addition, member states of EEA are pushed to develop national webs which guaranties cooperation with a need for EU and international reporting.

The requirements for environmental information are passing through an evolution. At the start EEA thought as an organisation providing environmental data and information, now it not only discusses about the position of environment with European Parliament, European Commission and member countries, EEA and EIONET at the same time with a direct help from European Union 6th Environmental Action Programme are working on 4 major topics. These topics involve solving the problems related to climate change, loss of biodiversity and understanding changes in the situation of environment, protection of human health and living standards, management and sustainable use of natural resources and wastes. Each one of these is affected by the sectoral activities like agriculture, chemicals, energy, communication and land use planning which need a consideration. Environmental topics and information particularly in the integration of communication with sectoral politics has shown a development, and studies on the energy-agriculture are continuing. They are fully cooperating on the submission of reports regarding the future of environment and effectiveness of basic environmental and sectoral politics. During the next five years EEA and EIONET will expand their activities for supporting a series of political aspects.

EIONET has made important contributions in the following fields and will continue its efforts in this direction.

- ✰ A true and timely evaluation of the measurement of greenhouse gases together with their progress within the target of Kyoto Protocol and a comparative demonstration of these between Europe and other areas in the world.
- ✰ Application of Nature 2000, development of performance indicators for biodiversity politics in connection with the sectoral politics.
- ✰ Support the structural needs at the European and global level on marine environment and soils together with the strategies related to these topics.
- ✰ Evaluation of the impact of air pollution on the public.
- ✰ Evaluation of environmental and economic effects of natural and technological dangers.
- ✰ Develop more the Eurowaternet -a European Water Tracking web- for supporting applications of water information system in Europe and

> preparing reports for water environment and other directives. EIONET has developed a lot during the first 10 years. It is now ready for adaptation to the changing topics and priorities. EEA member countries are working more openly and collectively on the environmental information in Europe through EIONET. EIONET is a successful story supporting the evaluation and developments in environmental politics, through partnership and information sharing, which can be exported to other parts of the world and will help in establishing connections between these parts. [41]

An important step taken by the Union within the environmental politics is the preparation of a New Environmental Action Programme after the Nice agreement (2000). EU Commission has announced its 6th Environmental Action Programme on 24 January, 2001, which gives basic and priority targets in the field of environment for the coming 10 years (2002- 2012) in Europe. In the programme called *"Environment 2000: Our Future, Our Choice"* 4 major topics have been selected as priority targets. These are; climate change, nature and biodiversity, environment and health, and natural resources and wastes. [42] Sixth Action Programme can be accepted as a framework study on the future 10 years activities of the EU with an aim to integrate environmental care in public politics for a successful sustainable development. EEA has made important contributions in the concretisation of this programme at the public level in particular from the point of sustainable development. [43] The Sixth Programme upon which the Council and the Parliament have reached to an agreement has been activated at the end of 2000; it embodies general strategies for the coming 10 years. Although a short strategic document, it provides an opportunity for preparation and application of detailed political targets and measurements. [44] At this stage Sixth Programme, stresses that the decisions to be taken on the environmental problems will be within the framework of mechanism of management not direct management, and is a binding for the organisations of the Union and the member countries. Moreover, this Programme will be a binding for the standing member countries while becoming full members due to the inclinations for expansion of the Union. It appears that decisions have been made within Sixth Action Programme towards increasing awareness on sustainable development through multifaceted discussions with standing member governments and cooperation with business sectors as well as non-governmental organisations working in the field of environment in the standing member countries.[45] The Action Programme too stresses propagation of understanding for more sustainable production and consumption on the way towards effective use of resources and stresses the need for eco-activities in Europe and the world, the most important aspect in this connection will be the steps taken towards environmental technology. [46]

The most important work related to the EU environmental politics is the acceleration gained lately regarding the attempts made to create e-Europe. The environmental politics of the Union is thus affected directly or indirectly during this period. An information web based on the effective use of information technology will increase the structural and functional effectiveness related to the environmental politics of the Union like other political fields. One of the important results from the EU meeting held in Lisbon in March, 2000 was development of e-Europe in order to adapt people and countries in EU on information age. The three basic aims of this development were: firstly to bring a digital age for all humans, homes, schools, business centres and government offices in the Union; secondly create a digital aged Europe with the support of innovative mind ready to finance and develop new ideas, and finally make arrangements in such a way that throughout this period the social aspects are not left out and confidence is developed among the humans on these concepts. In order to fulfil this aim, heads of the governments and states entrusted European Council and Commission to prepare a detailed action plan through highest level coordination coinciding with "e-Europe Introduction and Lisbon Strategy" keeping in view the national affords. Commission finished the e-Europe 2002 Action Plan draft, the first stage of e-Europe on 24 May, 2000. The draft with 64 targets was discussed by the member countries during the Commission meeting held at Feira on 19 and 20 June, 2002 and e-Europe 2002 Action Plan expected to be finished till 2002 was accepted. [47]

e-Europe plan prepared with the aim of arranging the passage of countries in EU towards an informative society and a e-Europe+ plan was also prepared for including standing member countries in 2001. This group including Turkey as well has prepared an action plan with an undertaking that at the end of 2003 they will finish the work grouped under 4 major articles. These major groups and their main headings are as follows: fasten the work on establishment of basic structural foundations of the informative society (provision of information services for acceptable cost to everybody, application and adaptation to the acquirements on informative society); make arrangements for providing cheaper, faster, safer internet communications (cheaper and faster internet communication, faster internet for students and researchers, safer webs and intelligent cards); investments on human resources (prepare European youth for the age of numbers, labour force in the information based economy, participation of everybody in information based economy); refreshing the use of internet (fasten e-trade, e-government-electronic communication in the public services, intern health, European numbered contents in global webs, intelligent communication systems, intern environment). The standing member countries will prepare three reports, 2 mid term and one final regarding the developments related to this work, and shall inform EU about the work done till that time. The mid term reports will be submitted in February/March 2002 and at the end of 2002, final progress report will be submitted at the end of 2003. [48] The standing member countries with the knowledge of major differences in economy, social and industrial conditions between them and the member countries have come to an agreement for a common date as 2003, in order to reach the goals set by e-Europe+ interference. In addition, in order to compare the data between e-Europe and e-Europe+ for a follow up and evaluation purposes, indicators set and accepted by 15 e-Europe countries have been selected. However, e-Europe has come into force at a time when in almost all countries in EU all houses had telephone communication. The situation in the standing member countries is not the same. Therefore, for these 3 elements in e-Europe + Action Plan a paragraph under the title "the activities related to the formation of basic structural foundations should be fastened in informative society" has been added. Additional indicators have been used for reporting and follow up of this part.

European Commission has invited South Cyprus Greek Administration, Malta and Turkey in February 2001 to join other standing member countries in the preparation of this joint action plan. Turkey has participated in this activity the same year. The most important project run by EU in the economic field has been e-Europe interaction. With a start given by the standing member countries too as e-Europe+ interaction, Turkey fastened the activities by carrying out studies on the way towards an informative society. The final progress report on the application of e-Europe+ Action Plan published by the countries on the way to join the Union and standing member countries was accepted with contentment in the European Ministers Conference on Informative Society held in Budapest during 26-27 February,2004, and the affords spent and results obtained were appreciated. It has been fully stressed that in order to reach the common goals defined in e-Europe+, the joint steps taken within e-Europe 2005 Action Plan will form a solid basis for progress made in the application of national action plans in the expanding EU members and related standing member countries. With the completion of e-Europe+, Turkey, Bulgaria and Romania have joined the e-Europe 2005 with an observer status, in addition to the 10 full members in May, 2004. [[49]] As a result of these developments, the meeting held by the European Council in March 2002 in Barcelona has entrusted the Commission to prepare a new action plan focussing on " widespread use and communication of broadband upto 2005 within the Union, development of Internet Protocol IPv6, net and web security, e-Government, e-Education, e-Health and e-Business." e-Europe 2005 Action Plan expected to be completed in 2005, was accepted during the meeting of European Council in Seville in June 2002. The aim within the framework of these goals is to develop on one side public services and application, services and contents embodying e-Business as well in the demanding part and on the other side develops a safe wideband communication infrastructure for a presentation of this work in the application part. Lisbon strategy not only contains productivity and development but also social adaptation and employment. e-Europe 2005 favours a human resources approach in this connection, with an aim to develop participation (e-inclusion) and provision of equal opportunity. Due to this provision of services through different channels is evaluated as an important tool in the success of this application. Moreover, the aim in e-Europe 2002 is to increase internet connections, but in e-Europe 2005 the goal is to change this infrastructure into economic activities. [[50]] A study increase in all these activities related to e-Europe is accepted as a start [[51]] of materialization of Information Society for European citizens.

This work and the politics put forth in the infrastructure, communication and public services by EU affecting the environmental politics as well sometimes directly, sometimes indirectly. The developments at the Society level related to the e-government and development of information technology in particular affect and direct the politics regarding environment at the very first level. During these developments affecting not only the member countries but also the standing members and other countries, Turkey which has started the integration discussions with the Union too is getting affected in many ways and is following many developments on different topics including environment.

CONCLUSIONS

There is a widespread feeling in both academic circles and public in the world that political aspects and living attitudes in the economic development, production and comfort in the West are the main reasons for the environmental degradation. Although such an important responsibility is attributed to the West, today the environmental problems due to their characteristics above the systems are standing in front of us as a common problem of the whole humanity. The necessity to keep under control the ecological changes due to their dynamic characteristics through preventive measures with new approaches is increasing and at the same time gradually becoming difficult. As such, a series of common preferences and goals should be determined above the national level and formation of effective politics is becoming must to solve environmental problems. Out of the efforts put forth for solving problems in this connection, both the due to the political standards put forth as well as its important position from the point of effectiveness, EU today is the focal point as an international actor in the standards on environmental politics, due to their cumulative characteristics. The truth that humans themselves are a part and parcel of the environment, environmental organisations and arrangements in the public sectors will continue to be one of the trio in the community with environmental awareness (like EU). [[52]]

European Union, which as a reconstruction has emerged in the context of global and regional constitutions. It is a concrete, influential and important model bearing a potential to generate policies, norms, values and standards on the subject of environment as it does in many other issues. Many technological, economic and social instruments have been put forward concerning environmental policies along with the progress and the experience of EU acquis and integration. Information and communication technologies have been one of these main instruments in carrying out environmental policies of EU.

The establishment of information net and information-communication infrastructure in the members and standing members by EU as in other political fields is an important structural formation related to the environment. The environmental informatics (Environmental Informatics- EI) holds an important position today in EU countries in the attitudes towards environmental protection and organisational activities. [[53]] The developments related to this are of great importance for solving the environmental problems both in the formation as well as effectiveness of environmental management and politics. Thus, the revised environmental politics by EC; as one of the Regional Organisations working on this with a parallelism in the development; is of importance for our country during the EU discussions. EU environmental legislation reflects the progressiveness from the point of environmental models applied in the present day world. As pointed out in the Union decisions summarised above real, effective and good examples for all countries in the world towards the protection and development of environment have been produced. It should be mentioned here that the shape of formation of environmental management in all countries and particularly in Turkey should be considered and applied within this perspective.

REFERENCES

1. Ökmen, M. Kent, Çevre ve Globalleşme, Alfa Yayını, İstanbul, Turkey, 2003; 134 pp.
2. Keating, M. Yeryüzü Zirvesinde Değişimin Gündemi, UNEP Türkiye Komitesi Yayını, Ankara, Turkey, 1993; 17 pp.
3. Evans, B.; Theobald, K. Policy and Practice: Evalating Local agenda 21 in Europe, Journal of Environmental Planning and Management, 2003, 46(5), 781-794.
4. Kerkhoff, L. V. Integrated Research: Consepts of Connection in Environmental Science and Policy, Environmental Science&Policy, 2005, 8, 452- 463.
5. Toprak, Z. Globalleşmede Çevre Faktörü ve Çevre Korumacı İdeolojiye Politik Bir Yaklaşım, Yeni Türkiye Dergisi, 1995, 5, 175- 189.
6. Durrschmidt, J. Multiple Agoras: Local and Regional Environmental Policies Between Globalization and european Pathways of Transformation", Innovation, 2002, 15 (3), 193-208.
7. Naisbitt, J. Megatrendler: Asya, (Çev. U. Kaplan), Altın Kitaplar Yayını, İstanbul, Turkey, 1997; 31pp.
8. Ökmen, M. Kent, Çevre ve Globalleşme, Alfa Yayını, İstanbul, Turkey, 2003; 137 pp
9. Karkkainen, B. C. Post-Sovereign Environmental Governance, Global Environmental Politics, 2004, 4 (1), 72- 96.
10. Trittin, J. The Role of the Nation State in International Environmental Policy, Global Environmental Politics, 2004, 4 (1), 23- 28.
11. Türkiye Çevre Vakfı, Türkiye'nin Çevre Politikası Nedir? Ne Olmalıdır?, T.Ç.V. Yayını, Ankara, Turkey, 1987; 5 pp.
12. Keleş, R. Çevre ve Siyaset, İnsan, Çevre ve Toplum, Eds: R. Keleþ, İmge Kitapevi, Ankara, 1992; 162- 178.
13. Keleş, R.; Hamamcı, C. Çevrebilim, İmge Kitapevi, Ankara, Turkey, 1993; 216 pp.
14. Quevauviller, Paul. Science-policy Integration Needs in Support of the İmplementation of the EU Water Framework Directive, Environmental Science&Policy, 2005, 8, 203- 211.
15. Ökmen, M. Uyum Sürecinin Ekoloji- Politiği: AB ve Türkiye'de Çevre Sorunları, AB Yolunda Türkiye: Müzakere Sürecinin Ekonomi Politiği, Ed: M. Dikkaya, Alfa Yayını, İstanbul, Turkey, 2006; 330- 380.
16. Öcal, M. İktisat, Çevre ve Avrupa Topluluğu'nda Çevre İdeolojisi, Türkiye Günlüğü, 1989, 3, 51- 62.
17. Ökmen, M. Uyum Sürecinin Ekoloji- Politiği: AB ve Türkiye'de Çevre Sorunları, AB Yolunda Türkiye: Müzakere Sürecinin Ekonomi Politiği, Ed: M. Dikkaya, Alfa Yayını, İstanbul, Turkey, 2006; 330- 380.

18. Johnson, S.P.; Corcelle G. The Environmental Policy of European Communities, International Environmental Law and Policy Series, London, UK, 1989; 16 pp.
19. Ökmen, M. Uyum Sürecinin Ekoloji- Politiği: AB ve Türkiye'de Çevre Sorunları, AB Yolunda Türkiye: Müzakere Sürecinin Ekonomi Politiği, Ed: M. Dikkaya, Alfa Yayını, İstanbul, Turkey, 2006; 330- 380.
20. İktisadi Kalkınma Vakfı, Avrupa Birliği'nin Çevre Politikası, p.10-12, http://www.ikv.org.tr/pdfs/4f3a608d.pdf (29.12.2005).
21. Avrupa Çevre Ajansı, Avrupa'da İklim Değiþikliği ve Nehir Taþkınları, EEA Briefing, 2005/01, s.3 http://reports.tr.eea.eu.int/briefing_2005_1/tr/briefing_2005_1-tr.pdf (25.12.2005).
22. İktisadi Kalkınma Vakfı, Avrupa Birliği'nin Çevre Politikası, p.13, http://www.ikv.org.tr/pdfs/4f3a608d.pdf (29.12.2005).
23. Avrupa Çevre Ajansı, AÇA İþaretler-2004, p. 22, http://reports.tr.eea.eu.int/signals-2004/tr/TR_Signals_web.pdf (30.12.2005).
24. Wettestad, J. The Rapid EU Process: Causes and Consequences", Environment, 2004, 46 (9), 43- 44.
25. Türkiye Çevre Vakfı, Türkiye'nin Çevre Politikası Nedir? Ne Olmalıdır?, T.Ç.V. Yayını, Ankara, Turkey, 1987; 38 pp
26. Egeli, G. Avrupa Birliği ve Türkiye'de Çevre Politikaları, Türkiye Çevre Vakfı Yayını, Ankara, Turkey, 1996; 58 pp.
27. Türkiye Çevre Vakfı, Avrupa Birliği'nde ve Türkiye'de Çevre Mevzuatı, T.Ç.V. Yayını, Ankara, Turkey, 2001; 47 pp.
28. European Commission, Environment 2010: Our Future, Our Choice The Sixth Environment Action Programme of the European Community, http://europa.eu.int/comm/environment/newprg/index.htm (01.01.2006).
29. European Commission, Environment Directorate General- 2005 Management Plan, p.1-15. http://europa.eu.int/comm/dgs/environment/management_plan_2005.pdf (03.01.2006).
30. European Commission, EU Environmental Policy in 2004: developments, new evidence and outlook for 2005. http://europa.eu.int/comm/environment/pdf/work_doc.pdf (03.01.2006).
31. Ökmen, M. Uyum Sürecinin Ekoloji- Politiği: AB ve Türkiye'de Çevre Sorunları, AB Yolunda Türkiye: Müzakere Sürecinin Ekonomi Politiği, Ed: M. Dikkaya, Alfa Yayını, İstanbul, Turkey, 2006; 330- 380.
32. Görmez, K. Çevre Sorunları ve Türkiye, Gazi Yayınevi, Ankara, Turkey, 2003; 97 pp.
33. Heldeweg, M. A. Towards Good Environmental Governance in Europe, European Environmental Law Review, 2005, January, 2- 24.

34. Turnock, D. The Role of NGO's in Environmental Education in South-Eastern Europe, International Research in Geographical and Environmental Education, 2004, 13 (1), 103- 109.
35. Williams, R. J. Ontologies for ecoinformatics, Web Semantics: Science, Services and Agents on the World Wide Web (2006), doi:10.1016/j.websem.(2006.06.002).
36. Hilty. L. M. The relevance of information and communication technologies for environmental sustainability - A prospective simulation study, Environmental Modelling and Software, 2006, 21, 1618-1629.
37. Bosch, P. The European Environment Agency Focuses on EU-Policy in its Approach to Sustainable Development Indicators, Statistical Journal of the United Nations ECE 19, IOS Pres, 2002, 19, 4- 13.
38. Avrupa Çevre Ajansı, Kimiz? Ne yapıyoruz?, Nasıl Yapıyoruz?, p.4. http://org.eea.eu.int/documents/brochure2004/general_brochure_web-TR.pdf (25.12.2005).
39. Avrupa Çevre Ajansı, AÇA Stratejisi: 2004- 2008, Kopenhag, 2004, p.3. http://org.eea.eu.int/documents/strategy-docs/strategy_web-tr.pdf (29.12.2005).
40. eEurope+ Action Plan, p. 26. http://www.bilten.metu.edu.tr/eEurope+ (01.01.2006).
41. AÇA, EIONET Nedir? http://www.Eionet.eu.int/reportnet.html (20.11.2006).
42. European Commission, Environment 2010: Our Future, Our Choice The Sixth Environment Action Programme of the European Community. http://europa.eu.int/comm/environment/newprg/index.htm (01.01.2006).
43 Bosch, P. The European Environment Agency Focuses on EU-Policy in its Approach to Sustainable Development Indicators, Statistical Journal of the United Nations ECE 19, IOS Pres, 2002, 19, 4- 13.
44. Roberts, P.; Colwell, A. Moving the Environmental to Centre Stage: A New Approach to Planning and Development at European and Regional Levels, Local Environment, 2001, 6 (4), 421- 430.
45. Budak, S. Uluslar arası Çevre Düzenlemeleri Bağlamında Politika, Adalet ve Katılım, Çevre Sorunlarına Çağdaþ Yaklaþımlar, Eds. U. Yıldırım, C.Marin, Beta Yayını, İstanbul, Turkey, 2004; 401- 415.
46. European Commission, EU Environmental Policy in 2004: developments, new evidence and outlook for 2005. http://europa.eu.int/comm/environment/pdf/work_doc.pdf (03.01.2006).
47. DPT, e-Avrupa 2002. http://www.bilgitoplumu.gov.tr/eAvrupa/euindex_2.html
48. eEurope+ Action Plan, pp. 2-25. http://www.bilten.metu.edu.tr/eEurope+ (01.01.2006).
49. DPT, e-Avrupa+. http://www.bilgitoplumu.gov.tr/eAvrupa/euindex_3.html

50. DPT, e-Avrupa 2005. http://www.bilgitoplumu.gov.tr/eAvrupa/euindex_4.html

51. Porrero, I. P.; Monteiro, M. *e*Accessibility and European Technological research and development, Technology and Disability, 2002, 14, 55- 88.

52. Özel, M. Avrupa Birliği: Ekonomik Topluluktan çevre Topluluğuna (mı)?. G.Ü İ.İ.B.F Dergisi, 2003, 1, 221- 238.

53. Pillmann, W. Survey of environmental informatics in Europe, Environmental Modelling and Software, 2006, 21, 1519- 1527.

Chapter 2

The Role of Communication Technologies for Access to Environmental Information and Policies: Perspective of Turkey in the Process of Adaptation to the European Union

***Serhat Baştan*[1] *and Mustafa Ökmen*[2]**
[1]*Celal Bayar University, School of Applied Sciences, Manisa, Turkey*
[2]*Celal Bayar University, School of Higher Education, Salihli-Manisa, Turkey*

ABSTRACT

European Union, which as a reconstruction has emerged in the context of global and regional constitutions. It is a concrete, influential and important model bearing a potential to generate policies, norms, values and standards on the subject of environment as it does in many other issues. Many technological, economic and social instruments have been put forward concerning environmental policies along with the progress and the experience of EU acquis and integration. Information and communication technologies have been one of these main instruments in carrying out environmental policies of EU. The computerized information and communication has taken on a dominant role on flexibility in manufacturing, productivity in the usage of production factors, activity in decision support and problem solving processes and advantage in access to knowledge for citizens of democratic societies. Thus environmental matters have become one of the primary titles of European information society objectives. In this context, applications relevant to use of information-technology infrastructure on solving environmental problems in member or candidate countries are gradually gaining priority in the scope of e-Europe 2002 and e-Europe+ 2003 action plans and their successor initiatives. These technologies have multi-dimensional effects from technical and operational viewpoint such as monitoring/measuring environmental happenings and empiric data exchange between scientific and expert organizations to political and social aspects of the subject such as public access to information and formation of public opinion concerning environmental agenda about environmental matters. Consequently, it is possible to propound that information and communication technologies have become the one of the main parameters of environmental issues. Depending on this suggestion, we investigated the Turkish and EU policies regarding the role of information and communication technologies in the creation of environmental information, its distribution and public access to environmental information sources. EU's policies and technology-based initiatives have importance at the great extent in this period of serious debates about Turkey's becoming a member of the Union. Furthermore, Turkey geopolitically and geostrategically stands in a focus of regional environmental problems due to its industrial development potential and its crossroads location in the center of Black Sea and the Mediterranean catchment area among the important energy and transportation routes. Therefore, potential of access to environmental information is vital to handle ecological problems of this region. The paper is based on analyzing these potential and possible future developments from the perspective of Turkish and EU affairs.

INTRODUCTION

It is possible to say that information and communication technologies have taken a position among the main parameters of environmental problems. In the light of this proposition, in this study role of information and communication technologies to track the environmental problems, their solution and pursuit of environmental agenda in the politics of Turkey and Europe will be evaluated on comparative basis. In this connection the political strategies of European Union and introductions related to the communication technologies are of great importance because of the serious discussions held for Turkey's membership to the Union. Moreover, geopolitically and geostrategically Turkey looks like a focal point in the regional environmental problems due to its industrial developmental potential, position on an important intersectional route of energy as well as transport and location in the heart of southeast Europe in particular Black Sea/East Mediterranean basin. The potential to reach the environmental information has thus immense importance in tackling with the regional environmental problems. The presentation is based on the questioning of this potential in the light of future developments within the context of European Union.

1. ENVIRONMENT AND ENVIRONMENTAL PROBLEMS AS A FACTOR SUPRANATIONAL

Industrialization and urbanization are the main source of environmental problems. By industrial revolution, the idea of dominating environment came into being. Industrialization at first by industrial wastes caused depletion of forests and other resources, extinct species, air, water, soil and noise pollution. Besides it brings some other problems as well.

Sensitivity to environmental problems in 1960s accelerated in 1972 by Stockholm Environmental Conference. Beyond developed-underdeveloped, socialist-capitalist, and similar dichotomies, the conference with many countries, except the Soviet Union, put forward emergency of environmental problems. Over-systems property of environmental problems was discussed first at the conference as well.

Environmental problems in different economic systems were discussed whether they have some characteristics or not. Although a controversial issue the two views, environmental problems are free from economic systems and the two phenomena are related to each other to a great extent, have prevailed.

For the first view, environmental problems have equal importance and characteristics both in liberal and command economies. They claim that political systems and economic systems have no sense something different pertaining to environmental problems. As a matter of fact, environmental problems of both systems corroborate their views. Regarding development and industrialization levels, population density and such issues have caused various dimension differences in various geographical and climatic conditions. However, economic systems, political regimes and dominant ideologies have not been influential concerning environmental matters. Environmental problems in a country may even emerge at various dimensions. Effective use of public enterprises emerges at various dimensions. Effective use of public enterprises comparing to that of private ones have affirmed that environmental issues are free from systems. [[1]]

Whoever responsible is and whatever the reason may be, a serious deterioration of ecosystems and the beginning of its hazardous effects against the life on the earth are turning into one of the main crisis of the humanity today. This situation menaces the whole world along with every aspects from developed to underdeveloped, the Easterner to Westerner, the Southerner to Northerner or socialists to capitalists. As the nature has increased the frequency of warning signals pertaining to the serious results of this fact which has been continuously being expressed by scientists, it has become a major national issue on which various expedients and attempts of solutions have been intensified. Today it has been understood that environmental problems with their cumulative characteristics are closely interrelated with many economic, social, cultural, political and administrative parameters. Ecological deterioration is one of the most important problems such as fast growth in population, consumption of natural resources, starvation, terrorism and regional conflicts being caused by occurrences which are more or less related to industrial and urban phenomenon which has been accelerating since the beginning of second half of the 20th Century. Moreover it is possible to claim that this is the most important problem among others threatening common future of the whole world. Consequently, environmental questions identically require global solutions in accordance with their global characteristics. There is a common thought in scientific community and public opinion in the world that Western policies, economic development models and lifestyles depending on consumption and prosperity are primary reasons of environmental exhaustion. Though the responsibility is imputed to the West at the great extent by this way, environmental problems stand as a common matter of the whole humanity with their supra-system characteristics. Moreover, the necessity to keep the ecological changes with their dynamic attributes under control is gaining more importance and constantly getting harder yet. Consequently, these facts require some common preferences and objectives to be determined and effective policies to be generated beyond the national level.

2. ENVIRONMENTAL PROBLEMS AND ENVIRONMENTAL POLICIES IN TURKEY

Population growth, industrial development and environmental pollution, which threaten natural resources in every country on earth, have made environmental problems one of the most important concerns of man in the last quarter of the Twentieth Century. Inventories of environmental problems are a working method in several developed and developing countries. In some countries, in fact, in addition to organizations concerned with the problems themselves, there are also offices whose special function is to conduct inventories of the problems. Such a need has also been felt in Turkey, where environmental problems have recently received increased attention-especially air, solid and water pollution.

Urbanization and industrialization are supposed to be the two fundamental causes of environmental problems. This two causes, though the characteristics and the dimensions are different, are also concerned with the environmental problems not only in developing countries -such as Turkey- but also similarly in developed countries. It consists of the problems and qualities especially on "non-industrialization" and "urbanization-not parallel with-industrialization". Environment is the main problem among the others.

Turkey came across with the environmental problems later in contrast with that of industrialized countries. Since the last period of 1970s, environmental problems have reached the level to threaten men's health especially in urban and coastal regions. Therefore, this point has been taken into consideration within the framework of the politics developed parallel with the rising of problems. Environmental problems have seriously been derived from industrialization and urbanization. [2]

Turkey faced the issues of industrialization and some environmental problems caused by industrialization at the beginning of 1960. However, in some rural areas (due to factors different than industrialization) environmental problems have lasted for years. Therefore, there are some fundamental problems in rural areas. In most agricultural areas there is the danger of erosion and in many regions forest areas are being removed for agriculture. Turkey began its industrialization process in the 1950s with the rise of urbanization in the same years. Thus, the tremendous problems of urban areas increased, especially in the 1970s. Now, Turkey has all sorts of environmental problems, except ones related to radiation. Some environmental problems are at lower levels, while others are at dangerously high levels.

Nowadays, there are increasing environmental problems in Turkey. Air pollution is one of the most serious problems in almost every city in Turkey. Also there is the issue of transportation, noise pollution and the increasing number of shanties in urban areas. One of the biggest causes of these increasing problems is unplanned urbanization. There is a faster increasing rate of urbanization in Turkey than in many industrialized countries. For the reason that most immigrants come to the big cities, these cities as a result have not been capable of compensating for the needs of their settlers. Thus the problems have increased over time.

In the early years of the Turkish Republic, environmental policies were rarely discussed. Of course, one could hardly expect environmental policies to be a very important issue for a nation just after the end of a war, with enormous problems waiting to be solved. In addition, after Ankara was selected as the capital city of Turkey, the bureaucrats were interested only in the environmental problems of Ankara. Yet, in the years after the Turkish War of Independence, the reconstruction of Anatolian towns corrupted during the war was one of the most important problems to be solved.

During the first years of the republic, both the establishment of the chairmanship of Public Improvement in Ankara and the Ankara Plan, prepared by H. Jansen, can be counted as the first arrangements related to the environment. Accordingly, in 1930s many important arrangements are made on that basis. In 1930, the Municipality Law (code 1580) and the General Healthcare Law (code 1593) came into practice. However, until 1960, except for the Forest Law, there were no major laws to protect the environment. Only a few minor regulations came into practice, none of which appear to be related to the problems that resulted from industrialization. However, at that times there were some important improvements in Europe.

If Turkey had applied some of the environmental precautions that Europe did, we wouldn't have the same environmental problems today. Turkey started a development programme in 1960 with progressive plans. Industrialization and development efforts were included in the 1961 Constitution. Starting with that period, Five-Year Development Plans were prepared. After the second Five-Year Development Plan, the "Environment Sector" took place according to these plans. Particularly after the 1972 Stockholm Environment Conference this issue became very significant. When we look at the third Five-Year Development Plan, we see that the main problem in solving environment problems was the negative effect on funds for development and for targets of developing with industrialization.

The Fourth Five-Year Development Plan gave a large place to environment. This plan emphasized man-environment relations and health in the period of urbanization, as well as saving the country's historical and natural wealth with economic integrity. The Fifth Five-Year Development Plan was more sensitive. The main approach of this plan with regard to the environment dealt not only with removing current pollution and preventing possible pollution, but also with using and improving resources in the best way so that the next generations could benefit from them as well. Despite the preceding plans, the Sixth Five-Year Development Plan had different characteristics in giving a wide and detailed place for environmental problems.

Turkey, which has not had any serious environmental problems until the 1970s, has not shown important developments in producing environmental policies. Some policies in a few problematical areas remained the same from the Ottoman Empire time to 1980s. Environmental pollution appeared in 1980s and sensitivity of the people for environmental pollution increased in the 1970s especially after the 1972 Stockholm Environment Conference. The Assembling Rank of the Environment and some other foundations are the results of these developments. The sudden and the rapid changing period of environmental politics started in 1980s.

This new period started with the 1982 Constitution, which has topics about the environment directly. In 1983 this period continued with the development of regulations such as the Environmental Law, the Culture and Natural Sources Saving Law, and the Straits Law. In a short time, new legal arrangements about every topic of the environment were made. Despite these positive developments, there were many inconsistencies and contradictions in the policies of this period. In addition, laws change often and mistakes are made, so legal permission is likely to be given for polluting the environment. There still has not been a consistent or positive environmental mentality at the administrative level. [3]

3. ENVIRONMENTAL POLICIES AND GLOBAL AND REGIONAL EFFECTS

The need and idea for national politics in Turkey has come to the front first time during the UN Environmental Conference organised in 1972, which played an important role in the EU environmental politics as well. As a matter of fact, the topic of environment was first time considered in the 3rd Five Year Development Programme prepared in 1973, stressing the need for inclusion of measures for environmental protection and regulation in legislations without hindering economic development. In this way, the plan to determine the environmental problems in Turkey was started, which entered an important stage with the articles in 1982 Constitution together with the Environmental Law that become effective in 1983. The changes published continued to formalise during 6th and 7th Development Plans with a backing from this law. [4] Particularly in the 7th Plan attention was drawn towards the global and regional developments, and collaborations in the economic globalisation together with fast developments in the regional integration activities were underlined. EU, NAFTA and Pacific Region are mainly becoming important within this framework and a three poled structure comes into being. Parallel to these developments the international norms and rules seem to gain weight with a support from the increasing force of sanctions. It is not possible now for the countries to prepare effective politics in this connection with abstraction from the developments in other countries. [5]

Other important development in the formation of environmental politics in Turkey has been National Environmental Strategy and Application Plan. The plan prepared with the financial sources of World Bank and a technical support from the Ministry of Environment and Secretariat of DPT (Government Planning Office) was published in 1998. One of the most important features of the Plan is efforts for an establishment of a balance between the development and environment. The Plan, in addition to its providing integration with the National Agenda 21, it has also been adapted to the approaches underlined in the UN development programme. The goals for the next 20 years in the environment have been determined together with this, at the same time a projection outlined for the 8th Five Year Development Plan. [6] In view of the global and regional developments same approaches with increments are continued in the 8th Plan. According to the Plan, the regional integral developmental activities are developing under an interactive influence with the globalisation in the economic field. Foreign trade is developing new norms and standards in the fields of environment and intellectual rights during the globalisation period and activity of international organisations is steadily increasing in this field. The organisations above the national level are steadily becoming determinative particularly in the economic politics of EU, and intensified joint interferences are following in the political and social fields as well. [7]

4. ENVIRONMENTAL INFORMATION AND ENVIRONMENTAL POLICIES OF TURKEY IN THE PROCESS OF ADAPTATION TO THE EU

4.1 Environmental Policies of Turkey in Process of EU

The environmental politics in Turkey are embodying many problems during the adaptation to the EU beyond this net and detailed approach by DPT. Environmental politics should be handled with an integral perspective due to its close relationship with all economic, social, political and administrative parameters. The work on the adaptation to EU Environmental Legislation is presently carried out by DPT. However, this work should not remain on the papers; Turkey should prepare its politics in a concrete form together with alternative scenarios during this transitory phase. It should not be forgotten that this topic has dimensions beyond the legal adaptation procedure. It can be said that these evaluations and goals regarding both environmental politics as well as environmental management are positive as compared to the earlier plans and are in accordance with the steps outlined above. But these should be handled in a more detailed and developed form in the application plans and should be applied effectively after preparing an organisational infrastructure. This is a must for Turkey; during this phase of fast developing informative society and technology together with new administrative thinking and developments; which has covered measure steps in its relation with EU. The cumulative characteristics of environment beyond the national and international levels, and international and regional organisations in this field together with the standards and rules originating in this connection too make it necessary. [8]

Environmental politics and organisational developments in this connection are important topics both in regional and global dimensional perspective for whole humanity, as in the group of countries named today as EU. Turkey has a close relationship with this topic due to its geo-economic, geo-strategic as well as political position as well as its requirement for local, regional and global situation. The steps followed within the framework of developments and expectations particularly during the entrance and adaptation steps to EU, the environmental politics in the Union and a similar approach in Turkey towards their joint evaluation is very important. As a reality, the acceptance of environment as a mere pollution problem has passed this level in parallelity with the regional-global reciprocal effects in the development, and speeded up the work on new concepts in the environmental governing with many regional and global organizational impacts coming to the forefront and similar other developments, both at UN, OECD and EU levels as well as international and above the national level organisations. European community as one of the effective organizations in leading the international developments on environment has been attempting to develop joint politics and actions for the members starting from 1970. These attempts have become more effective particularly after the enforcement of 5th Action Programme in 1993, in accordance with new global and regional developments and have become a concrete in the sense of communities environmental politics. The environmental politics of the Union together with other economic and social politics is closely related to Turkey both as a regional country as well as a country on its way to enter EU. Turkey is expecting structural changes in many fields by selecting socio-economic politics of the EU and its arguments as the goal for its integration and has submitted its stableness in this connection to the world community through Ankara agreement. It is openly known that Turkey's membership in EU requires a change in the present structural features or bring these to a suitable EU level in the fields covering economy to agriculture, industrial relation to infrastructural administration and environment. One of these changing fields will be environmental management and politics. [9] Particularly with the start of negotiation period this topic has become more important and environmental politics is one of the major fields seen in the EU standards.

The technology and production stages form one of the economic aspects of the environmental protection. In particular the steps taken by Turkey at the Customs Union level makes it necessary to undertake detailed studies on the economic dimension and cost/competition analysis of environment, due to its aim for a close connection with EC. The developmental model applied by the EC until now has in general been on the negative side. Insufficient use of labour force against the overuse of natural resources has resulted in a fast degradation of environmental conditions in Europe. As a result of these developments EC has underlined a series of goals in the member countries for development and spread of clean technology, their encouragement and application of environmental costs to the market within the framework of micro/macro-economic politics. The aims under question clearly point out that some solid measures are taken for the future developments in the field of environment. Turkey is a little below the European standards due to high rate of industrialisation as well as overuse of pollution producing technologies, but it is possible to say that it is more advantageous than most of the European countries as regards the measures to be taken for protection of environment. For a continuation in the industrial development and production of industrial products together with an entrance of the Turkish industrial products in the EC market, adaptation to the EC environmental and quality standards under question accepted by the producers of the EC are important for competition. The target for creation of full competitive surroundings at the same time within the Customs Union, Turkish Firms involved in the productions for industry particularly need investment in environmental infrastructure as well as standardisation, quality control and education in environmental protection for an easy trade with the EC. The protection and development of Turkish industrial efficiency both in quality as well as environmental safety will be possible through a production on European standards. [10] While upgrading the standards in quality and environment both conditions for competition and legislation for environment should also be adapted to the Community legislation and applied too.

The last and final steps in the responsibilities of Turkey regarding the adaptation to the EU environmental legislation have been laid through "Participation Joint Document" prepared after the 1999 Helsinki Summit, which entrusts Turkey with some short and medium range responsibilities within the framework of EU legislation and Copenhagen Criteria. [11] The "Participation Joint Document" (PJD) declared first in November 2000, and reviewed in April 2003, is a document determined by EU giving the conditions required to be fulfilled by Turkey for a full membership in EU. Turkey's responsibilities for adaptation to the EU environmental politics have been discussed on short and medium term basis in accordance with the reviewed form of PJD announced in April 2003. According to this, the short term topics are listed as; acceptance of a programme for a transfer of the acquisitions/expenses for adaptation of acquisitions and preparation of a plan for financial investments depending on real verified current public and private sources/legislation framework regarding the environmental protection legislation, legislation for water quality, starting transfer and application of legislation regarding the joint pollution control and management of wastes/enforcement and application of the directives for Environmental Impact Evaluation (EIE). The medium term responsibilities have been outlined as; integration of legislation into internal law for environmental protection and strengthening organisational, administrative and observatory capacities including data collection[12].

In addition, in the Reviewed National Programme related to the Turkey's acceptance of EU legislation and published in the Official Bulletin on 25 July, 2003; a list is given mentioning the achievements in the environment and the points which will be given a priority after this. According to this, the law regarding the participation of Turkey in the European Agency and European Information and Observation net has been put into force after its publication in the Official Bulletin of 28 January, 2003. At the same time, "Turkish Environmental Legislation Investigation Project" supported by the MEDA fund has been completed in 2002. A comparison of the present environmental legislation with EU legislation together with the determination of gaps and calculation of investment costs on the basis of directives needing investment have been made within this project. In the financial tables under this heading the results from "Turkish Environmental Legislation Investigation Project" the estimations related to the investment needs are given, but net investment needs will be produced exactly after the completion of "Integrated Environmental Adaptation Strategy Project for Turkey". First of all for the development of sectoral strategy followed by the finalisation of strategies under question, integrated adaptation strategy and general financial strategy will be prepared towards an adaptation and application of legislations.

The question of integration of principals of sustainable development in all other sectoral politics definitions and application is seen among the 2003 medium term goals of PJD. A project index has been prepared with this precaution and has been approved by the Commission within the framework of 2004 Financial Cooperation Programme. Moreover, the Draft Law pertaining to the changes in Environmental Law present in the Turkish Parliament includes the topic of communication on environmental information. The education of personal has been started in April 2003 for formation of national data bank and development of administrative structure within the " environmental information communication and development of administrative structure" component undertaken within the Project on Capacity Building in the field of Environment in Turkey (TR-362.03), which was accepted in the 2002 Financial Cooperation programme. Lastly in the National Programme it has been stressed that within the EU environmental legislation changes regarding the reporting, adaptation to the legislation, its application and infrastructure will be considered after the start of work. In the list of priorities in the National Programme; improvement of air and water quality, protection of nature, activation of waste management and strong environmental impact assessment, adaptation to the Strategic Environmental Assessment directives, management of environmental noise and chemicals, genetically modified organisms and nuclear safety problems have been taken into consideration[[13]].

4.2 Environmental Information, Communication Technologies and Environmental Policies of Turkey in Process of EU

It will be useful to open up a little more the relations on the environmental politics between EEA and Turkey as one of the topics in the National Programme, because the steps taken in this connection during the EU adaptation period and efforts spent have a special importance. Turkey applied for membership to the EEA in June, 1999; the discussions were started by the EU commission with a representation from Turkey during the meeting held on 27 March, 2000 at Brussels. The topic under question has been taken into consideration in the meeting held on 17-18 May, 2000 with the authorities of Environmental General Directorate of EU Commission visiting Turkey during the renegotiations on "Participation Strategy" towards Turkey's participation in the programmes and membership in EEA. In this meeting it has been openly stated by the EU Commission authorities that EEA is the first EU organisation opening the doors towards the full membership of Turkey in EU. In the meeting of "EU and Standing Member Countries Ministers for Environment" a participation agreement was signed between "Turkey and EU on EEA and European Information and Observation Net" on 9 October, 2000 at Brussels. A validation of the Agreement under question was ratified by Turkish Parliament on 23rd January, 2003 with the 4794 number law. Within this framework as in other countries, EEA requires that the data needed should be collected according to the EU standards, all required information on environment and other fields (industry, agriculture etc.) should be provided by Turkey during the period of adaptation to the European legislation, provision of all required information to the decision makers, provision of all information and data during the formation of politics, provision of information and data during the Environmental Impact Assessment for an acceptance of the projects prepared by the Standing member Countries of the EU as a prerequisite for getting financial support from EU, formation of a standard not attained due to different formats used by the public sectors/organisations, and private organisations related to the data on environment in Turkey, for this purpose in order to get support on the "Project on Structural Development and Communication on Environmental Information" was started by the "Ministry of Environment and Forests" with a participation from other organisations and sectors in this connection under a coordination of the Ministry, so as to create a National Data Bank for observation and reporting on environmental situation in Turkey, activation of EIONET in Turkey in order to reach on time the data of all member countries, and finally provision of data and information on environment to the public[14].

According to the progress report 2003 for Turkey, a limited progress has been made on such topics as air quality, nature conservation, chemicals, noise pollution, nuclear safety and protection from radiation, for the purpose of operation of EU legislation and strengthening of administrative capacity in Turkey. However there is a great need to spend more efforts in Turkey, both from the point of legislative adaptation and within the application aim of all subheadings in this section. In the case of air quality, steps are to be taken for applications on the adaptation of acquirements in legislation including development of observation system in air quality. The legislation on waste management although coinciding with acquirements, requires that more efforts should be spent on its application and transfer to the internal law. There is a need for allotting enough financial sources to this area. Due to this, the Twinning Project on Waste Sector which will run for 2 years as a part of 2003 Financial Cooperation "Project on Administrative Capacity Building in the Environment in Turkey" and expected to start on September, 2004, will allow an adaptation to the concerned legislation. According to the report the application of White Paper on Regulation of Wastes has been planned for last quarter of 2005, the Regulation on the Control of Waste Oils has been put into force after its publication on 21st January, 2004 in the official bulletin number 25353.

As regards the transfer and application of legislation on water quality to the internal law, in addition to the acceptance of a new framework law regarding the water sources, additional efforts should be spent on the adaptation of standards on drinking water and wastewater disposal with the acquirements of legislation. Due to this, the enforcement of Regulations on Water Products and Control of Water Pollution are planned for 2005. The regulations for prevention of waters from nitrate pollution originating from the agriculture were published on 18.02.2004 in the official bulletin number 25377. Although a series of regulations have been put in to force on nature conservation, legislative adaptation level is still low. Acceptance of a white paper on nature conservation and transfer of application of the legislative decrees to the internal law regarding the birds and habitat is of great importance. As such, changes are planned to be made in the new laws to be issued on conservation of biodiversity and nature, 2872 Environmental Law, 3167 Land Hunters Law, 2873 National Parks Law and 2863 Protection of Cultural and Natural Objects Law. The changes were made in the 3167 Land Hunters Law with the number 4915 Law and it was enforced after its publication in the official bulletin number 25165 on 11th July, 2003. Legislative changes are planned to be made in this field till the end of 2006. The efforts are needed to be spent on the chemicals and genetically modified organisms in order to achieve an adaptation in this connection. An evaluation of the Regulations on Dangerous Chemicals is planned till the end of 2006. [15]

One of the papers supplementing the development of environmental politics in Turkey during EU integration and prepared by the European Commission for Turkey, emphasizes the important topics under the Environment heading in the 2004 Progress Report. According to this, aims of the community for environmental politics are to protection of environment for the present and future generations and improvement of sustainable development. In the integration of environmental protection to the other political fields in the community basic thoughts are; preventive measures, rule of polluter pays, destruction of environmental pollution at the source and sharing of responsibility.

Acquirements embody more than 200 legislative regulations for side legislation, water and air pollution, management of waste and chemicals, biotechnology, nature conservation, industrial pollution and risk management, protection from radiation and noise. The adaptation to the acquirements needs important investments, but at the same time these provide important advantages for the human health, and reduce degradation of fisheries, landscape planning, buildings and forests. The presence of a well equipped strong administration at national, regional and local levels is a must for the effective application of environmental regulations. In the report developments achieved since the last progress report and topics not showing any development were also dealt. Accordingly, from the last progress report till now some advances have been made regarding the transfer to the internal law, but no progress has been recorded in the integration of environmental topics to other politics. In the side regulations limited advances have been made. A new regulation and law for EIA (environmental impact assessment) as well as regulations for reaching data information have been accepted. Turkey has also approved the UN Climate Change White Paper Agreement. In the field of air quality limited advances have been made, changes have been made in the regulations regarding the fuel quality of petrol and diesel, as well as CO_2 emissions and enlightening consumers on fuel economy in the new passenger buses. Moreover, in July 2004 a communication has been published on the quality of petrol and diesel fuels.

Some advances have been made in the field of waste management. Pouring wastes into the Mediterranean and a law related to the reduction of this pollution originating from their movements beyond the borders has been accepted. At the same time regulations for the wastes from packaging, soils from excavations, control of building and demolition wastes, waste oils, management of wastes from the establishments around the ports and bays, and for batteries and accumulators have been accepted. In the case of water quality, limited advances have been made. The regulation related to the protection of water sources from nitrate pollution has been accepted. A limited advancement has been made in the nature conservation. The changes have been made in the regulations regarding the application of CITES agreement. Advances have been made in the fields of industrial pollution and risk management as well as genetically modified organisms. In the case of chemicals, very limited advancement has been achieved. In this connection the changes towards a prohibition of marketing and use of pesticides and related products have been accepted. As regards the noise no advance has been made, but a limited progress has been achieved in the case of nuclear safety and protection from radiation. The regulation for the wastes originating from the use of radioactive substances has been accepted. As a result of merger of Ministries of Environment and Forest in 2003, Turkey has taken some measures to strengthening the administrative capacity. Together with this, limited advances have been made in the development of administrative capacity including the conflict on powers and application rules. A governing statute has been accepted for the formation of a special commission that will provide scientific support on environmental problems. [16]

The formation of environmental policies has come up as one of the important parameters during the discussions, and its measure aspect has been information and communication infrastructure regarding this topic. Although the steps taken by Turkey regarding EEA and EIONET have been dealt with above, it will be useful to open up this topic a little more, because as in other topics the importance of information and communication is great for Turkey and EU particularly in the adaptation of politics in environment. As such, the steps taken to form e-Europe within EU and affords in connection with the participation and supplementations from Turkey in E-Europe+ are very important. In view of the e-Europe+ planning, E-Turkey was started under the chairmanship of TUBITAK in Turkey and many working groups formed results were discussed at the 1.Information Council on 10-12 May, 2002. In the Urgent Action Plan prepared by the 58th Government, place was given to the e-Transformation Turkey Project and for this project Undersecretariate of GPO was entrusted with the responsibility for coordination, follow up, evaluation and leading. In addition, in the circular; number 2003/12 dated 27 February, 2003; issued by the Prime Ministry, aims, organisational structure, and application rules of the e-Transformation Turkey Project have been determined. An Information Society Section Head was established within the GPO in March, 2003 in order to fulfil this responsibility. The aims of this project are; first of all a reorganisation of the policies and legislation on information and communication technology within the framework of EU legislation, application of the action plan to Turkey under e-Europe + for candidate countries, development of mechanisms for the participation and decision making capacity of public with the help from information and technology, supplementation by official administration which should be open to accountability, make maximum use of information and communication technology while securing services from the government, application and supplementation of the good administrative rules, spreading the use of information and communication technology, in order to reduce the loss of source extravagancy merger of the official investment projects which resemble or show duplication, observation, evaluation and coordination among the investing official services, and enlightening the activities of private sector within the above rules. The work done with the coordinators from 28 April, 2003 created after the approach accepted for the execution of e-Transformation Turkey Project, e-Turkey interference started after its participation in the e-Europe+ interference with the use of work done till now, scientific knowledge and formations and ended up with the preparation of Short Term Action Plan (STAP) within this framework. In the STAP a total of 73 projects were outlined in the fields of strategy of the informed society, technical infrastructure and security of information, education and human resources, legislative infrastructure, standards, e-Government, e-Health, e-Trade, and e-Environment.

The affords to create this information and communication infrastructure at the national level is closely related to the environmental policies and its formation, although it has gained momentum in the affiliation to the EU there is still a long way to go. The collection, storage, functionalising, distribution and communication of all sorts of information are critically important for the economic, societal and technical progress in our country as well like the countries entering information age. But in our country the importance of information in the economic progress and societal development has not been fully perceived by the Government. Although "information" is the basic factor in almost all fields like production, education, health, research, public administration, and information services and information technology are closely related to these fields, but until now its integration in the 5 years development plans at the national level in the information management and organisation has not been dealt with. Still we do not have national information politics embodying strategies for effective and productive management of information technology, sources, systems and services. Although we do come across few developmental cases on national information politics in our country from the year 1925 onwards, but their incompleteness and lack of integrity has not made it possible to achieve continuity in these affords. Until now no cooperation has been achieved among the universities, research-developmental organisations in the public and private sectors, TUBITAK, Government Institute of Statistics, Institute of Turkish Patents, Institute of Turkish Standards, National Library, Directorate General of Government Archives, in the Universities, School and Public Libraries, organisational archives, information centres and museums in the production, management and organisation of information at the national level. [17] However, it can be said that particularly in the period following the discussions with EU, important steps have been laid towards the formation of information net and information communication which will affect different fields including the environmental politics in a positive way. It can be said that activities on the e-Government together with internet as well as covering formation of information communication net and efforts for the creation of information net for telecommunication infrastructure on a wide scale has fastened this type of work on the way to EU.

It seems that, environmental politics has taken an unforgettable place in the relations between EU and Turkey within the acceleration gained following the 1999 Helsinki summit. In the last five years with a gradual speed up and concretisation of the relations, topic of environment has acquired a position among the foremost topics and important steps have been laid in the Turkey- EU standards towards the environmental politics. The topic of environment is holding an important position among the discussion headings in the new period starting on October, 2005. The use of EU funds has brought the topic of environment to the forefront and this can be evaluated singly as an important indicator as regards the importance given to the topic in the Turkey-EU relations. [18]

CONCLUSIONS

EU environmental legislation reflects the progressiveness from the point of environmental models applied in the present day world. As pointed out in the Union decisions summarised above real, effective and good examples for all countries in the world towards the protection and development of environment have been produced. It should be mentioned here that the shape of formation of environmental management in all countries and particularly in Turkey should be considered and applied within this perspective. If we look at the conditions of environmental management applied in Turkey an important backwardness as compared to the system put forward by EU is openly visible. A long and troublesome period is required by Turkey in order to reach such an understanding on environmental management for a successful transition and adaptation on its way to become a member of EU. National legislation should be reorganised in this direction and the truth is that both national and local environmental focal points will face some difficulties for reaching a level to apply this legislation. However, an upperhand of EU political authority as compared to the member countries and acceptance for adaptation to the Union legislation by the member country at the start does not leave any other choice in his direction. In a way, most probably adaptation of Turkey to the EU on the topic of environment will be more difficult and painful than political and economic integration. The effectiveness in this case and in the final analysis the important catalyser and easiest factor for a formation and application of successful environmental politics will be an establishment of an information and information-communication infrastructure. This seems as a must for Turkey on its way to EU, particularly its adaptation to the environmental politics or from the point of view of achieving the standards put forward.

Although an important step is laid towards the EU standards through the formation of legislation regarding the environment, it should be accepted that this is not everything. The environment is a change in the mentality and needs willpower towards transfer and application of legislation beyond this. It should not be forgotten that an integrated approach is needed because environment is intermingled with technological, economic, social, cultural and political administrative factors. The approaches bringing to the forefront scientific technology and information should be transferred to applicability by tackling with the topic in national, regional and global perspective in the 7th and 8th Five Year Development Plans and particularly in the 9th Plan, and policies creating balance between the national requirements and regional and global realities should be created and transferred for application.

REFERENCES

1. Arslan, N.; Ökmen, M. The Economical and International Dimensions of the Environmental Problems, Environmental Problems in the Black Sea Region and the Role of the Voluntary Organizations. Building and Environment, 2006, 41 (8), 1040-1049.

2. Ökmen, M.; Parlak, B. An Agenda for Urban Air Pollution of Environmental Problems and Policies: The Turkish Case. Interdisciplinary Environmental Review Issue (IER), 2006, 6(1), 96-107.

3. Ökmen, M.; Parlak, B. An Agenda For Environmental Policies of Local Government in Terms of Globalization & Localization: The Türkish Case. Proceedings of ECOSUD 2001 Conference, Alicante, Spain, June 6-8, 2001.

4. Türkiye Çevre Vakfı, Avrupa Birliği'nde ve Türkiye'de Çevre Mevzuatı, T.Ç.V. Yayını, Ankara, Turkey, 2001; 93 pp.

5. DPT, 7. Beþ Yıllık Kalkınma Planı, DPT Yayını, Ankara, 1995, 1 pp.

6. Algan, N.; Mengi, A. Turkey's Sustainable Development Policies in the EU Accession Process", European Environmental Law Review, 2005, 95-108.

7. DPT, 8. Beþ Yıllık Kalkınma Planı, DPT Yayını, Ankara, Turkey, 2000; 2 pp.

8. Ökmen, M. Politika ve Çevre, Çevre Sorunlarına Çağdaş Yaklaşımlar, Eds.: U.Yıldırım, C.Marin, Beta Yayını: İstanbul, Turkey, 2004; 326-378.

9. Yaşamış, F. D. Çevre Yönetiminin Temel Araçları, İmge Yayını, Ankara, Turkey, 1995; 175 pp.

10. Türkiye Çevre Vakfı, Çevre Yönünden Gümrük Birliği ve Türkiye, Türkiye Çevre Vakfı Yayını, Ankara, Turkey, 1996; 18 pp.

11. Algan, N.; Mengi, A. Turkey's Sustainable Development Policies in the EU Accession Process", European Environmental Law Review, 2005, 95-108.

12. İktisadi Kalkınma Vakfı, Avrupa Birliği'nin Çevre Politikası, 29 pp. http://www.ikv.org.tr/pdfs/4f3a608d.pdf (29.12.2005).

13. İktisadi Kalkınma Vakfı, Avrupa Birliği'nin Çevre Politikası, 30 pp. http://www.ikv.org.tr/pdfs/4f3a608d.pdf (29.12.2005).

14. T.C Çevre ve Orman Bakanlığı, Avrupa Çevre Ajansı. http://www.cedgm.gov.tr/aca/aca.htm (08.01.2006).

15. Avrupa Komisyonu Türkiye Delegasyonu, Avrupa Birliği'nin Çevre Politikaları http://www.deltur.cec.eu.int/default.asp?lang=0&pId=3&fId=10&prnId=7&hnd=1&ord=6&docId=317&fop=0 (10.01.2006).

16. DPT, Türkiye'nin AB'ne Katılım Sürecine İlişkin 2004 Yılı İlerleme Raporu, DPT Yayını, Ankara, Turkey, 2004; 118 pp.

17. DPT, Sekizinci Beþ Yıllık Kalkınma Planı Biliþim Teknolojileri ve Politikaları Özel İhtisas Komisyonu Raporu, DPT Yayını, Ankara, Turkey, 2001; 15 pp.

18. Ökmen, M. Uyum Sürecinin Ekoloji- Politiği: AB ve Türkiye'de Çevre Sorunları", AB Yolunda Türkiye: Müzakere Sürecinin Ekonomi Politiği, Eds.: M. Dikkaya, Alfa Yayını, İstanbul, Turkey, 2006; 330- 380.

Chapter 3

Drinking Water System, Izolation Zones and Pressure Management Strategy in Denizli

Mahmud Güngör[1], Ümit Cantürk[2] and Şule Vardar[3]
[1] Denizli Municipality, Deputy Mayor DENİZLİ/Turkey
[2]TEKSU Teknolojik Su Kontrol Sistemleri ve Yönetimi A.Ş.
[3]Denizli Municipality Water and Sewerage Manager DENİZLİ/Turkey

ABSTRACT

Denizli, which is in the western part of the Aegean Region, is a city of industries and tourism. Province has an area of 11.868km^2 and a population of 480,200. After 2009 local elections, with the new law, 13 districts and 10 villages were merged with Denizli Municipality. Drinking water is supplied to the city from 3 springs and groundwater sources. After chlorination, water is disseminated to the 99 per cent of the municipality in a healthy and adequate manner. However, especially in the old distribution networks high amounts of loss are now taking place. Denizli has about 3,25 per cent of surface and 1,41 per cent of groundwater sources of Turkey. Due to poor pipe material used, poor construction and long term use of pipelines, financial losses are increasing. Denizli Municipality has recently received a loan of 30 million Euro from World Bank, which will be used for reconstruction and development of drinking water, rainwater, and sewerage systems. Within this scope, similar recent applications related to demand-based pressure and management around the globe reveal the fact that Denizli Municipality will be able to reduce leakage, resulting especially from poor customer connections, to bottom levels in existing pipe network. The Municipality has divided the network into 150+ District Meter Areas (DMA), almost 40 per cent of which will be pressure-managed based on demand patterns. All these DMAs shall be metered continuously and data shall be transferred to a central office for on-time reaction of the municipality to prevent further losses.

Introduction

Denizli, a city of industry and tourism in the southwestern part of the Aegean region is surrounded by the provinces of Afyon and Burdur in the east, Aydın and Manisa in the west, Muğla in the south and U°ak in the north. The province lies between 28° 30' - 29° 30' east longitude and 37° 12' - 38° 12' north latitude. It is located at the northern hillside of Akdağ and on the plateau slightly divided by branches of Aksu stream -a branch of Buyuk Menderes River. The province covers an area of 11,868 km^2.

It includes 18 counties namely; Acıpayam, Akköy, Babadağ, Baklan, Bekilli, Beyağaç, Bozkurt, Buldan, Çal, Çameli, Çardak, Çivril, Güney, Honaz, Kale, Sarayköy, Serinhisar and Tavas (Figure 3.1), covering an area of 11.868 km^2. The population density is 64 persons/km^2. Changes in the population during 1945-2008 are shown in Table 3.1. The population data of 2007 and 2008 has been taken from the results of "Census based on address" which is coordinated by Turkish Statistical Institute (TÜİK).

Table 3.1: Population Changes in Denizli Between 1945-2008 (Turkish Statistical Institute, 2009)

Years	*Central County Total (ind.)*	*Central County+ County Towns (ind.)*	*Towns+Villages (Central) (ind.)*
1945	65,424	20,162	45,262
1950	72,935	22,017	50,918
1955	84,965	29,909	55,056
1960	96,826	48,925	47,901
1965	118,744	64,331	54,413
1970	143,362	83,372	60,990
1975	171,521	106,902	64,619
1980	206,561	135,373	71,186
1985	220,206	169,130	51,076
1990	271,346	204,118	67,228
2000	400,719	275,480	125,239
2007	494,961	323,151	171,810
2008	508,870	479,381	29,489

After 2009 local elections, based on the Greater city law, 13 districts and 10 villages were merged with Denizli Municipality: Akkale, Bereketli, Gümüþler, Kayhan, Kınıklı, Servergazi, Hallaçlar, Üçler, Korucuk, Bağbaşı, Gökpınar Cankurtaran and Göveçlik Municipalities and Bozburun, Eskihisar, Goncalı, Hisar, Kadılar, Karakurt, Saruhan, Şirinköy, Güzelköy (Yiğenağa) Karakova villages were added, in accordance with the Municipal Law number 5393-article 11. The total population of towns and villages which are connected to Denizli Municipality is about 29.489, whereas the

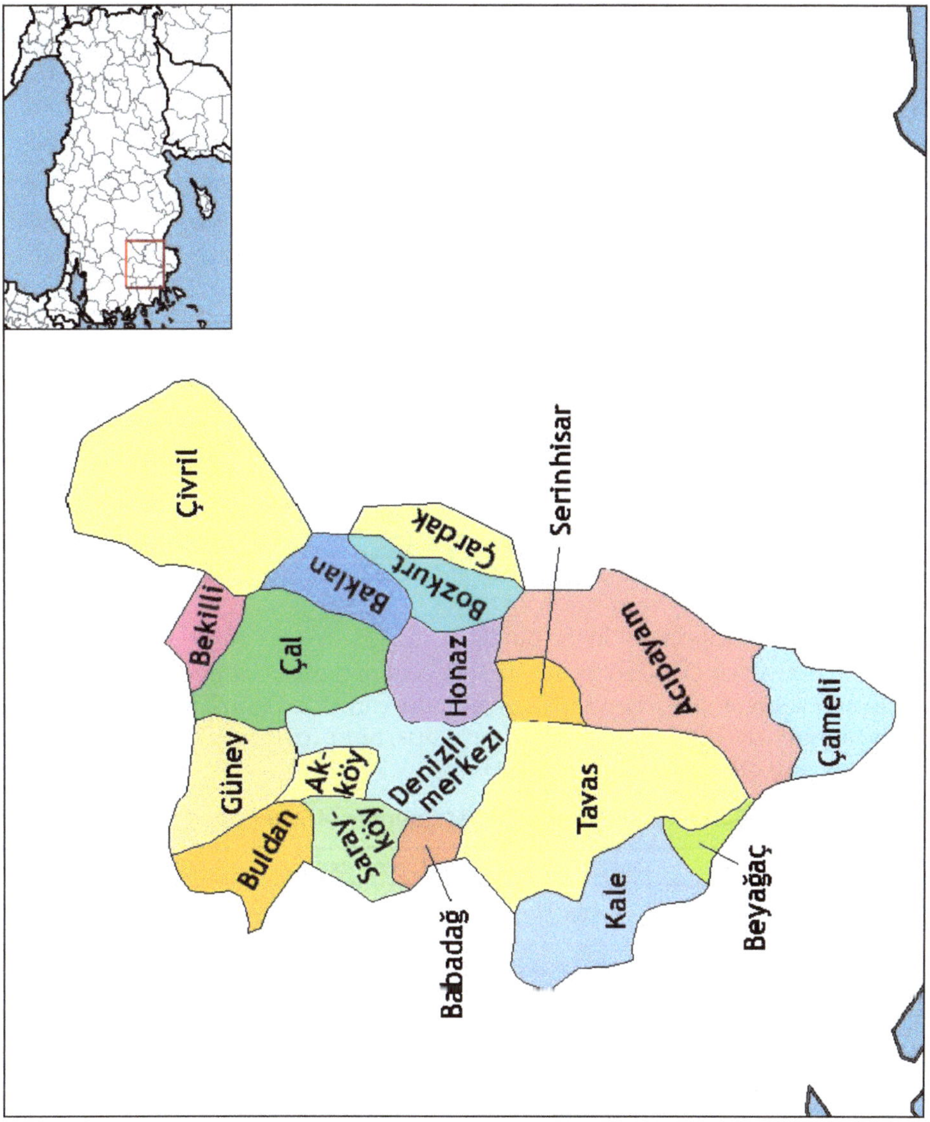

Figure 3.1: Location of Denizli Province in Turkey with its Districts

population of city center and surroundings is now about 508.870. The total population of the province is now around 917.836.

The economical development of the province surpasses Turkey's average migration from country side to Denizli Municipality. Besides, when increase of population in Denizli is analysed, it is clear that population increases linearly without showing any fluctuations. Census results from 2000 and 2007 are shown in Table 3.2.

Table 3.2: Population Change in Towns Between 2000 and 2007 (Turkish Statistical Institute, 2008)

Town Name	Population (ind.)	
	2000	2007
Bağbaşı	11,995	21,437
Bereketli	2,866	10,203
Güveçlik	1,990	2,064
Gümüşler	16,008	24,558
Hallaçlar	3,580	4,336
Kayhan	5,716	8,168
Kınıklı	19,699	25,413
Gökpınar	2,739	3,533
Servergazi	5,588	10,469
Akkale	4,198	6,753
Üçler	3,560	9,290
Korucuk	1,836	1,809
Başkarcı	2,782	2,433
Total	**82,557**	**130,466**

DRINKING WATER SUPPLY SYSTEM AND WATER SOURCES

Drinking water is supplied to the city of Denizli from 3 springs and groundwaters. After chlorination, water is disseminated to the 99 per cent of municipality in a healthy and adequate manner. However, high amounts of physical loss occur, especially in the old distribution networks. Denizli has the 3,25 per cent of the surface and 1,41 per cent of groundwater sources of Turkey. Usable water per capita in Denizli is about 2.874 m^3, which is higher than the average of Turkey. Thus, our province is in a better state in comparison with other provinces (Table 3.3).

Table 3.3: Surface and Ground Water Sources of Turkey and Denizli

Sources	Denizli (hm^3)	Turkey (hm^3)	Denizli/Turkey %
Surface Water Sources	3.188	98.000	3,25
Groundwater Sources	198	14.000	1,41
Total	**3.386**	**112.000**	**4,66**

Denizli municipality provides drinking and utilization water from different sources as shown below. After the merger of neighbouring municipalities to Denizli, province currently owns 70 drinking water depots and 103 drinking water wellbores.

Within the context of increasing amount of water obtained from these sources and wells; 150 L/sec water has been compiled and included in the system by drainage of Gokpinar Yukari Santral Source and 25 L/sec water has been compiled and included in the system from Gokcen Source. Also, 30 L/sec water of Israfil Gutter, which is in the boundaries of villages Hisar-Kadılar has been included in the system with the Project of İsrafil Gutter Source, completed in 2008 (Table 3.4). Extra source with 30 L/sec water has been added to the system since 2008. Additional storage capacity of drinking water make the network more safe. With the merger of neghbourhood municipalities, the number of water storage with different volumes increased from 21 to 70 and so, water storage volume has increased up to 73,355m^3.

Drinking Water Sources (2009)

Sources	*Water Flow (L/sec)*
Gokpinar Drinking Water	600
Derindere Drinking Water Source	183
Benlipinar Water Source	20
Wellbores	402
Yukari Santral Water	100
Israfil Water Source	30
Total	**1335**

Table 3.4: Drinking Water Sources and Capacities for 2006-2008

Source Name	*Unit*	*Water Quantity Produced*			*Maximum Water Catch Capacity*
		2006(1)	*2007(2)*	*2008(3)*	
Derindere drinking water Source	lt/sec	196	91	74	82.22
Gökpınar drinking water Source	lt/sec	674	579	554	615.56
Benlipınar Water Source	lt/sec	20	20	20	22.22
Wellbores	lt/sec	392	556.1	536	717.10
Water sources in towns	lt/sec			156.00	186.00
	lt/sec	1,282	1,246	1,340	1,623
	m^3/day	110,765	107,663	115,776	140,236
Total	**m^3/year**	**40,429,152**	**39,297,010**	**42,258,240**	**51,186,082**
Gökpınar Dam	lt/sec				900.00
Akbaş Damı	lt/sec				750.00
	lt/sec	1,282	1,246	1,340	3,273
	m^3/day	110,765	107,663	115,776	282,796
Total	**m^3/year**	**40,429,152**	**39,297,010**	**42,258,240**	**103,220,482**

Population in the Future and Forecasting of Related Water Requirement

It is expected that the most effective factor in forecasting population is the growth rate of Denizli in industry which is much higher than Turkey's average. So, classical projection techniques used for population forecasting must be reevaluated within the current economical vision of Denizli. In the light of these evaluations, predicted population forecasts are shown in Table 3.5.

Table 3.5: Final Population Projection

Belediye Isimleri/ yıllar	*Gerçekleşen*		*Projeksiyon*							
	2007	*2008*	*2010*	*2015*	*2020*	*2025*	*2030*	*2035*	*2040*	*2045*
Denizli	323,151	339,000	370,624	459,656	562,194	678,876	813,237	933,455	1,043,065	1,113,781
Başkarcı	2,433	2,468	2,698	3,346	4,093	4,942	5,921	6,796	7,594	8,109
Bereketli	10,203	12,232	13,373	16,586	20,286	24,496	29,344	33,682	37,637	40,189
Göveçlik	2,064	2,075	2,268	2,813	3,441	4,155	4,977	5,713	6,384	6,817
Hallaçlar	4,336	4,456	4,872	6,042	7,390	8,924	10,690	12,271	13,712	14,641
Korucuk	1,809	1,805	1,974	2,448	2,994	3,615	4,330	4,971	5,554	5,931
Servergazi	10,469	11,451	12,520	15,527	18,991	22,932	27,471	31,532	35,234	37,623
Üçler	9,290	10,654	11,648	14,446	17,669	21,336	25,559	29,337	32,782	35,005
Akkale	6,753	7,228	7,902	9,800	11,986	14,474	17,338	19,901	22,238	23,746
Bağbaşı	21,437	23,291	25,464	31,581	38,625	46,642	55,873	64,133	71,664	76,522
Gökpınar	3,533	3,664	4,006	4,968	6,076	7,337	8,789	10,089	11,273	12,038
Gümüşler	24,558	26,106	28,542	35,398	43,294	52,280	62,627	71,885	80,326	85,772
Kayhan	8,168	8,595	9,397	11,655	14,254	17,213	20,620	23,668	26,447	28,240
Kınıklı	25,413	26,355	28,813	35,735	43,706	52,777	63,223	72,569	81,090	86,588
Merkez Toplam	**453,617**	**479,381**	**524,100**	**650,000**	**795,000**	**960,000**	**1,150,000**	**1,320,000**	**1,475,000**	**1,575,000**
Köyler[(1)]	41,344	29,489	29,371	28,898	28,126	27,375	25,732	24,794	23,891	23,020
Kabul			29,400	28,900	28,150	27,400	25,750	24,800	23,900	23,000
Toplam	**494,961**	**508,870**	**553,500**	**678,900**	**823,150**	**987,400**	**1,175,750**	**1,344,800**	**1,498,900**	**1,598,000**

Denizli Municipality is responsible for supply, treatment and distribution of drinking water within its boundaries. Total current and future water consumption for Denizli Municipality is shown in Table 3.6.

Table 3.6: Average Daily Water Consumption in Denizli in 2008 and Future Projections

Types of Consumption	*Water Consumption (m^3/day)*		
	2008	*2025*	*2045*
Households	49,827	101,760	204,750
Corporate, Commercial and Industrial	7,075	36,706	71,588
Permitted consumption uncharged (Non-profit)	473	947	1,553
Görünen Kayıplar	58,400	61,909	76,381
Total	116,093	201,873	355,243

Numbers and types of consumers belonging to 12th period of 2008 are given in Table 3.7.

Table 3.7: Consumer Types and Numbers

Consumer Type	*Consumer Record Number*	*Active Consumer Number*
Houses	118,577	112,377
Houses (Town consumers)	6,934	6,565
Commercial	17,605	16,530
Tourism incentive	17	17
School	117	103
Bakery	57	52
Factory	395	352
Construction	2,814	2,270
Official	180	132
Mosque	134	128
Public utilities	26	21
Official Hospitals	13	11
Disabled	861	860
Houses (Own source)	20	2
Commercial (Own source)	76	13
Official public utilities	16	14
Veteran and Martyr families	187	183
Disabled (Town consumers)	58	54
Green area	336	336
Private Dormitories	17	17
TOTAL	**148,440**	**140,037**

Sectoral distribution of water consumers according to four main categories is given in Figure 3.2.

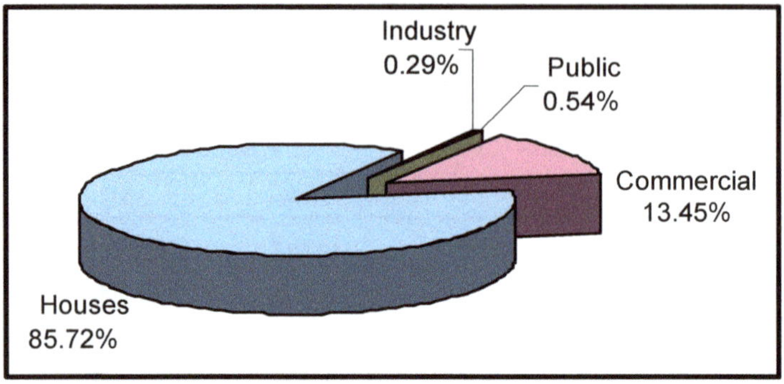

Figure 3.2: Sectoral Distribution of Water Consumers

The consumption of drinking water for 2008 is given in Tables 3.8 and 3.9 by categorizing consumer groups.

Table 3.8: Water Consumption by Consumer Category in 2008

Consumer Category	*Connection Number*	*Measured Consumption m^3*	*Unmeasured Consumption m^3*	*Total Consumption m^3*
Houses	113,476	10,648,854		10,648,854
Commercial	18,830	1,677,594		1,677,594
Official	298	737,506		737,506
Industry	404	79,166		79,166
Water supplied to towns	6,565	7,538,104		7,538,104
Garden utilization	336	88,263		88,263
Income-yielding water	139,909	20,769,487	0	20,769,487
Mosques	128	172,584		172,584
Water Loss			21,316,169	21,316,169
Non-profit water	128	172,584	21,316,169	21,488,753
Total	140,037	20,942,071	21,316,169	42,258,240

Table 3.9: Household Consumption per Capita in Denizli in 2007 and 2008

Years	*Consumer Category*	*Total Consumption, m^3*	*Population Ind.*	*Water Consumption per Capita lt/ind/day*
2007	Houses	10,648,854	339,000	86.06
	Water supplied to towns	7,538,104	140,381	147.12
	Houses+Towns	18,186,958	479,381	103.94
	Gross(1)	20,942,071	479,381	119.69
2008	Houses	10,872,245	323,151	87.87
	Gross(1)	15,156,338	323,151	128.50

With a population of 479,381, water requirement of Denizli Province is about 1335 L/sn. The water obtained from sources, reservoirs and wellbores as per the data available from the end of 2009 is only enough for Denizli Province. With growing population, in the year of 2012 water requirement will be 1628 L/sn. Therefore current water sources are insufficient and there is an immediate need for new sources to fulfill the future water requirements. A Master Plan is being developed to treat current sources and also to transform Honaz Akba° Dam into drinking water facility which is planned as irrigation facility. It will take a long time to finish the water treatment plant and the transfer line which will provide water from Akbaş Dam. Until the year the investment is realized; water requirement, which can't be obtained from Gokpınar and Derindere sources, has to be reinforced by the water from wellbores and new sources need to be found.

URBAN INFRASTRUCTURE AND GIANT INFRASTRUCTURE PROJECT

Within the scope of this project; drinking water, sewage, rain water integration systems of Denizli have been evaluated as a whole. The implementation areas within the scope of this project and priorities have been determined by the Municipality and loan demanded from World Bank. Suggestions by holistic approach are summarized under the main headings given below.

DRINKING WATER DISTRIBUTION NETWORK

The first water network of Denizli Municipality was constructed in 1952, improved in 1958. In this network, cast iron pipes were used with diameters ø60 - ø200 mm. Construction of drinking water system covering major part of residential area was completed by Provincial Bank, during the years 1975-1981. In this network, ø80-500 mm AÇB pipes were used. The economical life-span of these pipes has come to an end and on monthly basis an average of 600 water break-downs have been detected in the drinking water network meaning a water loss occurs of about 60 per cent. Operation of drinking water system is not automated as yet. When low pressure occurs in any zone, municipality authorities turn down the other line valves in order to supply more water to these zones. Operation system is not preplanned; it is run according to daily requirements.

In drinking water and sewage systems, maintenance is more oriented to fulfill urgent requirements. 90-95 per cent of breakdowns detected in the network is defined as branch breakdown (consumer linkage). This breakdown is caused by consumers since they use cheap materials instead of proper quality materials in the linkages which they install on their own. In order to assess the water loss, frequent and reliable measurements are necessary. Especially, water loss in high pressure networks reaches high values and it is known that more breakdowns occur in these networks.

Within the boundaries of Denizli Greater City, drinking water pipes (network) of residential area will be replaced (Figure 3.5). The aim of this endeavor is to construct drinking water infrastructure investment and to run it efficiently and effectively in order to get rid of water loss problems. For this reason, network replacement and

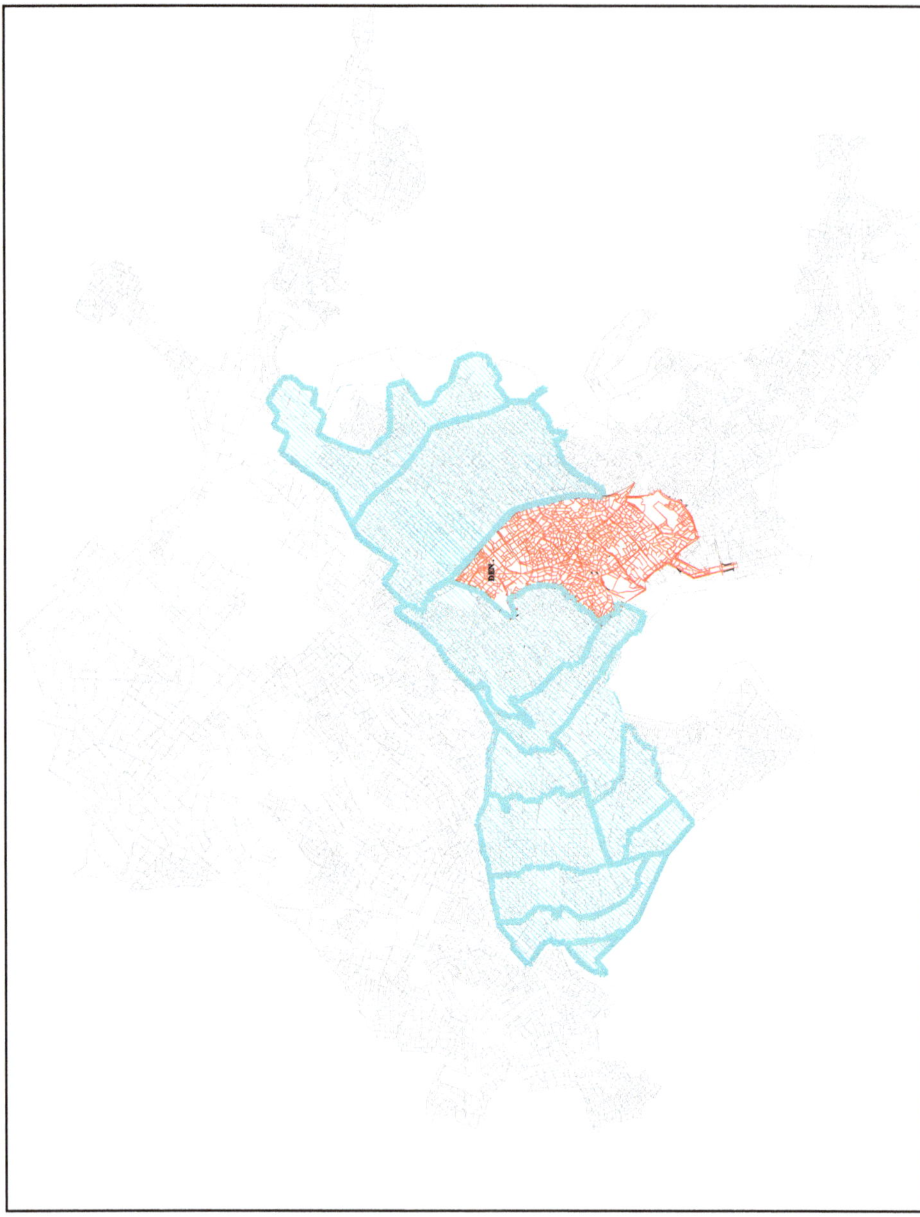

Figure 3.3: Project Area to be Invested

Table 3.10: Denizli Municipality Drinking Water System Components

Bölge adı	*Çapiara göre uzunluklar (m)*													
	100 mm	*150 mm*	*200 mm*	*250 mm*	*300 mm*	*350 mm*	*400 mm*	*500 mm*	*600 mm*	*700 mm*	*800 mm*	*900 mm*	*1000 mm*	*1500 mm*	*TOPLAM*
Denizli-1.2	104,428.50	2,489.00	2,030.00	1,030.50	2,314.50		2,874.00	11,007.50	1,089.00			121.50	2,175.50	9,811.00	**139,371.00**
Denizli-1.3	237,165.00	4,866.50	4,296.50	1,241.50	327.50	301.00	840.50	1,619.00	2,825.00	2,679.00	351.00		2,080.50		**258,593.00**
Denizli-1.4	248,024.50	8,374.50	3,129.00	2,609.00	2,068.00	450.50	821.00	5,989.00	8,101.50	13,282.00	408.00	1,845.00	233.50		**295,335.50**
Denizli-1.5	315,274.00	7,869.50	4,321.50	3,134.50	2,775.00		5,701.00	765.50	5,868.50	6,794.50	1,338.50				**353,842.50**
Denizli-1.6	19,143.50	633.00	1,192.50	1,621.50	1,109.50				5,550.00	6,349.00					**35,599.00**
Denizli-4.2	10,679.50	426.00	551.50												**11,657.00**
Denizli-4.3	28,575.00	1,286.50	565.50	122.00	1,048.00		73.50								**31,670.50**
Denizli-4.4	30,961.00	825.00	573.00	725.50	30.00		907.50								**34,022.00**
Denizli-4.5	14,683.00	185.50	5,007.00	128.50	3,034.00		1,844.00	756.00	749.50	1,939.00	9,008.00		254.00		**37,588.50**
Denizli-5.1.1	3,080.00	222.50	2,149.00	526.50	3,178.00		1,017.50								**10,173.50**
Denizli-5.1.2	21,423.50	867.50	2,388.50		4,267.50		381.00								**29,328.00**
Denizli-5.2	41,782.50	440.00	2,414.00	2,406.00	3,997.00		3,086.00	778.00	543.50						**55,447.00**
Denizli-5.3	58,458.50	845.00	376.50	483.50	131.50		790.00		543.50						**61,628.50**
Denizli-5.4	23,091.00	259.00	318.50	436.50			1,251.50			1,565.00					**26,921.50**
Denizli-6.1	51,361.00	829.50	3,164.00	3,574.50	476.00		9,835.00	1,617.00	2,084.00	2,383.50			1,561.00		**76,885.50**
Denizli-6.2	56,590.50	4,949.00	721.00	1,911.50	617.00		4,377.00								**69,166.00**
Denizli-6.3	58,457.00	5,256.00	667.50	1,236.00		1,230.00	2,949.00								**69,795.50**
Denizli-6.4	102,585.50	5,860.50	1,388.00	406.00	126.50	1,644.00	283.50	806.00							**113,100.00**
Denizli-6.5	14,583.00	3,447.00		1,075.00	2,143.00	1,378.00									**22,626.00**
TOPLAM (MERKEZ)	**1,440,346.50**	**49,931.50**	**35,253.50**	**22,668.50**	**27,643.00**	**5,003.50**	**37,032.00**	**23,338.00**	**27,354.50**	**35,113.50**	**11,105.50**	**1,845.00**	**6,304.50**	**9,811.00**	**1,732,750.50**

Table 3.11: Denizli Greater City Drinking Water System Components

Zone Name	*Lengths Due to Diameteres (m)*														
	100 mm	*150 mm*	*200 mm*	*250 mm*	*300 mm*	*350 mm*	*400 mm*	*500 mm*	*600 mm*	*700 mm*	*800 mm*	*900 mm*	*1000 mm*	*1500 mm*	*TOTAL*
TOTAL (CENTER)	1,440,346.50	49,931.50	35,253.50	22,668.50	27,643.00	5,003.50	37,032.00	23,338.00	27,354.50	35,113.50	11,105.50	1,845.00	6,304.50	9,811.00	1,732,750.50
Towns Total	869,872.50	45,593.00	21,705.50	16,631.50	12,979.50	3,084.50	10,854.00	7,759.00	9,277.50	822.50	168.00				998,747.50
TOTAL	2,310,219.00	95,524.50	56,959.00	39,300.00	40,622.50	8,088.00	47,886.00	31,097.00	36,632.00	35,936.00	11,273.50	1,845.00	6,304.50	9,811.00	2,731,498.00

renovation will be carried out. The projected lines with respect to their pressure zones are given in Tables 3.10 and 3.11.

SCADA SYSTEM

SCADA system is developed to have a central water management system to operate the water supply and distribution system properly and efficiently. This system, shortly named as SCADA (Supervisory Control and Data Acquisition), is designed to monitor the conditions of the water supply network system of the city, water level in the storage tanks and the pumps in the pumping stations, to gather information, audit, and control the system remotely. The main purpose of the project is, to monitor the production, storage and distribution of drinking water systems that are under the responsibility of the Municipality, and to run the water loss by a computer based central management system to be installed in Water Breakdown Unit which is considered by the management.

With this project, for Water Loss Management , isolated measurement zones will be developed with easy control areas. Flowmeters according to requirement, control valve for regulating pressure, electronical BKV controller, sewage holder, air valve, and equipment will be furnished into measurement chambers which are to be located at the entry of these zones.

Target purposes with Scada System are as follows:

- ✰ Ensure athe real-time coordination between stations through central monitoring, command and automation.
- ✰ Operation convenience and reliability with central monitoring, command and automation.
- ✰ Ensure time, energy and workpower savings of the personel who are responsible for maintenance and command of the plants through minimizing their daily tasks.
- ✰ Prevent the delay of interventions and to minimize the water outage and loss originated from this occasion by geting informed about breakdown instantly.
- ✰ Control storage levels and to prevent reservoir overflows.
- ✰ Ensure time and equipment savings through minimizing the breakdowns caused by negligence.
- ✰ Take the necessary precautions automatically by predefining the situations which tend to cause breakdowns.
- ✰ Include new stations and/or isolate zones in SCADA and Water Loss Control System easily.
- ✰ Keep the statistical data belonging to stations and isolate measurement zones in a reliable way properly.
- ✰ Ensure distribution of current water to consumers continuosly and sufficiently in accordance with quality standards.
- ✰ Realize more reliable, practical and economical operations.

INSULATED ZONES

Due to poor pipeline material used, poor workmanship and long term use of pipelines, current non-revenue water amount in Denizli has risen up to 60 per cent. The Municipality, until now, has not applied an active leakage detection program. However, management now aims to reduce and minimize the leakages in the city.

The Municipality is experiencing the benefits of this new method and needs a water loss reduction plan for future. Even though old pipes in the city will be replaced by ductile pipes with the credit obtained, pressure regulation connected to utilization quantity will be applied in order to minimize non-revenue water. Similar applications based on pressure regulation around the globe reveal the fact that it will be possible to reduce leakage, derived especially from poor customer connections to the midway of existing pipelines. Moreover, even the replaced pipes are sometimes exposed to extreme pressure, this is caused by minimum necessary pressure in critical points of each pressure management zone. With this project, the Municipality will separate the network over 150 Districts Metering Areas (DMA). These DMAs will be measured continuously and the data will be transferred to a central office to prevent more loss through instant intervention of the municipality.

Leakage reduction is one of the most important issues facing utilities, due to the increasing shortage of drinking water in large areas of the world. Control valves are the main components in preventing the supply of excessive pressure in order to reduce leakage. Pressure management has two main benefits in reducing water consumption. Firstly it reduces the amount of leakage escaping from the network via existing holes or cracks in the reticulation pipes and secondly it reduces the pressure induced stress in the pipes by minimizing the pressure surges. It helps to reduce the likelihood of formation of new leaks or bursts. As a utility owner, Denizli Municipality has appointed a network designer to select the optimum solution for each zone. There are a number of different options for a successful design of pressure management schemes for different design requirements.

The meter measures the flow of DMA, accurately giving the ability to review and analyze the flow in the zone and flow to the consumers (residential, commercial, and industrial measurements). The pressure reducing valves shall be located and adjusted to reduce pressure to meet the minimum requirements of water users without any surplus pressure. The dimensions of DMAs vary between 500 and 3000 service connections (Figures 3.4–3.6).

The calculated simple payback period of the City of Denizli pressure management area is 2 (two) years. With the added benefit of the reduction in water network breakdowns, the payback is further reduced to less than 1.5 years.

Municipality is expecting the following outcomes from pressure management process:

- ✰ Reduction in leakage.
- ✰ Reduction in the frequency of bursts; subsequent savings in repair costs due to reduced leakage.

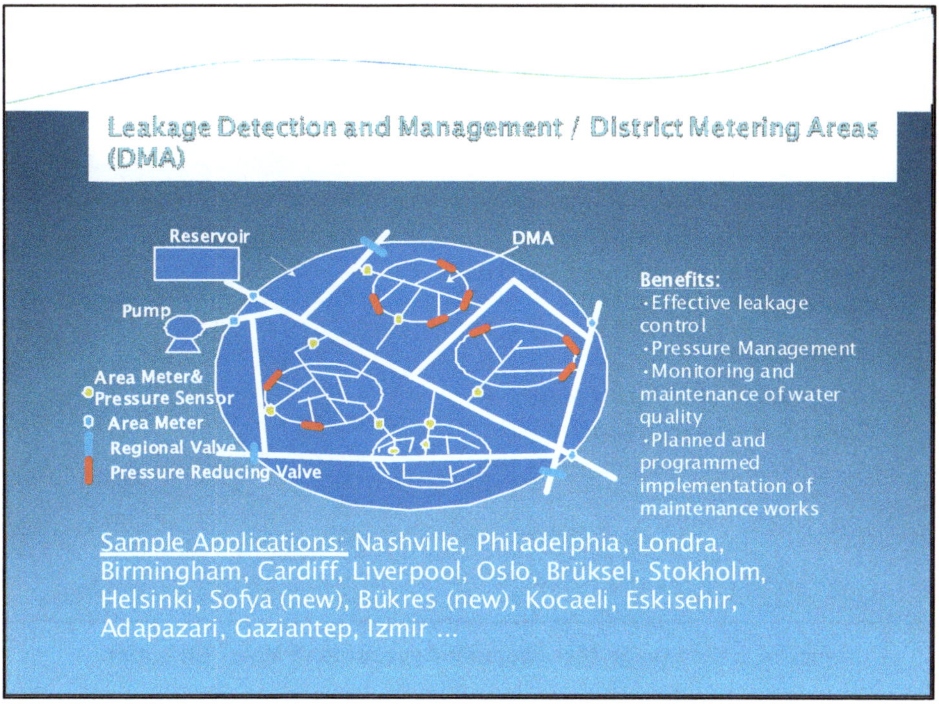

Figure 3.4: Schematic Display of District Metering Areas (DMA)

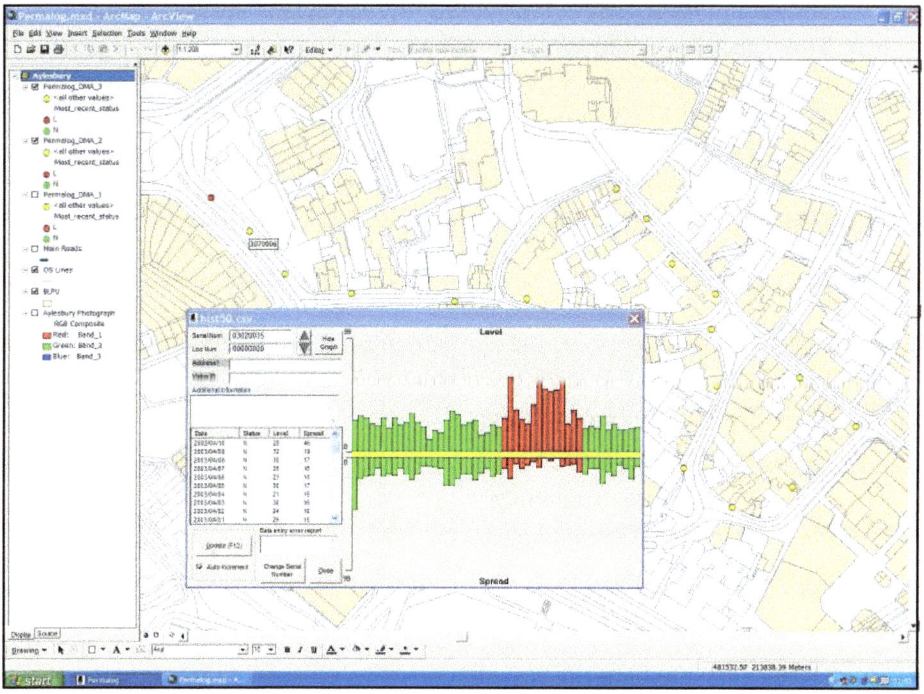

Figure 3.5: Display Screen of Leakage Management Program

Instruments Used In Meter Chambers:

- Isolation Valve
- Filter Strainers
- Meter or flowmeter
- Pressure Regulating Control Valve(BKV)
- Datalogger
- Electronic BKV Controller + Communication Module

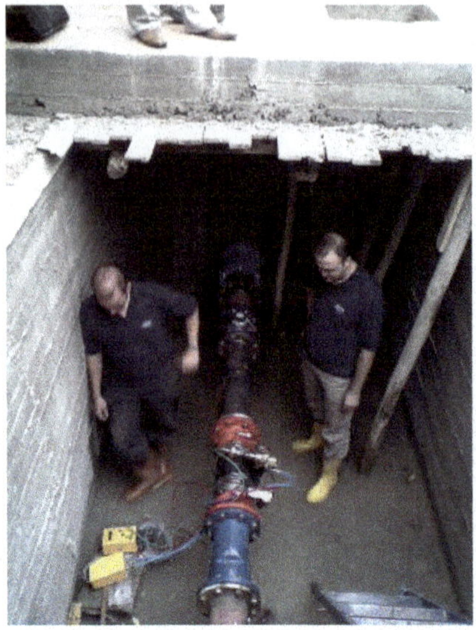

Figure 3.6: Leakage Management/Assessment: Meter Chamber

- ✰ Stabilization of the pressure, decreasing the possibility of pipeline movement and eliminating certain household plumbing problems.
- ✰ To be able to provide a more constant service to customers – large daily pressure variations may give customers an impression of a poorly managed service, and unnecessarily high pressures raise customers expectations and perceptions of what is adequate.
- ✰ Enable to standardize on pipes and parts which have a lower pressure rating, and are therefore cheaper.
- ✰ Assist demand management when flow restriction is necessary e.g. during drought.

In addition to these savings, the implementation and application of demand and leakage control through pressure management yields the following benefits:

- ✰ Reduction in background loss.
- ✰ Reduction in consumption.
- ✰ Reduction in burst frequency.
- ✰ Improved system performance.
- ✰ Extended asset life.

These benefits include:

Short term financial benefits associated with the costs of buying water from a bulk supplier, water treatment cost savings including the marginal costs of chemicals, power and sludge disposal and water reticulation costs. Long term benefits related to asset utilization life, reduction in pipe failures (burst frequency), savings in the costs of repairing main pipe bursts with extended asset life.

Indirect financial benefits also are created from more efficient use of existing water supplies. In particular, reduced water loss ensures that existing water supplies can meet future increases in demand. This can defer construction of new water infrastructure such as dams, treatment plants, reservoirs and water networks. A degree of drought security is also possible as long as the lower water demand maintains the security of water supply for longer periods.

System verification of pipes and cables provides improved information about the distribution system. This enables staff to become more familiar with the system, including the location of pipes and valves. This information assists staff to respond more quickly to emergencies such as network breakdowns and provides an early indication of any increases in water loss from leakages.

SEWAGE AND WASTEWATER TREATMENT

Since 1984, various sewage projects have been prepared and most of the city has been connected to sewage system. The collected wastewater is discharged from three points (Bozburun, Bağırsak Gutter, Ellez Gutter) and ø200-ø1000mm diameter pipes are used in the sewage network and ø1000-ø1600 mm diameter pipes are used in the collectors. The first level of Denizli Wastewater Plant was constructed in 2005 and finished in March 2007. Treatment process chosen is Active Sludge System. Denizli Province Bagirsak Gutter Basin Sewage Practice Project was completed by Provincial Bank and the project was approved in May 2001. Bagirsak Gutter Sewage Project involves Bagirsak Gutter Basin which includes Denizli (center) southeast residential areas existing in Gokpinar Dam water catch basin, constructed for drinking water supply (Figure 3.7).

One part of this basin was considered in the project limits which was approved in 1984 and included collectors with Ø600 mm, construction of a part of the network was completed. Project area was considered as large as possible, because project purpose is to protect Gokpinar Dam water catchment basin. In addition to east region of Denizli Center, towns like Tekkekoy (Gokpinar), Bağbaşı, Kınıklı and Kayhan were included in the project area. Besides the projects certified between 1986 and 2001, Bagirsak Gutter and Bozburun Region Sewage Network Project was carried out. The Municipality has connected parts with suitable elevation to 1600 mm Ellez collector in order to protect the environment by repairing Bagirsak Gutter Collector. Thus, the wastewater discharged without treatment is led to existing Eskihisar Wastewater Treatment Plant, and so discharging with treatment follows.

ESKIHISAR WASTEWATER TREATMENT PLANT

Eskihisar (center) wastewater treatment plant was planned with 2 levels according to required flows of the years 2005 and 2025 so as to treat the wastewater collected with Sewage Network of Denizli Sewage Practice Project which was approved in 1986. Wastewater comes to the plant designed according to the active sludge system, with an Ø 1600 mm Ellez Gutter collector. First level of wastewater

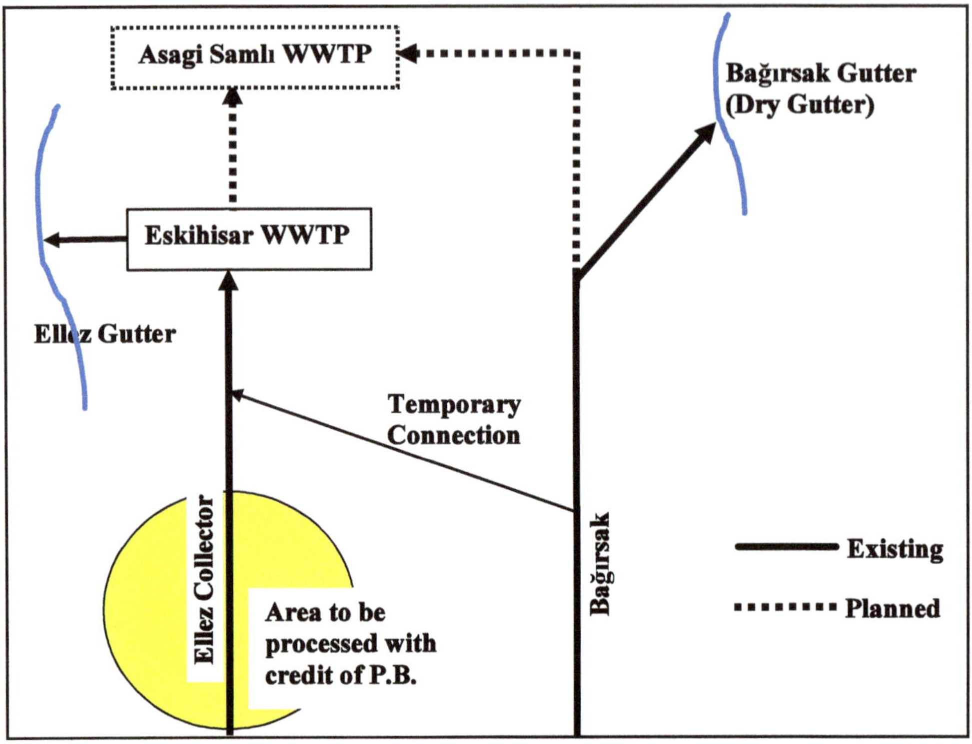

Figure 3.7: Schematic Display of Denizli Wastewater Collection System (Main collectors)

treatment plant was designed according to the year 2005. It was completed and taken into operation in March 2007 (Table 3.12).

Table 3.12: Eskihisar Wastewater Treatment Plant Project Values

Parameters	*Unit*	*2005*	*2025*
Population	Person	378,353	703,838
Q_{max}	lt/sec	2,123	3,609
Q_{proje}	lt/sec	1,692	2,806
Q_{ort}	lt/sec	1,333	2,138
Q_{min}	lt/sec	736	1,318
BOI_5 (total)	kg/day	24,369	40,855
BOI_5 (avg)	mg/lt	240	232
Nitrogen-N	mg/lt	46.00	43.74
Phosphorus-P	mg/lt	7.33	7.60
AKM	mg/lt	526	540

Plant is composed of automatic screen, grit chamber, primary settling, active sludge with surface aeration and sedimentation pools. Sludge which is to be formed in the plant is condensed in the pools with gravity, and then stabilized in unaerobic digesters. When the plant reaches its final capacity, the second level composed of same units will be constructed. Treated water coming from the plant is connected to Ellez Gutter with a 2.5 km collector line.

As can be seen from Table 3.12, first level design of WWTP project was prepared and constructed according to 115,171 m^3/day wastewater flow. After Denizli Municipality took over the responsibility of plant operation, 800 L/sec wastewater was treated in the plant with the measurements by the new flow meters.

The plant is composed of three main units:

1. Pre-treatment (entry, thin and thick screen, grit chamber with aeration).
2. Biological treatment (primary settling, aeration, secondary settling, chlorination).
3. Sludge removal (sludge thickener, unaerobic sludge digester, belt filter).

The exiting 20-25 tons/day sludge from the plant has an 18-20 per cent solid material and is sent to Solid Waste Storage Area of Denizli Municipality.

RAIN WATER (STORM WATER) COLLECTION SYSTEM

There are many illegal connections from building roofs to existing sewage system which serve as an integrated system and this causes the overloading of the capacity of sewage system during heavy rains. For these reasons, wastewater collection system has to be renewed and separated. Since there is insufficient rain water network, in case of heavy raining, water fills in wastewater lines and causes water floods with overflowing and blowbacks. In order to solve emergency situations in Denizli Province, a rain water network has been serving a limited area since last four years, with the municipality's own resources. Various rain water networks have been constructed with own resources of Municipality between 2004-2008 in different diameters and lengths. These are given in Table 3.13.

Table 3.13: Existing Rain Water Network Collection of Denizli

Years	*Total Length (m)*
2004	575
2005	7,200
2006	8,853
2007	5,584
2008	6,160
Total	28,372

Partial Rain Water Project including the areas of drinking water and sewage systems to be renewed with the World Bank's credit, was constructed by Denizli Municipality Water and Sewage Department in 2008. In this project, network calculation of Denizli Municipality partial rain water drainage system was performed and discharge spots were determined. According to Provincial Bank Technical Specifications, rain water lines were calculated as the diameters would be Ø 400, Ø 500, Ø 600, Ø 800 ve Ø 1000 mm. For the rain data used in the project, rain-intensity-recurrence curves belonging to Denizli published by the Department of Meteorology were utilized. Due to the project, total 78.062 m rain water line was planned in Denizli center. Rain water project of the whole province is being carried out and projects are about to be finished. Moreover, to protect Denizli Province from rain water floods coming from southern mountains, canals of strengthening project are applied by DSI (Public Waterworks Administration) and construction of some canals has been completed.

REFERENCES

Denizli Water Supply Rehabilitation Project Feasibility Report , KAYA Engineering and Consulting Co.Ltd. SNS Consulting and Engineering Co.Ltd Joint Venture

Korkmaz, N.1988. Yağıþ-yeraltısuyu seviye ilişkisi ve su kaynaklarının projelendirilmesine etkisi DSİ Ankara.

Yurdagül, M. 1983.Yukarı Çürüksu Havzası hidrojeolojik etüt raporu DSİ Ankara.

Denizli-Honaz Kaynakları karst hidrojeoloji etüt raporu. DSİ Ankara.

Gokalp, E.1971. Pamukkale (Denizli) bölgesinin jeoloji vr hidroloji etüdü.MTA derleme rap. 4561.

Gokalp, E.1971. Denizli Vilayeti Yenice-Gölemezli ve Karahayıtlı Kaplıcaları çevresi jeolojik etütleri ve jeotermik enerji imkanları; MTA Derl. Rap.4571.

Chapter 4

Integrating Environment into Land-Use Planning through Strategic Environmental Assessment: A Case Study on Environment Plans in Turkey

P.N. Say, D. Çolakkadıoğlu and S. Özyurt

Çukurova University, Department of Landscape Architecture, Balcalı, Adana

ABSTRACT

Strategic Environmental Assessment (SEA) can be defined as a systematic process in which environmental, economic and social consequences of proposed policies, plans and programs are evaluated at the earliest appropriate stage of decision-making. SEA has potential as an approach to achieve sustainable development. Current Project–level Environmental Impact Assessment (EIA) has been successfully established world-wide as an effective tool to address the environmental impacts of project development proposal. But, today it has been discussed that it is necessary to strengthen project-level EA and to practice SEA.

SEA was introduced in the Member States of the European Union (EU) when the *"Directive, 2001/42/EC of the European Parliament on the assessment of the effects of certain plans and programs on the environment"*. SEA came into force on June 27, 2001 and had to be enacted by the Member States until July 21, 2004. Spatial planning/land use planning is one of the main issues to be stated by the SEA-Directive. Currently SEA practice has been increasing; however, there are still defects about where and how to utilize it.

Project–level EA has been practiced in Turkey since 1993. SEA has gained more importance gradually and presently it is inevitable to utilize SEA in the early phases of land use plans in Turkey.

This paper aims to present the findings from a project on integration of SEA into Environment Plans in Turkey. The aim is to test the concept of SEA at the level of spatial planning and analyze its possible practice opportunities of SEA on Environment Plans.

Keywords: *Strategic Environmental Assessment (SEA), Environmental Plan, Adana.*

INTRODUCTION

SEA can be defined as a systematic process, evaluating the environmental, economic and social results of any suggested policy, plan or programme (Sadler and Verheem, 1996; Goodland, 1998; Bukley, 1998). Therefore; it has more advantages than project-level environmental assessment (EIA-Environmental Impact Assessment). Also, it provides a more holistic approach towards sustainable development. The aims of SEA applications are given in Table 4.1.

Table 4.1: Aims of SEA Applications (Rossouw *et al.*, 2000)

	SEA
✰	Determines the scope for development proposals.
✰	Makes a relation between environmental factors and development actions.
✰	Evaluates regions or sectors which are performing development areas.
✰	The process ongoing and renewable aiming at the provision of the necessary information in the required time
✰	Identify practices and issues which are necessary for sustainable development and cumulative impact assessment.
✰	Targets keeping environmental quality in a specific level.
✰	To create a wide network of information, the boundaries of the study is extensive. Details of the issues are lower.
✰	Creates an information network can be measured against the benefits and effects.

SEA can be defined as a legal means of decision making for competent corporations responsible for approving the plans, EIA projects. As seen in Figure 4.1, EIA and SEA are forming the main stages of Integrated Environmental Management (IEM) as two affect evaluation system implemented in different stages of development process and completing each other. As seen in Figure 4.1, EIA is carried out at project and practice levels but SEA can be carried out policy, plan and program levels.

SEA is developed as a more comprehensive evaluation system as a result of arising concerns related with the insufficiency of EIA carried at project level and it became more prevalent in 1990s. SEA especially legal in the developed countries became widespread after 2001/42/EC Directive of EU. During the accession process to EU, the preparations of Turkey about legal frame of the subject are going on. The draft SEA regulation was prepared in 2005, depending on the directive of EU about the subject.

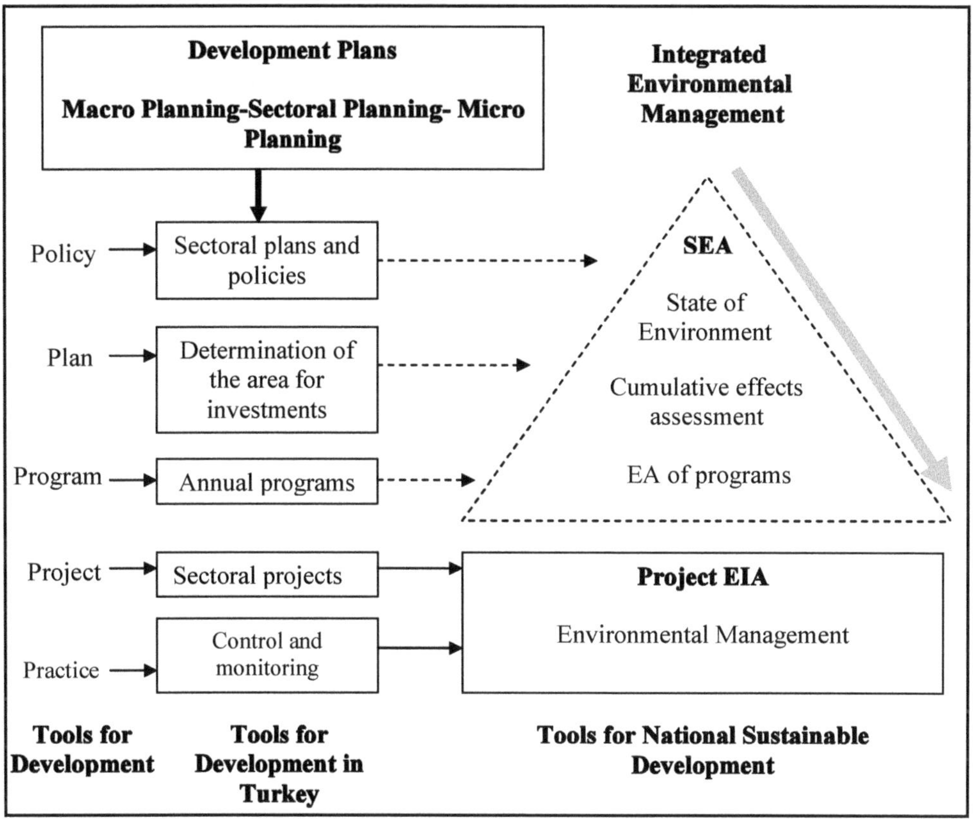

Figure 4.1: The Relationship of Development Tools with Integrated Environment Management (Say and Yücel, 2006)

SEA practices have mostly been observed in the developed countries; however, the practice of this subject is more important for the developing countries aiming sustainable development (Barrow, 1998; Say, 2000; World Bank, 1999; Abaza, 2000; Wood, 2003; Abaza *et al.*, 2004; Dalal-Clayton and Sadler, 2005). In these countries, the development process till now has been followed with an approach which aims economic income in the short term. This approach has often caused the environmental factors to be ignored. The evolving problems have been tried to be solved by policies which intend to decrease current pollution. The international relationships, organizations, the external financial sources for investments, non-governmental organizations and the public impact have been effective in order to form "Preventive policy tools" such as EIA, SEA.

The practices of SEA are closely related to the administration system and decision making procedure in the countries. For this reason, practices in the SEA scope and practice procedures may change from country to country. The subjects that required strategic decision-making process have been gathered in five groups for Turkey by taking into consideration national and international SEA systems explained by Sadler and Verheem (1996) and the SEA practice fields determined by Partidario (1996),

Bukley (1998), Goodland (1998), Say (2000) and SEA Directive by the European Union on 27 June 2001.

1. National Five-Year Development Plans (FYDPs) and Sectoral Plans: socio-economic plans including macro-economic targets and sectoral development policies prepared by Prime Ministry State Planning Organization (SPO).
2. Regional Development Plans: plans, aiming to improve the socioeconomically underdeveloped regions, to provide the effective use of sources in the region and to improve income distribution in the country such as Southeastern Anatolia Project—SAP and Eastern Anatolia Project—EAP.
3. Legal Regulations: laws and international convention for especially effective uses of natural sources such as the Forest Law, Tourism Encouragement Law, Coast Law.
4. Privatization Policies: privatization activities have been in progress with the effect of laissez-faire economy since 1985 in Turkey. Especially privatization policy in the energy and tourism sectors has great impact on environment.
5. Physical Plans: urban and rural developmental plans and other land use plans, environmental plans prepared by the Ministry of Environment and Forestry, urban master plans by municipalities, Infrastructure plans, and tourism plans.

This research project aims at analyzing the possible practice opportunities of SEA in Turkey and the practicability of SEA into Environment Plans which determine decisions of land uses. The research is conducted on two sections. In the first section, procedural approaches to SEA on environment plans are investigated and a framework for these approaches is adapted at the institutional level. In the second section, SEA practice form is investigated for environment plans in case of Adana. In this article, the findings related to the first part of the project are presented.

MATERIALS AND METHODS

The present study was carried out in three phases as seen in Figure 4.2. As seen in Figure 4.2, theoretical framework and scope determination of SEA was researched at the first phase of the study. At the first phase, current situation about SEA, importance and requirements of SEA for Turkey and topics may evaluate in SEA were carried out.

At the second phase of the study the investigation of environmental plan was researched. At this phase the process of preparing plans and policies tracked to determine decisions was carried out.

At the last phase, investigation of SEA implementation on environmental plans was examined.

EU SEA regulations, case studies, national draft regulations, country based development plans are used as materials in this study.

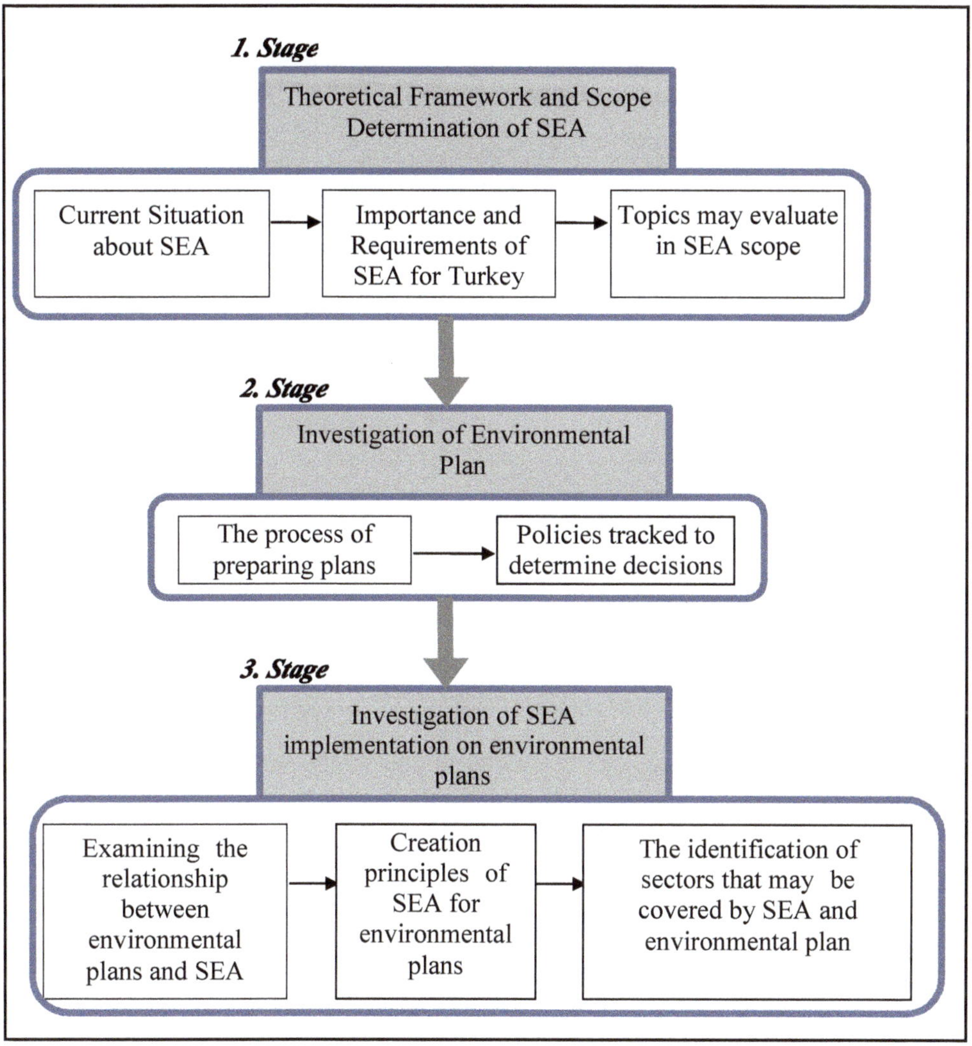

Figure 4.2: Flowchart of the Methodology

Planning Hierarchy and Environmental Plans in Turkey

Planning defined as process to settle the current problems or those will arise in future can only been effective means for the solution of problems with an integrated approach. The integrated planning approach addressing areas and locations may only be realized in a hierarchical system (Figure 4.3).

As seen in Figure 4.3, social-economic plans include national development plans, regional plans and master plans. National development plans, which can be carried out country plan scale, have macro-economic targets, sectoral aims, objectives and policies. Regional plans and master plans have superior physical plans or high level

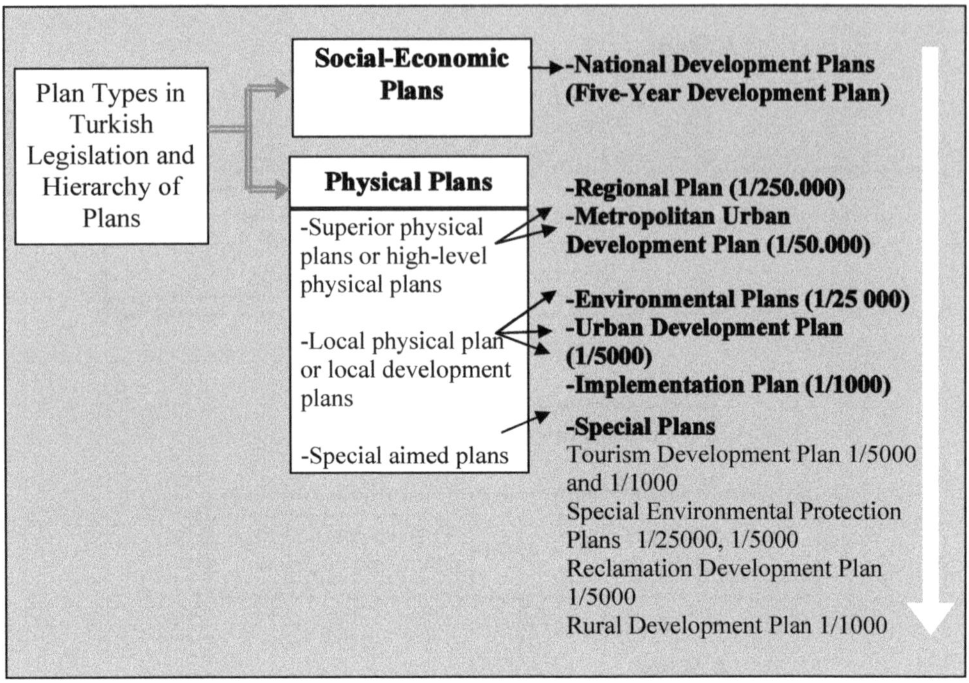

Figure 4.3: The Planning Hierarchy in Turkey (Say, 2004)

physical plans. Regional plans can be carried out 1/500 000 and 1/100 000 scale. Metropolitan plans can be carried out 1/50 000 scale.

Physical plans include environmental plans, master plans and implementation plans and also tourism development plans, reclamation development plans and rural development plans. Environmental plans, master plans and implementation plans are local physical plans or local development plans. Their frame scale can be performed respectively for sub region, province 1/2000, 50000, 100000; for urban 1/2000, 5000 and for rural 1/1000.

Tourism development plans, reclamation development plans and rural development plans are special aimed plans. Their frame scale can be performed in respectively for sub region 1/1000, 5000 for urban 1/5000 and for rural 1/1000.

ENVIRONMENTAL PLANS

In 1/100 000, 1/50 000 or 1/25 000 scales specifying the settlement and land uses decisions such as for residents, industry, tourism and transportation are high scaled plans which are integral with the plan notes and reports. In other words, these are plans determining the utilization decisions for protection of natural sources and their sustainable use. For example, preparation of Environmental Plans has importance for the generation of decisions to be the basis for lower scaled plans. Environmental Plans aim:

- ✰ To prevent of unplanned urbanization and industrialization generally in country,

- ☆ To ensure sustainable development,
- ☆ To manage or orientating urban and rural developments,
- ☆ To ensure the consideration of ecological, economical, cultural, social and physical values integrally,
- ☆ To establish protection-utilization balance,
- ☆ To avoid environmental problems by the generation of strategy, decisions and general land use decisions,

When the scope of environmental plans are considered they are forming important examples for impact assessment systems due to their forming the base for urban development plans where main land uses decisions are made.

INTEGRATION OF ENVIRONMENTAL PLAN AND SEA

As a result of the findings received at the first part of the study, the legal framework of SEA on the environment plans has been established for the administrative system in Turkey.

Our findings related to this part are presented as follows;

- ☆ The phases of the SEA process on Environment Plans
- ☆ The issues to be solved for SEA process
- ☆ Tools to be used on SEA process
- ☆ The main issues taken into account to reach these results:
- ☆ The theoretical frame of SEA
- ☆ The administrative system in Turkey
- ☆ Planning hierarchy
- ☆ The perspectives of the institutions responsible for the planning (questionnaire-interview)
- ☆ The content of the development plans
- ☆ The SEA Directive of the European Union

A procedure based on the basic principles of SEA's general stages is given in Figure 4.4.

SEA is the result of a systematic process. SEA implementation starts with a decision whether SEA will be carried out or not. Implementation of the plan will emerge effects monitoring is the final step of SEA. Practitioner of a plan or program (a ministry or other public institutions) is responsible for implementing most of the steps and decisions.

According to the draft SEA regulation in Turkey, the SEA process is based on 6 phases: *screening, the determination of the scope, the preparation of the SEA report, quality control, decision making* and *monitoring*. The very similar EU members were assigned to these six phases. They are given in the Figure 4.5.

As seen in Figure 4.5, the first phase of the draft SEA regulation in Turkey is "screening". In this phase the Ministry of Environment and Forests decides whether

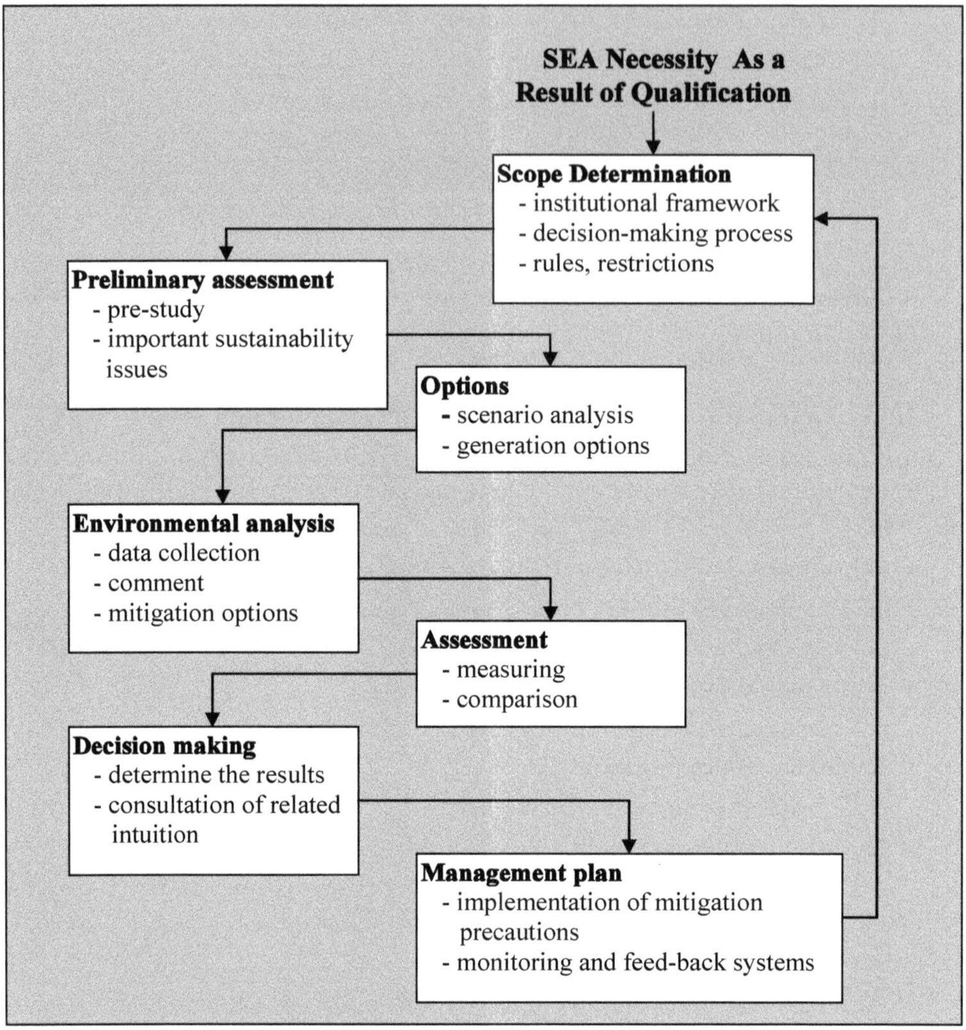

Figure 4.4: Basic SEA Application Template (Nilsson *et al.*, 2005)

SEA will be applied on a plan or program. In this phase the Ministry of Environment and Forests consultant with the other related institution.

In the second phase, the scoping document is prepared by considering the suggestions about public participation. The responsible of this phase is authorized institution. And also first public participation meeting is held in order to get views about the format of the SEA report. In the third phase, "SEA-Report", the format of the SEA report is determined and according to this format, the authorized institution prepares the SEA report.

At "quality control phase", the fourth phase of the draft SEA regulation in Turkey, Authorized institution completes the report by taking into account the views in the meeting and presents it to the Ministry of Environment and Forests. The Ministry of

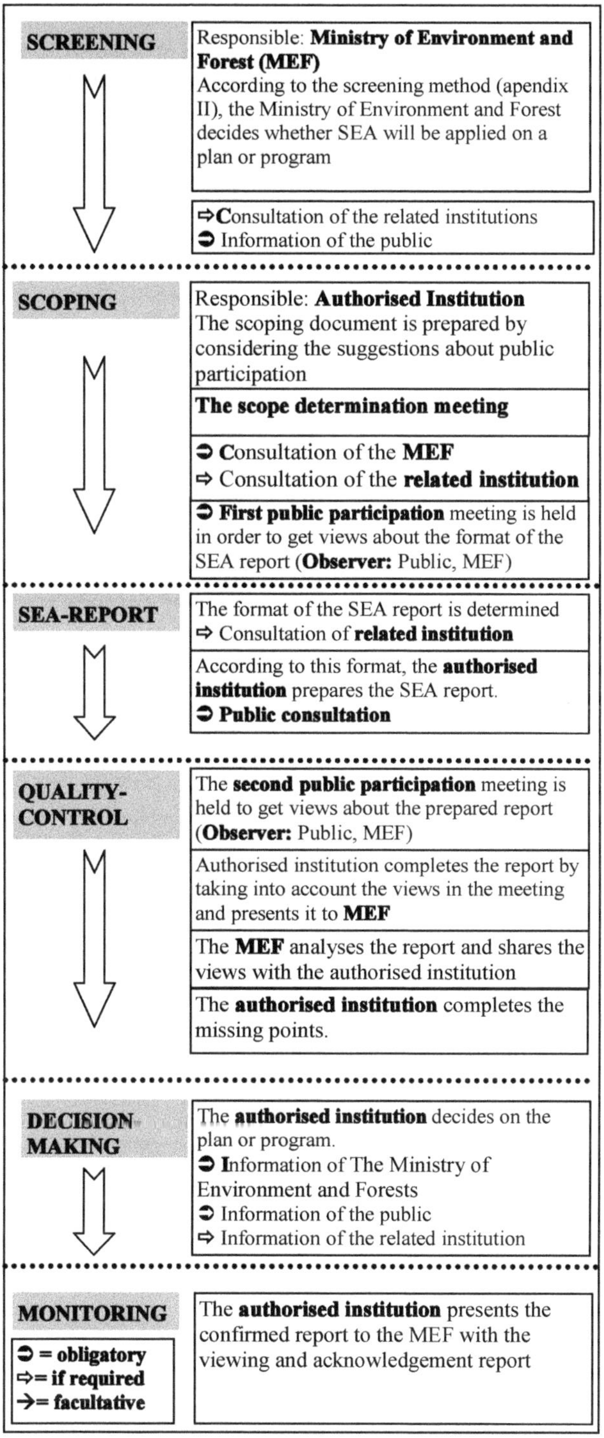

Figure 4.5: The Phases of the Turkish Draft SEA Regulation (Say *et al.*, 2008)

Environment and Forests analyses the report and shares the views with the authorized institution. The authorized institution completes the missing points. The second public participation meeting is held to get views about the prepared report.

In the decision making phase, the authorized institution decides on the plan or program and inform The Ministry of Environment and Forests, public and related institution. In the monitoring phase the authorized institution presents the confirmed report to the Ministry of Environment and Forests with the viewing and acknowledgement report.

In the present project, main stages of the regulative plus practical process of SEA are determined. Henceforth, practical phases of SEA can group into three main headings (Table 4.2).

1. Determination of the Present Situation
2. Proposed Scenarios for the Future
3. Evaluation applicability

Table 4.2: Main Stages of the Regulative Plus Practical Process of SEA (Adapted from Say, 2004)

	Phases		*Ancillary Tools*
TODAY	**1-Determination of the Present Situation**		
	Land uses	Sector production-consumption situation natural resources used by sector	Statistical data
	The Pollution situation of the Country	Pollution maps Pollutions caused by uses	Meteorological and topographic data GIS Emission measurements Scientific researches Statistical data Limit values
	Population	Population density, urban and rural population	
	Sensitive Regions Protected area	Protected areas Historical and cultural areas	Protection statutes Maps for protection areas
FUTURE	**2- Proposed Scenarios for the Future**		
	The Development of the Sector	Sector share in economic growth Change in the sector production and consumption	Economical analysis Statistical analysis
	Economic Development	Basic economic Indicators towards economic growth	Basic economic indicators in development plans
	Population	Population growth	Methods of population prediction
	Impact	Impacts caused by sector	

Contd...

Table 4.2–Contd...

	Phases		Ancillary Tools
PRECAUTION	**3- Evaluation applicability**		
	The results of the evaluation	Co-evaluation of subjects in the 1. and 2. phases	
	Basic evaluation criteria	Criteria determining the general frame of the policy Activity determination for the practice of criteria	
	Policy Tools	Sectoral Political Local	Local governments and their policies Administrative structuring National and international law

In first stage determination of the present situation includes land uses such as agriculture, industry, tourism and transportation. The pollution situation of the country includes such as sulfur dioxide pollution maps and noise pollution maps. Population of research area includes population density, urban and rural population and socio-economic structure of the population. Sensitive regions and protected area includes national parks, nature reserve areas and natural monument

In the second stage proposed scenarios for the future includes the development of the sectors economic development, projection of population and impact assessment. In the third phase evaluation applicability includes the results of the evaluation, basic evaluation criteria, policy tools such as sectoral (agricultural plans), political (environmental laws), local (municipalities).

In the present project, main stages of the regulative plus practical process of SEA are determined. Henceforth, practical phases of SEA can group into three main headings

1. The Analysis of the Present Situation
2. Expectations-Predictions
3. The Evaluation of Applicability
    - The first category is based on the sector, land uses and the analysis of the available situation of affected factors.
    - The following phases include the future suggestions and the scenarios.
    - Lastly, the findings from these two phases are evaluated and planning criteria are set.

ACKNOWLEDGEMENTS

This research project was supported by grants Çukurova University, Research Projects Unit (Project Numbers: ZF2007BAP40).

REFERENCES

Abaza H., 2000. Strengthening future environmental assessment practice: an international perspective. In: Lee N, George C, editors. Environmental assessment in developing and transitional countries: principles, methods and practice. Wiley Publishers.

Abaza H, Bisset R, Sadler B., 2004. Environmental impact assessment and strategic environmental assessment: towards an integrated approach. Geneva: The United Nations Environment Programme (UNEP).

Barrow AL., 1998. The environmental overview as a realistic approach to strategic environmental assessment in developing countries. In: Porter AL, Fittipaldi JJ, editors. Environmental methods review: retooling impact assessment for the new century, AEPI and IAIA, USA.

Buckley R., 1998. Strategic Environmental Assessment. Editors Alan L. Porter and John J.Fittipaldi, Environmental Methods Review: Retooling Impact Assessment for the New Century. AEPI and IAIA. USA.

Dalal-Clayton B., B. Sadler, 2005. Strategic Environmental Assessment: A Sourcebook and Reference Guide to International Experience. Earthscan Publications Ltd.

Goodland R., 1998. Strategic Environmental Assessment. Environmental Methods Review: Retooling Impact Assessment for the New Century. Editors, Alan L. Porter and John J. Fittipaldi. AEPI and IAIA. North Dakota, The Press Club, USA.

Nilsson, M, Björklund, A., Finnveden, G., Johansson, J., 2005. Testing a SEA Methodology for the Energy Sector: a Waste Incineration Tax Proposal, Environmental Impact Assessment Review, 25: 1-32.

Partidário M.R., 1996. Strategic Environmental Assessment: Key Issues Emerging From Recent Practice. Environmental Impact Assessment Review, Volume: 16: 31-55, Elsevier Science Inc. New York.

Rossouw N., Audouin, M., Lochner, P., Heather-Clark S,.Wiseman, K., 2000. Development of Strategic Environmental Assessment in South Africa. Impact Assessment and Appraisal, volume 18, number 3 September 2000, 217-223, Guildford, Surrey GU1 2EP, UK.

Sadler B., Verheem., R., 1996. Strategic Environmental Assessment, Status, Challenges and Future Directions. Ministry of Housing, Spatial Planning and the Environment of The Netherlands, International Study of Effectiveness of Environmental Assessment. The EIA Commission of The Netherlands

Say N.,P.,2000. Importance of strategical environmental assessment for developing countries in the case of Turkey. 20th Annual Meeting of the International Association for Impact Assessment IAIA'00, 19–23 June 2000, Hong Kong; 2000.

Say, N., 2004. Stratejik Çevresel Değerlendirmenin Beş Yıllık Kalkınma Planları ve Enerji Sektörü Örneğinde Araştırılması ve Bir Uygulama Modelinin Geliştirilmesi. Ç.Ü. Fen Bilimleri Enstitüsü (Doktora Tezi), Adana.

Say N., Yücel, M., 2006. Strategic environmental assessment and national development plans in Turkey: Towards legal framework and operational procedure. Environmental Impact Assessment Review 26 (2006) 301– 316

Say N., Herberg A., Köller J., 2008. Development and implementation of SEA: a comparative review of draft regulation in Turkey and Germany experiences (in press).

Wood C., 2003. Environmental impact assessment in developing countries. Int Dev Plan Rev 2003;25:301– 21.

World Bank., 1999.Environmental assessment. Operational policy, bank procedure and good practice. Washington (DC) World Bank.

Chapter 5
Land Degradation, Urbanisation and Biodiversity in the Gediz Basin-Turkey

Munir Ozturk[1], Mustafa Okmen[2], Salih Gucel[3], Aykut Guvensen[1] and Ali Celik[4]

[1]*Ege University, Science Faculty, Botany Department, Bornova-Izmir,Turkey*
[2]*Celal Bayar University, School of Higher Education, Salihli, Turkey*
[3]*Institute of Environmental Studies, Near East University, Lefkosa, Northern Cyprus*
[4]*Pamukkkale University,Science and Arts Faculty, Biology Department, Denizli, Turkey*

ABSTRACT

Gediz basin lies in the West Anatolian part of Turkey covering an area of over 1.7 million ha around Gediz river and its tributaries. It is an important land-farming river basin and agricultural centre from a financial point of view. The area has a typical Mediterranean climate. Forest area in the basin is around 470.000 ha., the agricultural area is 472.000 ha. and grasslands cover 101.266 ha. The basin has a rich plant diversity which includes 319 endemic taxa from 206 genera and 60 families. The endemics constitute 10 per cent of the total endemic plant cover of Turkey. The seaward fringe of the basin is an ecologically important wetland ecosystem forming a category A nature reserve. The delta is used by 211 bird species out of a total of 426 recorded in Turkey.

The basin includes 2.3 per cent of the total area of Turkey and 2 per cent of the irrigated land, supporting 3.7 per cent of the population. It is traversed for 401 km by Gediz river which irrigates 521.000 ha of land out of 2 million ha spread over four states. Approximately 386.000 ha of the total irrigated land area constitute the plains. Nearly 1.5 million ha (84.69 per cent) have different slopes. The land degradation problems pose a serious threat to the rich biodiversity and agricultural areas. Forests originally protected the soils on the mountains, but the trees have been cut for building and to provide firewood. During the last 20 years, the population of the basin has increased from 2 to 4 million inhabitants. This increase, together with unsustainable land use practices and migration have led to increased urbanisation resulting in land degradation problems as well as unhealthy development in the city outskirts. Annually tonnes of sediment are transported to the Aegean Sea due to erosion. Urbanisation is increasing and large parts of prime quality land are being permanently removed from agricultural use. These are the major environmental problem of the basin. An attempt has been made here to highlight this situation.

INTRODUCTION

Population growth, industrial development, land degradation and environmental pollution; which threaten natural resources in every country on earth;have made environmental problems one of the most important concerns of man in the last quarter of the twentieth century (PRB,2004). These problems are actually extending beyond the national boundaries which necessitates international cooperation, and such problems are going to be a determining factor in international relations in the 21st century (Arslan and Okmen, 2006; Okmen and Parlak, 2006). The situation in Turkey is not different from other corners in the world. One of the major problems we are facing in this connection is land degradation. It is believed that although land degradation is more important than environmental protection; because it includes additional aspects not considered in the concept of environmental protection; it has not received the global attention that it deserves (Ozturk *et al.*, 2002).

Land degradation in Turkey started centuries ago and has been on-going but has gained a greater momentum during the last few decades (Ozturk *et al.*, 2008). The factors responsible for this are urbanisation, salinization and alkanization, grazing, forest cutting, fires and logging (Ozturk *et al.*, 2010). In particular, urbanisation puts greater demand on arable land (Ozturk, 1995, 1999). In addition, large areas of arable land have been taken over by urban sprawl, while some areas are used for non-agricultural activities. Moreover, forest cover is becoming thinner and erosion exposes the surface and results in great deflation of the topsoil. Some environmental issues such as soil pollution, erosion, generation of dust, and soil compaction are emerging at a fast pace. Erosion alone is resulting in a loss of millions of tonnes of sediments, which are transported to the Aegean Sea (Alpaslan and Atýþ, 1995; Ozturk *et al.*, 1994). Wrong irrigation practices are leading to a dramatic increase in the salinization of arable lands due to an increase in the area of irrigated land. In Turkey nearly 1.5 million ha are facing acidity and 2.8 million ha show salinity alkalinity problems (Guvensen *et al.*, 2006).

Gediz Basin is getting an equal share from these unfavourable developments taking place in the country.It lies in the West Anatolian part of Turkey covering an area of over 1.7 million ha around Gediz river and its tributaries and is an important land-farming river basin as well as an agricultural centre from a financial point of view. The climate is typical Mediterranean. The mean annual precipitation is 609 mm, but varies between 492 mm(Salihli)-726 mm(Manisa). Rainfall is common in winter and scarce in summer. The temperature varies between 6.2–27.7°C. These climatic features are highly favourable for agricultural productivity.

The 401 km long Gediz river traversing through the basin is a backbone for the irrigation of nearly 522.000 ha of land out of 2 million ha spread over four states. More than 100 thousand ha in this green corridor are used for the production of several agricultural crops like grapes, cotton and tobacco. Overuse of water for irrigation results in a rise of the water table, water-logging, salinization and ultimately secondary salting of soil (Ozturk *et al.*, 2007). The patches of salinity are visible at the farm level. The basin is experiencing a decrease in productivity and biological diversity because of the decrease in nutrient-use efficiency, physical and chemical degradation of soil, and inefficient water use. The use of monocultures, mechanisation and excessive use of chemicals for plant protection are affecting the crops, plant, and animal diversity negatively. During the last 20 years the population of the basin has doubled followed by heavy urbanisation (SIS,2004). The demographic outburst together with water erosion and unsustainable land use have lead to land degradation problems in the basin and are also threatening its biodiversity. Present communication gives a synopsis of these degradation factors and their possible future impacts in the basin.

LAND DEGRADATION

The ruined cities in the wide, fertile plain of the Gediz basin were probably settled for the first time in the third millennium B.C. They experienced droughts, earthquakes, famines, fires, floods, invasions, landslides, and sieges (Ozturk *et al.*, 1994,1995). The area was the western terminus of the Persian royal road described by Herodotus. The splendor of ancient Sardis in the basin is still evident and gives reason to ponder over the future of this basin. Much of the history of this basin is reflected in the soil. The top layer of the soil at the interface between earth and atmosphere is the place where atmospheric attack is maximum. A layer of phytosphere lies between the soil and atmosphere, which depending on its density has a protective role. The processes of soil degradation are accelerated when the vegetation is destroyed by human activities (Figures 5.1 and 5.2). The data on land degradation in the Gediz Basin was summed up from the sources provided by the Directorate of Water and Soil Resources, Menemen, as well as the data published earlier by Ayyildiz (1983), Gul *et al.* (1995), Taysun and Uysal (1995), Ozturk *et al.* (1994, 1995, 1996a,b, 2007), Gokmen *et al.* (1997), Gecgel *et al.* (2000), Ozturk and Guvensen (2002), Guvensen and Ozturk (2003).

Gediz basin includes 2.3 per cent of the total area of Turkey and 2 per cent of the irrigated land, supporting 3.7 per cent of the population. Total area of land is 1.721.895 ha.,out of which agricultural areas constitute 643990 ha, grasslands 101266 ha, and forest and macchia 919423 ha. The soils are entisols and inceptisols on steep slopes or in alluvial valleys, affected by landslides, erosion, alluvial deposition and overuse through cultivation. Approximately 400.000 ha of the total irrigated land area constitute the plains. Nearly 1.5 million ha (84.69 per cent) have different slopes.The area of different classes of soils is; 134,578 (Class-I), 124,086 (Class-II), 127,349 (Class-III), 134,896 (Class-IV), 124 (Class-V), 315.537 (Class-VI), 845.623 (Class-VII), and 39.702 ha.(Class-VIII). These abound in mica flakes, which contribute to sliding, especially when they are irrigated. In earlier times farmers had designed settling pots at close intervals in certain water lines so that soil sediments could be removed periodically. This practice is not used now(Gul *et al.*, 1995). The percentage of the irrigated cotton area is about 50 per cent, however, the area has decreased from 50.000 ha (1980) to 36.000 ha (2000). On the other hand, the area of vineyards has increased from 10.000 ha to 25.000 ha during the same period, whereas the area of

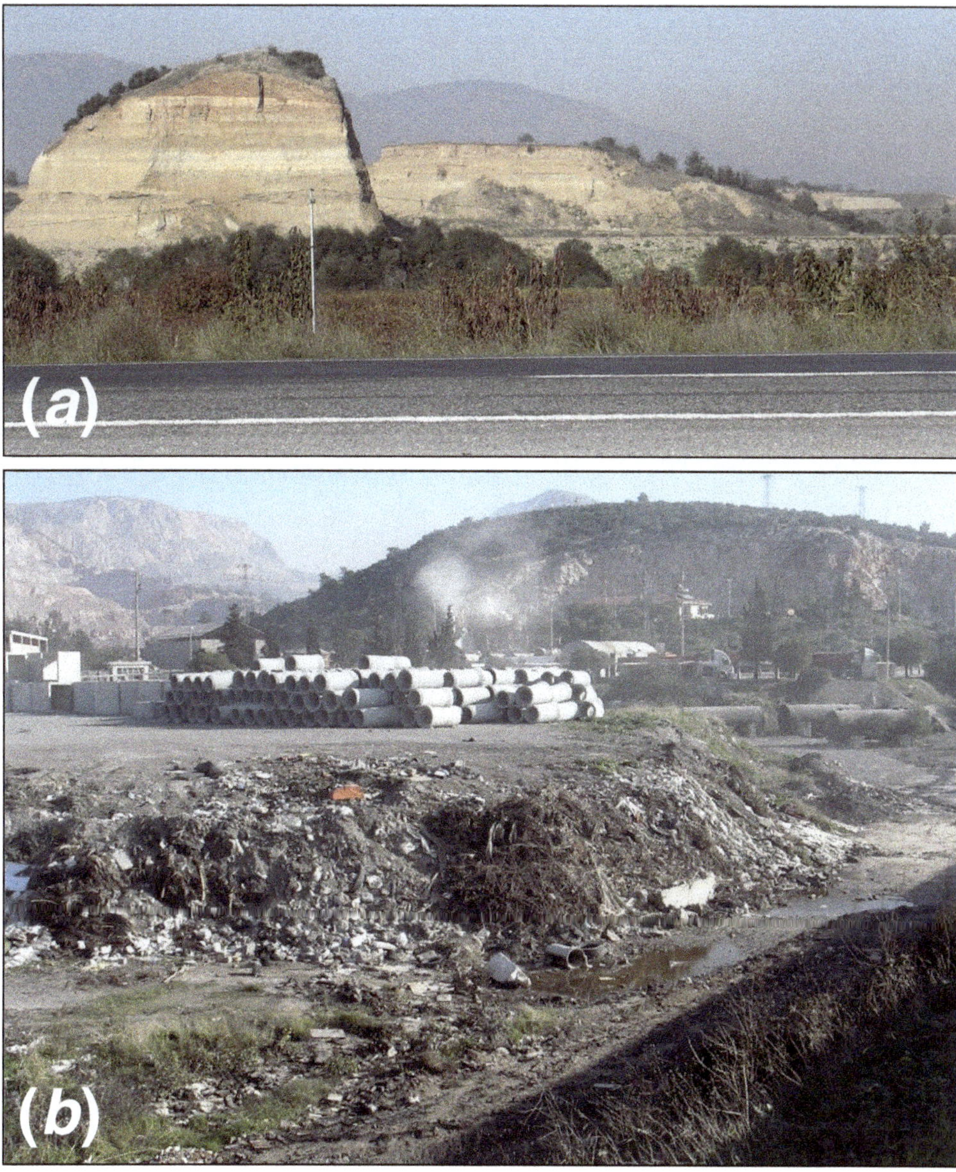

Figure 5.1: (*a*) Soil Extraction for Brick Factory (*b*) Industrial establishment for cement production and the together with a waste dump

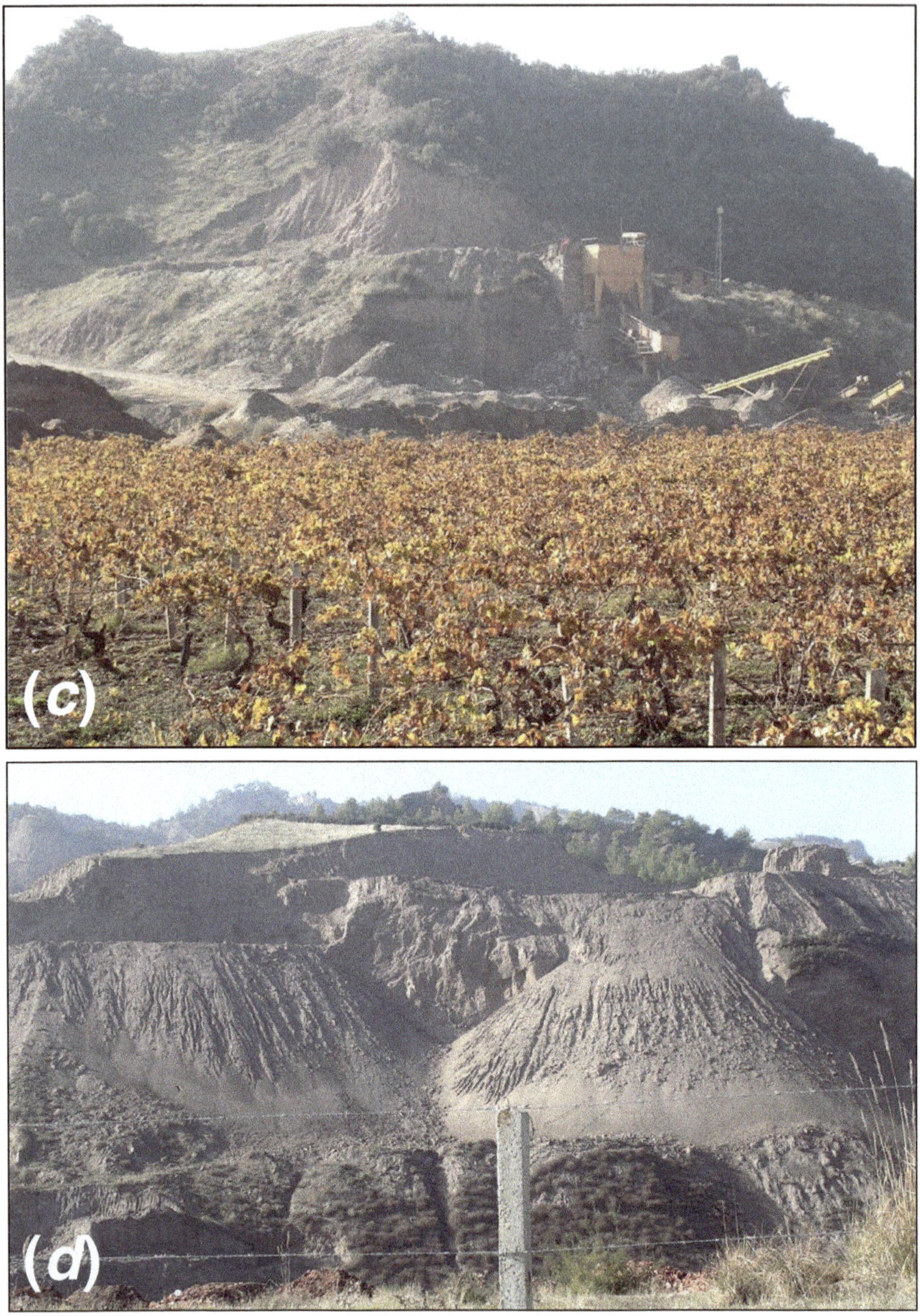

Figure 5.1: (c) Degradation of forest areas
(d) Soil extraction for brick factory
Visible alongside the highway crossing the study area.

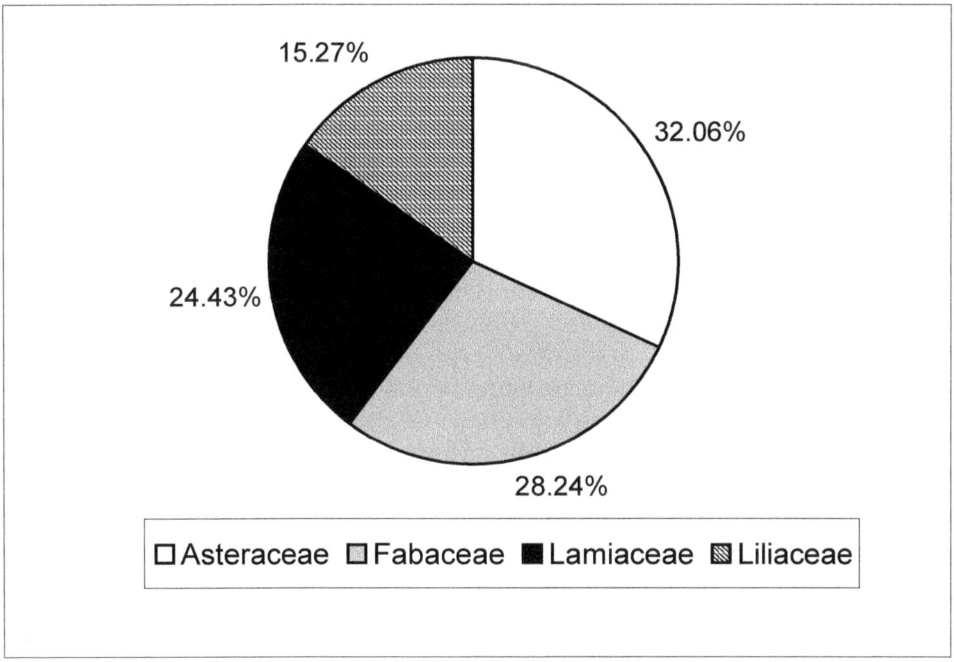

Figure 5.2: Percentage of Dominant Families on the Basis of Endemics in Gediz Basin

tobacco is decreasing (Gecgel, 2000). The erosive agents like rainfall and runoff are very effective in the soil erosion.The different forms of water induced erosion include geological and man-induced forms. A major part of the land, nearly 1.5 million ha (84.69 per cent) show slight, medium, steep or very steep slope. Soils on upper slopes have been overgrazed by goats and sheep for centuries. Forests originally protected the soils on the mountains, but the trees were cut for construction purposes and to provide firewood. Different degrees of water erosion are observed and these affect nearly 90 per cent of the land area (Taysun and Uysal, 1995). Only 265.000 ha of the land surface here are safe from erosion. Medium, strong and very strong erosion is observed on 220.000, 640.000 and 540.000 ha respectively (Gul *et al.*, 1995). Erosion and deforestation pose a great threat. Approximately 100.000 ha of agricultural land around Kemalpasa, Turgutlu, Salihli, and Alasehir has been used for non agricultural purposes. In addition to this loss today millions of hectares of land are awaiting planning due to wrong use or over use. During the last few decades the brick and tile factories are using large areas of the productive land from the basin. In general, 60 per cent of the land from the most productive class I-IV soil groups has been used unsustainably (Ozturk *et al.*, 1994,1995). More than 30 brick factories and some leather workshops are a major cause of pollution in this region.

URBANISATION

Human activities are the determining factor at all stages of degradation. Industrialisation and demographic explosion have been an important driving force in urbanisation (SIS,2004). Improved individual mobility of inhabitants has changed the pattern of urban growth. The latest trend in the region has been to use the best quality arable lands for housing schemes, highways, touristic establishments, sports complexes, universities, airports and other activities. An increase in the land of construction on a monetary basis results in a decrease in the productive value of cultivated areas. The causes, mechanisms of deterioration, and the impacts underlying this originate from population increase together with emmigrations due to favourable climatic conditions. For urbanisation, the information published by the State Institute of Statistics was used (SIS, 1995, 2003, 2004) together with the quesstionares collected from the muncipalities in the basin. The demographic developments have followed an increasing trend during the last 2 decades and the population of the basin has doubled goimng up from 2 to 4 million inhabitants. Urbanisation is increasing and large parts of prime quality land are being permanently removed from agricultural use. Although an overall rate of urbanisation in this area as compared to other regions of Turkey is lower, area of settlements and other related factors covers 57216 ha. including 16.000 ha of prime quality land. This area has been lost due to intensive construction of housing schemes during the last decade. This increase, together with unsustainable land use practices and migration have led to increased urbanisation resulting in land degradation problems as well as unhealthy development in the city outskirts. The noise pollution in the major cities of the basin is above the OECD standards (65 dB) and the waste disposal sites also cause problems (Ozturk *et al.*, 1994, 1995).

PLANT DIVERSITY

The productive forest area in the basin covers 120.000 ha, degraded forest 350.000 ha, agricultural fields 340.000 ha, orchards 132.000 ha, and grasslands 101,266 ha. The basin in general shows a thermo-mediterranean and mediterranean plant cover mainly represented by *Pinus brutia* forests and macchias, but in the supra-mediterranean parts forests of *Pinus nigra* ssp. *pallasiana* dominate and in the sub-alpine zone we come across the species of *Astragalus* and *Acantholimon*. In the grasslands, major taxa are; *Phleum, Hordeum, Cynodon, Alopecurus, Stipa, Avena, Trifolium, Vicia, Onobrychis, Medicago, Carex, Phalaris, Sinapis, Raphanus, Agropyron, Papaver,* and *Glycyrrhiza.*

The plant collections were done during spring, summer and autumn seasons of 1998-2000. These were identified with use of the Flora of Turkey and the East Aegean Islands (Davis,1965-1988, Davis *et al.*, 1988;Guner at al.,2000). Gediz basin is included in the mediterranean phytogeographical region with varying habitats which embody a rich plant diversity. In all 319 taxa belonging to 206 genera and covering 60 families were collected, all included in the division Spermatophyta. The dominating families are Asteraceae (32.06 per cent), Fabaceae (28.24 per cent), Lamiaceae (24.43 per cent) and Liliaceae (15.27 per cent). In all 77 taxa are from Mediterranean, 49 from Irano-Turanian, 2 from the Euxinic phytogeographical regions and 3 from the unknown group. Some of the taxa are listed in Table 5.1.

The endemics constitute 10 per cent of the total endemic plant cover of Turkey (Ozturk *et al.*, 1992; Ekim, 2009). Most important ones are; *Achillea nobilis spp. sipylea,*

Table 5.1: List of the Plant Taxa Collected from Gediz Basin

Family and Plant Name	Habitat	Flower Colour	Flowering Time	Habit	Altitude (m)	Element
			ASTERACEAE			
Doronicum reticulatum	Alpine slopes	Yellow	6-7	PR	1700-2000	MED
Senecio castagneanus	Rocky slopes, forest openings	Yellow	4-6	PR	1250-2200	MED
Anthemis cretica spp.*tenuiloba*	Steppe, ruderal, river bed	Yellow	5-6	PR	300-2100	IR-TR
A. cretica spp. *anatolica*	Steppe, ruderal, moist heaps	Yellow	5-7	PR	280-2285	IR-TR
A. cretica spp. *leucanthemoides*	Serpentine rocks, slopes, pine forests, hill tops	Yellow	5-6	PR	200-1600	IR-TR
A. cretica spp. *absinthifolia*	River beds, hill tops	Yellow	4-7	PR	700-1070	IR-TR
A. cretica spp. *candicans*	Rocky, calcareous slopes	Yellow	5-6	PR	800-2000	IR-TR
A. cretica var. aciphylla	Mixed subalpine forests	Whitish	5-7	PR	350-1600	MED
A. cretica var. discoidea	Subalpine forests	Whitish	5-7	PR	300-1600	MED
A. xylopoda	Schist rocks	Yellow	7	PR	1400-1600	MED
A. dipsacea	Slopes, mountain base	Yellow	6-7	AN	800-1500	MED
A. wiedemanniana	Moist slopes, steppe	Golden yellow	5-6	AN	400-1800	MED
Achillea phrygia	Steppe, moist slopes, oak associations	White or yellow	5-7	PR	700-1500	IR-TR
A. nobilis spp. *sipylea*	Forest, steppe, grassland, rocky slopes, volcanic areas		6-8	PR	400-2350	MED
Tripleurospennum hygrophilum	Base of mountains, grassland	White	4-5	PR	800-980	MED
T.conoclinium	Meadows	White	4-7	BI/ PR	0-1200	MED
Cirsium leucopsis	Edge of spring waters		7	PR	1400-1600	MED
Carduus olympicus ssp. *Hypoleucus*	Rocky slopes, black pine forests	Purple	6-9	PR	1400-1700	MED
Jurinea cadmia	Rocky mountains	Lilac-pink	6-8	PR	1070-2100	MED
J. pontica	Forest, steppe, grassland	Purple	6-8	PR	50-1450	IR-TR
Serratula lasiocephala	Steppe	Pink	6	PR	1700-2000	IR-TR
Centaurea amasiensis	Forest, slopes	Yellow	6-7	PR	100-1100	IR-TR

Contd...

Table 5.1–Contd...

Family and Plant Name	*Habitat*	*Flower Colour*	*Flowering Time*	*Habit*	*Altitude (m)*	*Element*
C. zeybekii	Open red pine forest	Pink	6	PR	600-1300	MED
C.a olympica	Forests	Rose purple	7-8	2	500-1200	MED
C. polyclada	Red pine forests, maquis, grassland	Rose purple	6-8	PR/BN	0-500	MED
C. sipylea	Higher elevations	Rose purple	7	PR/BI	*	MED
C.calolepis	Sandy heaps, grassland, ruderal	Rose purple	6-7	PR	50-1300	MED
C.aphrodisea	Steppe, hill tops	Rose purple	6-8	PR	1300-1900	MED
C. acicuaris var. urvillei	Rocky areas	Yellow	5-6	PR	0-1200	MED
C. lydia	Pine forests, grassland		6-7	PR	400-1600	MED
C. reuterana ssp. *reuterana*	Rocky slopes	White	5-7	PR	1300-2100	MED
C.pichleri ssp. *extrarosularis*	Steppe, rocky slopes		6-7	PR	750-3200	MED
Scorzonera suberoza ssp. *cariensis*	Cultivated areas	Lilac or purple	5-6	PR	1200-1650	IT-TR
S. acuminata	Grassland slopes	Yellow	5	PR	950-1400	IR-TR
S. eriophora	Pine forest, calcareous steppe	Yellow	5-6	PR	500-1900	IR-TR
S. tomentosa	Steppe, rocky slopes, hill tops	Yellow	6-8	PR	800-2600	IR-TR
Tragopogon olympicus	Arid habitats	Yellow	7-8l	PR	1500-2300	MED
T. subacaulis	Arid habitats, calcareous rocks	?	7-8	PR	1100-1500	MED
Picris olympica	Mountain slopes, subalpine meadows	?	7-8	PR	1500-2100	MED
Hicraceum leucutechum	Rocky habitats, steppe	Yellow	6-7	PR	1340-1960	MED
H. tumeloum	Mountain steppe	Yellow	7	R	1300	MED
Cicerbita variabilis	Rocky places, wet slopes, pine and fur forests	Blue	6-8	PR	1400-2000	MED
		LEGUMINOSAE				
Chronanthus orientalis	Oak forest, High slopes	Yellow	6	PR	Over 1000	MED
Genista involucrata	Maquis, Conifer forests, Wet rocks	Yellow	6	PR	600-1500	IR-TR
G. aucheri	Rocky slopes, Steppe, Meadow	Yellow	6-7	PR	500-1500	IR-TR

Contd...

Table 5.1–Contd...

Family and Plant Name	*Habitat*	*Flower Colour*	*Flowering Time*	*Habit*	*Altitude (m)*	*Element*
Colutea monocalyx sp.*davisiana*	Forests, Rocky slopes	Yellow	3-5	PR	0-1250	MED
Astragalus macroscepus	Calcareous steppe, Red pine forests		6-8	PR	700-1460	IR-TR
A. oxytropitolius	Pine forests, Wet slopes		5-7	PR	850-2100	IR-TR
A. papasianas	Oak and wheat associations	Yellow	5-6	PR	50-100	MED
A. ptilodes var. *ptilodes*	High plateau	Whitish	6-7	PR	Over 500	MED
A. ptilodes var. *cariensis*	High plateau	Whitish	6-7	PR	Over 500	MED
A. strictispinis	Rocky places, Pine forests	Pink	5-8	PR	850-2000	IR-TR
A. condensatus	Steppe, Forest	Pink	5-7	PR	900-3000	IR-TR
A.weidemannianus	Steppe, Meadows	Pink	5-7	PR	900-2200	IR-T
A. baibutensis	Mountainous areas	Purplish	6-7	PR	500-2130	IR-TR
A. mitchelianus	High altitudes	Pink	7	PR	700-1100	MED
A. squalidus	Pine forests, Calcareous steppe	Purple, rarely greenish yellow	5-7	PR	500-2350	IR-TR
A. gaebotrys	Pine forests, *Calcareous steppe*	Yellowish	6	PR	–	IR-TR
A. acmonotrichus	Rocky slopes	White or purplish Yellow	5-7	PR	1500-1900	MED
A. gymnolobus	Steppe, Meadows, Forest	White or Yellow	5-6	PR	1000-1800	IR-TR
A. trojanus	Steppe, Maquis	Pink	5-7	PR	0-1040	MED
A. consimilis	High altitudes	Red	6	PR	900-1200	MED
A. lydius	Steppe, Meadows	Purplish-blue	5-7	PR	100-1400	IR-TR
A. paecilanthus	Cedrus forests	White or Yellow	5-6	PR	940-2000	IR-TR
A. flavescense	Rocky slopes	Yellowish	7-8	PR	1750	MED
A. vulnerarie	Forest, Steppe slopes	Yellow	5-7	PR	750-2150	IR-TR
Trifolium caudatum	Forests	Purple, Pink, Cream	5-8	PR	600-1600	IR-TR
T. pannonicum ssp. *Elongatum*	Meadows, Steppe, Forest	White or Cream	6-8	PR	300-2350	IR-TR
T. chlorothricum	Open areas	White	5	AN	–	IR-TR
Melilotus bicolor	Dry, rocky places	White	4-5	AN/BN	1000-1200	IR-TR

Contd...

Table 5.1–Contd...

Family and Plant Name	Habitat	Flower Colour	Flowering Time	Habit	Altitude (m)	Element
Trigonella eretica	Waste grounds, Eroded areas	Yellow	4-6	AN	100-500	MED
T. rostrate	Rocky slopes, steppe, Black pine forests	Bright yellow	5-6	AN	850-1500	IR-TR
T. smymea	Rocky slopes, Wet rocks	Bright yellow	5-6	AN	0-1150	MED
T. cephalotes	Wet phrygana, River banks, Lower places	Yellow	4-5	AN	0-50	MED
Lotus macrothricus	Wet valleys, Riversides	Yellow	4-7	AN	50-1000	MED
Hedysarum cappadocicum	Wet rocks, Steppe, Rocky slopes	Pink, purple or Yellowish	5-8	AN/ PR	800-3500	IR-TR
Onobrychus podperae	Wet rocks, slopes	Pink	5-6	PR	300-1000	IR-TR
O. armena	Wet rocks, Steppe, Maquis, Ruderal forest	Dark Pink	5-8	PR	100-2000	MED
Ebenus plumose var. *plumose*	Arid places	Pink	6-7	PR	100-850	MED
		LABIATAE				
Lavandula stoechas ssp. *cariensis*	Red pine forest, Maquis, Phyrygana, sand dimes, Ruderal	Blackish-purple	3-6	PR	0-700	MED
Sentellaria salviifolia	Rocky areas, Maquis, Forests	Bright yellow	5-8	PR	400-1900	–
Phlomis russeliana	Conifer forests, Hazelnut shrubs	Yellow	5-9	PR	300-1700	EX
Phlomis armeniaca	Pine forests, cereal and fallow fields	Yellow	6-8	PR	800-2350	IR-TR
P. capitata	Steppe, Rocks, Oak forests	Yellow	6-8	PR	540-2400	IR-TR
P. nissolii	Rocky areas, Ruderal, Steppe	Yellow	6-8	PR	0-1550	IR-TR
Lamium pisidicum	Dry calcareous areas	Pinkish-purple or Lilac	5	PR	915-1830	–
Wiedemannia orientalis	Dry rocks, Steppe, Ruderal, Open areas	Pale or dark pink, Lilac, purple	4-6	AN	700-1650	IR-TR

Contd...

Table 5.1–Contd...

Family and Plant Name	*Habitat*	*Flower Colour*	*Flowering Time*	*Habit*	*Altitude (m)*	*Element*
Ballota nigra ssp. *anatolica*	Valleys, River banks, Heaps, Waste ground	Purple or pink	5-9	PR	0-1900	IR-TR
Marribium rotundifolium	Calcareous habitats	Whitish	5-7	PR	700-2000	MED
M. globosum ssp. *globosum*	Rocky slopes	Whitish	4-8	PR	800-2500	IR-TR
Sideritis sipylea	Calcareous places, Mixed forest	Yellow	5-8	PR	200-1600	MED
S. tmolea	Sloppy rocks	Yellow	7-8	PR	1900	MED
Stachys tmolea	Calcareous places, young forest	Pale lemonade yellow	5-8	PR	200-1900	MED
S. cretica ssp. *smymea*	Red pine forest	Rose pink	4-8	PR	5-1380	MED
S. cretica ssp. *anatolica*	Meadow, calca-reous slopes	Rose pink	5-9	PR	100-2900	IR-TR
S. setifera ssp. *lycia*	Stream banks, wet plains	Pink	5-8	PR	750-1400	IR-TR
Nepeta cadmea	Rocky slopes	White	7-8	PR	200-1900	MED
N. nuda ssp. lydiae	Rocky slopes	Purplish-blue	6-8	PR	600-1700	MED
N. viscida	Rocky slopes	Pale purple or white	5-7	PR	740-1800	MED
Origanum spyleum	Calcareous rocks, Forest, maquis	Pink	6-8	PR	100-1500	MED
Micromeria cristate ssp. *phyrgia*	Calcareous rocks, steep places	White, Lilac or Purple	7-8	PR	1000-2000	MED
Thymus zygioides var. *lycaonicus*	Red pine forest, open maquis	Pink	4-6	PR	0-1600	MED
T.sipyleus ssp. var. *sipyleus*	Mountain steppe, Rocky slopes	White or pink	5-8	PR	400-2700	MED
T. chaubardii ssp. *chaubardii* var.*anatolianus*	Open conifer forests	Pale pink	5-7	PR	250-2200	MED
Zizophora taurica ssp. *eleonioides*	Steppe, open habitats	Reddish-purple, Lilac	5-8	AN	100-2060	MED
Salvia cedronella	Shrubby ground	Yellow	5-6	PR	–	MED
S. wiedemannii	Calcareous areas, Steppe, Ruderal	Lilac or Blue	5-7	PR	500-1400	IR-TR
S. pisidica	Oak maquis, Field borders	Purplish-blue	5-7	PR	950-1750	IR-TR

Contd...

Table 5.1–Contd...

Family and Plant Name	*Habitat*	*Flower Colour*	*Flowering Time*	*Habit*	*Altitude (m)*	*Element*
S. tehihatcheffii	Calcareous areas, Red pine forests, Vineyards	Light purple or White	5-6	PR	400-1200	IR-TR
S.cadmica	Rocky slopes, Steep slopes, Oak shrubs	White	5-7	PR	900-1810	?
S. smymaea	Rocky habitats	Purplish-blue	5	PR	920	MED
S. dichroantha	Steppe, Forest, Ruderal	–	7-9	PR	810-1810	IR-TR
		LILIACEAE				
Allium sibthorpianum	Grassland, Rocky slopes	Lilac pink or Purple	8-9	PR	700-2500	MED
A. phyrgium	Artemisia steppes, ruderal	Purplish-Brown	8	PR	900-1850	IR-TR
A. olympicum	Forest, Grassland, Shrubby ground	Purple, Red, Lilac, Pink	7-8	PR	1300-2800	EU
A. huber-monathii	Forests, Shrubs, Maquis, Rocks, Dry areas	Pale pink or white	6-8	PR	1000-2000	IR-TR
A. pictistamineum	Maquis, Shrubs, Steppe, Rocks	Pale green, yellowish-green	5-6	PR	200-980	MED
A. proponticum var. *proponticum*	Maquis, Steep slopes, Coast, Rocky places, Vineyards	Pale Purple or Purple	5-7	PR	0-1600	MED
A. stylosum	Forest, Shrubby, Rocky places, Steppe	Dark purple	5-7	PR	100-1800	MED
A. reuterianum	Slopes, Rocky habitats, Calcareous places	Pink, Red or Wine red	7-8	PR	1800-2100	MED
Chionodoxa lucilae	Open slopes	Lavender, Blue	5-6	PR	1600-2000	MED
C. forbesii	Open slopes, forests	Bright blue	3-4	PR	1000-2500	MED
C. sardensis	Pine forests, Wet North slopes	Bright blue	3-4	PR	550	MED
Ornithagalum nivale	Snowy slopes	White	6	PR	–	MED
Muscari aucheri	Wet places, forests	White or Pale blue	4-6	PR	1000-3000	–
M.latifolium	Pine forests	Purplish-Black	4-5	PR	1100-1800	MED
M. bourgaet	Wet places, Calcareous areas	Light blue	5-7	PR	1500-3000	MED
Hyacenthalla lineata	Shrubs, Forests, Grasslands	Dark blue	3-5	PR	200-1500	MED

Contd...

Table 5.1–Contd...

Family and Plant Name	*Habitat*	*Flower Colour*	*Flowering Time*	*Habit*	*Altitude (m)*	*Element*
Fritillaria buhynica	Forest, Shrubs	Greenish-yellow	3-5	PR	100-1200	MED
F. flerschenana	Clayey heaps, Steppe, Shrub	Dark pink	2-5	PR	1000	IR-TR
F. canca ssp. *canca*	Forest, Rocks, Calcareous places	Yellow or orange	3-5	PR	200-1500	MED
Tulipa armena var. *lycia*	Rocky slopes	Pale red, pink or yellow	4-6	PR	1000-2750	MED

MED: Mediterranean; IR-TR: IranoTuranian; EU: Euxin; AN: Annual; BN: Biennial; PR: Perennial.

A. phrygia, Anthemis cretica spp.absinthifolia, A. cretica spp.anatolica, A. cretica spp.candicans, A. cretica spp.leucanthemoides, A. cretica spp.tenuiloba, A. aciphylla var. aciphylla, A. cretica var.discoidea, A. dipsacea, A. wiedemanniana, A. xylopoda, Carduus olympicus ssp. hypoleucus, Centaurea acicularis var. urvillei, C. amasiensis, C. aphrodisea, C. calolepis, C. lydia, C. olympica, C. pichleri ssp. extrarosularis, C. polyclada, C. reuterana ssp. reuterana, C. sipylea, C. zeybekii, Cicerbita variabilis, Cirsium leucopsis, Doronicum reticulatum, Hieraceum leucotechum, H. tmoleum, Jurinea cadmea, J. pontica, Picris olympica, Scorzonera acuminata, S. eriophora, S. suberosa ssp. cariensis, S. tomentosa, Senecio castagneanus, Serratula lasiocephala, Tragopogon olympicus, T. subacaulis, Tripleurospennum conoclinium, T. hygrophilum (ASTERACEAE); Astragalus acmonotrichus, A. baibutensis, A. condensatus, A. consimilis, A. flavescenses, A. gaeobotrys, A. gymnolobus, A. lydius, A. macroscepus, A. mitchelianus, A. oxytropifolius, A. paecilanthus, A. papasianus, A. ptilodes var. cariensis, A. ptilodes var. ptilodes, A. squalidus, A. strictispinis, A. trojanus, A. vulneraria, A. weidemannianus, Chronanthus orientalis, Colutea melanocalyx sp.davisiana, Ebenus plumose var. plumose, Genista aucheri, G. involucrata, Hedysarum cappadocicum, Lotus macrotricus, Melilotus bicolor, Onobrychis armena, O. podperae, Trifolium caudatum, T. chlorothrichum, T. pannonicum ssp. elongatum, Trigonella cephalotes, T. cretica, T. rostrate, T. smymea (FABACEAE); Ballota nigra ssp. anatolica, Lamium pisidicum, Lavandula stoechas ssp. cariensis, Marrubium globosum ssp. globosum, M. rotundifolium, Micromeria cristate ssp. phyrgia, Nepeta cadmea, N. nuda ssp. lydiae, N. viscida, Origanum sipyleum, Phlomis armeniaca, P. capitata, P. nissolii, P. russeliana, Salvia cadmica, S. cedronella, S. dichroantha, S. pisidica, S. smyrnaea, S. tchihatcheffii, S. wiedemannii, Sideritis sipylea, S. tmolea, Stachys cretica ssp. anatolica, S. cretica ssp. smyrnea, S. setifera ssp. lycia, S. tmolea, Thymus chaubardii ssp. chaubardii var.alternatus, T. sipyleus var. sipyleus, T. zygioides var. lycaonicus, Wiedemannia orientalis, Ziziphora taurica ssp. cleonioides (LABIATAE); Allium huber-morathii, A. olympicum, A. phyrgium, A. pictistamineum, A. proponticum var. proponticum, A. reuterianum, A. sibthorpianum, A. stylosum, Chionodoxa forbesii, C. luciliae, C. sardensis, Fritillaria bithynica, F. carica ssp. carica, F. fleischeriana, Hyacinthella lineata, Muscari aucheri, M. bourgaei, M. latifolium, Ornithagalum nivale, Tulipa armena var.lycica (LILIACEAE), Chaenorhinum litorale ssp. pterosporum, Digitalis cariensis, Linaria genistifolia, L. corifolia, Scrophularia cryptophila,

S. depauperata, S. floribunda, S. scopolii var. smyrnaea, Verbascum antinori, V. cheiranthifolium var. asperulum, V. coronopifolium, V. exuberans, V. kastamunicum, V. lobatum, V. luciliae, V. lydium var. lydium, V. parviflorum, V. salviifolium, V. serratifolium, V. smyrnaeum, V. splendidum, V. stenotachyum, V. uschakense, V cuneifolia ssp. cunefolia, Veronica elmaliensis, V. multifida (SCROPHULARIACEAE); Arenaria acerosa, A. macrosepala, A. sipylea, A. tmolea, Bolanthus minuartioides, B. spergulifolius, Dianthus anatolicus, D. brevicaulis ssp. setaceus, D. leucophaeus var. leucophaeus, D. lydus, Gypsophila tubulosa, Minuartia juressi ssp. asitacia, M. mesogitana ssp. lydia, M. saxifraga, Saponaria chlorifolia, S. prostrata ssp. prostrata, Silene phrygia, S. sipylea, S splendens, Velezia hispida (CARYOPHYLLACEAE); Astrantia maxima ssp. haradjianii, Bupleurum sulphureum, Eryngium bithynicum, E. thorifolium, Ferula anatolica, Ferulago humilis, F. macrosciada, F. aucheri, Heracleum platytaenium, Muretia aurea, Peucedanum chryseum, Pimpinella cretica var. cretica, P. tragium ssp. polyclada, Tordylium macropetalum (APIACEAE); Consolida aconiti, C. phrygia ssp. phrygia, C. raveyi, Delphinium fissum ssp. anatolicum, Ranunculus heterorhizus, R. reuterianus (RANUNCULACEAE); Corydalis solida ssp. solida, Papaver argemone ssp. davisii, P. strictum, P. virchowii (PAPAVERACEAE); Alyssum davisianum, A. fulvescens var. stellatocarpum, A. oxycarpum, Erysimum kotschyanum, Hesperis balansae, H. buschiana, H. kotschyi (BRASSICACEAE); Fumana paphlagonica, Helianthemum nummularium ssp. lycaonicum (CISTACEAE); Amygdalus balansae, Alchemilla bursensis, Crataegus dikmensis, Potentilla buccoana, P. sublaevis, Pyrus anatolica (ROSACEAE); Rosularia chrysantha, Sedum hispanicum var. planifolium (CRASSULACEAE); Valerianella glomerata (VALERIANACEAE); Pterocephalus pinardii, Scabiosa hololeuca, (DIPSECACEAE); Herniaria olympica, Paronychia anatolica ssp. balansae, P. angorensis, P. chionaea (ILLECEBRACEAE); Hypericum adenotrichum, H. aviculariifolium ssp. aviculariifolium var. aviculariifolium, H. aviculariifolium ssp. aviculariifolium var. bourgaei (GUTTIFERAE); Malope anatolica (MALVACEAE); Linum aretioides, L. cariense, L. hirsutum ssp. anatolicum var. anatolicum, L. hirsutum ssp. pseudoanatolicum, L. tmoleum (LINACEAE); Erodium absinthoides ssp. absinthioides (GERANIACEAE); Haplophyllum megalanthum, H. myrtifolium (RUTACEAE); Acer hyrcanum ssp. keckianum (ACERACEAE); Asyneuma limonifolium ssp. pestalozzae, A. rigidum ssp. sibthorpianum, A. virgatum ssp. cichoriiforme, Campanula betonicifolia, C. lyrata ssp. icarica, C. lyrata ssp. lyrata, C. raveyi, C. teucrioides, C. tomentosa, Jasione supina ssp. timolea (CAMPANULACEAE); Vincetoxicum canescens ssp. pedunculata (ASCLEPEDIACEAE); Convolvulus galaticus (CONVOLVULACEAE); Alkanna tinctoria ssp. gladulosa, A. tubulosa, A. areolata var. areolata, A. areolata var. sublaevis, Moltkia aurea, Nonea macrosperma, Onosma armenum, O. isauricum, Paracaryum aucheri, P. calycinum, Symphytum anatolicum (BORAGINACEAE); Acanthus hirsutus (ACANTHACEAE); Acantholimon acerosum var. brachystachyum, Limonium effusum (PLUMBAGINACEAE); Aristolochia hirta (ARISTOLOCHIACEAE); Euphorbia anacampseros var. tmolea, E. erythrodon, E. falcata ssp. macrostegia (EUPHORBIACEAE); Quercus vulcanica (FAGACEAE); Asperula daphneola, A. lilaciflora ssp. lilaciflora, A. lilaciflora ssp. phrygia, A. nitida ssp. hirtella, A. tenufolia, Crucianella disticha, Galium brevifolium ssp. brevifolium, G. campanelliferum, G. dumosum, G. incanum ssp. centrale, G. pendiliflorum. G. stepparum (RUBIACEAE); Phoenix canariensis (PALMAE); Arum balansanum, Biarum tenuifolium (ARACEAE); Sternbergia schubertti (AMARYLLIDACEAE); Crocus biflorus ssp. nubigenus, C. flavus ssp dissectus, C. fleischeri, C. oliveri ssp. balansae, Gladiolus anatolicus, Iris purpureobractea (IRIDACEAE); Juncus sparganiifolius (JUNCACEAE); Carex distachya var. phyllostachioidea, C. divulsa ssp. coriogyne (CYPERACEAE); Alopecurus davisii, Bromus cappadocicus ssp. sclerophyllus, B. macrocladus, B. sipyleus, Festuca pinifolia var. phrygia, Helictotrichon pubescens ssp. longifolia, Pseudophyleum gibbum (POACEAE).

FAUNAL DIVERSITY

The seaward fringe of the basin is an ecologically important wetland ecosystem; forming the A category nature reserve which used to receive excess water from the Gediz River, but since 1990, with restrictions on irrigation releases, the reserve suffers from water shortages during summer months. There are four freshwater lagoons, namely; Kýrdeniz Dalyan (400 ha), Homa (1824 ha), Chilazmak (725 ha) and Ragýppasha (500 ha). In addition Cigli swamp (140 ha) and Sazlý lake (30 ha) are two spots worth noting. These form rich habitats for waterbirds, fish and other organisms.The delta is used by 211 bird species out of a total of 426 recorded in Turkey, 46 being local, 54 summer migratory, 43 winter migratory, and 30 transitory (Sýký, 2002). The important species are; Great Cormorant *(Phalacrocorax carbo), Pygmy* Cormorant *(Phalacrocorax pygmeus), Dalmatian* Pelican *(Palecanus crispus), Greater* Flamingo *(Phoenicopterus ruber),* Ruddy Shelduck *(Tadorna ferruginea),* Lesser Kestrel *(Falco naumanni), Eurasian* Oystercatcher *(Haematopus ostralegus), Black-winged Stilt (Himantopus himantopus), Pied* Avocet *(Recurvirostra avosetta), Eurasian Thick-knee (Burhinus oedicnemus), Collared Pratincole (Glareola pratincola), Kentish Plover (Charadrius alexandrinus)*, Spur-winged Lapwing *(Vanellus spinosus)* and Common Redshank *(Tringa totanus).*

Livestock number 1.3 million heads. The land degradation problems pose a serious threat to the rich biodiversity and agricultural areas. Forests originally protected the soils on the mountains, but the trees have been cut for building and to provide firewood.

CONCLUSIONS

The phenomenon of degradation cannot be arrested by physical or technical means alone. Guiding human activities towards consideration of ecology and natural landscape variations would be a positive step in assuring that future civilizations will enjoy greater success and persist longer than did past civilizations. Interdisciplinary efforts to better understand the environment and its relationships to the welfare of inhabitants would contribute towards a more optimistic view of the future.

There is a need to evaluate land resources to assess the desertification threats and potential for degradation to manage the resilience characteristics of the systems and select technological options in the framework of the resilience properties (Ozturk, 1995; Feoli *et al.*, 2003).

Although Agenda 21 (UNCED, 1992, Ch.12) emphasizes desertification with activities set-up for preventive actions, by mentioning that "to strengthen regional and global systematic observation networks linked to the development of national systems for the observation of land degradation and desertification caused both by climate fluctuations and by human impact, and to identify priority areas for action", but land degradation is receiving very little attention. A comprehensive new agenda is needed to protect the sustainability of the ecosystem and prevention of land degradation. The new agenda should indude a science based approach, to understand the land quality and its changes by management alternatives (Sakcali and Ozturk,2003). In the preparation of the new agenda, all sectors including stakeholder and non-governmental organizations should work together.

Tonnes of sediments are transported to the Aegean Sea due to the overgrazing and subsequent soil erosion as well as other land degradation activities (Ozturk *et al.*, 1994, 1996a,b; Alpaslan and Atis, 1995). Erosion control is particularly critical for maintenance of soil fertility for food production as well as environmental quality considerations in the basin in general. There is an urgent need for a mobilization of the scientific community to start an integrated program for data collection, sample areas for assessment, monitoring of land degradation and development of mitigation technologies; develop methods for integrated and sustainable monitoring and management of biodiversity; land use models including natural and human-induced factors; information systems that link environmental monitoring, accounting and impact assessment; policies that encourage sustainable land use and management; and finally develop economic instruments in the assessment of land degradation to encourage the suitable use of land resources. In order to avoid an induced impoverishment and the desiccation of land, it would be more plausible to start taking measures now through well organised ecological land use planning.

With demographic outburst and economic growth, the demands on the functions of the water resources system are increasing at an alarming rate in the basin. An effective utilisation of land and water inventory data for land use planning is needed vis-a-vis the re-charging of groundwater reservoirs in areas of water table decline related to the agro-industrial watershed based planning. For example, until 1980 deep wells were almost unknown in the basin. The picture of the challenges facing irrigated agriculture is becoming clear in the context of increasing water scarcities. There is no option other than the improvement of efficiency in order to achieve high production annually. Government should double grants for water development, in order to reduce by half the number of muncipalities lacking water treatment systems. For a protection of biodiversity, water management will play a critical role by optimizing the performance of the water resources system in order to satisfy the socio-economic as well as ecological requirements. This can be acheived through an interstate cooperation.

The rich biodiversity of Gediz Basin flourishes on different ecological habitats such as alpine/subalpine zone, forests, macchias phrygana and wetlands in the deltaic zone. These are mainly dominated by the plants belonging to the mediterranean phytogeographical region. The area shows an endemic ratio of 20-25 per cent. These endemics constitute 10 per cent of the total endemic plant cover of Turkey (Ozturk *et al.*, 1992; Ekim,2009). The plant diversity in general is under great pressure, in particular nearly 50 species of endemics. Overexploitation can result in a decrease of genetic diversity (Ozturk *et al.*, 2008) together with the degradation or destruction of the wetlands around the deltaic zone due to excessive use of Gediz river, which leads to a rise in the water table, water-logging, and salinization. The herbaceous and long rooting taxa like species of *Medicago, Agropyron, Bromus inermis, Festuca ovina, Lotus corniculatus,* shrubs like *Capparis* spp., *Pistacia* spp., *Rhus coriaria, Rosmarinus officinalis, Crateagus monogyna, Paliurus spina-christii, Vitex agnus-castus,* trees like *Populus tremula, Robinia pseudo-acacia* and *Pinus brutia* can be used for reforestation and reclamation of problematic areas, and prevention of erosion. The salinized lands can be reclaimed and evaluted by planting economically important halophytic taxa or salt resistant species.

Greater pragmatism is required to subscribe to a new land ethic, to safeguard ecological, biological and genetic diversity as well as the economic productivity potential of the basin. We believe that this study will add greater value to the concept of land degradation and provide a forum to identify the research gaps and needs concerning the assessment, monitoring and development of mitigation technologies, indicators for early warning of degradation and determination of land quality. There is an urgent need for the establishment of a sustainable land management institution, to carry out the above mentioned tasks.

REFERENCES

Alpaslan, A. H., Atis, I. 1995. Erosion Problems in Gediz Basin and Projects of DSI. I Gediz Basin Erosion and Env. Symp., 111-118.

Arslan, N., Okmen, M., 2006. The Economical and International Dimensions of the Environmental Problems, Environmental Problems in the Black Sea Region and the Role of the Voluntary Organizations, Building and Environment, Volume 41, Issue 8, August, pp. 1040-1049.

Ayyildiz, M. 1983. Irrigation Water Quality and Salinity Problems. Ankara Univ., Fac. of Agric., Publ.No. 879.

Davis, P. H. Davis.1965-1985. Flora of Turkey and the East Aegean Islands, Vols. 1–9, University Press, Edinburgh,UK.

Davis, P. H., Mill, R. R., Tan, K.1988. Flora of Turkey and the East Aegean Islands (Supplement), Vol. 10, University Press, Edinburgh, UK.

Ekim,T. 2009. The Rare Endemics of Turkey. Turkiye ÝÞ Bankasý, Yayýnlarý. pp:537.

Gecgel, G., Girgin, A., Beyazgul, M., Defraiture,C., Murray-Rust, H. 2000. Performance assesment of Gediz Basin Using Secondary Data. Proc. of Intern. Symp. on Desertification, Konya-Izmir, pp. 96-102

Gokmen, D., Durmus, H., Ozturk, M.1997. Problems Facing Biodiversity in the Coastal Zone of Soke. Great Menderes Delta Agric. And Env.Problems Symp.,Soke,149-158.

Gul, A., Avcioðlu, R., Acar, M. I., Ozel, N. 1995. Use of alternative land use systems on the marginal agricultural areas under the effect of erosion in Gediz Basin. I Gediz Basin Erosion and Env. Symp., 66-83.

Guner,A., Ozhatay,N., T. Ekim, T., K. H. C. Baser,K.H.C. 2000. Flora of Turkey and the East Aegean Islands (Supplement 2), Vol. 11, University Press, Edinburgh, UK.

Guvensen, A.,Ozturk, M., 2003.Halophytic Plant Diversity of South Aegean Coastal Zone in Turkey. Pakistan Journal of Botany, 35,5,853-864.

Guvensen, A., Gork, G., Ozturk, M., 2006. An overview of the halophytes in Turkey. Sabkha Ecosystems Vol. II, West and Central Asia (Eds. Khan *et al.*), Springer, The Netherlands, pp: 9-30.

Okmen, M., Parlak, B., 2006. An Agenda for Urban Air Pollution of Environmental Problems and Policies: The Turkish Case, Interdisciplinary Environmental Review Issue (IER), Vol VIII, Number 1, June, pp. 96-107.

Ozturk, M., Pirdal,M., Uysal, I.1992. Ecology of Some Turkish Endemics. Agriculture and Village, 74: 20-21.

Ozturk, M., Erdem,U., Alparslan,E. 1994. Ecological restoration of human disturbed areas -A case study from Gediz Basin-West Anatolia Turkey. In: Urban Ecological Development, Research and Application (Wang and Lou, eds.). China Environmental Science Press, Beijing, 216-223.

Ozturk, M.A., 1995.Recovery and rehabilitation of mediterranean type ecosystem- A case study from Turkish maquis. In: Evaluating and Monitoring the Health of Large-Scale Ecosystems, NATO-ARW (Eds. D.Rapport, C.L. Gaudet, P.Calow), Springer- Verlag, pp: 319-332,

Ozturk, M., Erdem, U., Butuner, H., Dalgic, R. 1995. Non-agricultural use of land in Turgutlu-Gediz basin. I Gediz Basin Erosion and Env. Symp., 354-369.

Ozturk, M., Celik, A., Alparslan, E., Erdem, U.1996a. Land Degradation in Relation to Urbanisation and Industrialisation in the Aegean Region of Turkey.Assessment and Monitoring of Soil and Land Degradation(Eds.Kapur,S.,et al.), Cukurova University Press, Turkey, pp.305-317.

Ozturk, M., Guvensen,A., Gokmen, D., Caliskan, I., 1996b.Drought and Plant Interactions in the Soke Plain. Great Menderes and Delta, Agric. and Env.Problems Symp., Soke, 24-36.

Ozturk,M.,1999.Urban Ecology and Land Degradation.In:Perspectives in Ecology (Editor:Almo Farina). Backhuys Publishers, Leiden, NL,115-120.

Ozturk, M., Guvensen, A., 2002. Studies on the Halophytic Vegetation of Saline-Alkaline Habitats in West Anatolia-Turkey. Int. Symp. on Optimum Resources Utilization in Salt-Effected Ecosystems in Arid and Semiarid Regions, 8-11-April, Cairo-Egypt.

Feoli, E., Giacomich, P., Mignozzi, K., Ozturk, M., Scimone, M., 2003. Monitoring desertification risk with an index integrating climatic and remotely-sensed data: An example from the coastal area of Turkey. Management of Environmental Quality: An Intern. Journal, Vol. 14, No.1, pp. 10-21.

Ozturk,M.,Celik,A.,Yarci,C.,Aksoy,A.,Feoli,E., 2002. An overview of plant diversity, land use and degradation in the Mediterranean region of Turkey. Environmental Management and Health, 13,5, 442-449.

Sakcali,M.S.,Ozturk, M., 2003. Eco-physiological Behaviour of Some Mediterranean Plants as Suitable Candidates for Reclamation of Degraded Areas. Journal of Arid Environments, 57, 141-153.

Ozturk, M., Guvensen, A., Gork, G., Gork, C. 2006. An overview of the coastal zone plant diversity and management strategies in the Mediterranean region of Turkey, In: Biosaline Agriculture and Salinity Tolerance in Plants (Eds. Ozturk *et al.*), Birkhauser Verlag, Basel Switzerland, pp:89-100.

Ozturk, M., Sakcali, S., Bahadýr, H., Gork,G., Beyazgul, M.2007. Salinity-Alkalinity and Pollution Problems in the Gediz Basin-Turkiye. International Conference On Environment: Survival and Sustainability 19-24 February, Near East University, Nicosia-Northern Cyprus, Vol. 7: MT-11: Environmental Science And Technology, pp: 2935-2941.

Ozturk,M., Guvensen, A., Sakcali, S., Celik, S. 2008. Agriculture and Ecological Interactions. In: Ecological/Organic Agriculture and Environmnt (Edit. İbrahim AK). F.Özsan Press, EKODER, Bursa, pp:14-26.

Ozturk,M., Gucel,S., Sakcali,S., Gork,C., Yarci,C., Gork,G.2008. An overview of plant diversity and land degradation interactions in the eastern Mediterranean. Chapter 15, In: Natural Environment and Culture in the Mediterranean Region (Eds. Efe *et al.*), Cambridge Scholars Publ.,UK, pp:215-239.

Ozturk,M.,Gucel,S.,Kucuk,M., Sakcali, S. 2010. Forest,Climate Change and Forest Fires in the Mediterranean Region of Turkey. Journal of Environmental Biology-Special Issue,31,1-9.

PRB. Population Reference Bureau. 2004. World Population Data Sheet. Washington D.C.

Sýký,M.(2002) Birds of Gediz Delta. Ekoloji, 44, 11-16.

SIS 1995. State Institute of Statistics -The Population of Turkey, 1923–1994 Demographic Structure and Development. Ankara: SIS.

SIS. 2003. State Institute of Statistics-Census of Population: Social and Economic Characteristics of Population. Ankara: SIS.

SIS. 2004. State Institute of Statistics-Population and Development Indicators. Web site: http://nkg.die.gov.tr, September 7, 2004.

Taysun, A., Uysal, H.1995. Impacts of Water Erosion in Gediz Basin and its Significance. I Gediz Basin Erosion and Env. Symp., 7-15.

Chapter 6

Land Degradation in Sri Lanka and Necessity of Mitigation

M.A.K. Munasinghe

Research Officer, Department of Agriculture,
Natural Resource Management Center, Peradeniya, Sri Lanka

ABSTRACT

Sri Lanka with a population of 20 million people in an area of 65.5 000 Km^2 is one of the countries with high population density. Problem of land degradation has been evident throughout its history. This has manifested itself in several ways; heavy soil loses, high sediment yields, soil fertility decline and reduction of crop yields, marginalization of agricultural lands, land slides, salinization, alkalinization, acidity including both desertification, and formation of acid sulfate soils, iron toxicity, toxicity associated with high nutrient contents, water logging and indiscriminate disposal of waste. Major causes for land degradation are both natural and human related. Main natural causes are geomorphological conditions and rainfall characteristics. Human induced causes are : high demand for agricultural lands under increasing population density, which leads to encroachment of forests and marginal lands; high dependency on land for living; inadequate attention on soil conservation particularly in annual cropping; inadequate legislation or non-implementation of available legislation towards proper use of lands; land fragmentation, ownership and unfavorable tenancy condition; lack of initiatives for proper land use; lack of proper land use policies; adhoc planning of infrastructure development or lack of holistic approach, etc.

The most significant of these are heavy soil losses and high sediment yields, thus soil erosion should receive the highest attention. There has been no continuous monitoring of erosion status/process or sediment transport. However attempts have been taken to asses the risk of erosion under the present land use in the Central Province. Available evidence suggests that the frequency and magnitudes of landslides has increased in recent years causing serious damage to life and property. Occurrences of salinity and water logging are associated with coastal areas and irrigated areas within the dry zone. With development of recent irrigation facilities, it was anticipated that salinity would become a problem over the years, but because the country is not as arid as most of the areas of the world affected by salinity, salts that accumulate in surface flushes with the high intensity rains occurring on seasonal basis. Studies show that saline/alkaline soils, waterlogged soils and potential acid sulphate soils are the major problems in Sri Lanka. Saline and alkaline soils have been classified as Solodized Solonetz (Natraqualfs) and Soloncladeschacks (Salothids). These soil groups cover an area of about 16,000 ha. There is no systematic monitoring programme except few site specific observations. Therefore, there is a great neccessity for a proper monitoring programme coupled with related database.

INTRODUCTION

Sri Lanka with a population of 20 million people in an area of 65.5 thousand square kilometers is one of the countries with high population densities. This tropical island is comparatively less affected by natural calamities and blessed with climatic conditions which are favourable for year round crop production. In order to harness the agro-climate, heavy investments on irrigation and other agricultural projects have been made, but projected returns for those investments were not achieved during the past few decades. There exists a great potential in land productivity enhancement through proper planning and efficient project implementation. The land and water resources management in the country has been identified as one of the priority areas for action to achieve sustainable food security like other Asian and Pacific region countries. There is a need to raise the land productivity by reversing land degradation and water loss, increasing biodiversity and the quality of the environment. In the above effort the necessity of an integrated approach is well accepted in order to improve the countries ability to plan and monitor the use of its land resources for better use and management. In this context fundamental understanding of land water resources potentials, their limitations, possibilities of their use for better food and agricultural productivity is very important.

Sri Lanka has made a considerable progress in understanding its resource potentials and limitations, because these are necessary to address the questions related to soil inventory, land evaluation, land use planning and management, land degradation assessment and land use mapping. In this paper an attempt is made to describe the achievements in land degradation assessments, and highlight the issues that should be focused in order to use the management tools effectively when implementing activities related to land resources management under present organizational context.

GEOGRAPHY AND CLIMATE

Based on elevation and nature of the terrain five geomorphic regions can be identified: the coastal fringe, the central highlands, the south-west country, the east and southeast, and the north central lowlands. The central massif from which headwaters of all Sri Lanka's major rivers originate, is a compact physiographic unit bounded on the south by a high mountain, the *Worlds End*. The topography of the southwest; the other part of Sri Lanka's Wet Zone is characterized by long parallel ridges cut by the rivers beginning in the hill country. The eastern and southeastern lowlands are characterized by rolling hills, undulating plains and isolated residual hills. The coastal fringe has important fisheries and the region contains large parks and wildlife refuges that are home to wild elephants. The north central lowlands are somewhat similar to those in the east. In this region, the country's hydraulic civilization flourished.

The spatial distribution of mean annual rainfall over the island varies between 750-5500 mm. approximately, 2/3rd of land area of the country receives less than 2000 mm of rain per year. However, this amount is not well distributed. Temporal rainfall variation shows bimodal distribution pattern in many parts of the country.

- *The convectional convergence period: (March to mid April):* This period is characterized by the bright clear mornings, formation of clouds by the convectional activity in the early afternoon and thunderstorms in the late afternoon.
- *The pre-monsoonal period (mid April to late May):* Characterized by transitional weather pattern. During this period convectional activity is reduced and monsoon rains begin.
- *The southwest monsoon period (late May to late September):* This season begins with the largest amount of rainfall to the southwestern lowlands and the windward side of the central highlands.
- *The convectional cyclonic period (late September to late November):* Begins to weaken the monsoons and can include cyclones.
- *The northeast monsoons (November to February):* This period begins with rainfall to the large extent of the lowland located in the eastern, northern and southern parts of the central highlands.

Due to this seasonality of rainfall, two dry periods with varying lengths can be clearly identified in many parts of the country. Generally shorter dry spell continues from mid January to mid March and longer dry period extends from June to September. The two rainy periods are related to the two monsoons namely northeast and southwest. The two seasons are locally called "Maha" season and "Yala" season respectively. The Maha season extend from October to February and Yala season from March to September. Even if the total annual rainfall is comparatively high, year-to-year variation make rainfall very unreliable.

The solar radiation varies between 275-400 cal/cm^2/day during different months in different stations. The highest solar radiation is experienced in February and March and the lowest is in November and December.

The mean monthly temperature of the island differs slightly depending on the seasonal movement of the sun with some modifying influence caused by rainfall. In the lowlands (below 300-m msl) the mean annual temperature is 27°C with mean daily range of 6°C. Temperature changes with the changing elevation in the central highlands. Generally temperature decreases by about 6°C per each rise of 1000 m from mean sea level (msl). The temperature regime is favorable for plant growth in most parts of the country. Diurnal range of temperature is relatively small.

Relative humidity (RH) closely follows rainfall pattern of the region with diurnal differences caused by the warming effect of the sun during daytime and cooling effect of the night. RH values in the nights remain about 10 per cent higher than the daytime RH, but is above 70 per cent throughout the year in all parts of the country. Generally dry months record low values. High RH is conducive for growth of plant pathogens.

High wind speed is damaging some crops. The wind speed varies over the year and over the country. May, June, July and August are windy months in most of the dry districts. During these months wind speed of over 20 km/hr is not uncommon.

SOILS OF SRI LANKA

The soils have been classified at great group level for the whole country and series level for some parts. The mapping has been done mainly at great group level. Fourteen great soil groups have been described (Moorman and Panabokke, 1961). Generalized soil map of Sri Lanka is given in Figure 6.1. Extents of major great groups of the country are presented in Table 6.1. The chemical and physical properties of soils have been discussed by De Alwis and Panabokke (1972). The physical properties (except for water holding capacity) of major soil groups such as Reddish Brown Earths, Red Yellow Podzolic soils, Red and Yellow Latasols, Reddish Brown Latasolic soils are favourable for plant growth. The chemical fertility of wet zone soils is poor because these soils are extensively leached due to high rainfall. The CEC values of most of the soils are low,therefore there is a need for the use of organic manure and special fertilizer management practices on these soils. The base saturation of the dry zone soils remains at a higher range. Solodized solonetz, Bog and Half Bog soils are the major problematic soils. The agricultural potential of soils has been discussed by De Alwis and Panabokke (1972). In general, soils do not pose major problem for crop production.

Table 6.1: Extents of Different Soil Mapping Units and Major Soil Groups of Sri Lanka

Sl.No.	*Soil Mapping Units*	*Extent of Soil Mapping Unit (ha.)*	*Extent of Major Soil Groups (ha.)*
1.	Reddish Brown Earths and Low Humic Gley soils; undulating terrain	1800129	
2.	Reddish Brown Earths with moderate amount of gravel in subsoil and Low Humic Gley soils; undulating terrain	55034	
3.	Reddish Brown Earths with high amount of gravel in subsoil and Low Humic Gley soils; undulating terrain	89648	
4.	Reddish Brown Earths and Solodized Solonetz; undulating terrain	177456	
5.	Reddish Brown Earths, Noncalcic Brown soils and Low Humic Gley soils; undulating terrain	247938	
6.	Reddish Brown Earths and Immature Brown Looms; rolling, hilly and steep terrain	313500	2683705

Contd...

Table 6.1–Contd...

Sl.No.	Soil Mapping Units	Extent of Soil Mapping Unit (ha.)	Extent of Major Soil Groups (ha.)
7.	Noncalcic Brown soils and Low Humic Gley soils; undulating terrain	67507	
8.	Noncalcic Brown soils, soils on old alluvium and Solonetz; undulating terrain	65281	132788
9.	Red-Yellow Latosols; flat to slightly undulating terrain	263380	263380
10.	Calcic Red-Yellow Latosols, flat terrain	35152	35152
11.	Solodized Solonetz and Solonchaks; flat terrain	149963	149963
12.	Grumusols; flat terrain	14094	14094
13.	Soils on Recent marine calcareous sediments; flat terrain	26675	26675
14.	Alluvial soils of variable drainage and texture; flat terrain	501978	501978
15.	Regosols on Recent beach and dune sands; flat terrain	153249	153249
16.	Red-Yellow Podzolic soils and Mountain Regosols; mountainous terrain	261737	
17.	Red-Yellow Podzolic soils; steeply dissected, hilly and rolling terrain	821754	
18.	Red-Yellow Podzolic soils with strongly mottled subsoil and Low Humic Gley soils; rolling and undulating terrain	157265	
19.	Red-Yellow Podzolic soils with soft or hard laterite; rolling and undulating terrain	247037	
20.	Red-Yellow Pdzolic soils with dark B horizon and Red Yellow Podzolic soils with prominent A1 horizom; rolling terrain	12901	
21.	Red-Yellow Podzolic soils with semi-prominent A1 horizon; hilly and rolling terrain	25212	1525906
22.	Reddish Brown Latosolic soils; steeply dissected, hilly and rolling terrain	62326	62326
23.	Immature Brown Loams; steeply dissected, hilly and rolling terrain	54088	54088
24.	Bog and Half-Bog soils; flat terrain	57423	57423
25.	Latosols and Regosols on old red and yellow sands; flat terrain	54607	54607
26.	Alluvial soils of variable drainage and texture; flat terrain	51892	51892
27.	Regosols on Recent beach sands; flat terrain	7852	7852

The most recent major contribution for soil resources inventorization was implemented in 1998–2008 through a project conducted by the Sri Lanka Soil Science Society with the financial assistance of Canadian Soil Science Society. The first phase of the project concentrated for the wet zone of Sri Lanka and soils were identified on series level based on soil chemical and physical properties (Wet Zone Soils of Sri Lanka, 1998). Subsequently, soils in Intermediate and Dry zone were completed

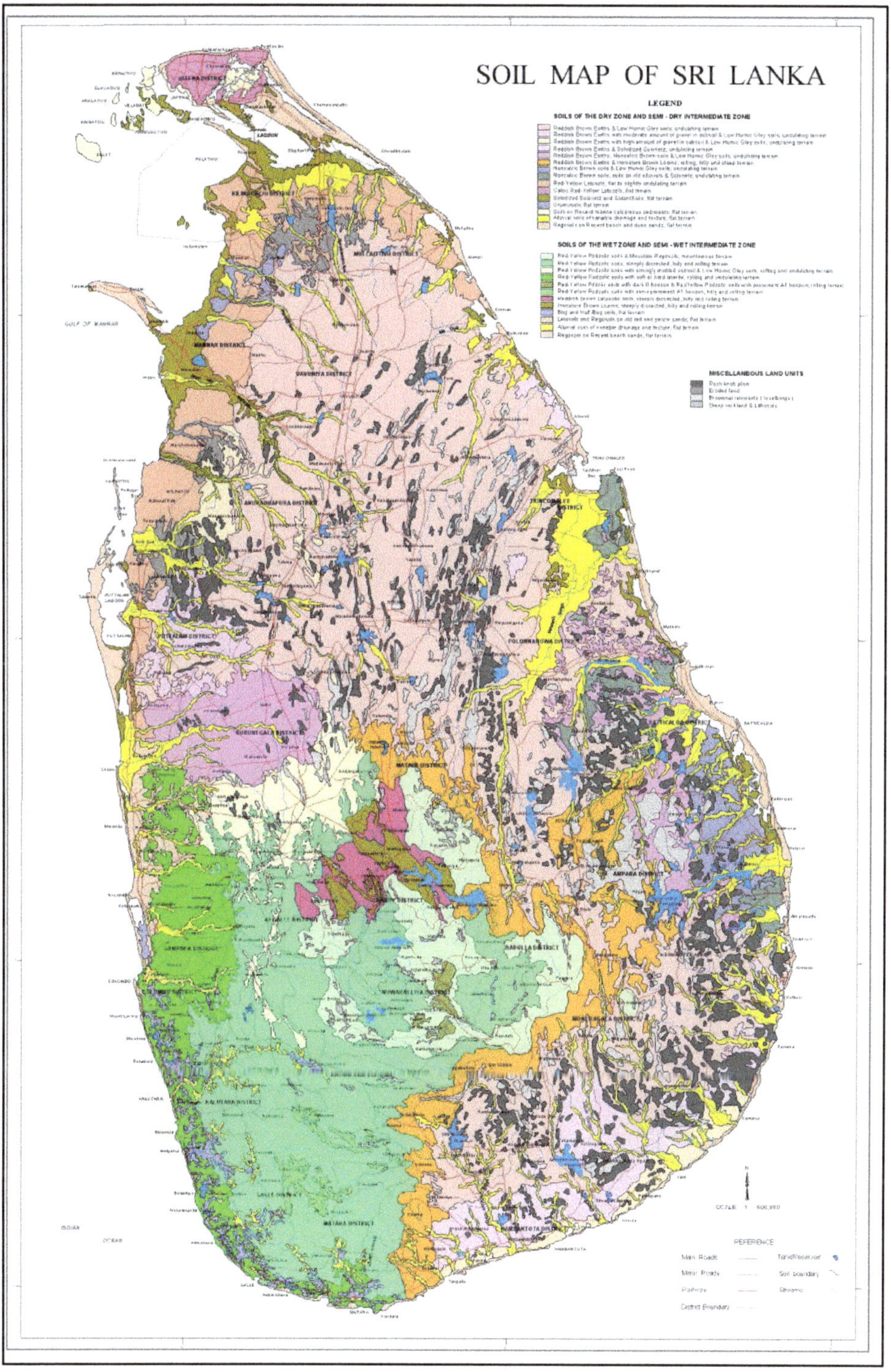

Figure 6.1: Generalized Soil Map of Sri Lanka

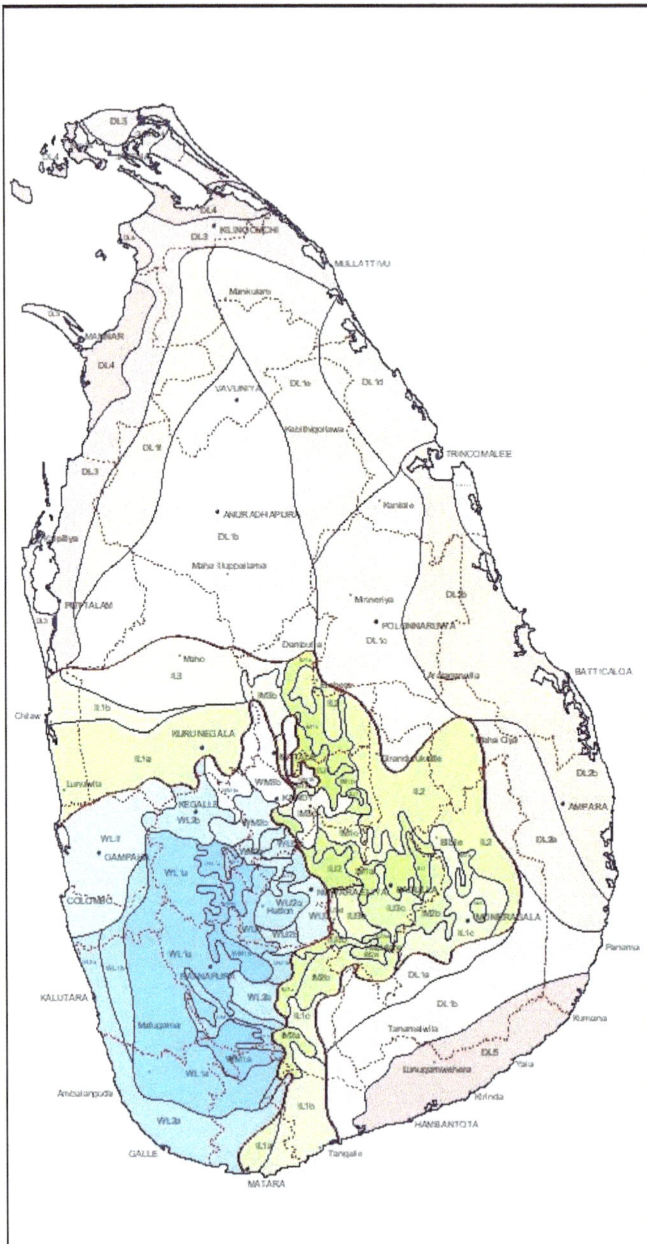

CLIMATIC ZONE	AGRO - ECOLOGICAL REGION		75% EXPECTANCY VALUE OF ANNUAL RAINFALL (mm)
WET ZONE	UP COUNTRY	WU1	> 3,100
		WU2a	> 2,400
		WU2b	> 2,200
		WU3	> 1,800
	MID COUNTRY	WM1a	> 3,300
		WM1b	> 2,900
		WM2a	> 2,200
		WM2b	> 1,800
		WM3a	> 1,600
		WM3b	> 1,400
	LOW COUNTRY	WL1a	> 3,200
		WL1b	> 2,800
		WL2a	> 2,400
		WL2b	> 2,200
		WL3	> 1,700
INTERMEDIATE ZONE	UP COUNTRY	IU1	> 2,400
		IU2	> 2,100
		IU3a	> 1,900
		IU3b	> 1,700
		IU3c	> 1,600
		IU3d	> 1,300
		IU3e	> 1,400
	MID COUNTRY	IM1a	> 2,000
		IM1b	> 2,000
		IM1c	> 1,300
		IM2a	> 1,800
		IM2b	> 1,600
		IM3a	> 1,400
		IM3b	> 1,200
		IM3c	> 1,100
	LOW COUNTRY	IL1a	> 1,400
		IL1b	> 1,100
		IL1c	> 1,300
		IL2	> 1,600
		IL3	> 1,100
DRY ZONE	LOW COUNTRY	DL1a	> 1,100
		DL1b	> 900
		DL1c	> 900
		DL1d	> 900
		DL1e	> 900
		DL1f	> 800
		DL2a	> 1,300
		DL2b	> 1,100
		DL3	> 800
		DL4	> 750
		DL5	> 650

Figure 6.2: Agro-ecological Map of Sri Lanka (2003)

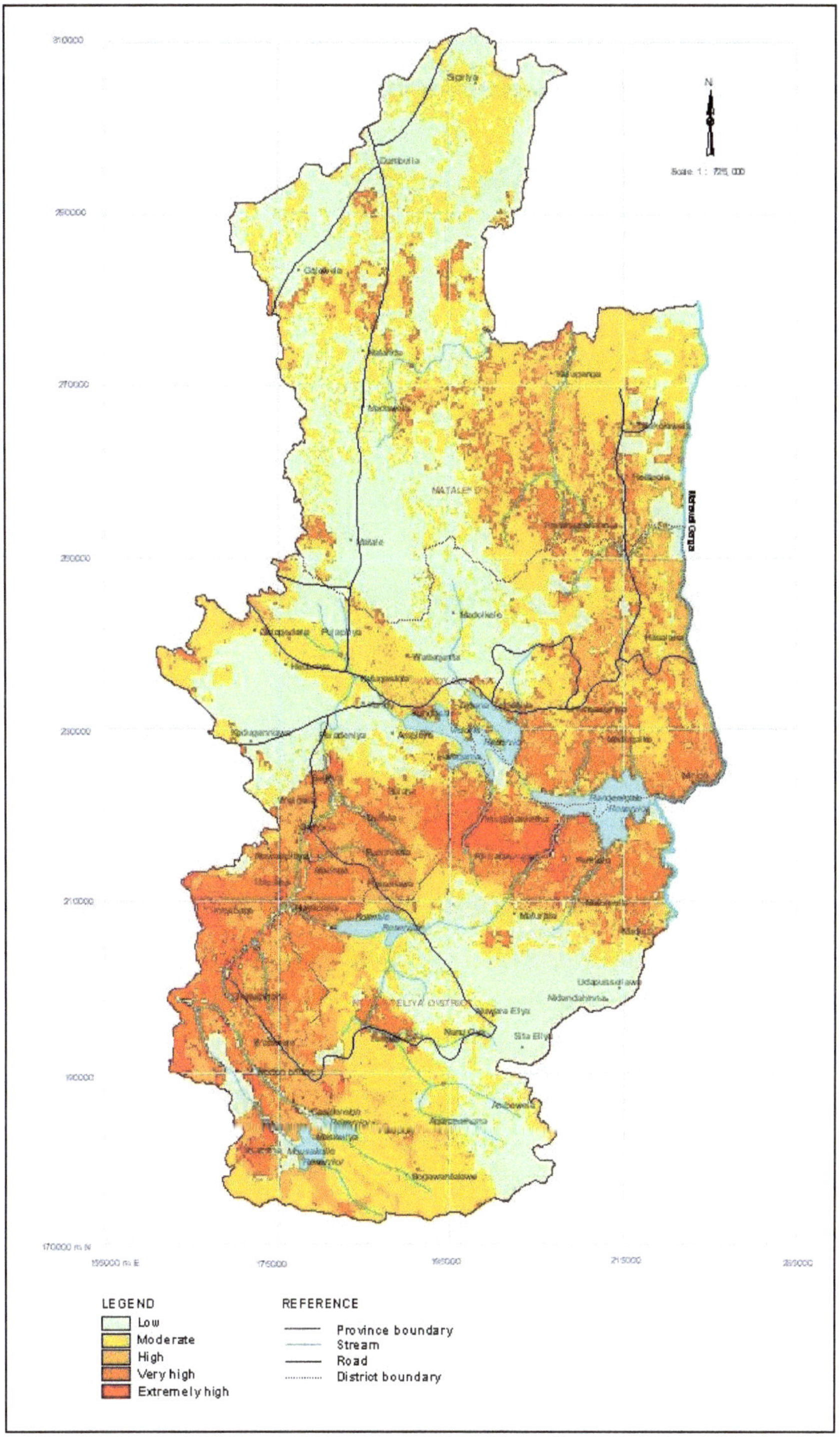

Figure 6.3. Soil Erosion Hazard Map: Central Province of Sri Lanka

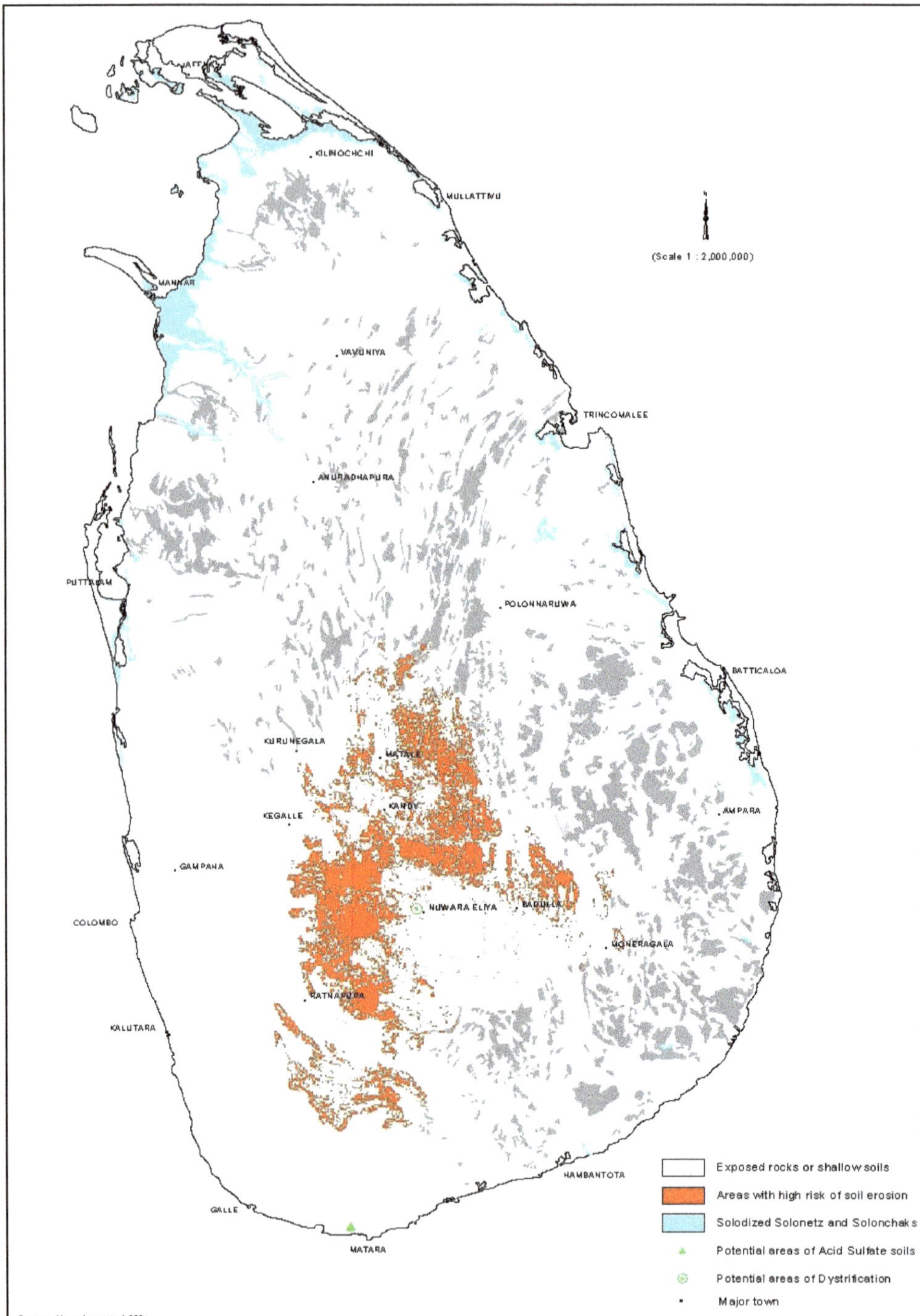

Figure 6.4: Map Indicating Land Degradation

through the same project (Soils in the Intermediate Zone of Sri Lanka 2003, Soils in the Dry Zone of Sri Lanka, 2007). The information generated by the project is already in digital form and has been developed into an information system. Thus information is being used for number of research and development planning activities.

AGRO-ECOLOGICAL ZONING

An agro-ecological region represents a particular combination of the natural characteristics of climate, soil and relief (Panabokke, 1996). When an agro-climatic map, which can be considered as areas where the integrated effect of climate is uniform throughout the area for crop production, is superimposed on soil and terrain the resulting map identifies agro-ecological regions. Thus, each agro-ecological region represents somewhat uniform agro-climate, soils and terrain conditions and as such would support a particular farming system where certain range of crops and farming practices find their best expression. Regionality of the rainfall distribution has traditionally been generalized in to three climatic zones in terms of "Wet zone" in the southwestern region including central hill country, and "Dry zone" covering predominantly, northern and eastern part of the country, being separated by an "Intermediate zone," skirting the central hills except in the south and the west. In differentiating aforesaid three climatic zones, land use, forestry, rainfall and soils have been widely used (Panabokke, 1996). The Wet zone covers the area, which receives relatively high mean annual rainfall over 2,500 mm without pronounced dry periods. The Dry zone is the area, which receives a mean annual rainfall of less than 1,750 mm with a distinct dry season from May to September. The Intermediate zone demarcates the area, which receives a mean annual rainfall between 1,750 to 2,500 mm with a short and less prominent dry season. As low temperature is an important climatic factor affecting plant growth in the Wet and Intermediate zones, a sub-division based on the altitude takes into account the temperature limitations in these two climatic regions.

In this delineation, the Low-country is demarcated as the land below 300 m in elevation and the Mid-country with elevation between 300-900 m while the Up-country is the land above 900 m elevation. Both Wet and Intermediate zones spread across all three categories of elevation while the Dry zone is confined to the Low-country resulting in seven agro-climatic zones covering the entire island.

Based on many decades of work, 24 agro-ecological regions have been identified (Panabokke and Kannangara, 1975). The differentiation of the Wet zone into its distinctive agro-ecological regions was governed primarily by differences in rainfall and elevation. In the Dry zone, on the other hand, it was the nature of the soil that primarily determines the identity of individual agro-ecological region. In the Intermediate zone, it was observed that rainfall, elevation and soil play an equally important role. According to this map, there were 10 agro-ecological regions in the Wet zone; 9 in the Intermediate zone and 5 in the Dry zone.

Agro-ecological boundaries cannot be considered as fixed. They are best treated as variable, not only because of global and local changes in the environment but also, and often more significantly, because of the means we have at our disposal to estimate and represent them (Wood and Pardey, 1998). Moreover the impacts of climate change are visible in the form of increased variability of seasonal rainfall in the country (Punyawardene 2003), which necessitates updating existing agro-ecological models.

Even if the delineation criteria remain unchanged, the physical boundaries of agro-ecological regions may change with more spatial and temporal coverage of data, improved data interpolation and interpretation algorithms and reduction of the scale. Thus, in view of the environmental change, availability of more spatial and temporal data and advancement of GIS technology led to the sub-division of 24 agro-ecological regions into a map with 46 agro-ecological sub-regions (Figure 6.2) on an enhanced scale (Punyawardena *et al.*, 2003). The main distinguishing characteristic of each agro-ecological region is denoted by a 4-character code consisting of letters and a number. Three major climatic zones are indicated by the first upper case letter of the code (W, I and D). The second upper case letter of the code (L, M and U) denotes three categories of elevation. The numerical character in the third place of the code represents a more detailed moisture regime (rainfall and evaporation combined) with a degree of wetness on the scale of 1 to 5 where 1 being the most favorable. The lower case letter in the fourth place indicates a sub-region as determined by rainfall distribution and other physical environmental factors where degree of wetness decreases $a \rightarrow f$.

In the Wet zone, there are 15 agro-ecological sub-regions. Four sub-regions found in the Up-country wet zone show a distinct variation in the distribution of the South West Monsoon (SWM) rains. Being in the most effective area of the SWM rains, WM1a, WL1a and WU1a sub-regions receive the highest amount of rainfall. Apart from the amount and distribution of SWM rains, relative effectiveness of North East Monsoon (NEM) rains has also played a vital role in distinguishing 6 sub-regions in the mid-country wet zone. The four months period from December to March is relatively "dry" in WM3a agro-ecological sub-region while there are two distinct dry periods in the WM3b due to reduced effectiveness of SWM rains over this sub-region. In the Low-country Wet zone, amount and distribution of SWM as well as First Inter Monsoon (FIM) rains were important in identifying the 5 agro-ecological sub-regions. Meanwhile, the months of July, August and December in WL3 agro-ecological region do not receive adequate amount of rainfall and hence cannot be considered as wet months. As such, 4 months period extending from December to March is relatively "dry" in this region.

The Intermediate zone consists of 20 agro-ecological sub-regions out of which 15 sub-regions lie in the central hills. Varying degree of effectiveness of different rainfall governing mechanisms across the central hills has caused variety of growing environments in this region. There are 7 agro-ecological sub-regions in the Up country Intermediate zone out of which IU1 is reported to receive the highest annual rainfall among all sub-regions of the entire Intermediate zone. Being in the Knuckles range, this region receives ample amount of rains from NEM while the contribution from SWM rains is also substantial. Complex geographical settings of the IU3 agro-ecological region which encompasses almost whole of the so-called "Uva basin" have resulted in 5 agro-ecological sub-regions due to high spatial variability of intermonsoonal and NEM rains in this region. Meanwhile, due to its location in the rain shadow area of the SWM, this region does not receive adequate rains during June to September resulting in dry and windy environment.

The Mid country Intermediate zone has 7 agro-ecological sub-regions. Most of these sub-regions also do not receive adequate rains from SWM and, hence, 4 months from June to September are relatively dry. Low country Intermediate zone consists of 5 agro-ecological sub-regions. Other than IL2, all other agro-ecological sub-regions in the Low country Intermediate zone show a bi-modal rainfall distribution. Since Second Inter Monsoon (SIM) and NEM rains are the only effective rainy seasons in the region, the IL2 agro-ecological region exhibits a distinctly uni-modal rainfall distribution along with a long and pronounced dry period from April to September.

In the Dry zone, there are 11 agro-ecological sub-regions with distinct differences in rainfall distribution. The DL3, DL4 and DL5 agro-ecological regions of the Dry zone receive the lowest annual rainfall along with some limitations of soils that are found in these regions. Out of 11 agro-ecological sub-regions, only DL1a and DL1b are characterized by two discernible peaks in the rainfall distribution and thus, support crops in both *Maha* and *Yala* growing seasons. Those agro-ecological sub-regions found in the eastern sector of the Dry zone, *i.e.*, DL1c, DL1d, DL1e and DL2a and DL2b, exhibit a distinct uni-modal rainfall pattern, and support only the crops in *Maha* season. The rest of the agro-ecological sub-regions of the Dry zone also support only the *Maha* crop since *Yala* rains in those sub-regions are not adequate to meet the evapotranspiration requirements (National Atlas of Sri Lanka, 2007)

LAND AND LAND DEGRADATION

The total land area of Sri Lanka is approximately 6.56 million hectares. Out of this, nearly 300,000 ha. are covered by water bodies. Present per capita land availability of the country is 0.3 ha. There is a decreasing trend of this value as the population continues to increase. Moreover 2/3rd of the land area in the country falls within flat to undulating landform. The rest comprised of hilly and mountainous terrain where higher percentage of land area consists of highly dissected, steep slopes and narrow valleys. Most recent scientific landuse mapping exercise was undertaken by the Survey Department of Sri Lanka during 1983-1988 periods (Survey Department, 1988). The uses shown in these maps were grouped into 7 categories namely agricultural lands (arable), residential and buildup lands, forest lands, rangelands, water bodies, wetlands and barren lands.

ARABLE LANDS

Arable land extent is nearly 2.9 m hectares which are about 45 per cent of the total land area of the country. The main agricultural land uses include paddy (27 per cent) and plantation crops (tea, rubber and coconut, 24 per cent). Nearly 44 per cent of the agricultural lands are sparsely used which means that there remains a great potential for these lands to be properly developed/used. This is about 20 per cent of the land area of the country.

DYNAMICS OF LAND USE IN SRI LANKA

Traditional landuse has been rice lands in lowlands, and chena (slash and burn) and the homesteads in the uplands. Rice lands were irrigated from reservoirs or streams. These lands provided the staple food rice while chena provided coarse grains, pulses, oil seeds, vegetables and condiments. Homestead consists of perennial trees such as fruit and coconut.

This traditional landuse system in the country gradually changed with the influence of the western world starting from 15th Century. An interest was developed to convert highlands to plantation crops. Cinnamon and Coconut were first established in coastal areas of the wet zone. Rapid change took place following introduction of coffee, tea and rubber plantations. Most of the forest in the wet zone was converted into plantations. This change of landuse resulted in heavy soil erosion.

In the dry zone, during the latter part of the 19th century, renovation of ancient irrigation schemes which had been abandoned was commenced. This resulted in clearing of forests for creating paddy fields and homesteads for new settlement schemes. Rapid increase of the irrigation extents in the country took place since 1930, which affected the traditional shifting cultivation where unavailability of forest lands made it impossible for farmers to have a sufficient fallow lands. This led farmers to do settled farming in the uplands. However low input farming practice continued in these lands and the result was an eroded upland with low production levels. This situation has posed challenges for land and water resources related researchers to put forward suitable/viable cropping systems. Rice was the main crop grown in paddy fields. However in early seventies upland crop cultivation under irrigation was introduced in these lands. This was the start of rice based cropping systems; paddy grown in the main rainy season while it was followed by upland crops during minor rainy season.

Land is one of the major natural resources that have a greater bearing on sustaining the economy of Sri Lanka. It provided the space and resources for expansion of development and other social activities so far, but the pressure to meet and continue to accommodate development needs is however becoming untenable. This is the major challenge for resources planners and there is a need of innovative reforms concerning land resource use.

LAND DEGRADATION AS A CONSTRAIN FOR SUSTAINABLE PRODUCTION

Land degradation can be defined as the process of weakening or impairing land productivity due to natural or anthropogenic activities. Problem of land degradation has been evident through out the Sri Lankan history. However, significant land degradation began during the past 150 years, particularly after the introduction of commercial plantation agriculture. In 1900, with a population of only 3.5 million, Sri Lanka had approximate forest cover of 70 per cent. By 1953, when population reached 8.1 million, natural forest cover had diminished to 44 per cent, and when the population doubled by the mid 1980's the forest cover was around 25 per cent. At present with an estimated population of 19 million the per capita land availability is 0.3 ha. As the arable land is almost half of the above figure, at present a tremendous pressure on land is eventually degrading land resources. This has manifested itself in several ways; heavy soil loses, high sediment yields, soil fertility decline and reduction of crop yields, marginalization of agricultural lands, land slides, salinization, alkalinization, acidity including both desertification, and formation of acid sulfate soils, iron toxicity, toxicity associated with high nutrient contents, water logging and indiscriminate disposal of waste. The most significant of theses are heavy soil losses and high sediment yields, thus soil erosion should receive the highest attention.

SOIL EROSION

Serious concerns about soil erosion have been expressed since 1873, when Sir J.D.Hooker stressed dangers of indiscriminate plantation agriculture. One result was the restriction of land cultivation above 1,525 meters msl. In 1931, a Committee on Soil Erosion documented the damage caused by plantation crops. The Soil Conservation Act of 1951, which followed devastating earth slips in the hill country, in particular in the Kothmale valley, has been largely neglected since. As Stockings (1986, unpublished reports, Mahaweli Development Program) observed, the reality of soil conservation in Sri Lanka today, is one of much rhetoric, little action; the concern of some, the disdain of others.

In one of the earliest estimates of sediment transport based on measured sediment yields Joachim and Pandithesekara (1930), estimated 132-833 thousand tons/year (115/tons/hectare/yr) for the upper Mahaweli catchment, 55 per cent of which was under tea. The studies have shown an average of 15 million tons of sediment, for the period of 1952-1982 passed in the upper Mahaweli watershed through the Peradeniya gauging station. TAMS/USAID (1980) reports soil losses in Maha Oya catchment (Mid Country) in lands under Tobacco on slopes ranging from 45 to 60 per cent. Estimates range from 388 to 913 t/ha/yr. Other estimates on soil erosion under different land uses are available on adhoc basis. There has been no continuous monitoring of erosion status/process or sediment transport. However attempts have been taken to asses the risk of erosion under the present landuse in the Central Province where most of the central massif exists. Figure 6.3 indicates the susceptibility for soil erosion or erosion hazard under the existing land uses. The extents falling under each erosion hazard classes are given in Table 6.2. (Munasinghe *et al.*, 2001). This study helps to prioritize erosion prone areas within the Central Province and identify areas where landuse changes are necessary. However, it is necessary to carry out assessments of soil erosion to answer the questions of 'where it is occurring","what will happen under the present management", "what changes are necessary especially interms of landuse and where it should be?" Therefore, studies related to monitoring of soil erosion on regional or watershed basis is essential in the near future rather than concentrating on plot or point measurements which are incorporating biasness to the result if one generalizes.

Table 6.2: Extents Under Each Soil Erosion Hazard Class for Each District within the Central Province of Sri Lanka

Soil Erosion Hazard	Nuwaraeliya		Kandy		Matale	
	Extent (ha)	Per cent	Extent (ha) (ha)	Per cent	Extent (ha)	Per cent
Low	49,610	29	46,792	25	94,500	47
Moderate	53,635	31	54,618	29	75,929	37.4
High	28,185	17	36,171	19	12,450	6.2
Very high	32,890	19	47,790	25	18,900	9.3
Extremely high	6,255	4	3,790	2	160	0.1
Total	170,575	100	189,161	100	201,939	100

LAND SLIDES

Available evidence suggests that the frequency and magnitudes of landslides has increased in recent years causing serious damage to life and property (Jayewardene, 1980). However, records before 1970 appear sketchy and many landslides have gone unrecorded, like the Kothmale landslide of 1947. Small landslides that do not cause much damage to life and property are not recorded by the Geological Survey or the Social Services Department. The National Building Research Organization (NBRO) which is the mandated institute for land slide hazard assessments recorded around 136 landslides during 1986/87 period. In early June 1988, Sri Lanka experienced a devastating landslide damage, which caused over 300 deaths. According to Parliament Reports approximately 225,000 people in ten Districts were affected by floods and landslides that destroyed over 15,00 homes. Cost for rehabilitating was estimated at about 120 million rupees at 1988 prices (Madduma Bandara, 1989). It must be noted that in the year 2003, there were nearly 500 landslides and cutting edge failures occurred within the southwestern slope of the country claming nearly 200 human lives along with colossal amount of property damage. The situation was similar in 2007, where series of major land slides were reported. Increasing damage to life and property from landslides obviously related to the spread of human settlements into unsuitable areas prone to landslides. Needs for mapping and identification of hazardous areas is a high priority and the activity is in progress. Most of the vulnerable areas have been mapped at 1:10,000 scale by the NBRO and have been submitted to the relevant authorities for the implementation through relocation of settlements. However the implementation needs collaboration of stakeholders and should be lead by Provincial Administration where the progress is not agreeable.

SALINITY AND WATER LOGGING

Occurrences of salinity and waterlogging are associated with coastal areas and irrigated areas within the dry zone. With development of recent irrigation facilities, it was anticipated that salinity would become a problem over the years, but because

Sri Lanka is not as arid as most of the areas of the world affected by salinity, salts that accumulate in surface flushes with the high intensity rains occurring on seasonal basis. According to Lathief and Nayakakorale (1993), saline/alkaline soils, waterlogged soils and potential acid sulphate soils are the major problematic soils found in the country. Saline and alkaline soils have been classified as Solodized Solonetz (Natraqualfs) and Soloncladeschacks (Salothids) (De Alwis and Panabokke, 1972), containing Na and Ca as dominant cations. These are found in the areas where soil is affected by coastal salinity and cover an area of about 16,000 ha. Weerasinghe (1994) reported that the affected paddy extent due to acid sulphate soil in Matara District (Main district of the Southern Part) was about 1000 ha.

Desertification is mainly concentrated in the up country where intensive temperate vegetable cultivation is practiced. Salinity is mostly reported in the southern Dry zone where country's driest places exist and as pockets in the major irrigation schemes of the other part of the Dry zone especially where drainage is impeded. Sodicity or alkalinity has become a constraint to some paddy lands located along the coastal regions. Acid sulphate soils are being formed in the Nilvala Ganga Flood Protection Scheme in the southern province due to drainage improvement to the lands under paddy which were below the sea level, may be million years ago.

OTHER FORMS OF LAND DEGRADATION

It is accepted that soil erosion, landslides, salinity and alkalinity, formation of acid sulfate soils and dystrification are the most threatening land degradation processes in Sri Lanka. In addition, as land includes both soil and water, deterioration of water quality due to human interference cannot also be neglected and hence, it could be a major form of land degradation in years to come (Figure 6.4).

Moreover, it was observed that the conversion of agricultural lands to other uses such as residential, trade, industrial, infrastructure development mining are taking place at an increased rate over the last two decades and will continue even in the future. Adhoc paddy land filling also creates a problem of drainage, resulting in considerable extent of paddy lands uncultivable especially in low country wet zone where urbanization is rapid. Flash floods and heath hazards due to mosquitoes are a frequent dilemma in these areas. Integrated or system level development according to a well prepared plan is essential to avoid above situations.

MAJOR CAUSES FOR LAND DEGRADATION

Natural

- ☆ Geomorphological conditions and rainfall characteristics are conducive to severe land degradation in the form of soil erosion.

Human Related

- ☆ High demand for agricultural lands under increasing population density leads to an encroachment of forests and marginal lands.
- ☆ Farmers do not see the short term benefits of implementing soil conservation and are not concerned with the soil conservation.

- ☆ High dependency on land for living.
- ☆ Inadequate attention on soil conservation particularly in annual cropping.
- ☆ Inadequate legislation or non-implementation of available legislation towards proper use of lands.
- ☆ Land fragmentation, ownership and unfavorable tenancy condition (necessity of land tenure).
- ☆ Lack of initiatives for proper landuse.
- ☆ Lack of proper land use policies.
- ☆ Adhoc planning of infrastructure development or lack of holistic approach
- ☆ Decision makers/Politicians still use land alienation among poor unemployed citizens as a means of employment generation.
- ☆ Lack of efficiency in plan implementation.
- ☆ Lack of collaboration in related institutes, departments etc.

MAJOR STEPS TO BE TAKEN TO COMBAT LAND DEGRADATION

Implementation of Soil Conservation Act

Soil Conservation Act of Sri Lanka first came into effect in 1951 as Act No. 25 with the objective of conserving soil resource, prevention or mitigation of soil erosion and protection of land against floods and droughts. It was amended in 1996 by Soil Conservation (Amendment) Act No. 24 of 1996 to include wider scope thus addressing land degradation as well. Department of Agriculture has been implementing agency for the Soil Conservation Act from the start. However from the very beginning the implementation of Soil Conservation Act was restricted only to awareness creation among stakeholders and technical assistance as and when necessary, but no attempts were taken to take legal action against the violators for obvious reasons. Among those ambiguity or lack of clarity in legal provisions, inadequacy, inconsistence and vague legal expressions or provisions, inadequacy of institutional framework and infrastructure can be quoted.

In order to overcome these drawbacks another amendment to the existing Act is on the way. Followings points will be taken into consideration in the proposed amendment.

Gazetting New Regulations

Recent development activities other than agriculture have resulted in accelerated erosion. Thus there is a necessity of identifying such activities and widen the legal frame work to act against such activities.

Identification and Declaration of Conservation Areas

Declaration of conservation areas under the Soil Conservation Act has been done mostly based on expert judgments, but amendments are proposed based on a study conducted (Munasinghe, 2002) to identify such areas and it suggests that a geographical area nearly 10, 618 km^2 should be declared as a Conservation area. The proposed area was gazetted as new "Conservation Area" in Government Gazette 2008.

Institutional Strengthening

The Authority of implementing Soil Conservation Act vests on the Department of Agriculture (DOA) and the Natural Resources Management Centre (NRMC) is the responsible Unit for implementing the Act. However at present institutional framework of the NRMC is not adequate enough to handle the Act effectively. Human resources and infrastructure development, establishment of an operation network at field level are essential for its successful implementation.

Review Existing Legislation

Legal provisions pertaining to the protection and sustainable use of land recourses are contained in several major enactments other than the Soil Conservation Act (Paddy Lands Act, Act of Agrarian Services, National Environmental Act etc.) and these are enforced by a number of agencies. The presence of large number of Acts do not provide sound basis for efficient land management. Hence there is a need to review existing laws and regulations and develop appropriate legal frame work for effective implementation of land management strategies. In this effort one can use the Soil Conservation Act as an umbrella framework.

Appropriate Landuse

This plays a key role in determining the vulnerability of a land for soil erosion or land degradation. Therefore steps need been taken to ensure land is being utilized based on its suitability for a particular use. The Ministry of Lands has already prepared the landuse policy. Once the Policy is accepted and approved, the Land Use Act will be prepared. However, while implementing such Acts promotion of watershed and river-basin planning to address the issues of soil erosion in critical catchments is essential.

Land Policies

Successive governments have used land resources to provide employment to the rural population. This policy of providing land in lieu of employment opportunities should be discouraged. In addition encroachments on state lands especially in environmentally fragile areas have become a problem. Hence there is need to review current land alienation and encroachment regulation policies and to formulate a land alienation policy that would encroach into sensitive area. Reviewing land tenure systems in relation to their impact on land productivity also should be considered. On the other hand large numbers of people who use land without a proper title often do not take any interest in conserving soil due to their uncertainty in ownership. Land policies should be formulated to handle the problem of land fragmentation by providing minimum land extents for different uses in different regions.

Establishment of Appropriate Institutional Framework and their Coordination

There are number of government organizations dealing with land management. However under local context coordination among these institutes are minimum or not existing at all. Hence there is need to promote and strengthen coordination among different organizations dealing with land resources management.

Create Public Awareness

Public seem to be unaware of the seriousness of the soil erosion problem and need for soil conservation. Holding education and awareness creation programmes on soil conservation focusing on schools, government officers and the general public should draw the attention.

Providing Alternative Income Generating Opportunities

It is widely accepted that environmental degradation and poverty are closely linked. Rural community rely on natural resources mainly on land for survival but do not have the ability to invest on improvements or conservation of resource base. This degrades the resource base. A policy of weaning rural people away from land based employment should be adopted by promoting and establishing agro-based and other industries in rural areas for efficient use of agricultural products and generation of additional income. Livestock farming should also be encouraged and marketing and value addition should received high priorities.

Building up a Database

There is no systematic monitoring program on land degradation in Sri Lanka. Efficient monitoring program will help to understand the present situation and provide information for decision making. However a relevant, updated database is a prerequisite for efficient monitoring. In addition, maintenance of useful information in public domain is essential, hence planners, researchers, extensionists and public would be able to carryout there land management activities based on a sound footing.

CONCLUSIONS

Although understanding of land resources in terms ecology and soils, have been completed on a national scale, monitoring of degradation processors in spatial and temporal terms are yet to be carried out. Therefore, there exists a great necessity of a proper monitoring programme coupled with a related database. It must be noted that agro-ecological zonation is providing a meaningful platform for making stratification of point/region observations for land degradation assessments.

REFERENCES

De Alwis K.A. and Panbokke C.R. Handbook of Soils of Sri Lanka 1972, Journal of Soil Science, Sri Lanka 2:1-9.

Detric P.Weerasinghe, K,D,N. Gunaratne, D.A.B.N. Lexa and K. Vlseal 1982, Rice production on Acid Sulphate soils of Sri Lanka, Paper presented at the International Conference on Acid Sulphate Soils, Vietnam.

Lathief M.A. and Nayakakorale H.B. Problem soils of Sri lanka, Country paper proc. Workshop of SAARC countries on problem soils 23–24 Nov. 1993 Sri Lanka, Department of Agriculture, Sri Lanka.

Moorman F.R. and Panabokke C.R. 1961 Soils of Ceylon, Tropical Agriculturist, 117–169.

Panabokke, C.R., 1996. Soil and agro-ecological environments of Sri Lanka. Natural Resources Series -No.2. Natural Resources, Energy and Science Authority of Sri Lanka. No. 47/5, Maitland Place, Colombo–7, Sri Lanka.

Panabokke, C.R. and R.P. Kannangara, 1975. Agro-ecological regions of Sri Lanka. Land-use division, Department of Agriculture, Peradeniya, Sri Lanka.

Punyawardena, B.V.R., T.M.J. Bandara, M.A.K. Munasinghe, Nimal Jayaratna Banda and S.M.V. Pushpakumara, 2003. Agro-ecological regions of Sri Lanka. Sri Lanka Atlus, 2007

Punyawardena, B.V.R., 2003. Impact of climate change on crop production in Sri Lanka. Proceedings of the international conference on Monsoon Environments: Agricultural and hydrological impacts of seasonal variability and climate change. 24–28 March, 2003. International Center for Theoretical Physics, Trieste, Italy.

Munasinghe M.A.K., Pushpakumara V, Bandara T.M.J and Herath H.M.B. (2001) Use of Geographical Information Systems for Soil Erosion Hazard Assessment of the CENTRAL PROVINCE OF SRI LANKA, VOL. 3, 2001, P 148.

Munasinghe M.A.K and Pushpakumara V. (2001), Re-demarcation of Soil Conservation Areas of Sri Lanka based on Soil Erosion Potential, Journal of Soil Science

Wood, S. and P.G. Pardey, 1998. Agro-ecological aspects and evaluating agricultural research and development. Agricultural Systems, 57(1):13-42

Chapter 7

Land-Use Conflicts in Kenya: Drivers and Potential Solutions

Joyce W. Ngure*, Mwangi Githiru and Urbanus Mutwiwa

**Chief Research Officer,*
Ministry of Higher Education Science and Technology,
Directorate of Research Management and Development,
Utalii House, Kenya

ABSTRACT

Land in Kenya is the most important resource from which the country generates goods and services for the people. The national economy is primarily agro-based with 90 per cent of the population living in rural areas deriving livelihood directly from land. Land resources are the means to a livelihood and determine the levels of prosperity or poverty, fulfilling social obligations, and also conferring social status. The main economic activities in the high and medium potential areas is intensive agriculture and livestock husbandry while in the low potential areas it is pastoral, ranching and wildlife based systems and dryland farming. In urban areas, land use is determined by various zoning laws, which prescribe the areas to be used for commercial, residential, industrial, recreational and public purposes. Specific sites along mountains, river valleys, and the unique savannah and grassland ecosystems have been set aside for conservation as well as indigenous forests, wildlife sanctuaries, water catchments, marine life, monuments and cultural sites, protected as tourist attraction sites.

Kenya has not had a clearly defined or codified National Land Policy since independence. This together with the existence of many land laws, some of which are incompatible, has resulted in a complex land management system. There have been conflicts arising in conservation like conflict of human beings and wildlife involving damage by wildlife on human beings, crops and livestock as well as man poaching wildlife and destroying habitats. There have been challenges in past land distribution injustices as well as political interference. The practice of inheritance has caused land fragmentation and over exploitation. This together with increasing population has lead to problems of squatters that is widespread in forests. This paper discusses the current challenges of land use in Kenya and proposes recommendations.

Keywords: *Agriculture, Environmental conservation, Land-use guidelines, Land-use policy.*

INTRODUCTION

Kenya covers a total area of 587,000 Km^2 of which 11,000 Km^2 is covered by water (State of Environment Report Kenya, 2003). The country boasts an almost-complete repertoire of land cover and habitats from the snow-capped top of Mt. Kenya (5199 m a.s.l.) to the Indian Ocean coastline including forests, savannahs, grasslands, wetlands, fresh and saline water bodies, alpine zones and moorland, as well as deserts. In terms of moisture availability and temperatures, most of the country is very dry (classified as arid to very arid zones with less than 25 per cent moisture content, and covering about 60 per cent of the land surface (Figure 7.1). In contrast, high potential areas for agriculture (humid and sub-humid zones with more than 65 per cent moisture content cover only about 10 per cent of the land surface) (UNEP, 1982). Indeed, agro-ecological zonation of Kenya based on the length of growing period (LPG) shows the arid to very arid areas having a very short LGP of up to about 50 days per year, whereas very humid areas have upto about 344 days (FAO, 2007). To a large extent, land use is influenced by climate, but with additional impact of socio-economic factors as well as government policies. In all, agriculturally high and medium potential areas (humid to semi-humid) cover about 18 per cent of the total land area, while the rest (semi-arid to very arid) are classified as low potential, predominantly only suited for ranching and nomadic pastoralism (State of Environment Report Kenya, 2003).

Land is without doubt the most important resource in Kenya, from which the country generates goods and services for her people. Land resources are the means to a livelihood and determine the levels of prosperity or poverty, fulfilling social obligations, and also conferring social status and political power (Mwagore, 2002). The main economic activities in the high and medium potential areas are intensive agriculture and livestock husbandry, while in the low potential areas it is pastoral, ranching and wildlife-based systems; dry-land farming is becoming increasingly common in some of these areas. Agriculture remains the backbone of the national economy. It contributed 25.3 per cent of real gross domestic product (GDP) in 2003, 24.3 per cent in 2004 and 24.2 per cent in 2005 (CBS, 2006). Agriculture supports over 80 per cent of the population, and directly or indirectly employs 70 per cent of the country's labour force (Ogendo and Kosura, 1995).

Land use is currently determined by various zoning laws and policies that fall under the mandate of various Government bodies (Ministries and State Corporations). In urban areas for instance, zoning laws prescribe the areas to be used for commercial, residential, industrial, recreational and public purposes (Waiganjo and Ngugi, 2001). In addition, there are protected areas which largely comprise of specific sites along mountains, river valleys, and unique savannah and grassland ecosystems that have been set aside for conservation of indigenous forests, wildlife, water catchment areas, marine life, monuments and other culturally-significant sites; in addition, other areas have been set aside as mining zones. Most of these designated areas are interspersed within the diverse agriculturally potential areas with human habitation in rural and urban areas. Of the total land cover however, only about 2 per cent is forest which is much below than the recommended 10 per cent (Akotsi *et al.*, 2006).

What Policies Drive Land Use in Kenya?

The country has not had a clearly defined or codified National Land Policy since independence. This, together with the existence of many land laws, some of which are incompatible or in direct conflict, has resulted in a complex (sometimes not well-coordinated and difficult to manage) land management and administration system (KLA, 2000). Some of the key laws related to the land use are:

- The Local Government Act.
- The Physical Planning Act.
- The Agriculture Act.
- The Forest Act.
- The Wildlife Conservation and Management Act.
- The Water Act.
- The Environmental Management and Coordination Act.

Policy-wise, Kenya has been grappling with the land use problems since independence in 1963, which subsequent government regimes have been unable or unwilling to solve. During this period, partly owing to the lack of a cohesive land use policy/guidelines, land abuse has manifested itself in many ways including fragmentation, breakdown in land administration, disparities in land ownership, pollution, erosion and other forms of unsustainable use caused by or leading to poverty (KLA, 2000). This has resulted in environmental, social, economic and political problems including deterioration in land quality, squatting and landlessness, disinheritance of some groups and individuals, urban squalor, under-utilization and abandonment of agricultural land, tenure insecurity and conflict.

Shifts in policy, legislation and institutional land-use practices have over the years brought about changes in the way land-based resources have been used in space and time. Such changes can roughly be grouped into three groups each with unique landmark changes. These groups include the period of hegemony during the colonial time (pre-1963); the benevolence period immediately after independence (1963-1983) and the dark period (1983-2000) (KLA, 2000).

Pre-1963

This period could be called the "golden era" for natural resource conservation in Kenya because there was little destruction of the resources. Clearing of indigenous forests was minimal. Wildlife protection was probably at its best. There was no economic pressure to reclaim wetlands and other protected areas.

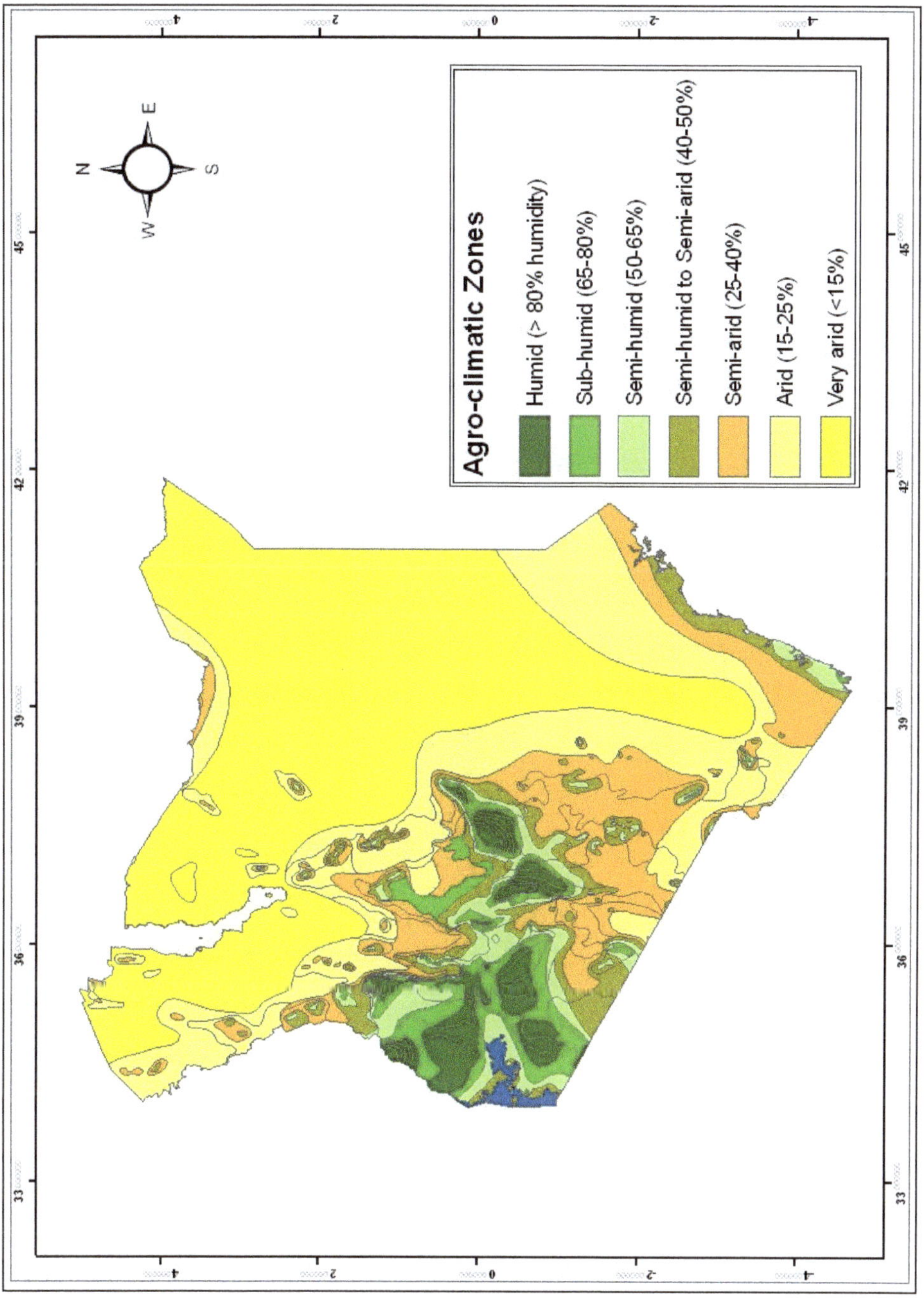

Figure 7.1: Agro-climatic Zones in Kenya

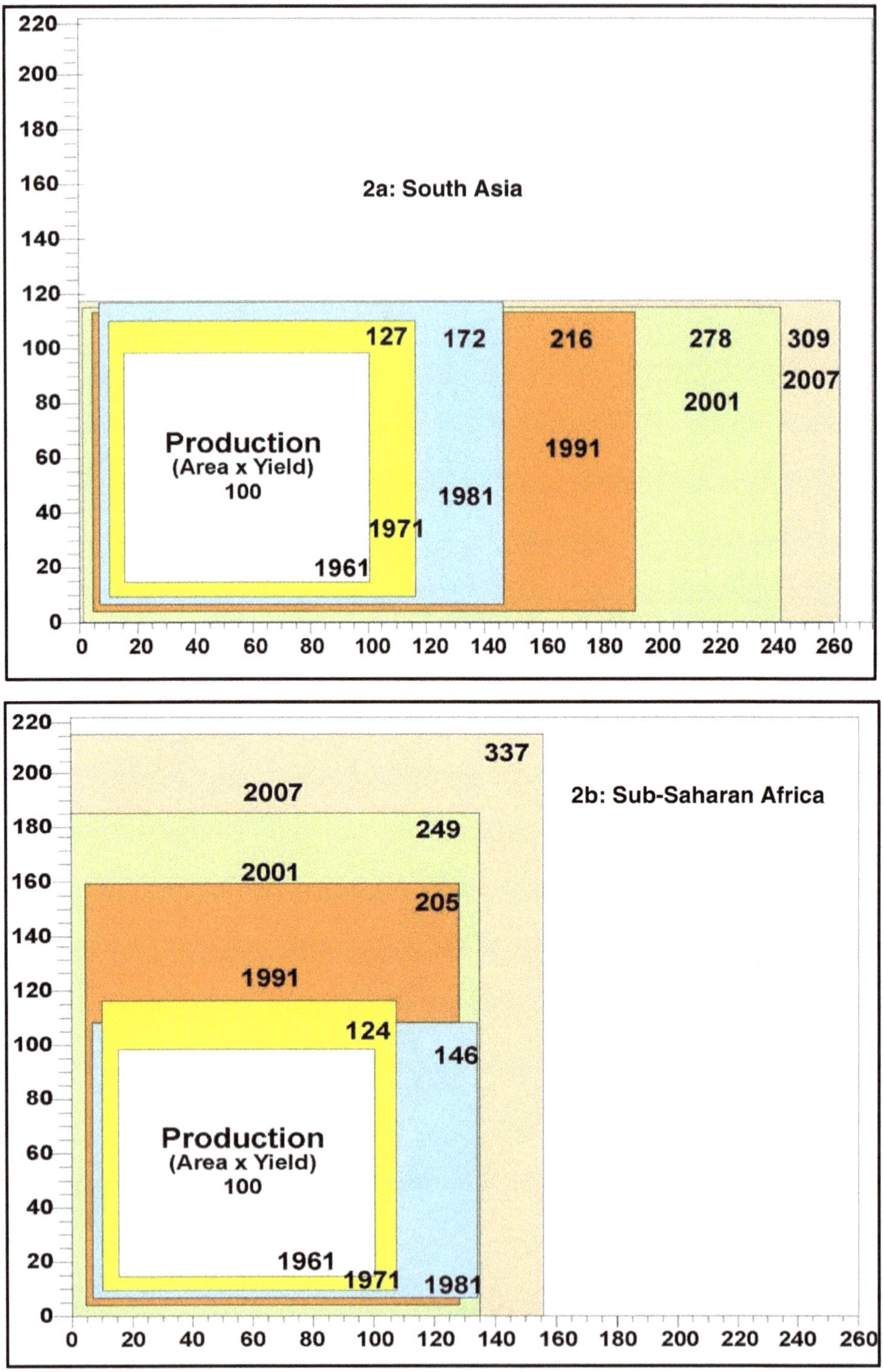

Figure 7.2: Trends in Agricultural Production and Land Area Used in Production (Data source: FAO)

1963-1983

During this period, natural resources management and environmental protection enjoyed unprecedented political goodwill and support form the highest office in the land. Resource conservation was part of the political agenda. This enabled technocrats to introduce some meaningful changes within the sectoral polices and international protocols.

1983-2000 (and to present)

Despite all the gains at technical and policy level, there was no political will to support conservation of natural resources. Some individuals and private citizens destroyed forests, wildlife and wetland areas with impunity because of the short-term political survival interests, and high-level corruption in the power structures and government system. All the policies, laws and institutions charged with sustainable utilization of the natural resources were ignored. Forests were excised and converted into private land without regard of their role as a public good.

CHALLENGES AND CONFLICTS FACING LAND USE IN KENYA

Policy Driven Conflicts

A poor environmental regulation is one of the important among various issues noted as being behind land abuse in Kenya (including population growth, natural and man-made disasters) (KLA, 2000). A concise policy framework from which to generate holistic and integrated strategy and action does not exist (KLA, 2000). Sectoral policies, *e.g.*, forests, soils, wildlife and water, have not been effective on the ground. The multiplicity of tenure regimes and in-fighting between government agencies has contributed to uncertainty of tenure and unsustainable land use, especially in rural areas.

Land tenure falls under three categories *viz.*, Government, trust land and private ownership. Under these tenures the absolute ownership of land together with rights and privileges are vested in the owner. This has posed some problems for management of natural resources such as wetlands. For wetlands, tenure and property rights commonly fall into all categories leading to their abuse (Odote *et al.*, 2008). For the large part, wetlands are open access resources resulting in uncontrolled and unregulated exploitation. Since wetlands lack proper recognition as providing crucial ecological services apart from water; in line with the tragedy of the commons (Hardin, 1968); they have been subjected to drainage for agricultural use, burning, dumping of waste from urban sewerage and industries, introduction of exotic invasive fauna and flora (Kibwage *et al.*, 2008).

Formulation of numerous policies for related resources has also led to differences in policy implementation. For instance, the Wildlife Act led to the formation of the Kenya Wildlife Service (KWS) which enjoyed a stronger mandate than the defunct Forest Department (FD) in enforcement of this Act (as well as other Acts related to wildlife and biodiversity such as the Forest Act). This led to considerable in-fighting between the two government agencies when implementing conservation in forests with each clamouring for superiority on the ground. For instance, there is evidence from Kakamega forest, western Kenya that under the management of FD from 1963 to 1986 there was massive logging but after 1986 when KWS joined in the management there has been considerably less logging due to change in protection status to National and Nature Reserve (KWS) from Forest Reserve (FD) (Bleher *et al.*, 2006). This indicates a need to harmonize conservation and enforcement practices within ecosystems regardless of institutional mandates.

Moreover, major policy gaps have arisen in important aspects of land relations such as the role of the state as owner or guardian of land resources. There remains a lack of clarity in the definition of 'public interest' in relation to expropriation. The Acts that govern land typically provide loopholes for corruption. Survey Act which governs the demarcation of land states "on all tidal rivers a reservation of not less than 30 m in width above the high water mark shall be used for government purposes". This gives the implementing officer the discretion to alter the width where 'public interest' demands and has thus created a major legal loophole that has been abused, and majority of land titles do not indicate a riparian reserve. The Physical Planning Act, Agriculture Act and Environmental Act have not addressed the riparian reserve issue comprehensively, given that they still use the Survey Act as a standard (Mwathane, 2009).

A clear and common problem caused by the above issue has been in the privatisation of riparian reserves. In the case of Nyando Basin, for instance, the land adjudication process made no allowances for the riparian reserves-the river was drawn as a thin line demarcating the pieces of land during demarcation. This would mean that if your land does not reach the river you have no access to water. Ideally the local authority should be the custodian of riparian reserves but this is rarely the case. On the other hand, where riparian reserves are public land, they have been turned into high density slums through (illegal) encroachment, such as those along Nairobi River, in Nyando basin Muhoroni and Ahero town where the landless find a home (Onyango *et al.*, 2007).

Historical and Politically Driven Challenges

The Kenyan traditional land tenure by indigenous people was communal and took cognisance of the existing physical and biological environment. The communities had ample land and practiced shift cultivation that provided a cycle allowing natural regeneration. This was practiced by the cultivators and pastoralists, both of which by then were sustainable as population density was still low (Hermunen, 2004). Land ownership disputes started after the European Berlin conference held in 1885 which marked the decision to colonize Kenya as well as other African countries by the Europeans. Britain gave money to build the railway and after seeing the immense resources and to recoup the costs incurred after completing the railway in 1901, they decided to move from Mombasa a coastal town, to Nairobi the current Capital city. The government alienated about 1.1 million ha in the most arable land and referred to it as the "white highlands" which were used to accommodate more settlers. They then moved the indigenous people to native reserves that were too small for their needs. Their pleas to the colonial masters to address the unequal land distribution were ignored. The settlers were given leases of 9 years which were reviewed to 99 years and 999 years, thus further aggravating the landless natives (Mungazi, 1999).

After independence in 1963 there was an attempt at Africanising the economy which included transfer of land from settlers to Africans. The first transfers were made of large scale farms to African individuals, partnerships, companies and cooperatives. Then there were resettlement schemes for about 180,000 families by 1974. These resettlement schemes resulted in land parcelisation and latter further subdivision, often to uneconomical and ecologically unsustainable levels (State of Environment Report Kenya, 2004). During the colonial era, colonialists determined the land policy and after independence it was the interest of the public officials, influential politicians and the rich that determined the land policy. This precipitated the problem of Kenya's land policy being influenced by (and consequently land being owned by) an elitist minority since only 5 per cent of the population benefitted from the post-independence resettlement (Republic of Kenya, 1999). In urban centres such as Nairobi, Mombasa and Kisumu over 60 per cent of the population live in informal settlements (largely slums) that occupy a paltry 5 per cent of the total urban land. The same scenario can be witnessed in the rural hinterland where most people do not own land but are squatters in environmentally sensitive areas such as forest reserves, riparian reserves, flood prone areas and water catchment areas (State of Environment Report Kenya, 2004).

In addition to this historical injustice, politicians in Kenya have been practicing "land grabbing", an irregular privatisation of public lands. Corrupt leaders have grabbed land for themselves, or to settle landless people in conservation sites (largely water catchment areas) in order to gain votes before elections, or as gifts to political allies (Klopp, 2006). This was evident in the well-publicised case where top-politicians tried to develop a housing project on half of the 1,063 ha of Karura Forest in 1996, the only remaining forest around Nairobi, besides de-gazetting of more than 67,000 ha of forest land from 12 forests countrywide in 2001. Both were only stopped after a series of violent and well-publicised civil society-driven protests organised by the Nobel Laureate Prof. Wangari Maathai. Similarly, in 1995 there was a proposal to de-gazette parts of the internationally famous coastal Arabuko Sokoke Forest, rich in the diversity of bird species and other biodiversity. This allocation could have opened the way for deforestation on 5,000 ha of the forest and powerful individuals would have profited from the land either by selling it or gaining political capital. Furthermore, other forests listed for excision were two water towers Mount Kenya and the currently disputed Mau complex of forests which encompasses the watershed for nearly half of Kenya's major rivers west of the Rift Valley. Stopping these actions has taken all the effort of several civil society groupings and other NGOs, including the Greenbelt Movement, the East African Wildlife Society, and the Environmental Liaison Centre International, the Kenya Alliance of Residents Association and the Law Society of Kenya and halted development in the area. (Gachanja, 2003).

Culturally-Driven Land Issues and Conflicts

Cultural habitats and habits have changed for some communities like the aboriginal Ogiek who virtually covered the central highland zones of the country but have been slowly squeezed out by invading tribes, the colonial settlers and now by settling communities. Ogiek did hunting in the Mau Forest. Traditionally for the Ogiek three-quarters of their diet consisted of game meat while honey contributed a significant proportion as well. They cared for the wild animals in their forests and only killed to meet their subsistence and cultural needs, and predominantly only those species with buoyant populations such as warthog and tree hyrax (Kamau, 2000). The colonial government used the Ogiek as labourers to destroy their own habitat by clearing the forest and replacing the indigenous trees with exotic ones to meet the colonial needs of timber. They also lost their land through declaration of their ancestral land as protected Forest Reserves. The destruction of these forests saw the Ogiek turn to small-scale arable farming and livestock production in the forests which are not as compatible as their erstwhile livelihood means (Ngece, 2003).

The cultural practice of inheritance where normally a father shares his piece of land to his adult sons (and increasingly to unmarried daughters) has led to extensive subdivisions of agricultural lands. Apart from increasing the number of boundary related disputes, the subdivisions have created intrinsically small parcels, some of which can hardly sustain a family unit even for subsistence needs. The small parcels are over-exploited and lose their fertility quickly as a result of over-cultivation, overgrazing, soil erosion and other poor farming practices (Waiganjo and Ngugi, 2001). This, together with increasing population, has lead to problem of squatters that is widespread especially in forests where there are constant incursions by not only the poor and landless but also the 'high and mighty' for farming or expropriating high valued timber and other forest products. This has contributed to massive soil erosion and its disastrous consequences downstream, displaced wild animals from their natural habitat together with other negative environmental consequences (KLA, 2000). With time the squatters demand to be allocated the occupied lands and have in the past, out of political expediency, succeeded thus encouraging even more incursions (Waiganjo and Ngugi, 2001).

Sustaining the livelihoods of such a high population with finite land resources places a high demand for access to resources, making use and ownership of land fundamental issues ((Mwagore, 2002). The lack of appropriate technology for maximum utilisation of the raw materials, low recovery, poor silvicultural practices, and low budgetary provision for recurrent forest operations have all hampered effective management. As a result, timber, wood fuel, pole wood and carving wood are-being exploited at unsustainable levels (Ochego, 2003). The same can be said of many other natural resources (KLA, 2000). This is also a key cause of human-wildlife conflicts involving both damage by wildlife on human beings, crops and livestock as well as humans causing damage to wildlife through poaching, retaliation (after wildlife attacks) and destruction of habitats for communities which formally co-existed peacefully with wildlife (Mwagore, 2002: Marekia, 1991).

SOME SOLUTIONS TO LAND–USE PROBLEMS IN KENYA

Grounds for a National Land-Use Policy

It is clear that the current land use systems in Kenya portend a gloomy future if the status quo is maintained. The pollution will continue, destruction of water catchments will increase leading to cessation of flow of major rivers and limiting water and power supplies, land-use conflicts over arable land will increase, water and pastures will escalate, the savannahs and grasslands will be subjected to severe desertification due to expanding agricultural activity, overgrazing and the search for fuel wood (KLA, 2000). Indeed, many of these effects are being felt in Kenya today with the increasing rationing of water and electricity across the country including major cities. Besides, loss of biodiversity will accelerate the collapse of the life-supporting ecosystems.

There is a clear need to create an institutional framework for land laws and land policy advocacy in Kenya given that the policy, legal and institutional framework created largely in 1950s is inappropriate because of the many changes in the social, political, economic and cultural fronts in the country (KLA, 2000). For instance, the increase in population has resulted into a heightened competition for access to land and natural resources, besides changes in the global environment combining to create a significantly different reality. There is a need for development of an all-inclusive and coherent land use policy that will consider all major options and stakeholders to ensure efficiency in land use. The system should also be responsive to various changes in demand. Equity should be established to ensure that the tenure system provides reasonable access to all especially those of low income or vulnerable groups. Land use system should be comparable with other policy instruments dealing with economic development and must not contradict existing legislation and should adequately cover for freehold, leasehold, customary or other interests (Waiganjo and Ngugi, 2001).

It is worth noting that developing countries' quest for food security through agricultural expansion which often leads to deforestation and land degradation. The main challenge for much of sub-Saharan Africa is how to design agricultural landscapes to resolve livelihood-environment conflict and maintain forest ecosystem benefits such as water storage, erosion control, biodiversity conservation and soil rehabilitation. The use of sustainable agricultural systems such as agroforestry has worked in various parts of the region including Machakos Kenya (Boffa, 1999). Integrating trees into agricultural landscapes on a massive scale would create an effective carbon sink while ensuring sustainable food production, and would help adapt to climate change in other ways too (Akinnifesi *et al.*, 2009). Agroforestry is also considered to be environmentally suitable for the mountainous areas as the rate of soil erosion under such system is considerably less (O'kting'ati *et al.*, 2004; Rasul and Thapa, 2006). Such crucial land use strategies such as agroforestry and other similarly useful ones (*e.g.*, zonation, terracing etc.) need to be explicitly entrenched in such a policy. It should also give policy guidelines on how to undo some of the mistakes undertaken in the past in the fairest of means to all stakeholders, *e.g.*, in revoking of illegal forest allocations especially by politicians, for political ends (Ochego, 2003).

RATIONALE FOR NATIONAL LAND-USE GUIDELINES

There is a further need for the development of a national land use plan besides a national land use policy (KLA, 2000); to provide coherent, rational and coordinated guidance on the use of land resources in Kenya. This should lead to the development of national land use guidelines which then form the standards for future land use in the entire country. This process will need to undertake the following key steps:

- ✰ Undertaking a comprehensive review of land use guidelines.
- ✰ Identification of gaps in the existing land use regulations.
- ✰ Identification of overlaps in sectoral laws related to land use.
- ✰ Identification of conflicts in land use regulations.
- ✰ Development of harmonization strategies so as to avoid conflicts in future implementation.

There are good reasons for better land use planning in Kenya.

First, it is needed to identify changes required in land use practices which will increase productivity and opportunities.

Second, it shall be used to decide where any changes are possible and should be directed *e.g.*, in the face of climate change, and other changing circumstances like refugee influxes. This will ultimately enable avoidance of misuse of land resources.

Some of the issues that require attention include:

harmonizing utilization within a given area,

allocating land use according to suitability,

reducing on land use conflicts,

harmonizing land use technologies according to their appropriateness and

educating the public to support sustainable land utilization (Kenya National Environmental Action Plan, 1994; KLA, 2000).

The figure below illustrates the impact of improving technologies where between 1961 and 2007 there has been an increase in yield with minimal increase in land area under production in Southern Asia (Figure 7.2a). It depicts an excellent land use system. In contrast, under the same period, there has been an increase in land area for agriculture with minimal increase in yield in sub-Saharan Africa (Figure 7.2b), which is also the case in Kenya. In this respect, there is a need for Kenya (and other countries in this region) to strive to optimise the sustainable use of land by increasing efficiency, which is the ideal situation with proper land use planning and ensuring appropriate land uses in all ecological zones.

CONCLUSIONS

The absence of a concise national land-use policy has had a direct (and largely adverse) impact on the way natural resources have been used in Kenya. Most of the issues, challenges and conflicts outlined in this paper, whether political, historical or cultural can be, at least partly, attributed to the lack of robust, cohesive land use policy in Kenya. The multi-disciplinary, multi-stakeholder processes needed to develop an appropriate land use policy would address many of these issues such as correcting historical injustices, dealing with human-wildlife conflicts under different scenarios, and land subdivision, whilst bearing in mind people's culture and values.

Fortunately, this process has, at least to some extent, been undertaken in Kenya and a draft National Land Use Policy is in place. It however needs refinement, further discussion and eventually cabinet approval.

Introduction of multiple land use will depend on formulation of a national land use policy enacted and enforced to guide and harmonize land-use planning activities (KLA, 2000). Controlled and well-regulated agricultural intensification retains a key role in Kenya's natural resources conservation and management by ensuring extra land is not sought for food production. Other techniques including agro-forestry and eco-agriculture (McNeely and Scherr, 2003) will also play an important role and need to be explicitly included in the land use policy.

The impacts of unsustainable land use in Kenya are many. The decline in the supply of pastures and potable water are creating conflicts among pastoralists and between small and large scale irrigation farmers. Poor quality of air and water increases disease risks in human beings and extinction of other life forms. Destruction of water catchments causes shortage of water and electricity supply necessitating rationing (KLA, 2000), as is indeed being currently experienced across the country.

The loss and degradation of land resources are to be seen in the context of policy, socio-economic conditions and natural and man-driven environmental changes. Their impact on agriculture and food production, as well as on the ecological and protective functions of natural and managed ecosystems is becoming increasingly manifest. The current trend of deterioration and attrition by continued land abuse can be curbed through the entire Kenyan society opting for rational land use through the application of wise-use principles. Striking a balance between satisfying the human livelihood needs and sustainable use of natural resources *i.e.*, sustainable development remains a daunting challenge.

REFERENCES

Akinnifesi, F. K., B. Muys, and O.C., Ajayi 2009. Growing trees among crops could sustain both environments and livelihoods in Africa, 8th July © 2009 SciDev.Net

Akotsi, E. F. N., M., Gachanja, and J. K., Ndirangu 2006. Changes in forest cover in Kenya's five 'Water Towers' 2003-2005. DRSRS and KFWG, Nairobi, Kenya.

Bleher, B., D., Uster, and T., Bergsdorf 2006. Assessment of threat status and management effectiveness in Kakamega Forest, Kenya. Biodiversity and Conservation 15:1159–1177

Boffa, J. M. 1999. Agroforestry parklands in sub-Saharan Africa Agroforestry Systems 52: 169-170. Kluwer Academic Publishers. Netherlands.

Central Bureau of Statistics, 2006. Kenya facts and figures. Ministry of Planning and National Development, Nairobi, Kenya.

Elias, T. O. 1956. The Nature of African Customary Law, Manchester University Press.

The Kenya National Environmental Action Plan (NEAP), 1994. Ministry of Environment and Natural Resources Nairobi, Kenya.

FAO 2007. AEZ-Agro-ecological Zoning System. http://www.fao.org/ag/agl/agll/prtaez.stm

Gachanja, M.K. 2003 Public perception of forests as a motor for change: the case of Kenya. Unasylva 213, Vol. 55.

Hardin, G. 1968. The tragedy of the commons. Science 162:1243-1248

Hermunen, T. K. 2004. Land use Policy in Kenya–Experiences from Taita Taveta District. University of Helsinki, Kumpula Science Library.

Kamau, J. 2000. Ogiek: The Ongoing Destruction of a Minority Tribe in Kenya rightsfeatures@alphanet.co.ke. Rights News and Features Service.

KLA, 2000. Land Use in Kenya: The case for a national land use policy. Kenya Land Alliance: Land Reform Vol. 3, Nakuru, pp. 88.

Kibwage, J. K., P. O., Onyango, and H., Bakamwesiga 2008. Local institutions for sustaining wetland resources and community livelihoods in the Lake Victoria basin. African Journal of Environmental Science and Technology Vol. 2 (5), pp. 097-106.

Klopp, J.M. 2006. Pilfering the Public: The Problem of Land Grabbing in Contemporary Kenya. Winter, Vol. 47, No. 1, pp. 6-26.

Marekia, E., Njeri 1991. Managing wildlife in Kenya. In Kiriro, Amos and Calestous Juma (Ed) Gaining ground institutional innovations in land use management in Kenya 155-176. African Centre for Technology Studies (ACTS) Environmental Policy Series: 2 ACTS press and African Centre for Technology Studies. Nairobi.

McNeely, J., and S., Scherr 2003. Ecoagriculture: strategies to feed the world and save wild biodiversity. Island Press, Washington D.C.

Mungazi, A. D. 1999. The last British liberals in Africa: Michael Blundell and Garfield Todd. Greenwood publishing group.

Mwagore, Dali(Ed) 2002. Land use in Kenya the case of a national land use Policy. Kenya Land Alliance. Land reform Vol. 3. Print fast Kenya limited Nairobi.

Mwathane, I., 2009. Tread carefully on land issues. http://www.nation.co.ke. (Accessed Aug. 15th 2009)

Ngece, K., 2003. Challenges in forestry conservation in East Africa. Is community based forestry the key to forest survival? http://www.cbnrm.net.

Ochego, H. 2003. Application of Remote Sensing in Deforestation Monitoring: A Case Study of the Aberdares (Kenya) 2nd FIG Regional Conference Marrakech, Morocco, December 2-5, 2003.

Odote, C., B., Ochieng, and O., Makoloo 2008. Implication of property rights for wetlands Management in Kenya. Institute of Law and Environment Governance, Nairobi. Glos.ac.uk/conference.

Ogendo, O., and O.W., Kosura 1995. Land Tenure and Agricultural Development in Kenya: Ministry of Agriculture, Livestock Development and Marketing.

O'kting'ati, A., J. A., Maghembe, E. C. M., Fernandez, and G. H., Weaver 2004. Plant species in the Kilimanjaro agroforestry system. Agroforestry Systems Volume 2, No. 3 Pp. 177-186.

Onyango, L., B., Swallow, J. L., Roy, and R., Meinzen-Dick 2007. Coping with History and Hydrology: How Kenya's settlement and Land Tenure patterns shape contemporary water rights and gender relations in water. (Eds) Koppen, B. V., Giordano, M., Butterworth, J. www.cabi.org pp. 179-182.

Republic of Kenya, 1999. The Contemporary Political History of the Republic of Kenya 1998-1999. www.hartford-hwp.com/archives/36/index-bbb.html. (accessed 10th Aug. 2009)

Rasul, G., and G. B., Thapa 2006. Financial and economic suitability of agroforestry as an alternative to shifting cultivation: The case of the Chittagong Hill Tracts, Bangladesh Agricultural Systems Volume 91, Issues 1-2, Pp. 29-50.

State of the Environment Report Kenya, 2004. Land Use and Environment. National Environment Management Authority (NEMA). Nairobi, Kenya.

State of the Environment Report Kenya, 2003. National Environment Management Authority (NEMA). Nairobi, Kenya.

UNEP, 1982. GRID database derived from the Exploratory Soil Survey. Kenya Soil Survey, Nairobi.

Waiganjo, C., and P. E. N., Ngugi 2001. The Effects of Existing Land Tenure Systems on Land Use in Kenya Today. International Conference on Spatial Information for Sustainable Development Nairobi, Kenya.

Chapter 8

Country Status Report on Land Use, Land Degradation and Environment in Mauritius

Cecily Cyparsade
Environment Officer,
Ministry of Environment & NDU,
2nd Floor, Ken Lee Tower, Line Barracks Street, Port Louis, Mauritius

ABSTRACT

Mauritius is a Small Island Developing State (SIDS) and, like all SIDS, it has certain unique characteristics and vulnerabilities that create additional constraints for its sustainable development. These include its small size, ecological fragility, susceptibility to natural disasters and climate change, narrow resource base, limited local capital for productive investment, remoteness from and dependence on international markets, high costs of transportation, limited diversification in production and exports and vulnerability to economic shocks. Effects of climate change and sea level rise are already being felt in numerous sectors namely on the coastal zone, strategic infrastructures, agriculture and freshwater resources and health among others. Limited land resources result in major challenges with regard to land use and planning for activities that are not compatible. Waste management is a key challenge given the limited options that are available in SIDS for waste disposal and recycling. However, over the years, Mauritius has made sustained efforts to build resilience to face its numerous challenges. Mauritius has been able to build resilience to adapt to unfavourable national, regional and international conjectures as well as to the effects of climate changes including cyclones, seasonal droughts and coastal erosion. These challenges have called for innovative approaches and major reforms in a number of critical sectors including education and training, agriculture and the financial sector. Many important initiatives have been taken to strengthen national institutions and administrative capacity covering environmental policy, land use, legislation, education, transport and biodiversity. Technical measures taken by the government towards strengthening of development control and sustainable land management amongst others are implementation of the National Development Strategies (NDS), Environmental Information System (EIS), identification and delineation of ESAs, the establishment of an Integrated Coastal Zone Management (ICZM) mechanism and plan, capacity building and mainstreaming of Sustainable Land Management (SLM) into national policies, plans and legislation as well as the Maurice Ile Durable project.

Keywords: *Mauritius, Land use and resources, Environment.*

INTRODUCTION

The island of Mauritius is a small tropical volcanic island of about 1,865 km^2 situated in the West Indian Ocean, 800 km off the east coast of Madagascar. In addition to the mainland island, the Republic of Mauritius also comprises of the outer islands: Rodrigues, Agalega, St. Brandon, Tromelin and some small islets. The Exclusive Economic Zone (EEZ) is a marine area of approximately 1.9 million km^2 which is nearly a thousand times the size of the land area itself. The main island has no continental shelf, the ocean depth reaching 3000 metres within 20 km of the coastline. The coastline of the island is almost completely surrounded by coral reefs, enclosing a lagoon area of 243 km^2. It has many sandy beaches, protected bays and calm lagoons. These natural resources have allowed for agricultural development, lagoon fishing and intensive capital development in tourism.

The country enjoys a mild maritime climate with summer extending from October to April and winter from May to September. Trade winds are prevalent throughout the year but are stronger in the winter when strong anticyclones pass to the south of the island. On average one cyclone passes within 100 km of Mauritius each year. Mauritius has a high annual rainfall of 3,900 mm^3 on average, with maximum occurring in summer. Maximum mean temperature reaches 31°C in summer, while mean minimum temperature on the central plateau reaches 14°C in July and August.

CHARACTERISTICS OF MAURITIUS

It is a Small Island Developing State (SIDS) and, like all SIDS, it has certain unique characteristics and vulnerabilities that create additional constraints for its sustainable development. These include its small size, ecological fragility, susceptibility to natural disasters and climate change, narrow resource base, limited local capital for productive investment, remoteness from and dependence on international markets, high costs of transportation, limited diversification in production and exports and vulnerability to economic shocks. Effects of climate change and sea level rise are already being felt in numerous sectors namely on the coastal zone, strategic infrastructure, agriculture and freshwater resources and health among others. Limited land resources also, results in major challenges with regard to land use and planning for activities that are not compatible. Waste management is yet another key challenge given the limited options that are available in SIDS for waste disposal and recycling. However, over the years, the country has made sustained efforts to build resilience to face its numerous challenges. It has been able to build

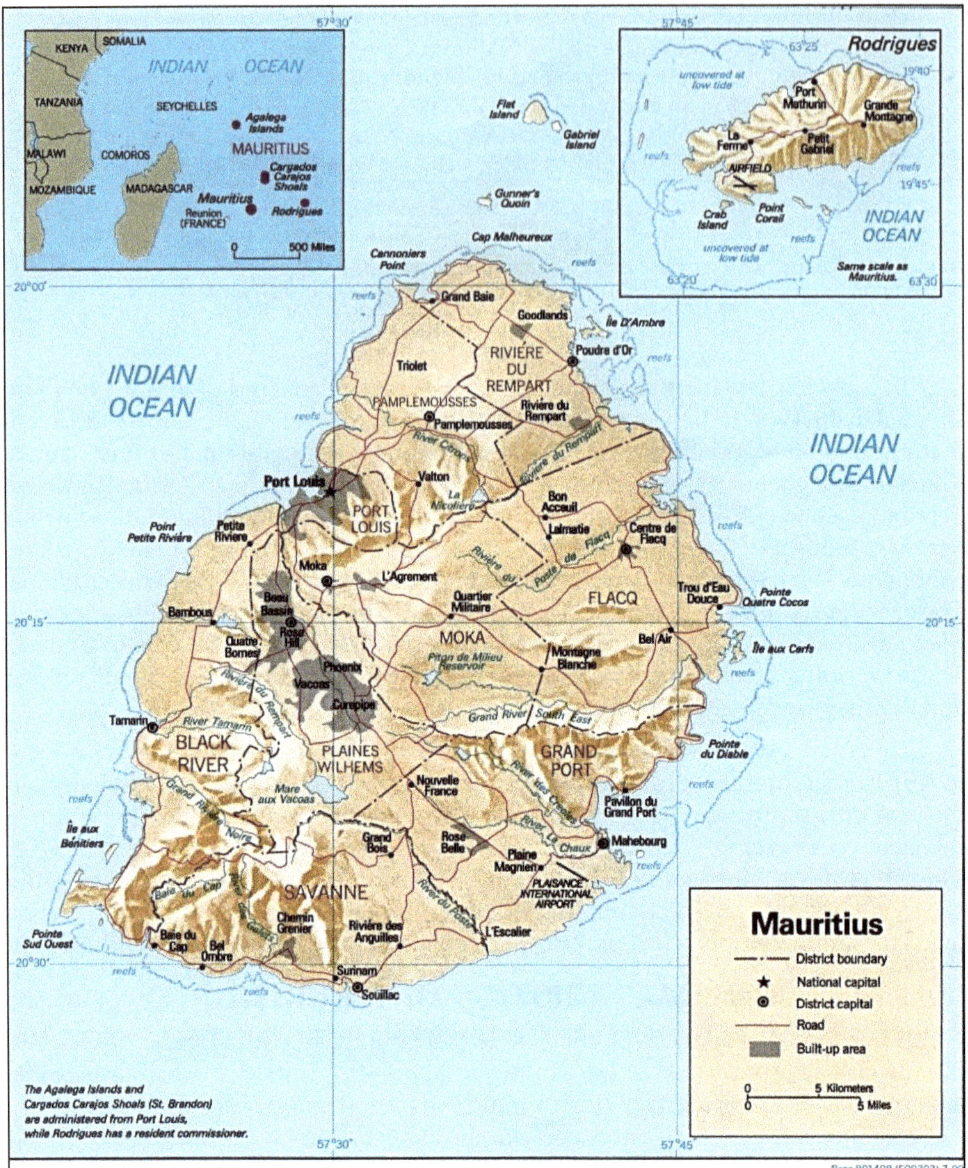

Figure 8.1: Map of Mauritius

resilience to adapt to unfavourable national, regional and international conjectures as well as to the effects of climatic changes including cyclones, seasonal droughts and coastal erosion. These challenges have called for innovative approaches and major reforms in a number of critical sectors including education and training, agriculture and the financial sector.

Many important initiatives have been taken to strengthen national institutions and administrative capacity covering environment policy, land use, legislation, education, transport and biodiversity. For example, a National Environmental Commission was set up chaired by the Prime Minister, with a broad membership including Ministries responsible for economic planning, finance, education, science, technology, natural resources, tourism, industry, health and justice. A key mandate of the commission is to ensure coordination and cooperation between public departments, local authorities and other governmental organizations engaged in environment protection policies.

These include:

- ✰ Adoption of new legislation to provide a sound and updated legal framework for the pursuit of sustainable development,
- ✰ Formulation of a National Environmental Strategy (NES) that provides a blue print for sustainable development.

SOCIO-ECONOMIC CONTEXT

Mauritius is a multi-cultural society comprising people of European, African, Indian and Chinese descent, with a population of over 1.2 million inhabitants and a density of approximately 591 inhabitants per km^2. Ranked 64 according to the UNDP Human Development Index (HDI), the Republic has experienced an average annual GDP growth of 5.5 per cent over the past two decades. Until very recently, the main sectors of the economy contributing to the GDP were: Manufacturing 23 per cent, Agriculture 5, 7 per cent, and Tourism and Services 62 per cent. It is a middle-income country with a per capita income of about US$3,890. Poverty prevalence on the mainland of Mauritius is around 9.7 per cent.

The economy has done remarkably well over the past three decades. Since independence in 1968, it has experienced significant transformation from an economy that was almost entirely dependent on sugarcane cultivation and manufacturing to one that has benefited from an export led strategy. Industry, in particular, clothing and textile manufacturing, constituted a cornerstone of the economy until recently. The sugar sector has long benefited from guaranteed prices and preferential access into European markets under the Sugar Protocol, the Lome Convention and the Cotonu Agreement. However, with the phase-out of the protocol, the country is now negotiating new agreements under the ACP/EU partnership.

Despite its impressive performance, the economy, like other Small Island Developing States (SIDS), is confronted to a number of short and medium term challenges, at both internal and external fronts. The main challenge of sustainability facing the country today is to adapt to increasingly free trade in the global markets. In the wake of increased global competition from low-wage countries and limited future opportunities through preferential arrangements, the Mauritian economy becomes even more vulnerable. At the same time, the Government is faced with the rising costs of social obligations, unemployment and environmental pressure from conflicting demand for land use.

URBANISATION

The number of households has been estimated at 287,700 in 2000, an increase of 59,000 since 1990, which implies a growth rate over double that for population-a direct consequence of declining fertility. Emerging trends show a decreasing family and household size: the average household size has decreased from 4.5 persons to 4.0 since 1990. These characteristics are likely to lead to changes in demand for house types, for example for smaller sized apartments and town houses, for more housing/ apartments/secure accommodation for the elderly and which could in turn produce higher densities in and around town centres. Housing has grown in most main settlements including those in rural areas, with a trend towards further increasing urbanisation in and around the capital city Port Louis and the adjacent Plaines Wilhems urban conurbation. Nearly 50 per cent of the increase during 1990-2000 occurred in the conurbation, with between 14 and 16 per cent in the north, east and south of the island. The introduction of the Cybercity project east of Rose Hill at Ebene will consolidate this trend.

STATE OF ENVIRONMENT

Biodiversity Resources

The Republic harbours a unique collection of some of the rarest plants and animals of the world. Indeed, the climatic, geological and topographical regimes prevailing in the island have resulted in an evolution of diverse biota with a high degree of endemism both in fauna and flora. However, this biodiversity has been adversely affected by development, especially human habitation, direct exploitation and invasion by alien plant and animal species. Mauritius is high on the world's list of countries for plant and bird species threatened with extinction. Native forests cover less than 2 per cent and are isolated to the south west area of the island. There are nearly 700 species of indigenous plants in the republic, out of which 300 are endemic and 70 already thought to be extinct. Furthermore, of the 44 known species of endemic land vertebrates that existed on the island and adjacent islands, 21 species are now extinct and many are still critically endangered.

The principle challenges addressed by the country are to protect and restore the wealth of biodiversity being lost for future generations, to promote valuable economic and social uses for these natural resources and to strengthen regional and international collaboration essential to the success of such activities in SIDS. Indeed, over the last decade, there has been growing awareness for the conservation of the indigenous flora and fauna and tremendous progress has been achieved in this field, with various organisations fully involved in the conservation and sustainable use of biodiversity resources.

Coastal and Marine Resources

The coastline of the country is 322 km long and almost entirely surrounded by a fringing coral reef enclosing a lagoon area of 243 km^2. It has a rich but vulnerable coastal and marine biodiversity which include coral reefs, mangroves and wetlands. The establishment of an approximately 1.9 million km^2 exclusive economic marine zone has vastly extended the fisheries and other marine resources available. The coastal zone is under growing pressure from both marine and land based activities and most of the pristine coastal sites have been exploited. Development associated with the expansion of tourism has resulted in the construction of a large number of structures on the coastal zone which is a major challenge for management of coastal zone.

Major initiatives to protect the coastal zone have been undertaken as part of an integrated coastal zone management plan, including the banning of coral sand mining, development of methods for beach nourishment, national sewerage programme and daily beach cleaning. Closer attention is also being given to the 49 islets surrounding the main island, whose quality has been depleted through uncontrolled waste discharge and alien trees, shrubs and animal invasion. Creating a balance in areas of conflicting interests is a major aspect of the future development in the management of the coastal zone. Another equally important requirement is professional and technical development in this field which is of crucial importance for a coastal state.

Freshwater Resources

Mauritius has a high annual average rainfall of 3,900 mm^3 and is also served by a well developed network supply with many river sources, boreholes, aquifers and reservoirs with a capacity of nearly 100 mm^3. While high population density, heavy use of water for irrigation as well as increasing demands from industrial and tourism sectors place growing stress on water resources, there has been considerable improvement in the water supply through the construction of run-off diversion schemes and dams, maintenance of networks, and replacement of obsolete potable water infrastructure. Inspite of all these measures water is becoming a scarce resource. Development has brought increasing demand nearly outstripping the increased capacity for supply. Indeed, the country is in water stress and water scarcity category by international standards and better integrated water management is a priority. The major challenge in this sector is to have a better demand management strategy with greater efficiency in use and the development of recycled water for industrial and agricultural use.

Energy Resources

The Republic imports all its fossil fuels, which is the main source of primary commercial energy. In 2006, 75 per cent of the total energy requirement was met from imported fuels, while the remaining 25 per cent was supplied from indigenous sources, hydro and bagasse principally. The transport sector with 49 per cent of energy consumption by sector is the largest and growing user of energy in the country. The country has completed its National Rural electrification Programme more than 25 years ago. It has a stable electricity supply through a national grid system based on a diversity of fuel mix, which has evolved with time, gradually shifting from diesel to coal and increasing use of bagasse (a by-product of sugar cane). However, in recent years, electricity demand has risen sharply due to sustained economic growth and a general improvement in the standard of living. Keeping pace with the expanding demand from all sectors whilst promoting the development of new renewable sources of energy is a major challenge. Mauritius has to focus on a few options in this direction which should be technically, economically and financially sound. With regard to the transport sector, the challenge is to provide the country with a reliable, energy efficient, environmentally friendly and fairly priced transport with solutions for road congestion.

Land Use and Resources

The Island was uninhabited until as late as 1598 and has been successively a Dutch, French and British colony, becoming independent in 1968. It is now one of the most densely populated islands in the world with a density of 591 inhabitants per km^2. On the main island, 43 per cent of the land is used for agriculture, 25 per cent for built up, 25 per cent for forests and grazing lands, 4 per cent for infrastructure and other services and 3 per cent as abandoned cane fields. High quality land, particularly in prime coastal areas is scarce and is sought after by a variety of competing users. Land degradation, increasing urbanisation, competition for land, management of environmentally sensitive areas and ensuring sound agricultural practices are among the principal concerns that are being addressed to ensure sustainable land use in the country. Major progress has been made in wider ownership of land and in the provision of housing for all people. Over 85 per cent of families live in homes which they own and which are predominantly robust and cyclone-proof. Three key land use trends have been identified:

- ✰ Continued urbanization of the conurbation by growth of housing, retail, commercial and industrial development,
- ✰ Dispersal and decentralization of building to the urban fringe and to the countryside locations where new housing is being developed through the various land conversion schemes, and
- ✰ Continued expansion of tourism and hotel resorts mainly on the coast.

Table 8.1: Evolution of the Land Use Pattern on the Island of Mauritius, 1995 and 2005

Land Use	*2005*		*1995*		*Change*	
	Hectares	*%*	*Hectares*	*%*	*Hectares*	*%*
Sugar cane plantations	72,000	38.6	76,840	41.2	–4,840	–6.3
Tea plantations	674	0.4	3,660	1.9	–2,986	–81.6
Other agricultural activities	8,000	4.3	6,000	3.2	2000	33.3
Forests, shrubs and grazing lands	47,200	25.3	57,000	30.6	–9,800	–17.2
Infrastructure	4,500	2.3	4,000	2.1	500	12.5
Inland water resource systems	2,900	1.6	2,600	1.4	300	11.5
Built-up areas	46,500	24.9	36,400	19.5	10,100	27.7
Abandoned cane fields	4,726	2.5	—	—	—	—
Total	186,500	100.0	186, 5000	100.0		

Source: Stocktaking and Stakeholders Consultation Climate Change Activities, May 2006.

LAND TENURE AND DEGRADATION

Most of the productive land in Mauritius, including all agricultural land, is privately owned. State lands consist primarily of "wild" lands including natural forests, degraded or scrub forest, forest plantations, and grasslands on degraded dry mountain slopes. Land degradation is one of the major environmental problems in the country. Less than 2 per cent of the original vegetation cover remains today. Many mountains are no longer covered by trees which lead to the soil erosion, leaving barren gullies on the mountain slopes. Other factors are essentially deforestation, unsustainable agriculture, overgrazing and recurring wildfires on grass-covered mountain slopes.

Table 8.2: Main Environment Indicators, 1998 and 2007

Sl.No.	*Environment Indicators*	*Units*	*1998*	*2007*
1.	Total land area	000 ha	186.5	186.5
2.	Irrigated land	Ha	18,379	21,321
3.	Forest area (as a % of total land area)	%	30.6	25.3
4.	Land protected areas	Ha	13,973	14,579
5.	Marine protected areas	Ha	7,190	7,190
6.	Threatened plant species (IUCN Red List)	Number	—	88
7.	Threatened animal species (IUCN Red List)	Number	—	65
8.	Total fish catch	Tons	9,786	6,952
9.	Mean catch per fisherman day	Kg	5.3	4.4
10.	Total carbon dioxide emission	000 Tons	—	3,450
11.	Per capita carbon dioxide emission	000 Tons	—	2.7
12.	Mean annual rainfall	Millimetres	1,676.80	1,914.00
13.	Annual fresh water abstraction	Mm^3	639	630
14.	Daily per capita domestic water consumption	litres	168.4	162
15.	Daily per capita solid waste generated (estimate)	Kg	0.7	0.9
16.	Total electricity generated	GWh	1,538.90	2,465.00
17.	Per capita primary energy requirement	toe	0.9	1.1
18.	Per capita final energy consumption	toe	0.6	0.7
19.	Energy Intensity	100,000 GDP	2	1.0

POLICY AND INSTITUTIONAL CONTEXT FOR LAND MANAGEMENT

Internal policies for land use have been guided by the National Physical Development Plan (NPDP–developed in 1994 and updated in 2003); the Tirvengadum Report; the Sugar Industry Efficiency (SIE) Act of 1988, 1998 and 2002; the Finance Act of 1994; the Environmental Protection Act (EPA), the "Morcellement" Act as well as the Municipal and District Council Regulations, the Zoning Committee of the Ministry of Housing and Lands, and the Sugar Productivity and Efficiency Unit of the Ministry of Agro Industry, Food Production and Security. Furthermore, the Land and Development Act, 2004 provides an updated framework for land use planning and decision making at national and local level. A new awareness is emerging for the value of architectural style, removing eyesores and promoting a built environment that is in harmony with the natural environment of the country.

Key policy elements from the above that are related to sustainable land use include the;

- ☆ Conservation of prime agricultural land as far as possible, in order to maintain sugar production quotas, and self-sufficiency in food crop production;
- ☆ Development in compact settlements (industry, tourism and infrastructure);
- ☆ Provision for future land requirements for housing, industrial, tourist and ancillary services;
- ☆ Agricultural diversification;
- ☆ Provision for land requirements for major field crops such as tea,tobacco, food crop and for forestry;
- ☆ Promotion of sustainable agriculture and environmental protection;
- ☆ Analysis of the suitability of land for irrigation;
- ☆ Improvement of land utilization;
- ☆ Land for livestock production and availability of suitable pasture lands;
- ☆ Rehabilitation of abandoned lands (about 900-1000 ha of arable land is presently being abandoned);
- ☆ De-rocking of land to make it more productive and amenable to mechanization, and
- ☆ Land subdivision and land conversion legislation and amendments.

In 2002, Government commissioned the National Development Strategy (NDS) for Mauritius and Rodrigues, following a review of the NPDP. The NDS (2003) has identified a number of gaps in the development framework, which have bearing on the processes underpinning land degradation.

With regard to the environment, the NDS has identified the following strategic priorities, to underpin sustainable development interventions:

- ☆ Safeguard valued elements of the natural and built environments;
- ☆ Use natural resources in a sensitive and sustainable manner;
- ☆ Promote land and property development and management practices which will benefit the environment, and
- ☆ Ensure that development makes a positive contribution to the environment.

The government is also placing a high emphasis on land management through the implementation of the recommendations made under the National Environmental Strategies (NES) and the second Environment Investment Program (EIP II). Three programs have been completed recently at the Ministry of Environment and National

Development Unit (MoE and NDU), which will directly influence sustainable land planning policies. These are the compilation of a basic Environmental Information System (EIS), studies to identify and quantify ESAs and the establishment of an Integrated Coastal Zone Management (ICZM) mechanism and plan.

At the global level, Mauritius has also signed and ratified a number of conventions, which are of relevance to sustainable land management. The most relevant ones include the UNCCD, UNFCCC, UNCBD, and the Ramsar Convention on Wetlands.

The Table 8.3 gives an indication of the various actors involved in land management in Mauritius. It also gives an indication of the level of institutional fragmentation characterizing the execution of land management functions in Mauritius.

Table 8.3: Institutional Fragmentation Characterizing the Execution of Land Management Functions in Mauritius

Institution	*Areas of Intervention and Relevant Issues*
☆ Ministry of Housing and Lands ☆ Min. of Tourism ☆ Ministry of Local Government ☆ Local Authorities	Planning Issues
☆ Min. of Local Government ☆ Forestry Service ☆ National Park and Conservation Service Mauritius Wildlife Foundation (NGO) Private Forest Owners ☆ Sugar Estates and others	Degraded forest lands in Mauritius, Rodrigues and Islets
☆ Agricultural Services (Min. of Agro Industry and Fisheries) ☆ Agricultural Research Extension Unit ☆ Mauritius Chamber of Agriculture ☆ Mauritius Sugar Industry Research Institute (MSIRI)	Degraded Agricultural Lands and improvement in soil fertility
☆ Ministry of Environment and National Development Unit ☆ Ministry of Tourism ☆ Beach Authority ☆ Min. of Local Government ☆ Forestry Service ☆ Ministry of Agro Industry and Fisheries and others	Coastal Zone/Belt
☆ Irrigation Authority ☆ Water Resources Unit ☆ Central Water Authority ☆ Agricultural Services ☆ Min. of Public Utilities ☆ Forestry Service	Dams, rivers, canals, etc.
☆ Min. of Education ☆ Ministry of Environment and National Development Unit ☆ Min. of Youth and Sports ☆ Mauritius Institute of Education	Awareness Campaign

Contd...

Table 8.3–Contd...

Institution	Areas of Intervention and Relevant Issues
☆ University of Mauritius ☆ MSIRI ☆ Mauritius Research Council	Research on affected areas
☆ Ministry of Environment and National Development Unit ☆ Remote Sensing Unit (Min. of Agro Industry and Fisheries)	Mapping of sensitive areas
☆ State Law Office	Legislative matters

INSTITUTIONAL AND LEGAL FRAMEWORK FOR ENVIRONMENT

The Ministry of Environment and NDU, through its technical department, the Department of Environment, has the primary responsibility of ensuring environmental protection, planning, monitoring, coordination and awareness rising. Other key Ministries too are concerned with environmental protection, mainstream environmental concerns within their work at both planning and implementation stages and deal with environmental complaints together with Local Authorities. The Ministry of Environment and NDU ensures coordination at different levels. The Police de L'Environment also plays a key role in the enforcement mechanism especially for the control of littering, illegal dumping, vehicular emissions and abatement of noise and eyesores. The Environment Protection Act (EPA) 2002 provides for the protection and management of the environment assets of the country.

STATUTORY BODIES AND ENVIRONMENTAL NETWORKS

National Environment Commission

As per the provisions of EPA 2002, the National Environment Commission (NEC), chaired by the Prime Minister, is the supreme environmental body and comprises all Ministers dealing with environmental issues. The NEC is the highest policy-making body mandated to monitor and review the work of public departments dealing with the protection and management of the environment and make recommendations accordingly.

Environment Coordination Committee

The Environment Coordination Committee (ECC) is steered at the level of the Ministry of Environment and NDU. Public departments and enforcing agencies dealing with environmental protection meet on this platform to ensure that policies are being implemented and that environmental laws are being complied with. The committee also advises the Minister on matters related to pollution control measures and codes of practice for environmental protection.

Environmental Liaison Officers Meeting

Under the Environment Protection Act, Enforcing Agencies play a key role in pollution control and environmental protection as per the mandate of their respective institutions. The Ministry of Environment and NDU ensures coordination at the level of the Environmental Liaison Officers (ELO) meeting. Furthermore, desk officers of all Enforcing Agencies and Local Authorities deal with environmental complaints in a concerted approach for pollution problem solving.

National Network for Sustainable Development

The National Network for Sustainable Development under the chairmanship of the Ministry of Environment involves all stakeholders including non governmental organizations, local authorities, community based organizations, and the business community. The network is a forum for discussions and consultations on sustainable development issues.

OTHER STATUTORY COMMITTEES

The Environment Impact Assessment (EIA) Committee is a platform where concerted actions are initiated by pertinent public organizations during the review of new projects. Recommendations made by all authorities concerned are listed in the promoter's EIA license and EIA post-monitoring exercises ensure compliance thereto.

The Integrated Coastal Zone Management (ICZM) Committee has over the years established itself and is active in coastal zone management through the active collaboration of all stakeholders involved in coastal zone management.

SUPPORT SYSTEMS

Environment Information System (EIS)

The Ministry of Environment (MOE) has developed the EIS project to strengthen environmental information management in the country. The project consists of two major components, namely: the development of a core set of Environmental Indicators for Mauritius and the development of a computer based EIS that makes maximum use of information technology for strategic management of the environment. In total, 28 different ministries, agencies and NGOs are involved in collecting and recording details of the parameters required for the core set of 32 indicators that have been developed through the project.

MUELEX

With the support of UNEP, Mauritius developed a National Legal Database for capacity-building to enhance access to environmental law information. This project is currently being implemented by the Ministry of Environment and NDU with collaboration of the Attorney General's Office and the Ministry of Information Technology and Telecommunications (Central Information System Division and Central Informatics Bureau).

PROGRAMMES FOR SUSTAINABLE DEVELOPMENT

Over the years Mauritius has shifted from a sectoral approach to a more integrated approach for environmental management and is moving on the path of sustainability. The Government is fully committed to protecting the environment, health, welfare and quality of life of all its citizens. A new National Environment Policy aims at providing a comprehensive flexible policy framework with a built-in system for environmental management, monitoring, review and competitiveness as well as to foster sustainable production and consumption patterns.

With the rapid growth of economy, consumption patterns have changed towards higher value, added industrial production and higher levels of consumption. In 1999 a second ten year National Environmental Strategy (NES 2) and National Environmental Action Plan (NEAP 2) for the Republic of Mauritius covering the period 2000-2010 was prepared. It consisted of an investment and implementation framework that included project proposals, activities, costs and responsibilities for 72 projects. EIP 2 was geared towards achieving sustainable development with the following programmes: management and coordination, resource management, sector programmes and a programme for Rodrigues. Some of the major projects completed include; legislative reform (Environment Protection Act, 2002), setting up of the Environmental Information System and Development of Environmental Indicators, strengthening of policy and strategic land planning (National Development Strategy), Islet recovery and restoration, National pest control strategy, Tourism Master Plan, Development of a Local Plan for Rodrigues, upgrading of existing dump sites and improvement of waste collection and storage.

These projects have been funded by the government through revenues raised by taxation, grants and loans from donor community, sponsorship and investment by the private sector as well as fund-raising by groups. As a small island state with a fragile ecosystem and limited resources, we are also fully conscious of the serious consequences resulting from the degradation of the global environment and its impact on the quality of life, health and on the economy. As the causes and effects of environmental factors cross-national boundaries, pollution problems have to be addressed in a spirit of cooperation from the world community, Mauritius has thus signed some 38 environment related Conventions and Protocols, which are legally binding. Actions have been undertaken by different stakeholders namely Ministries, Non-Governmental Organisations among others.

SECTORAL PROJECTS FOR SUSTAINABLE ENVIRONMENTAL AND LAND MANAGEMENT

National Sewerage Programme

Proper sanitation is of crucial importance to ensure the health and quality of life of our people. With the assistance of the European Commission and other funding agencies, Government has thus embarked on a National Sewerage Programme (NSWP), which aims at improving sanitation in the country. The NSMP provides a complete scheme for the development of wastewater sector in the country and targets above 80 per cent connection to the sewerage system by the Year 2030. Pursuant to the NSMP, the Wastewater Management Authority is currently installing new sewerage facilities in various regions. The overall purpose of the wastewater projects is to prevent further degradation of the environment (marine, coastal and inland) and attain a positive impact on the health and sanitary conditions of the Mauritians.

To-date, 24 per cent of households that is 75,000 families are connected to the sewerage network. It is projected that by end of 2009, 80,000 households will be connected to the network. As part of an Independent Audit exercise and with the contribution of different stakeholders, the MOE has worked out a plan to monitor the quality of ground, surface and coastal waters and treated effluents discharged from the treatment plants so as to ensure that the wastewater projects were serving their purpose.

DEVELOPMENT OF AN INTEGRATED COASTAL ZONE MANAGEMENT FRAMEWORK (ICZM) FOR THE REPUBLIC

The ICZM approach is considered as the most suitable framework for sustainable development planning in coastal and marine areas. In this context, the Ministry of Environment and NDU has completed the Development of an Integrated Coastal Zone Management Framework (ICZM) for the country. The framework comprises the following components; an ICZM Strategy for Mauritius in line with government policies and strategies for overall sustainable development and also in accordance with the requirements of various regional and multilateral agreements to which Mauritius is party and which are pertinent to ICZM, and review and develope an appropriate policy and regulatory framework to promote ICZM Development of ICZM plans for pressure zones. The area plans will guide future development and intervention to promote sustainable development in the coastal zone. The number of area plans will be elaborated following the diagnosis and prioritisation phases and consultation with key stakeholders.

ENVIRONMENTALLY SENSITIVE AREAS IN MAURITIUS AND RODRIGUES

Protection and conservation of environmentally sensitive areas has been identified as being a priority project in the National Environmental Strategies 1998. In this context, the Ministry of Environment and NDU has completed the study of Environmentally Sensitive Areas in Mauritius and Rodrigues with the main objectives to;

- ✰ Identify, classify and demarcate all environmentally sensitive areas (ESAs) in Mauritius and Rodrigues
- ✰ Prioritise the ESAs for protection and create a database for all ESAs to support and enhance decision-making
- ✰ Develop a comprehensive policy and legislations for the protection, conservation and sustainable development of ESAs in line with the NDS, and
- ✰ Prepare a comprehensive management plan for ESAs which will be financially and institutionally sustainable in Mauritius and Rodrigues

CAPACITY BUILDING FOR SUSTAINABLE LAND MANAGEMENT IN MAURITIUS INCLUDING RODRIGUES

The agriculture, pasture, forest and other terrestrial land management systems in Mauritius and Rodrigues are sustainable productive system that maintain ecosystem productivity and ecological functions while contributing directly to the environmental, economic and social well-being of the country. Capacity building programme for sustainable land management has already been completed for Mauritius and Rodrigues.This project is part of the UNDP/GEF LDC and SIDS Targeted Portfolio Approach for Capacity Development and Mainstreaming of Sustainable Land Management. The objective of this project was to build capacities for sustainable land management (SLM) in appropriate government and civil society institutions/user groups in Mauritius and Rodrigues. SLM is mainstreamed into government planning and strategy development. It will benefit a land surface area estimated at 50,000 ha and will contribute towards the achievement of the following long-term goal.

National Forest Action Programme for the UNCCD

Preparation of a National Forest Action Plan for UNCCD by the Forestry Services of the Ministry of Agro Industry, Food Production and Security is underway.

National Working Group on Wetlands

The National Parks and Conservation Service has set up a national working group on wetlands as per the RAMSAR Convention for evaluating and monitoring the wetlands.

Land Information System

The Ministry of Housing and Lands (MoHL) has plans to develop a comprehensive and elaborate land information system for Mauritius. A pilot project covering 20 km^2 of the northern part has been completed and it is envisaged that the LIS for the whole island would take 6 to 7 years. The LIS includes the boundaries of each parcel of land, ownership and information on land condition. The LIS database would be shared among relevant ministries, municipalities and other approved bodies. The Government is presently debating on the level of funding to be invested in the MoHL LIS. At a minimum, State lands will be digitized and integrated into the system.

National Remote Sensing Centre

The National Remote Sensing Centre (NRSC) has an annual budget for the purchase of satellite imagery, and their data are available to approved users. The NRSC has done an informal pilot study on local deforestation. In general, however, the use of remote sensing applications for SLM, natural resource management and monitoring and evaluation remain largely untested.

Maurice Ile Durable project

In the Budget speech 2008-2009, Government announced their vision to build a Green future for Mauritius through the Maurice Ile Durable project. The aim of the project is to reduce our reliance on fossil fuel by shifting to local renewable energy sources. This will be achieved through a number of strategies as follows:

- Revamping of the National Energy Fund into a Maurice Ile Durable (MID) Fund.

- ☆ Wind energy project at Bigara.
- ☆ 2 new hydro units at Midlands and La Nicoliere.
- ☆ Land based oceanic project.
- ☆ Waste to energy project.
- ☆ Tapping of landfill gas at Mare Chicose to produce electricity.
- ☆ Solar water heater loan scheme.
- ☆ Introduction of summer time (pilot project in 2008).
- ☆ Setting up of a solid waste recycling programme.
- ☆ 50 per cent reduction of taxes on hybrid vehicles, and
- ☆ Presentation of an "Energy Efficiency Bill".

CONCLUSIONS

The Republic of Mauritius has stepped into the 21st century with important economic challenges. The 36 per cent reduction in the guaranteed price of sugar, the dismantling of the multi-fibre agreement and the rising price of petroleum products have cumulated in increasing our economic vulnerability. Government has prepared an economic restructuring plan to consolidate the existing pillars of the economy and to develop new areas of growth such as Light Engineering, the Integrated Resort Scheme, Pharmaceuticals, Information and Communication Technology, Marine Industry and Medical Tourism. Government is fully conscious that the long-term socio-economic success of the country is not possible without environmental sustainability and is therefore putting environmental concerns high on its agenda. A sustainable programme or project today needs to focus on growth, poverty alleviation, environmental management, gender and governance issues. There is ample evidence to show that whenever economic objectives of projects have solely been preferred, these projects have proved to be unsustainable ones.

REFERENCES

Mauritius–Staking Out the Future (2005)–Ministry of Environment and National Environment Unit

Capacity Building for Sustainable Land Management in Mauritius (2007)–Forestry Services, Ministry of Agro Industry, Food Production and Security

Meeting the Challenges of Sustainable Development (2002)- Ministry of Environment and National Environment Unit

National Environment Policy 2007- Ministry of Environment and National Environment Unit Central Statistics Office, Mauritius

Landell Mills (2008). Development of an Integrated Coastal Zone Management Framework for the Republic of Mauritius. Draft final report.

Chapter 9

Land Degradation, Land Use, Urbanisation and Environment–Country Report for Myanmar

May Thae Su Moe
Associate Professor,
Department of Technical and Vocational Education,
N(21), Ministry of Science & Technology,
Nay Pyi Taw, Myanmar

ABSTRACT

Myanmar comprises of 7 states and 7 divisions. The population is around 51 million with 2.02 per cent growth rate in the 135 national races. Nearly 14 million (27 per cent) of the total population live in urban areas. Availability of portable drinking water in rural areas is a key problem. Arsenic content in the drinking water in many aquifers lying beneath the surface of Ayeyarwady's meandering alluvial plains is one of the country's major issues. One fourth of total area of the country's land is cultivable. The area sown during 2008-2009 was 11.88 million hectares. Expansion of agriculture land into the remaining 0.25 million hectares of fallow and 5.68 million hectares of waste land is being encouraged. The most common soil problems are water erosion, wind erosion, depletion of soil fertility, salinization, alkalization, and water logging. According to the reconnaissance survey done by Land Use Division 72 per cent of the total land is affected by water erosion, 11 per cent by wind erosion and severe sheet erosion is observed in the upper Myanmar. The thickness of soil is very thin and the fertility is very low compared to undisturbed soils. The alkalization occurs mainly in the dry zone. According to Land Use Division the area of saline and alkalinen soils in the dry zone is about 79234 ha. Myanmar has rich forest resources, but the demand for fuel wood and charcoal for cooking is rising with increase in population.The forests are overcut for timber as well as fuel wood. Moreover illegal loggings in some areas add to deforestation as well as environmental degradation. In the mountain areas shifting cultivation is the main cause of deforestation and soil erosion problem. The country is trying hard to combat land degradation. The technical collaboration and sharing of information through regional network can prove beneficial.

Keywords: *Land use, Land degradation, Urbanization.*

INTRODUCTION

Myanmar has rich resources like favorable weather conditions, a long coastal zone, minerals, petrol, fuel, forest, and natural areas. The land is undoubtly one of the most important resources. It enjoys a tropical and subtropical climate with rainy, dry cold and hot seasons. The average annual rainfall varies between 762-5080 mm. The coastal and hilly regions receive 2540 to 5080 mm and central region receives 762 mm to 1016 mm. The temperature also differs according to the season and region. In Central Myanmar, seasonal temperature varies between 40-43°C in hot season and 10-15°C in cold season. In the hilly region like Chin Hill and Shan plateau, maximum temperature is 29°C and minimum 7 °C. Myanmar is an agro-based country, the national crop being rice.

There are 135 ethnic races, main ones are Kachin, Kayah, Kayin, Chin, Bamar, Mon, Rakhine and Shan. The population of the country is around 51million (2002-03) with 2.02 per cent growth rate. Out of this around 14 million (27 per cent) live in urban areas. The population is increasing every year.

GROWTH OF YANGON CITY

In 2006 Naypyitaw was declared as the capital city. It lies in the centre of the country. Yangon was transformed into the main commercial city. Latter has grown and expanded under the physical constraints and limited affordable resources, by extending and implementing sites and services scheme. The three satellites of South Okkalapa, North Okkalapa and Thaketa were developed for more than 3 million inhabitants under squatter resettlement program. People from satellite towns and suburban areas mostly rely on the circular railway transport and pre-war lines concentrated in the central area.

Yangon is located between Hlaing River and Pazundaung Creek. The city has been expanding northward leaving CBD in the south, forming an elongated slope. The location of CBD in the southern part of the city created transportation problems. Therefore Yangon city authority developed Hlaing tharyar on the west side of Hlaing River with a plan to construct bridge over the river. First of all around 5000 fire victim families were resettled. The northern part of the city Shwepyithar was also developed at the same time in order to resettle the squatter areas displaced due to airport extension project.

Government initiated new town development in 1989 in the east, west and north of Yangon. The basic trend was the extension or improvement of the east-west growth axis of Yangon rather than north-south one. To develop the eastern side of Pazundaung in order to retard the north ward urban sprawl and to fulfill the housing

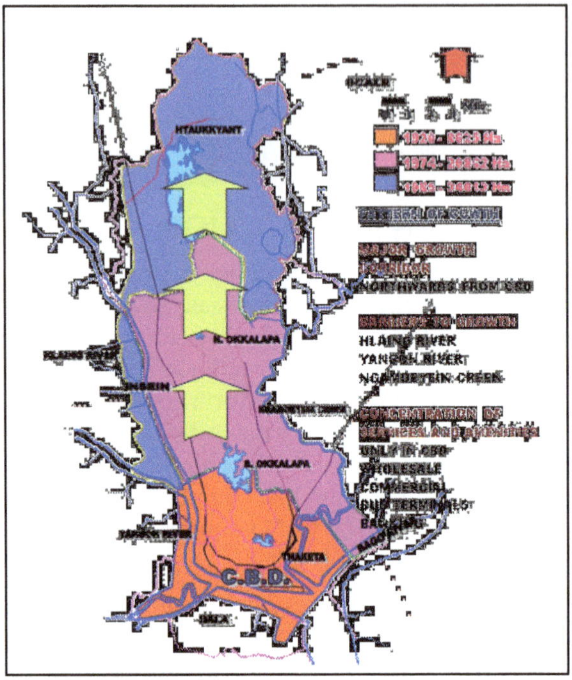

Figure 9.1: Yangon Between 1920 and 1980

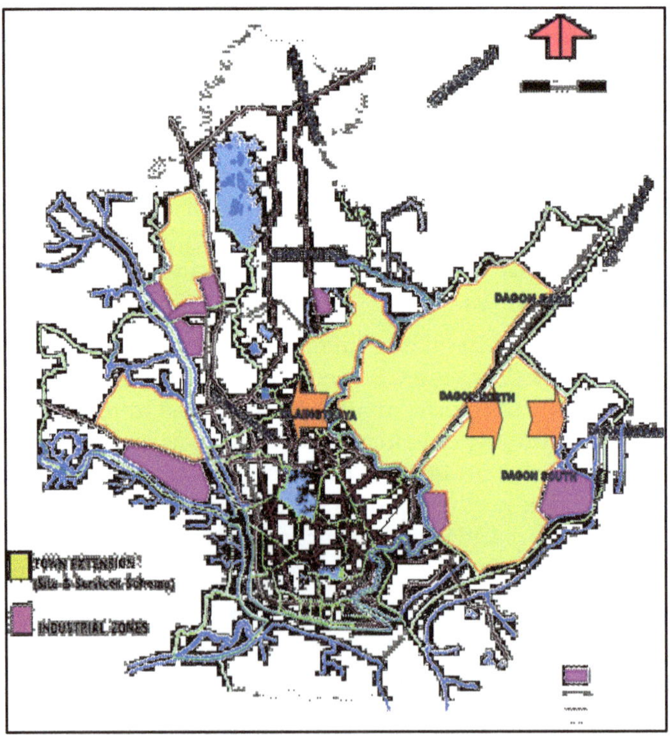

Figure 9.2: Planned Urban Extension After 1989

needs of the urban population Dagon north and Dagon south were developed in 1989. The whole development idea of Dagon was to integrate the connection with Thanlyin, North Okkalapa and Htauk-kyant to develope the city's communication line. At the west side of Yangon Hlainthayar and Shwephitha new satellite towns were developed to plan a cross pattern having an east-west axis and north-south axis. In all 8 industrial zones were developed by the "Department of Human Settlement and Housing Development" in the satellite towns and proximity of Yangon city and job opportunities for about 60000 people were created. It reduced the work trips from new township to the city centre. In the past ten years with implementation of sites and services scheme physical boundary of the city has expanded from 203 miles in 1993 to 233 miles in 2003. Urbanization has also increased job opportunities and decreased the number of dwelling units close to job places leading to an increased demand for public services and transportation The city has wide range of population density variation over 300 persons per acre in some CBD area and only about 30 persons per acre in the new suburban areas. The area of city has expanded significantly with a dramatical increase in the city population reaching about 4.5 millions in 2005. Due to population outburst, sanitation problems and environmental depletion occurred.

AIR QUALITY

Air pollution levels in the cities in developing countries have steadily increased due to an increase in the urbanization, population, industry, and transportation facilities. In Yangon air pollution is encountered. With increasing vehicular density and lack of appropriate pollution abatement devices in vehicles, this problem is likely to aggravate with the passage of time. In Ambient Air Quality monitoring in practice, particulate matter, sulphur dioxide and nitrogen dioxide are important parameters. The particulate matter is a problem for industrial, residential and commercial zones at Yangon. The concentration of sulphur dioxide and nitrogen dioxide are less than the maximum allowable levels of air quality standards.

ARSENIC CONTAMINATION

Safe drinking water is another important problem in Myanmar. The presence of arsenic was first detected in early 2000 mostly in shallow tube wells. These tube wells tap the groundwater from the upper strata. Arsenic testing under the Arsenic Mitigation Project showed the quantitative dominance of shallow tube well water source in Nyaungdone Township. It is bounded by Danuphyu and Pantanaw Townships to the west, Yangon Division to the east and north and Ma-u-bin Township to the south. A total of 3603 drinking water sources, attributing 2856 shallow tube well, 302 deep tube well, 307 dug wells, 127 river/streams and 11 ponds were tested for arsenic content. Arsenic can be removed by most suitable coagulant such as wood charcoal, laterite, rusted ion nails, etc.

LAND USE OF MYANMAR

As the population in Myanmar is rapidly increasing, proper land use planning and management is of utmost importance. The total land area of the country is about

261200 sq miles (67.65 million ha), 1/4 of total area of the country is cultivable. In 2008-2009, out of the total land area 11.88 million hectares were cultivated. Expansion of new agriculture land on remaining 0.25 million hectares of fallow land and 5.68 million hectares of waste land is being encouraged. Land utilization for the year 2008-09 in the country is given below.

	Million Hectares
Net Sown Area	11.88
Fallow Land	0.25
Culturable Waste Land	5.68
Reserved Forests	16.78
Other Forests	16.46
Others	16.60
Total	67.65

Ministry of Science and Technology has mostly constructed the new main buildings for Universities, Colleges, and Institutes in almost every State of Myanmar. In these universities, civil, chemical and environmental courses are taught together with the courses on land use, land degradation and environmental problems. Students now are investigating the reasons for the land degradation, urban population density, land use and physical infrastructure.

LAND DEGRADATION

Land degradation problems in the country include water erosion, wind erosion, soil fertility depletion, salinization, alkalinization, and water logging. In upper Myanmar, severe sheet erosion is seen, because the thickness of soil is very thin and the fertility is very low compared to undisturbed soil. Wind erosion is observed in 11 per cent of Myanmar and is one of the major factors of severe land degradation.

In the central dry zone areas major degradation problems are rapid depletion of woody biomass, expansion of farming into marginal and steeper slopes, extensive and uncontrolled use of limited pastures in erosion prone areas, and increased numbers of small livestock. The Dry Zone areas generally receive low rainfall, however, sometime; the rainfall intensity is very high. A less crop cover, undulating topography, and wrong cultivation practices result in serious soil erosion by water in the rainy season and by wind in dry season. In some areas, the undulating land composed of sand and sandy loam with low fertility is also subjected to wind erosion. As a result of this depletion of soil fertility becomes worse especially in arid regions and plateau. It is estimated to be about 12.3 per cent of the total area. In hilly regions factors like shifting cultivation, lack of proper agronomic practices, forest burning, fuel extraction, timber extraction, road making, mining etc. cause depletion of the natural forest cover resulting in water erosion.

Total water erosion affected area in Myanmar is approximately 72 per cent. According to "Land Use Division" saline and alkaline soils cover about 79234 ha in the dry zone. Approximately 1.4 per cent of the whole country is affected by both salinization and alkalinization ending up with land degradation. Salinization occurrs in the Coastal and Delta Regions due to sea water intrusion. Although heavy rainfall and flooded rice cultivation washes the deposited salts, but high tide and high evapotranspiration in summer makes the soils more saline. Salinization in central dry zone area is mainly due to saline groundwater evapotranpiration which results in the formation of salt crust on the soil surface. Use of saline water for irrigation also causes salinization. Scanty rain is not enough to wash out the accumulated salts. Alkalinization is confined only to certain areas of the arid region

DEFORESTATION

Myanmar has rich forest resources. These forests are over cut for timber as well as fuel wood purposes because of population pressure. In mountain areas shifting cultivation is the main cause of deforestation and soil erosion problem. The demand for fuel wood and charcoal for cooking has been rising with increase in the population. In addition to these factors, illegal loggings in some areas add to the deforestation as well as environmental degradation.

The tidal forests in delta and coastal region are the ideal ground for breeding of fish and prawns. Myanmar has an area of about 812,000 hectares of tidal forests. The mangrove trees in the tidal forests serve as important biomass for the villagers in terms of timber, fish, prawn, fuel wood, charcoal and some valuable forest products and herbs. However, the tidal forests are cleared for salt industries, fish and prawn breeding farms. It is estimated that about 3,000 hectares of tidal forests are cut every year which leads to a great hazards of biodiversity.

CONCLUSION

Like other developing countries Myanmar is facing several environmental problems arising mainly from underdevelopment and poverty. It is facing problems of deforestation, loss of biological resources, land degradation due to wind and water erosion, arsenic in drinking water, urbanization, waste management due to population outburst. Natural hazards like cyclones and earthquakes are few and frequency of occurrence is not very high. The degree of air and water pollution caused by industry or agriculture has been minimal due to still low level of industrialization and relatively small amount of chemicals used in the agriculture. This overview stresses the fact that for a healthy development of Myanmar precautionary measures should be adopted more carefully for a sustainable living of future generations in the country.

ACKNOWLEDGMENTS

I would like to express my thanks to the following persons for sharing information and letting me use their data for reproduction; Dr.Theint from Ministry of Science and Technology and U Aye Maung Sein, Director, Settlement and Land Records Department, Ministry of Agriculture and Irrigation, Nay Pyi Taw and U Khin Maung Nyunt, Senior Engineer, Ministry of Agriculture and Irrigation, for their information in corresponding reports. My thanks are also due to Dr. Than Than Thwe for the information on the Urbanisation of Yangon City.

REFERENCES

Dr.Theint, 2008. Combating land degradation in Myanmar. IAEA Research Paper.

Kyaw Yee,General Manager, Land Use Division, Myanmar Agriculture Service, Ministry of Agriculture and Irrigation.

Sandi April Maung, 2009. Assessment of the Effectiveness of Arsenic Removal Media used in Reduction of Arsenic Contamination in Groundwater Sources.

Urban and Regional Division, Department of Human Settlements and Housing Development Report, Ministry of Construction, 16th Feburary, 2009.

Chapter 10

Soil Protection Techniques from Erosion in the Sloppy lands of Vietnam

***Ngo Thi Loan*[1] *and Luu Nam Phuong*[2]**

[1]*Vice Head of Project Department, Centre for Regional Research and Development, Ministry of Science & Technology, 70, Tran Hung Dao Street, Hanoi, Vietnam*

[2]*Officer, International Cooperation Department, Ministry of Science & Technology, 39, Tran Hung Dao Street, Hanoi, Vietnam*

ABSTRACT

In Vietnam, slopy lands and mountainous areas cover three-fourth of the total territory. Soil fertility in slopy land is threatened by soil erosion which is main the cause to narrow the cultivated land. Soil erosion will be exacerbated by attempting to cultivate slopes that are too steep, cultivating up–and –down hill, continuous use of the land without proper crop cover, inadequate input of organic materials, heavy machinery used for tillage. These soil changes, with time, reduce agricultural productivity and hence income and household food security. The off-site of affects of soil erosion often have broader economic and environmental implications including flooding, sedimentation, water quality reducing, resulting in poorer living conditions for the people. Land and water management research has provided a range of technologies, which can reduce soil loss to acceptable levels. In this report, We introduce some techniques that prevent the loss of soil fertility and recommended mechanical measure *i.e.,* contour bunds, contour cultivation, contour rock walls, bench terracing and agronomic/fertility management measure, such as crop rotation, multiple cropping, manuring, fertilizer use, composting in Vietnam.

INTRODUCTION

Vietnam is a long and narrow country covering eastern portion of the Indochina Peninsular, along the Gulf of Tonkin and the East Sea. It is an S-shaped peninsula and shares borders with China, Laos and Cambodia. The country's land area is 330,991 km^2 lying between eastern longitudes of 102° 08′ and 109° 28′ and northern latitudes of 8°02′ and 23° 23′. Vietnam has a 3,260 km coastline and its inland borderline is 4,510 km. From north to south, the country stretches for about 1,650 km. administratively; the country is divided into 63 provinces.

Vietnam is characterized by high demographic pressure coupled with major dependency on land for livelihood. Uplands cover three-fourth of the territory and shelters one–thirth of the population. The uplands are a fragile environment characterized by sloppy lands that are prone to soil erosions. With increasing population pressure, these marginal lands have been brought into cultivation and the productivity is declining over the years as these lands are less fertile and vast majority of them are rain fed with uncertain and erratic rainfall. Farmers cultivating these lands are poor and don't take risk to use costly inputs such as fertilizers, because of uncertain response returns. This subsistence crop productivity and increasing poverty, but has also led to further degradation of natural resource.It is urgent to develop a new production system to bring economy up and to protected natural resources and living environment.

A risk of accelerated erosion exists on cultivated land from the moment trees, bushes, grass and surface litter are removed particularly on the sloppy lands. Soil erosion will be exacerbated by attempting to cultivate slopes that are too steep, cultivating up-and-down hill, continuous use of the land without proper crop cover, inadequate input of oganic materials, heavy machinery used for tillage etc. Soil erosion control depends on good management, which implies establishing sufficient crop cover and selecting appropriate soil and water conservation practices. The appropriate soil and water conservation measures generally include the agronomic and mechanical measures.

MECHANICAL MEASURES

Contour Farming

The traditional method of cultivation on sloppy lands is up and down the hill in most of the areas. This is the primary cause of accelerated erosion. Contour farming reduces soil erosion. Under this measure field operations such as ploughing and planting are performed on the contour. It helps in reduction of the surface runoff by impounding water in small depressions and decreases the development of rills (Figure 10.1).

Some times it is difficult to establish all crop rows on the true contour because of no uniform slopes in most of the fields. In order to establish row directions a true contour is laid out at one or more elevations in the field. For small fields of uniform slope only one contour maybe laid. The contour lines can establish using surveying equipment dumpy level or A-frame (A simple level made from local materials). The use of A-frame requires two persons one manipulating the A-frame and other marking the identified point on the contour. When farming is done on the contour instead of up and down cultivation, much less power is needed to be exerted by men, animals and machines. Moreover, the wear and tear of the machines and implements is less and work is done efficiently in less time.

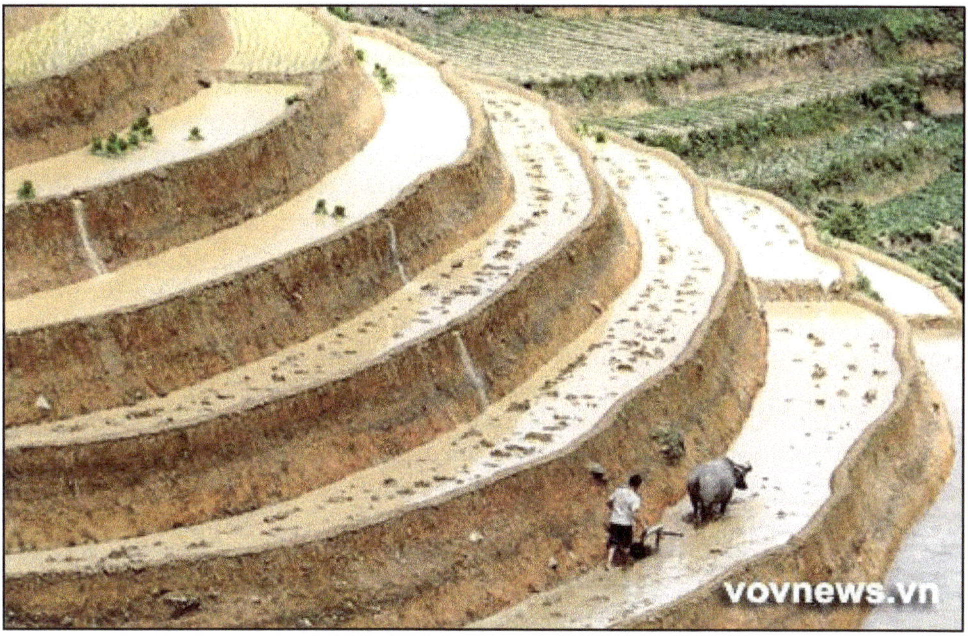

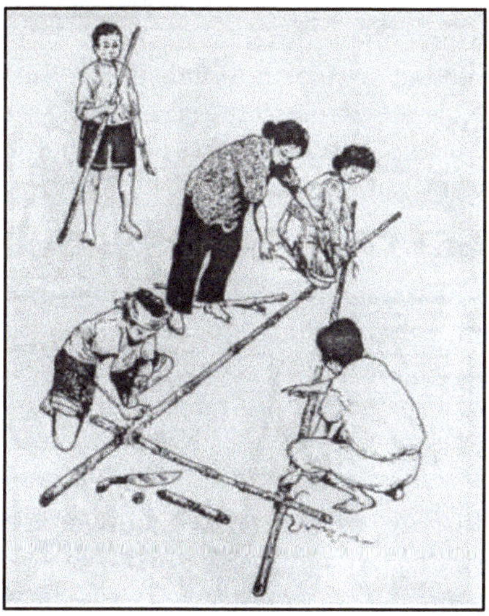

Figure 10.1: The Traditional Method of Cultivation on Sloppy Lands

The effectiveness of contour farming varies with land grade, type of soil and crop cover. Maximum effectiveness of contour farming is on medium slopes and on deep permeable soil. The relative effectiveness decreases as the land grades become very flat or very steep. On long slopes, where bunding is done to decrease the slope length, the bunds will act as guidelines for contour cultivation. All farming operations should be done parallel to these bunds.

Contour Bunds

Contour bunds are embankments of stone, or compacted solid (or combinations of these materials), which are constructed along the contour and serve to slow run-off and trap eroded soil. Establishment of contour bunds involves ploughing or digging along identified contour lines, the loosened soil then being used to form the bund. Removal of the soil from the contour line also creates a contour drain, which helps to reduce overland flow, increase infiltration, and divert excess water.

The hedgerows were typically planted on top of the bunds to reinforce them, as well as provide a source of biomass for green manuring and/or forage in a cut-and-carry system for penned livestock. Various hedgerow species were used, including grasses Guinea grass (*Pennisetum maximum*), Napier grass (*Pennisetum purpureum*), and leguminous shrubs (*Gliricidia sepium, Flemingia macrophylla*). The design of contour bunds include determination of the spacing between the bund, the cross sectional elements of a bund and the type and dimensions of the surplus system. The spacing between the two successive bunds is expressed in terms of horizontal and vertical intervals. This would depend on the type of soil, rainfall, grade of the land and vegetative cover during high rainfalls and farming practices.

Contour Rock Walls

Collecting the rocks and using them to construct a 0.5–1.0 m high rock wall running along the contour, sometimes in association with bench terraces. Contour rock walls are constructed at regular intervals down the slope, the spacing being calculated in relation to the steepness of the slope, soil type, and rainfalls. Walls are stabilized by planting grasses, shrubs or trees along their base, these plants being a source of green manure and/or forage, as with hedgerows.

VEGETATIVE BUNDS/BARRIERS

In many situations the vegetative bunds/barriers are better option for the soil and water conservation. Several times the vegetative bunds are combined with mechanical bunds. *Gliricidia sepium* has been grown on rice field bunds in several countries but mainly for the green manure, rather than for soil and water conservation. Erosion controls barriers using *Gliricidia* is a low-cost, simple and effective technology. *Gliricidia* can survive extreme climatic conditions. It is less sensitive to photoperiod and grows year-round. It generally dose not harbor pests and diseases. Most importantly, *gliricidia* unlike *leucaena* don't become a weed and not invasive. In tropical environments, such as Vietnam, *gliricidia* grows fast and allows 6-8 times pruning and provides 80-100 tons ha^{-1} of N rich biomass for in-situ application.

In cases where farmers do no like uneconomical biological barriers, like grasses or leguminous shrubs, pineapple could be used as a vegetative barrier. At Thanh Ha watershed the pineapple as a vegetative barrier was very much liked by the farmers and large number of farmers have taken up its cultivation on the contours and reaped multiple advantages such as economic benefit, soil conservation, limited or no competition by pineapple to cultivated crop and less labor costs as regular pruning is not required unlike *Gliricidia.*

OTHER PHYSICAL MEASURES

Several other physical soil conservation structures were also promoted by the project, which included contour canals, staggered trenches, drainage canals, silt traps, check dams and percolation tanks. These technologies were implemented in association with the physical and vegetative soil conservation measures already mentioned. Contour canals were promoted in association with contour bunds or contour hedgerows. It was found that the contour canals slow the flow or water, thereby increasing infiltration and soil moisture, and can also be used to drain excess water from the field. The latter function requires construction of a drainage canal, built down one side of a farmer's field to collect and remove diverted water from the field.

Drainage canals can also perform this function in association with contour bunds and hedgerows. Soil traps are pits in a gully or drainage channel used to slow runoff and trap eroded soil before it leaves the field. Check dams are physical or vegetative barriers placed in drainage canals and eroded gullies to slow runoff, prevent gully erosion, and trap eroded soil before it leaves the field. They slow down the water flow and allow heavier soil particles to settle. It prevents further widening and deepening of gullies and promotes the deposition of nutrients-rich sediments. The size of the check dam depends on the size of the drainage area or the gully size to be protected. Check dams can be built using bamboo, loose rocks, logs or other locally available materials.

AGRONOMIC MEASURES

These measures utilize the role of vegetation in helping to minimize the soil erosion by increasing soil surface cover, surface roughness, and surface depression storage and infiltration. Some measures adopted in the watershed are given below.

1. Strip Cropping/Alley Cropping/Hedgerow Intercropping

Contour hedgerow systems using *Gliricidia* was widely promoted in the watershed to minimize soil erosion, restore soil fertility, and improve crop productivity. Hedgerows are grown at intervals of 4-6 m along the contours. The strips or alleys between the hedgerows are planted with food crops: the pruned biomasses are used as green manure or as mulch *in situ,* or as fodder. Through time, natural terraces can form at the base of the hedgerow trees, and thereby minimize soil erosion and surface runoff. On sloping land, the improvement of soil fertility, stabilizing crop production may, in principle, pay off the labor inputs, but real farmer interest probably requires that the soil fertility accumulated in the hedgerows be used for profitable trees, regular or fodder crops.

2. Natural Vegetative Strips (NVS)

The use of Natural vegetative strips (NVS) has proven to be an attractive alternative, because they are simple to establish and maintain. When land is ploughed along contour lines, certain strips of 40–50cm wide are left unplugged, across the field on the contour. These strips are spaced at desired intervals down the slope and can be marked beforehand. The recommended practice for spacing contour buffer strips has been to place them at every one-meter drop in elevation, but a wider spacing may be acceptable. The contour lines can be determined using a *frame*. The natural vegetation of the strips filters the eroded soils, slows down the rate of water flow, and enhances water infiltration, marking them effective for soil and water conservation. Researchers found that these natural vegetative contour strips have many desirable qualities. They hardly need pruning maintenance compared with fodder grasses or tree hedgerows and give little competition to adjacent annual crops.

3. Minimum/Zero Tillage

In this system, simple farm equipment such as hoes abd digging sticks is used to prepare land and plant food crops. Farmers traditionally are familiar with minimum tillage practices. While more intensive tillage generally increases porosity of the topsoil and reduced barriers to infiltration of the soil surface, it normally interrupts the continuity of the macro-pores in the soil and can reduce deep infiltration, especially if a "plough-pan" is formed.

4. Crop Rotation

In this system various crop species are grown in sequence in the same field. For example: maize grown in the first and groundnuts in the second season. Groundnuts can replenish N (via N fixation) which can be used by maize. The different rooting pattern of different crop species planted may help soil structure and improve water percolation.

5. Monitoring Runoff and Soil Loss

The information on surface runoff and soil loss are needed for several purposes in the soil and water conservation. For example it is considered essential for designing the soil and water conservation practices and for planning supplemental irrigation activities. In the sloping lands of Vietnam, there exist very few runoff and soil loss at the watershed scale.

6. Farmers Perceptions on Soil and Water Conservation

Most farmers felt that soil erosion had decreased and soil fertility and soil moisture had increased. Decreasing soil erosion was primarily attributed to the direct effect of conservation measures. In the field, control of soil erosion resulted in the gradual formation of terraces. In sites where ploughing of the alleys occurred, terrace formation was relatively faster. Generally, terraces were perceived favorably as they were more easily and rapidly cultivated than sloping lands, particularly on very steep slopes. The adoption of hedgerows was seen to reduce the runoff and soil erosion, thus affecting soil fertility as well as increasing soil moisture. The formation of terrace directly increased the ease of cultivation. This, reduced runoff and erosion also increased the retention and hence the effectiveness of applying both organic and inorganic fertilizers. Where hedgerow trimmings were applied as green manure and mulch to the cultivated alleys, this has not only helped to reduce erosion but also was seen to increase the organic matter content of the soil, affecting both soil fertility and the soil's moisture-holding capacity. The combined effect of these changes on soil productivity was seen in increased maize yields (Figure 10.2).

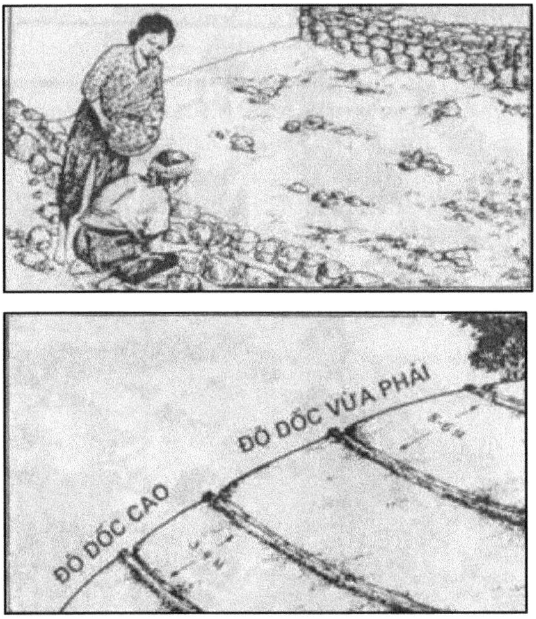

Figure 10.2: Soil Conservation Measures Exercised in Vietnam

CONCLUSIONS

Capacity building and awareness creation should invariably precede the planning activity on soil and water conservation. Farmers' active involvement is absolutely necessary. The soil conservation work must be developed in consultation with them. Balancing should be done between natural resource inventory and farmers socioeconomic conditions through appropriate technological interventions. The farm household should be the focus of every soil conservation program, as they take the daily decisions that shape the landscape. Farmers cannot ignore the short-term benefits of the land use decisions they make. Only the production oriented soil conservation strategies have a chance to be adopted that will provide a reasonable return on the labor a farmer has to invest.

Lack of the secure land tenure could be a major cause of less interest of farmer in the soil conservation works. Soil conservation programs have often led to 'pseudo-adoption' if strong social pressure, subsidies or other government incentives (including tenure security) were used to support adoption of practices that required substantial labor and other resource investment. In upland systems, plant yields are reduced more by a shortage or excess of soil moisture or nutrients rather than by soil losses *per se*. Therefore, there should be more emphasis on *rainwater management*, particularly water conservation, and *integrated nutrient management*, and less on soil conservation per se. Agronomic practices such as tillage and mulching that maintain infiltration rates are more easily adopted by farmers.

The implementation of programme of soil and water conservation should start from the ridgeline of the watershed to the valley–not on piecemeal basis in isolated patches. Forest, pasture, and cultivable land should be treated as inter-linked units of hydrological entity. Conservation works must be completed together with improved crop management in minimum time lag. In case of delay the benefits will not be adequate and client will get disillusioned with the usefulness of technology. Monitoring at each and every stage should be in-built in the system for quality and transparency, in a participatory mode.

Chapter 11

Erosion, Pollution and Urbanization in Agricultural Lands of Turkey

H. Samet[1] *and Y. Çıklı*[2]

[1]*Kocaeli University, Vocational School of Ihsaniye, 41040 Kocaeli/Turkey*

[2]*Duzce University, Vocational School of Cilimli, 81750 Duzce/Turkey*

ABSTRACT

The twentieth century witnessed a rapid urbanization of the world's population. Agricultural lands were sacrificed for the sake of industrialization and people were broken the soil off. According to United Nations population projections, approximately 5 billion people are expected to be urban dwellers in 2030. Particularly in developing countries, this expectation will cause the loss of the fertile soils in rural lands. Turkey has approximately 28 million hectares of arable agricultural land. The 10.8 million hectares of this land (1st Class and 2nd Class) have good quality for agricultural activities. The 10.9 million hectares are medium quality agricultural lands (3rd Class and 4th Class). As well as all over the world, agricultural areas in Turkey have three main problems: Erosion, pollution and inappropriate use. Especially in big cities, rural areas are under pressure coming from residential areas and cities are expanding gradually towards the edges. But, with increased migration to industrial areas, agricultural lands began to be filled with houses and small enterprises. For example, in Bursa, one of the largest industrial cities in Turkey, there were not any residential areas on 1st Class agricultural land. In the years of 1960, 825 hectares 1st Class agricultural land were used for industry and housing. This amount reached 1186 hectares in 1995. In this review, current status of the territory of Turkey was examined. Also erosion, pollution and inappropriate uses in agricultural lands and the precautions should be taken were scrutinized.

***Keywords**: Agricultural lands, Industrialization; Land degradation, Urbanization.*

INTRODUCTION

At the beginning of twentieth century, people living in rural areas migrated to the modern industrialized cities. These cities were a center of economic and scientific activities and provided a more stable livelihood, a greater opportunity for success, and a closer dependable accommodation with people [1]. Today, more than 50 per cent of population lives in urban areas, and by 2030, approximately 4.9 billion people (the 60 per cent of the world's population) are estimated to live in urban areas [2]. The accumulation of people had an impact on environment in terms of unplanned urbanization, resource consumption, and waste discharge. The urban population is expected to grow by 2 per cent per year during 2000-2015 and to reach an overall 65 per cent by 2050 [3]. Before 1950's, the arable lands that were available for intensive farming were adequate for the population in Turkey. With the mechanization in agricultural sector, the labor demand for agricultural activities declined and the rate of unemployment people in rural areas started to increase gradually. This significant social change increased the attraction of cities for villagers. As a result, a large scale migratory movement started to urban areas. This uncontrolled accumulation caused a distorted and unplanned urbanization [4].

LAND AND WATER POTENTIALS OF TURKEY

Turkey has a very central geography that bridges East and West, North and South, Europe and Asia and the Middle East. This unique geographical position implies rich biodiversity, sensitive ecological habitats and a wealth of cultural and historical resources [5].

Turkey's total area is 77.9 million hectares. The area of 28.1 (36 per cent of the total area) is suitable for many agricultural products. Forest and wooded lands (including shrubbery and maquis) are 23.2 million hectares (29.8 per cent of the total area) and the permanent meadow and pasture lands are 21.5 million hectares (27.6 per cent of the total land) (Table 11.1). The rest is non-agricultural lands.

The 1st Class and 2nd Class agricultural lands that are ideal for crop production are the sum of 11.9 million hectares. The 3rd Class and 4th Class agricultural lands that require some expenses for production are 14.71 million hectares. The total amount of 6th and 7th Class lands those are extremely important for biodiversity are 46.7 million hectares (59 per cent of the total area) (Table 11.1).

Table 11.1: Land Presence of Turkey (x 1000 ha) [6]

Usage Form	*Land Capability Class*								*Total*
	I	*II*	*III*	*IV*	*V*	*VI*	*VII*	*VIII*	
☆ Cultivable Lands	4825	6041	6036	4877	8	3965	2301	–	**28053**
☆ Meadow and Pasture Lands	149	444	738	1641	90	4163	14280	–	**21505**
☆ Forest and Wooded Lands	13	179	420	846	28	2624	19118	–	**23228**
☆ Non-Agricultural Lands	98	109	89	61	2	73	138	4543	**5113**
Total	**5085**	**6773**	**7283**	**7425**	**128**	**10825**	**35837**	**4543**	**77899**

Turkey's annual total water potential (including underground and surface sources) is 112 billion m^3. According to data from census of 2007, per capita annual water consumption is 1,586 m^3. In some countries those are rich in water resource such as USA, Canada, and Germany, this amount is above 10,000 m^3. According to scientists, in the years of 2030, per capita water consumption will fall below 1,000 m^3 and Turkey will be a country poor in water resources [7].

Turkey is known as one of the world's eight major genes center. More than 10,000 plant species live in Turkey of which one of three is endemic. Additionally, approximately 192 fish species that live inland waterway, 18 amphibian species, 83 reptile species, at least 426 bird species and 120 mammal species also live in this geography. Invertebrates are not included in this number [8].

All of this bio-diversity provides raw materials to medicine, pharmacy, agriculture, forests, animal husbandry, fisheries and tourism industries.

THE FACTORS THREATENING AGRICULTURAL LANDS IN TURKEY

Soil Erosion

Erosion is the most important soil problem in Turkey. Recent studies has estimated that roughly 500 million tons of topsoil is being lost every year and the lost of fertile soil causes a decrease in the storage capacity of reservoirs. For example; Firat, Yesilirmak, Dicle and Kizilirmak, which are the biggest rivers of Turkey, carry 73 million tons, 1,521 tons, 1,085 tons and 929 tons of soil every year, respectively. The amount of 1521 tons of soil carried by Yesilirmak is equal to 0.6 mm topsoil on the watershed [9].

The percentage of eroded land in Turkey is almost 86 per cent in which 59 per cent is severely and very severely eroded, 20 per cent is moderately eroded, and 7 per cent is slightly eroded (Table 11.2) [9]. There are a lot of factors to accelerate the soil erosion in Turkey.

Table 11.2: Degree of Erosion and the Eroded Lands in Turkey [10]

Erosion Risk	*Land (ha)*	*Rate (%)*	*Classification Criteria*
Slight	5,611,000	7.22	Less than 25 per cent of top-soil eroded
Moderate	15,503,000	20.04	Between 25 per cent and 75 per cent of top-soil eroded
Severe	28,334,000	36.42	All of the top-soil and the 25 per cent of sub-soil eroded
Very Severe	13,321,000	22.32	All of the top-soil and between 25 per cent and 75 per cent of sub-soil eroded
Total	**66,906,000**	**85.98**	

Tiredness and inappropriate use is the first reason accelerating the soil erosion. The Anatolian soils have been used intensively by various civilizations for 9-10 thousand years. Uncontrolled activity on these soils caused various problems such as loss of plant cover and organic matters in topsoil. The shortage of organic matters in soils reduces the particle stability and causes accelerated erosion. In Turkey, the rate of organic matter is less than 1 per cent in the 64 per cent of total lands [11]. Additionally, water contents and nutrition is generally not enough for germination in semi-arid eroded lands [12]. According to the Global Desertification Vulnerability Map, the middle region of Turkey is facing severe and very severe desertification problem.

The second reason of erosion is topographic factors. Turkey is a mountainous and hilly country. The average altitude is approximately 1,250 meters and 65.5 per cent of the total land has more than 15 per cent slope [13]. In Turkey, only 8.5 per cent of the total land has less than 8 per cent slope steepness. About 12.8 per cent and 16.2 per cent of the total land has the slope steepness of 5-10 per cent and 10-15 per cent, respectively [14].

The third reason of erosion is climatic factors such as rainfall, temperature, humidity and wind. The effect of these factors is generally seen on evaporation and transpiration process. On the other hand, a large part of Turkey has an arid and semi-arid climatic regime. Consequently, the precipitations take shorter time and they are downpour style.

The other reason of erosion is vegetation cover that plays a major role in reducing soil erosion. Despite all amelioration activities, wooded land and meadows are gradually cultivated for agriculture. Furthermore, early grazing and over-grazing during the period of vegetation causes the loss many kinds of plant species and less plant density. These open areas become vulnerable for soil erosion.

According to scientists, if the forest destruction and erosion continues with same speed without taking measures, desert climatic conditions will be experienced in the 85 per cent of Turkey's surface area.

Inappropriate Land Uses

Urbanization occurs as a result of the self revolution of a city or a society which completed industrialization in the developed countries [15]. But uncontrolled migration to cities and unplanned settlements causes very serious environmental, social and economic problems. Most important environmental effect is the inappropriate use of agricultural lands. Thus, the rural area surrounding the city turns to urban area rapidly and decreasing of the arable lands affects the bio-diversity and the ecologic balance negatively.

After 1950's, the migration to industrial cities from rural areas started to increase gradually. This accumulation caused illegal and unplanned constructions and sprawl process towards plains and plateaus. As seen on Table 11.3, Antalya an important tourism center in Turkey, was preferred by unemployment people in every period. Similarly, a lot of people or family migrated to Istanbul, which is the biggest city with its roughly 15 million populations. Kocaeli is very close to Istanbul and an important industrial city. Its net migration rate is relatively higher than the other city in every period. But in the period of 1995-2000, the number of migrant is only 211 since lots of people who came here with huge hopes dramatically lost their goods after the last earthquake in 1999. The unique reason of the damage is setting up factories and houses on valuable agricultural lands.

Some big cities are selected as a center of industry and foreign capital in Turkey, such as Istanbul, Izmir, Bursa, Kocaeli, Antalya and Adana. The cities had lost their agricultural lands because of being enterprise center. For instance, in Istanbul, while the settlement areas were 11 per cent in 1990, this rate reached to 16 per cent in 2005 and the agricultural areas decreased to 23 per cent from 27 per cent within the same period [16].

Table 11.3: Provincial Net Migration and Rate of Net Migration (1975-2000) [17]

Province	*1975-1980*		*1980-1985*		*1985-1990*		*1995-2000*	
	Net Migration	*Rate of Net Migration (‰)*	*Net Migration*	*Rate of Net Migration (‰)*	*Net Migration*	*Rate of Net Migration (‰)*	*Net Migration*	*Rate of Net Migration (‰)*
Antalya	17,142	26.5	25,339	32.8	87,737	89.7	90,457	64.3
Istanbul	288,653	73.4	297,598	60.5	656,677	107.6	407,448	46.1
Bursa	58,120	61	47,434	41.1	83,641	61.6	85,325	45.1
Izmir	119,896	73.7	82,173	41.9	146,208	63.8	120,375	39.9
Mersin	40,273	57.5	49,593	56.5	74,717	68.3	18,429	12.4
Kocaeli	53,640	112.9	41,287	67	83,362	108.2	211	0.2

The Bursa city is the best example of inappropriate land use in Turkey. The total areas of the city are 1,081,900 hectares and 17 per cent of this land is covered by plateaus [18]. The Bursa plateau is about 40,000 hectares and about 11,224 hectares of this land are under protection [19]. High added valued agricultural products are grown in the fertile lands of the plateau, such as fig, olive and chestnut.

Aksoy and Ozsoy [20] scrutinized the interaction of territory and urbanization in Bursa with using the Landsat satellite outputs of 1984, 1993 and 1998. Urbanized areas which was 5,089 ha in 1984 have reached 12,343 ha 1998 with the rise of 142 per cent.

Turkey is an industrializing country and as a result of this, speedily and unplanned urbanization occurs in big cities. In Bursa city, industrial areas that were established on 2nd Class lands were only 71.1 hectares in 1960. But as a result of uncontrolled expansion, it dramatically reached 756.4 hectares in 1976 and 1732 hectares in 1995 (Table 11.4).

On the other hand, while the settlement that was established on 1st and 2nd Class agricultural lands were 8.6 hectares in 1960, it increased swiftly and reached 3,054.1 hectares in 1976 and to 4092 hectares in 1995 (Table 11.4). As seen in Table 11.4, totally 9,876.8 hectares arable lands (Class I and II) that have got huge agricultural potential were sacrificed for the sake of industry and urbanization in 35 years period.

Table 11.4: The Loss of Agricultural Lands in Bursa Province [22]

Urban Texture	*Years*	*Land Capability Class*						*Total (ha)*
		I	*II*	*III*	*IV*	*V*	*VI*	
Industrial	1960	—	71.1	55.4	—	—	—	126.5
areas	1976	94.4	756.4	93.1	16.3	75.7	31.6	1,067.10
	1995	76.7	1,732.00	347.1	72.4	115.9	40.6	2,384.70
Residential	1960	5	3.6	238.5	613.1	—	13.2	873.4
areas	1976	731.4	2,322.70	802.2	1,005.10	163.9	321.6	5,346.90
	1995	1,110.10	2,973.80	1,007.30	1,046.40	276.9	701.8	7,116.30
Total (ha)		**2,017.20**	**7,859.60**	**2,543.60**	**2,753.30**	**632.4**	**1,108.80**	**16,914.90**

In 1977, in order to prevent the spread of the city towards the plains, "*The Plain Protection Protocol*" was signed. But unfortunately, plains could not be preserved and the city developed towards the north side [21].

In another investigation, the total amounts of small industrial estate in Turkey are 18,000 hectares and the 62 per cent of this land (approximately 11,150 ha) were located in agricultural lands (between Class I and Class IV) [23].

Between the years 1970-2000, Ozbek and Oztas [24] studied the use of agricultural lands for settlement in Erzurum Plain. According to these researchers, agricultural lands occupied in the first ten years (1971-1980) were calculated as 1,422 hectares. This amount was increased by 408 per cent in the second ten years (1981-1990) and has reached 7,237 hectares. In the last 10-year period (1991-2000) the loss of agricultural land has increased 1200 per cent and calculated as 18,491 hectares (Table 11.5).

Table 11.5: Inappropriate Use of Agricultural Lands in Erzurum Plain [25]

Periods	*The Loss of Agricultural Lands (ha)*			
	Class I	*Class II*	*Class III*	*Total*
1971-1980	847	575	—	1,422
1981-1990	493	619	1,683	7,237
1991-2000	11,185	4,228	2,378	18,491

Similarly, Alpan [26] studied on land use/land cover in Adana using satellite data of 1984 and 2000 and reported that the urban and built-up area has increased about 30 per cent on agricultural lands and 70 per cent on previously semi-natural lands by a factor of urbanization in the period of 16 years. Permanent immigration and urban development strategies were the main driving forces.

Soil and Water Pollution in Turkey

Soil and water pollution is a very serious problem in Turkey, as well as all over the world. Depending on industrial developments, mechanization in agriculture that is also called "Green Revolution" is developing rapidly in Turkey. Farmers and agricultural enterprises that closely followed these developments use excessive fertilizers and pesticides to increase their agricultural productions. All of these practices such as household wastes, industrial discharges, excessive fertilization and various disinfection materials are always problems for the environment.

Today soils, mineral waters, rivers, streams, lakes and coastal zones are all under the threat of heavy pollution. Another serious problem is air pollution. For example, a lot of plant living in the diameter of 37 km in Yatagan was affected badly by flue gas coming from Thermic Central [27].

CONSERVATION PRECAUTIONS

The History of Environmental Legislations

Ottoman Empire had some institutions for the protection of natural sites, though these were admittedly rare and benefited mainly elite people. The Turkish Republic replaced the Ottoman Empire and the founders of modern Turkey were overwhelmed with the problems of a war-torn country in the early years. However, by the 1930s, a number of organizations had become interested in environmental issues, including the Animal Conservation Association, the Turkish Association of Foresters, the Reconstruction Association, etc.

Some environmental problems such as the inadequacy of infrastructures in big cities, the rise of squatter housing because of rapid industrialization, and migration to the large cities, began to show their face around the same time. One of the first achievements of Turkish environmentalism, in 1955, was established as named "the Turkish Association for the Conservation of Nature and Natural Resources". In similar organizations, the founder's goal was to prevent forest destruction and to preserve soil and water resources [5].

With the influence of the conference named "United Nations Conference on Environment", held in Stockholm in 1972, environmental problems were dealt with for the first time in the third Five Years Development Plan (1973-1977) which was prepared in 1973.

Turkey's major environmental law was incorporated into the national canon in 1983. It was followed by the National Park Law, the Law for Protection of Cultural and Natural Wealth, etc. In the wake of all these new legislations, the Ministry of Environment was founded in 1991. Despite its small budget, the Ministry of Environment has organized some major conferences and invited different environmental groups and sectors to discuss the current environmental problems threatening Turkey, and how to deal with these problems.

In 2005, *"The Soil Conservation and Land Use Law"* was adopted in Turkish Parliament. The main purpose of this law is to prevent soil loss with natural and artificial ways, to preserve the quality of the soils and to identify the principles for planned land use and sustainable agriculture. In addition, all definitions such as land, agricultural land, absolute agricultural land, land capability classes, wooded lands etc. related to soil have been made in this law.

According to the law, *"Land Protection Board"* was created in all provinces. The committee's task is to control all the activities those will be done on land except agriculture and to allow them if suitable.

Non-Government Organizations (NGOs) and Environmental Awareness in Turkey

Thanks to growing importance of civil society in the world, traditional Turkish political life also has been changing in recent decades and leaving more space for public participation. This is the main reason for the advent and rapid growth of NGOs in Turkey since 1980s [5].

There are a few conservationist organizations in Turkey. One of them is "The Environment Foundation of Turkey (TCV)". It was established in 1978 as a non-governmental, non-profit, voluntary and independent organization. The other one is the Turkish Foundation for Combating Soil Erosion, for Reforestation and the Protection of Natural Habitats (TEMA). It was established in 1992 with the main objectives of trying to prevent soil erosion, protecting the natural heritage, and preserving biodiversity. Today, TEMA is an exemplary non-political and civil organization which provides environmental in-service courses to public. Also its major concern is to raise the public consciousness to environmental problems such as soil erosion, deforestation, conservation of the surface of the soil, the decline of productivity in farming areas, and threats to the biodiversity of the land [5]. Additionally, public education on these issues is vital importance for TEMA. And today, TEMA is continuing its own way with 375,000 volunteer.

RESULTS AND DISCUSSION

Urban growth, road construction, mining and industrial activities are the major factors in land degradation in different regions. Such activities have caused serious environmental problems. Vulnerable agricultural lands are often lost and it is not possible to gain them back. Besides the dwelling areas in the city, industry and trade functions also continue to spread on arable lands and this sprawl fact threaten the sustainability of natural sources.

To minimize the negative effects of these problems, the government and the local administrations must take some radical precautions.

Firstly, the result of migration to urban areas from rural areas should be examined carefully. Agricultural activities in rural areas should be strengthened and encouraged. Farmers should be aware of the new production techniques, should be informed about alternative crops and mechanizations. Thus, the population of rural areas will be held at the location and the migration will be prevented.

Secondly, the environment, nature and natural sources (*i.e.* water and soil sources, forests, mines and biodiversity) should be protected by means of direct or indirect educations, researches and applications. Furthermore, fertilization should be done according to scientific principles and agricultural disinfection materials (pesticide, herbicide, fungicide etc.) should be chosen according to the types and their following effects and should be applied properly.

In conclusion, the source of the problems lies in the socio-economic level of development of the country. We, as a young generation, have to take positions of leadership and look at the larger global picture to fix the problems.

REFERENCES

1. R. Hakim and M. Marshall. Urbanization Outlook. Our Task Inc. USA. (2008)
2. World Urbanization Prospects (http://www.unccd.int/knowledge/docs/migration.pdf) Last Visit 25th November 2009.
3. UNEP. Global Environment Outlook 3: Past, Present and Future Perspectives. Earthscan Publications Ltd. UK. (2002)
4. A.Z. Gurler, N. Kizilaslan, H. Kizilaslan and S. Dogan. Rural-Urban Migration in Turkey and the Socio-Economic Characteristics of the Immigrants (Tokat Case). Research Journal of Social Sciences, 2: 60-71.(2007).
5. I. Ozdemir. The development of Environmental Consciousness in Modern Turkey. Environmentalism in Muslim World. Ed. Richard Foltz, New York: Nova.
6. General Directorate of Rural Services. (2002). (http://www.khgm.gov.tr/arazi.xls) Last visit: 26th November, 2009.
7. A.Tomar. Toprak ve Su Kirliliği ve Su Havzalarının Korunması. TMMOB Izmir Kent Sempozyumu. 8-12 Ocak 2009. (2009).
8. K.Isık. Çevre Sorunları, Biyolojik Çeşitlilik ve Orman Gen Kaynaklarımız.TEMA Yayınları No:25. (1999).
9. M. Kışlalıoğlu, F.Berkeş. Çevre ve Ekoloji. Remzi K.evi. 4. Baskı (169-170). Ist. (1993).
10. H. Dursun. Türkiye Arazi Varlığı ve Sorunları, 1. Çevre ve Ormancılık Şurası 21-24 Mart 2005. Antalya. (2005).
11. H. Karaşahin. Yurdumuzda mera sorunları. Standart Çevre, 84-87. Ankara. (1995)
12. S.İ. Ürgenç. Ağaçlandırma Tekniği. Istanbul Üniversitesi Yayınları. Yayın No.3994. Orman Fak. Yay. No:441. İstanbul. (1998).
13. I. Bayramin, O. Dengiz, O. Başkan, M.Parlak. Soil Erosion Risk Assessment with ICONA Model; Case Study: Beypazarı Area. Turkish Journal Agriculture and Forestry. 27 (2003) 105-106. Ankara. (2003).
14. M. Reis, A. Yüksel, O. Erdaş, F. Tonguç, A.E. Akay. Afforestation practices in Preventing erosion in Watersheds of Turkey. International Congress on River Basin Management, 115-120.
15. H. Suher. Kentleşme-Kentlileşme, Gebze Yük. Tek. Enst., Kent-Kentleşme-Kentlileşme Sorunları ve Politikaları Paneli. (1996).
16. A. Geymen and I.Baz. İstanbul Metropolitan Alanındaki Arazi Kullanım Değiþimi ve Nüfus Artışının Izlenmesi. TMMOB Har. Kad. Müh. Odası Ulusal CBS Kon. (2007).
17. Turkey Statistical Yearbook. (2008). (http://www.tuik.gov.tr)

18. S.S. Eryılmaz, H.Cengiz, Y.Eryılmaz. The Urban Sprawl Model for an Affected Metropolis: Bursa–Istanbul Example, 44th ISoCaRP Congress (2008).
19. F. Eke and E. Ataç. Kentsel Mekanlarda Kaybedilen Doğal Değerler. Bursa Ovası ve AOÇ Örnekleri. Uluslararası Bursa Kent Sağlık Sempozyumu. (2006).
20. E.Aksoy and G. Özsoy. Bursa Kenti ve Yakın Çevresinde Tarım Arazilerinin Sanayi ve Kentsel Yerleşimle Kayıplarının CBS Teknikleri Kullanılarak Belirlenmesi, Nedenleri ve Çözüm Önerileri. GAP II. Tarım Kongresi, Sanlıurfa. II. Cilt s.1053-1060. (2001).
21. R. Arslanoğlu. Kent, Kimlik, Küreselleşme. Asa Kitapevi. Bursa. (1998).
22. E. Aksoy and G. Özsoy. Kentleşme ve Arazi Etkileşimleri. TMMOB Bursa Kentine Çözümler Sempozyumu. (2007).
23. Anonymous. T.C. Çevre Bakanlığı. 2000'li yıllara doğru çevre. S.169. Ankara.
24. A.K. Ozbek ve T. Öztaþ. Tarım arzilerinin amaçdışı kullanımı; Erzurum örneği. Ekoloji dergisi, 13:52(1-6). (2004).
25. C. Balabanlı, M.Türk, O.Yüksel. Erozyon ve Çayır Mera İlişkileri. SDÜ Orman Fak. Dergisi. Sayı: 2(23-34). (2005).
26. H.Alpan. Land Use Change and Urbanization in Adana. Land Degradation and Development. Vol.14, Iss. 6, p.575-586. (2003).
27. S. Somerson. Türkiyede Çevre ve Siyaset. Metiþ Yeşil Kitaplar (s.61). İstanbul. (1993).

Chapter 12

G-EIS: Global Environmental Information System and Egypt's Lands Degradations

Wafik Noseir
Environmental Engineering Consultant,
AUC, Abda Basha Abasia,
Cairo, Egypt

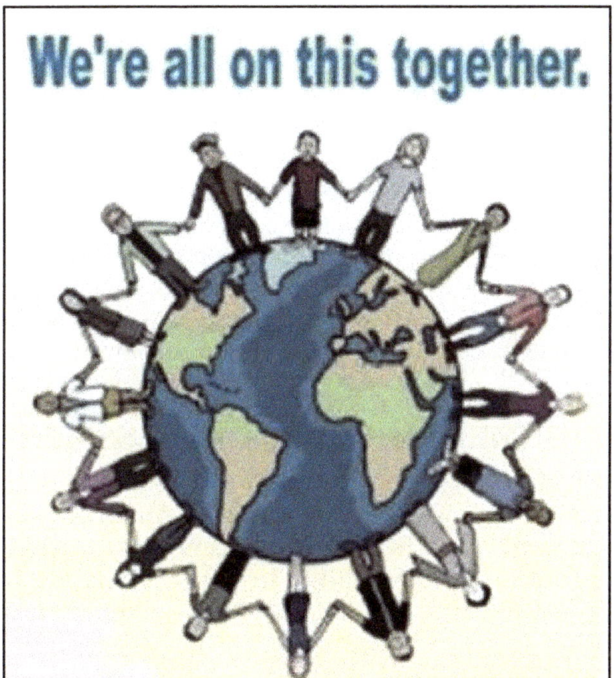

ABSTRACT

The aim behind this presentation is to establish a Global G-EIS (Global Environmental Information System) in a simple way that can be connected to the Goggle Earth program or any other map. The system will have all the Environmental parameters of the life on the Earth at all places. Each country can do its effort to collect all the parameters at each place around the country with a help or guidance of an Environmental global department. This could be established by the United Nations "EUN".

Each country can be divided into areas, varying from place to place according to their importance to the Environment and the Environmental parameters. G-EIS can have levels of security for each country to view the information that it wants to view publicly. However, all the information should be available to the EUN (Environmental-department of the United Nations). Therefore, the system can be a great help to each project to do its EIA (Environmental Impact Assessment), as well as for the ministries of Environment to do the needful. Moreover, the system can be a great help for EUN (Environmental-department of the United Nations) to approve and comment on the strategic projects affecting globally, not just the country where it is located. For example nuclear plants and factories of nuclear weapons, dams, major projects leading to global air, water and soil pollutions can be sorted out.

INTRODUCTION

World's environment is facing a very important challenge, a dilemma that should be solved as soon as possible to protect and preserve the environment of our earth and find the right answer to this question : "To be or not to be?" (Shakespeare). Since our environment on the planet Earth is connected globally, Soil is connected through the Earth, Water is connected through the surface or underground and Air through our atmosphere. We as humans can disagree about things like politics, economics, religion, social life, etc., but we are compelled to come to an agreement on our Environmental issues, in order to preserve and protect the environment of the only planet that we live on, at a GLOBAL LEVEL. Some facts in this connection are; we don't know what we don't know, if we can't measure it, we really don't know much about it, if we don't know much about it, we can't control it, and if we can't control it, we are at the mercy of chance. So let us look at this in a schematic representation (Figure 12.1). The global warming problem and energy interactions are presented in Figure 12.2.

"EIA" ENVIRONMENTAL IMPACT ASSESSMENT IN EGYPT

The total surface area of Egypt is 1 001 449 km^2, estimated population in 2009 is 85 million. The country is one of the driest parts in Africa with an average annual precipitation of 51 mm/year. There are hardly any forests present in the country. Almost 100 per cent of the cropland is irrigated. There are many important environmental issues like urbanization, pollution, soil erosion, land degradation and threats to the Biodiversity.

EIA shows how projects are approved in Egypt and everywhere on our planet with a slight difference in each country. The problem here is that all projects are

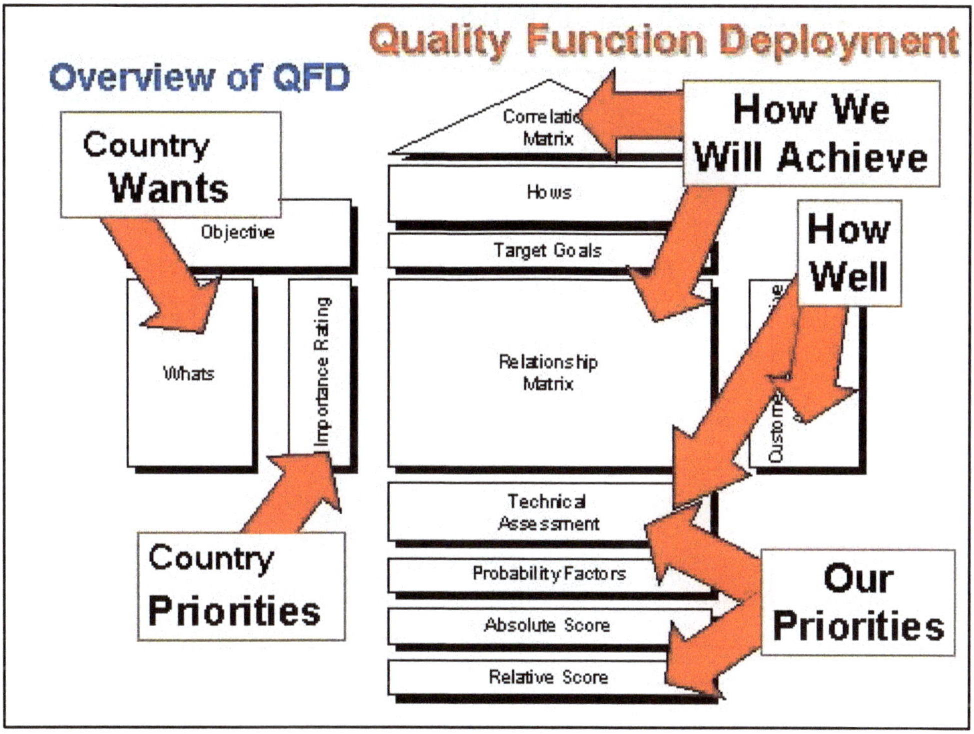

Figure 12.1: Schematic Representation of Quality Function Development

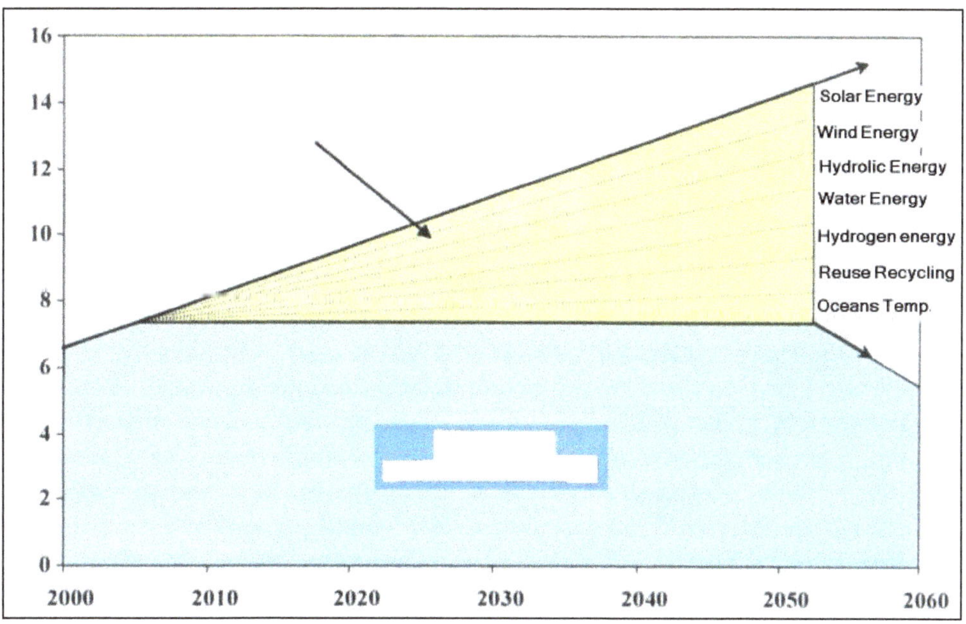

Figure 12.2: Global Warming Problem and Energy Situation

ENVIRONMENTAL INFORMATION SYSTEM (EIS)

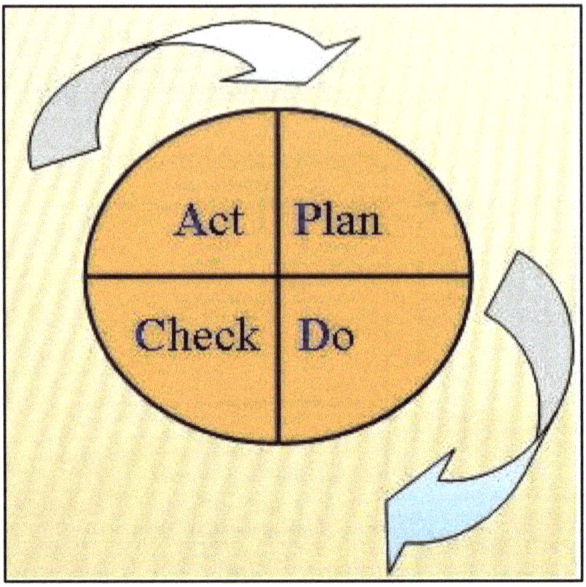

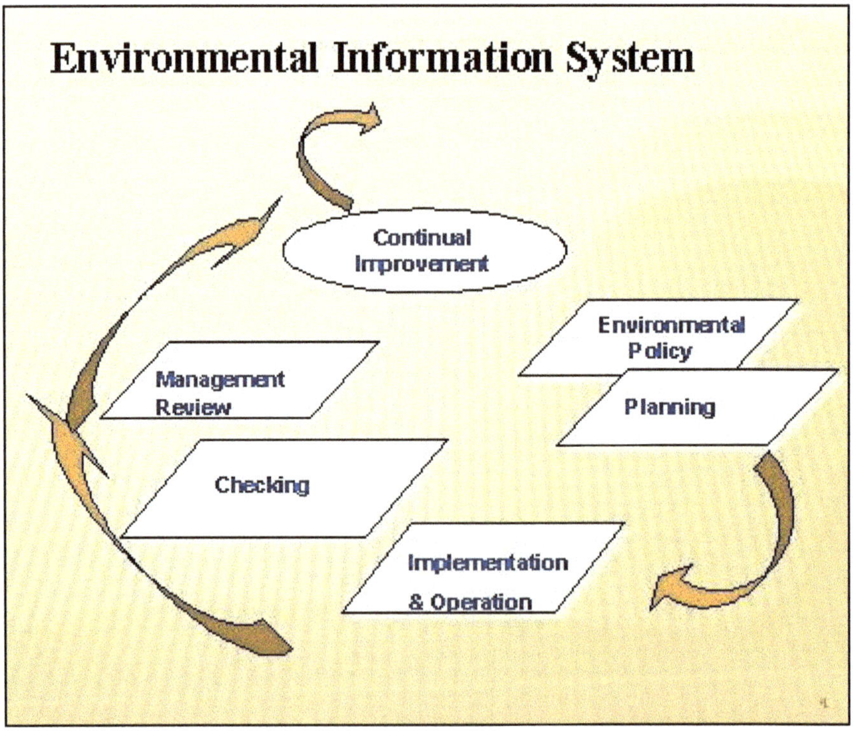

GLOBAL ENVIRONMENTAL INFORMATION SYSTEM
G-(EIS) Organization Overview

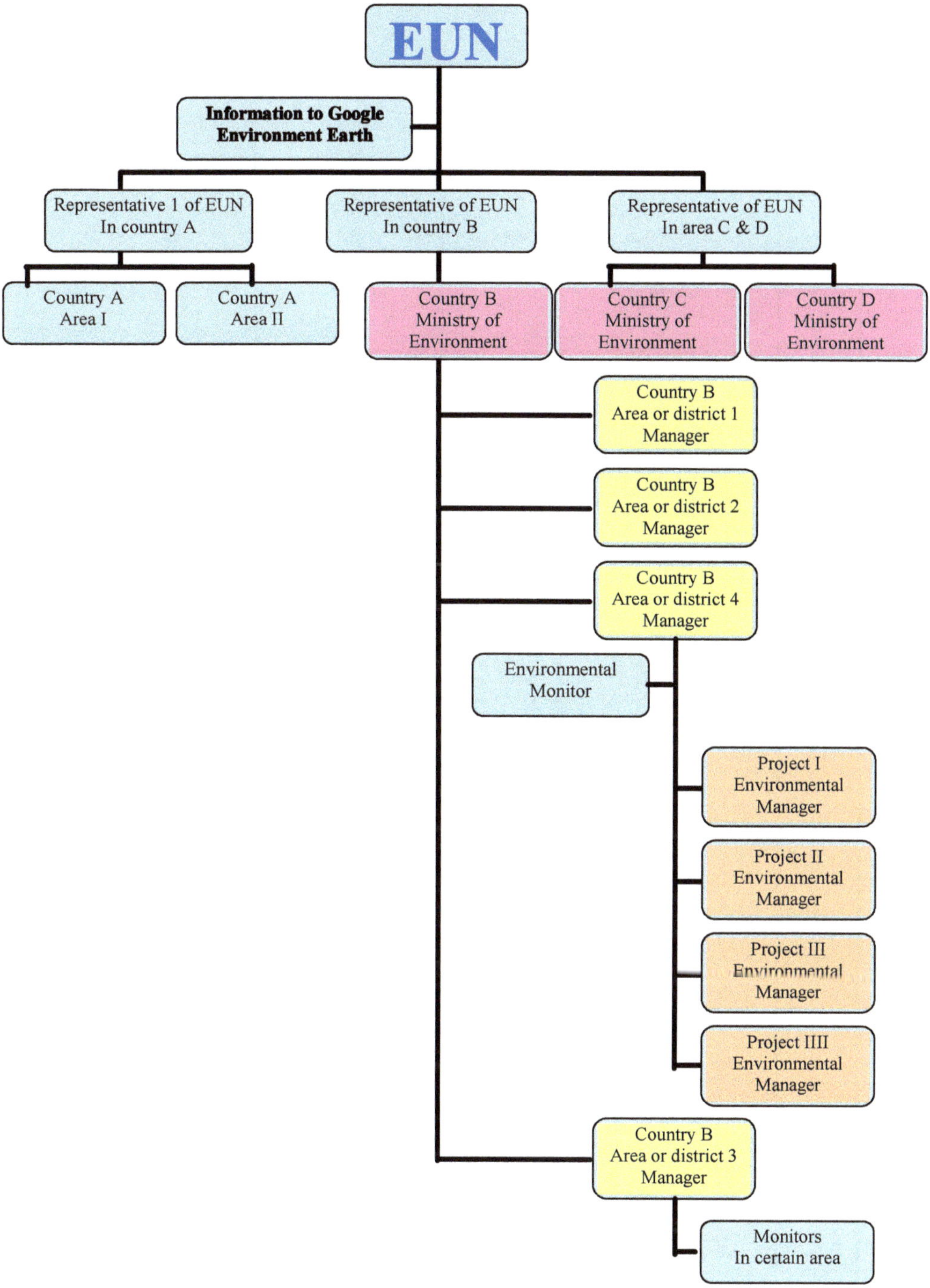

ORGANIZATION MATRIX

Environmental MatrixG-EIS	*UNEP Main Office*	*UPEP Represent in the Country*	*Country Environmental Ministry*	*Country Environ. Rep. in an Area*	*Area Represent.*	*Project Owner or Chairman*	*Project Environ. Manager*
UNEP Main Office	World Strategy +awareness	Coordination+ Methodology	Country Strategy+ EI Aapprovals				
UPEP Represent		Coordination+ Methodology	Country Strategy+ EI Aapprovals				
Country Environmental Ministry			Country Strategy+ EI Aapprovals		Social Habits and awareness		EIA Establishing+ follow-ups
Country Environ. Rep.				Mythology+ Monitoring			
Area Represent				Mythology+ Monitoring	Social Habits and awareness	Project targets+ approvals	
Project Owner or Chairman					Project targets+ approvals	Project targets+ approvals	
Project Environ. Manager				Mythology+ Monitoring			EIA Establishing+ follow-ups

STRATEGIC PLANNING

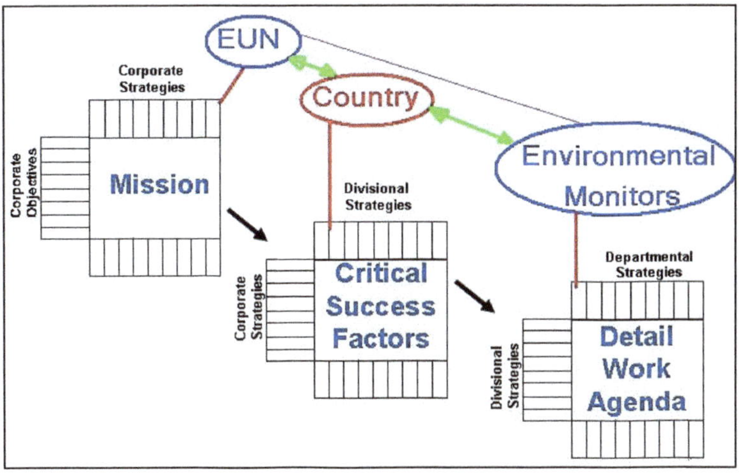

PROJECT MODULE G-EIS

Inputs	*Tools*	*Outputs*
Initial Environmental Information+ Impact on the area in all the Information elements mentioned before	*Scientists measures and Technical Corrects+ Social Studies scientists advises and Corrections+ NetworksSoftware*	*Reports and Decisions and Advises and Orders*

approved separately without any global planning that takes the wide picture into consideration. The Environmental laws also varies from a country to country, as if each country is separated from the other and has its own planet to live on.

Therefore, we need a Global Environmental planning and Global LAW.

A) "Land degradation due to global warming > rising sea level"

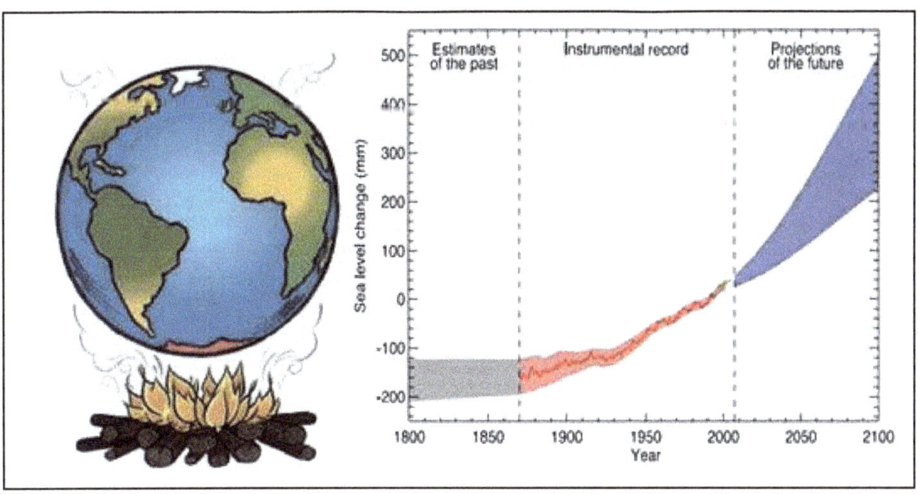

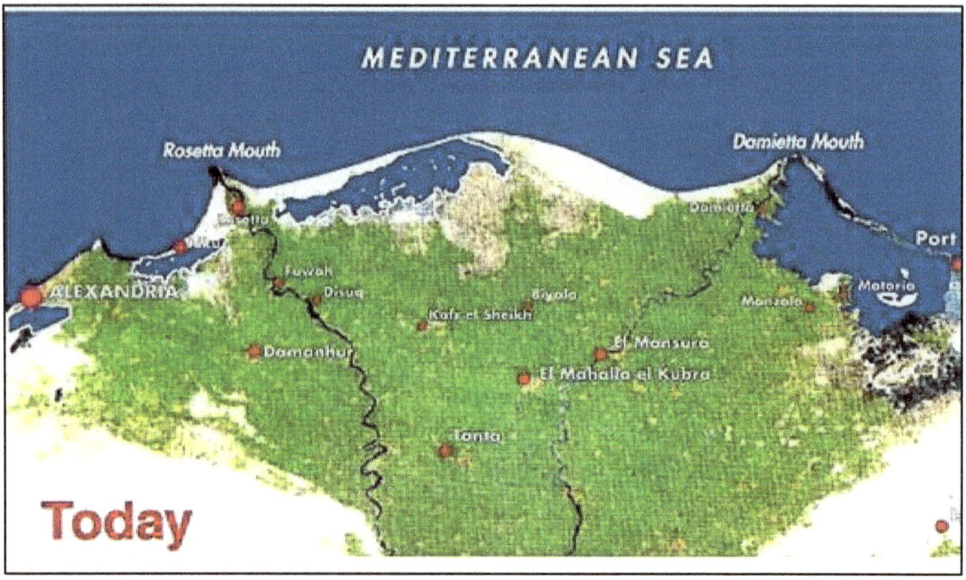
MEDITERRANEAN SEA
Rosetta Mouth
Damietta Mouth
Port
Fuwah
ALEXANDRIA
Disuq
Biyala
Kafr el Sheikh
Mataria
Manzala
El Mansura
Damanhur
El Mahalla el Kubra
Tanta
Today

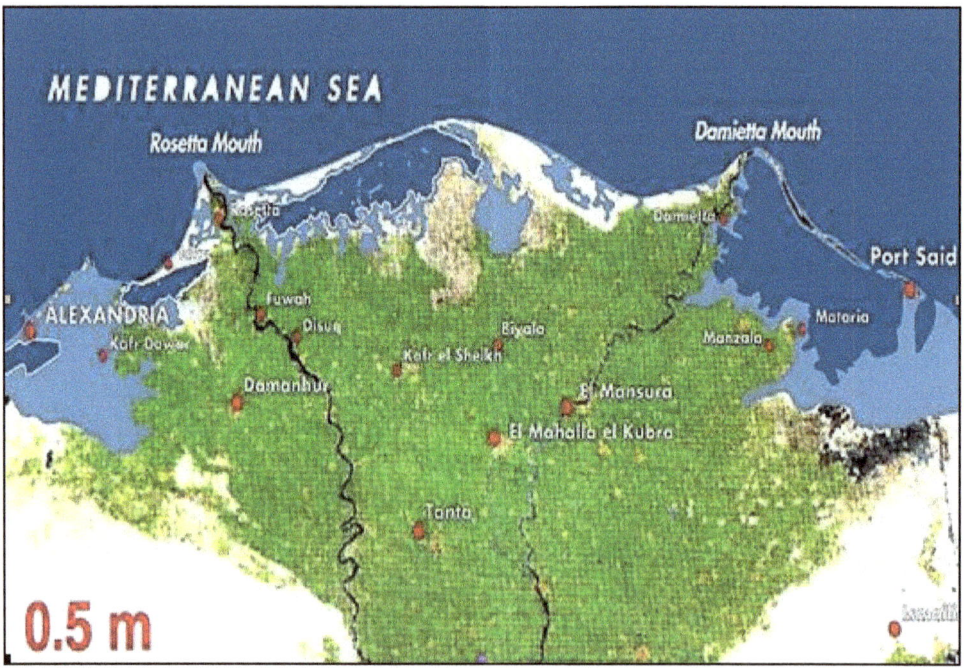
MEDITERRANEAN SEA
Rosetta Mouth
Damietta Mouth
Port Said
Fuwah
ALEXANDRIA
Disuq
Biyala
Kafr el Sheikh
Mataria
Manzala
El Mansura
Damanhur
El Mahalla el Kubra
Tanta
0.5 m

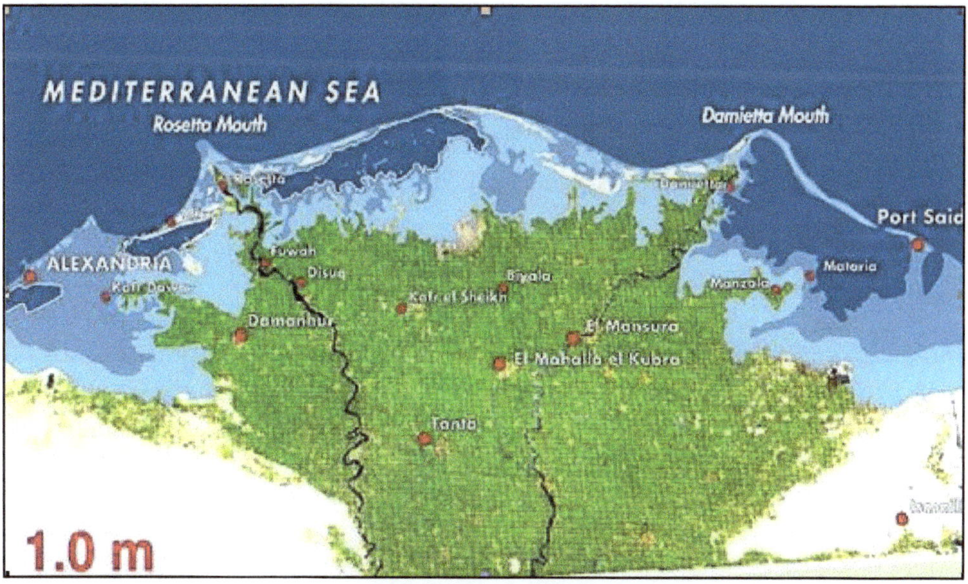

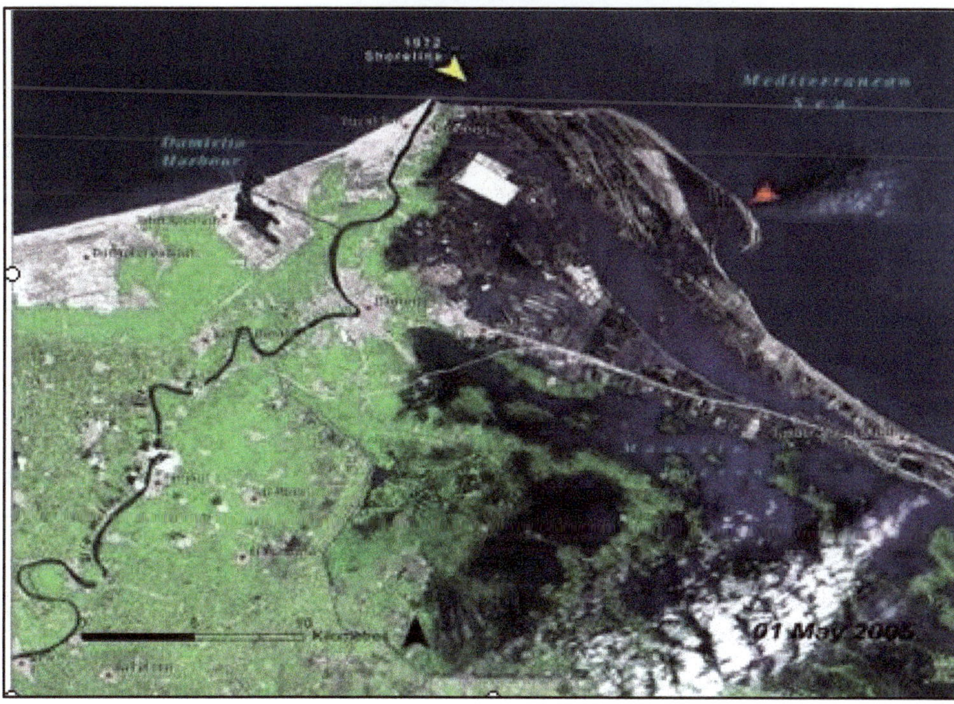

At several points along the coast, the delta is now receding, Damietta Promontory has eroded dramatically as waves and currents have stripped its sands faster than the river can replenish them (yellow arrow), while there are local areas of accretion such as the Damietta spit (red arrow), the delta is shrinking.

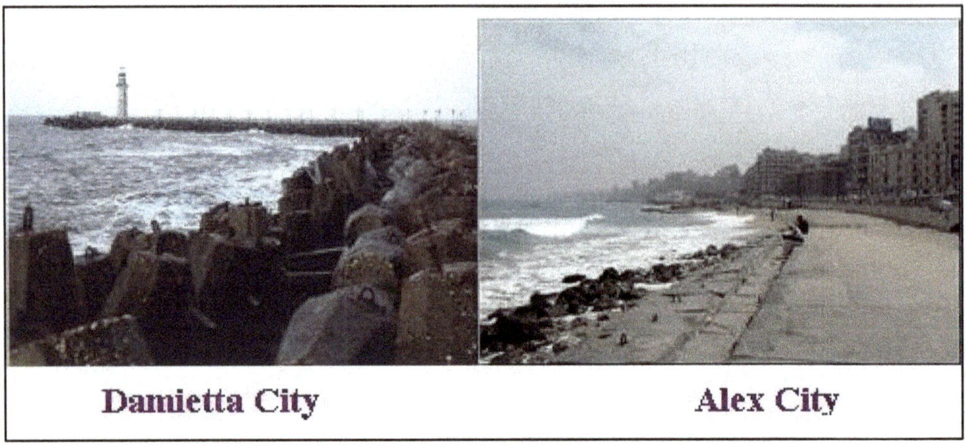

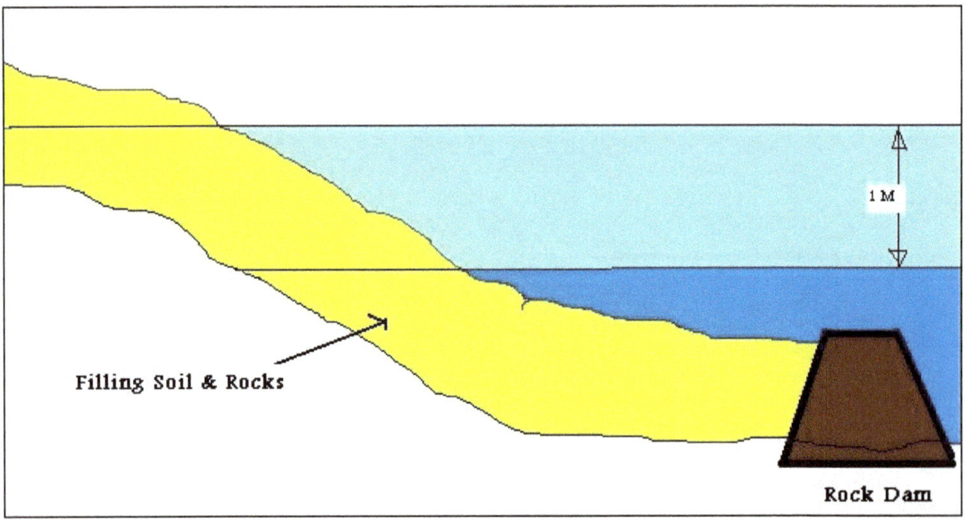

B) "Land degradation due to population increase in Egypt > use of agriculture lands for housing"

The population of Cairo has increased from less than six million in 1965 to more than ten million in 1998 and the city's current estimates are around 20 million. It is growing continuously.The urban extent of Cairo (gray areas) has expanded dramatically between the 1972 and 2009 (Images).

Much of the needed infrastructure is already in place and crops are already being produced on irrigated land including fruits and wheat (green around the lakes–2007 image).

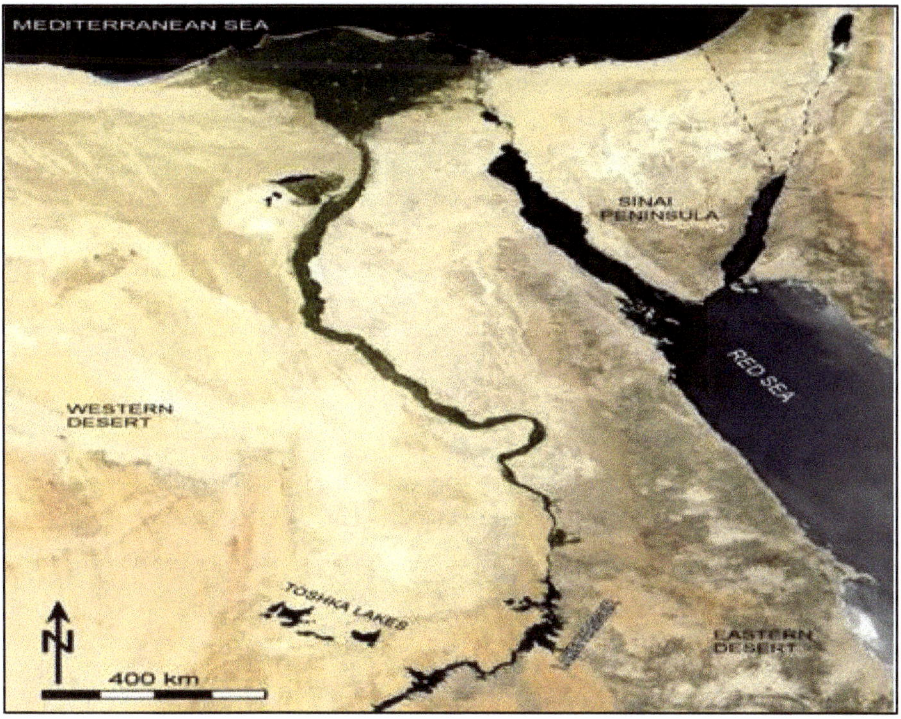
MEDITERRANEAN SEA
SINAI PENINSULA
RED SEA
WESTERN DESERT
TOSHKA LAKES
EASTERN DESERT
N
400 km

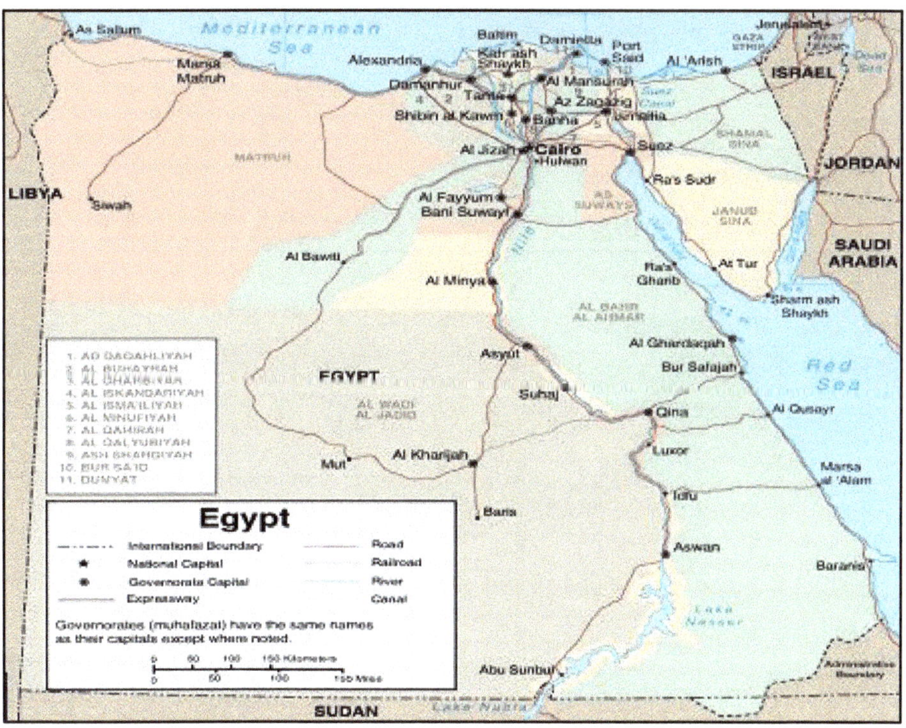
Mediterranean Sea
As Sallum
Marsa Matruh
Alexandria
Damanhur
Tanta
Shibin al Kawm
Banha
Al Mansurah
Az Zaqaziq
Port Said
Al 'Arish
ISRAEL
JORDAN
LIBYA
Siwah
MATRUH
Al Jizah
Cairo
Hulwan
Suez
Ra's Sudr
Al Fayyum
Bani Suwayf
SAUDI ARABIA
Al Bawiti
Al Minya
At Tur
Sharm ash Shaykh
Asyut
EGYPT
Suhaj
Qina
Luxor
Al Kharijah
Mut
Baris
Idfu
Aswan
Al Qusayr
Al Ghardaqah
Bur Safajah
Red Sea
Marsa al 'Alam
Baranis
Abu Sunbul
SUDAN
Egypt
International Boundary
National Capital
Governorate Capital
Expressway
Road
Railroad
River
Canal
Governorates (muhafazat) have the same names as their capitals except where noted.

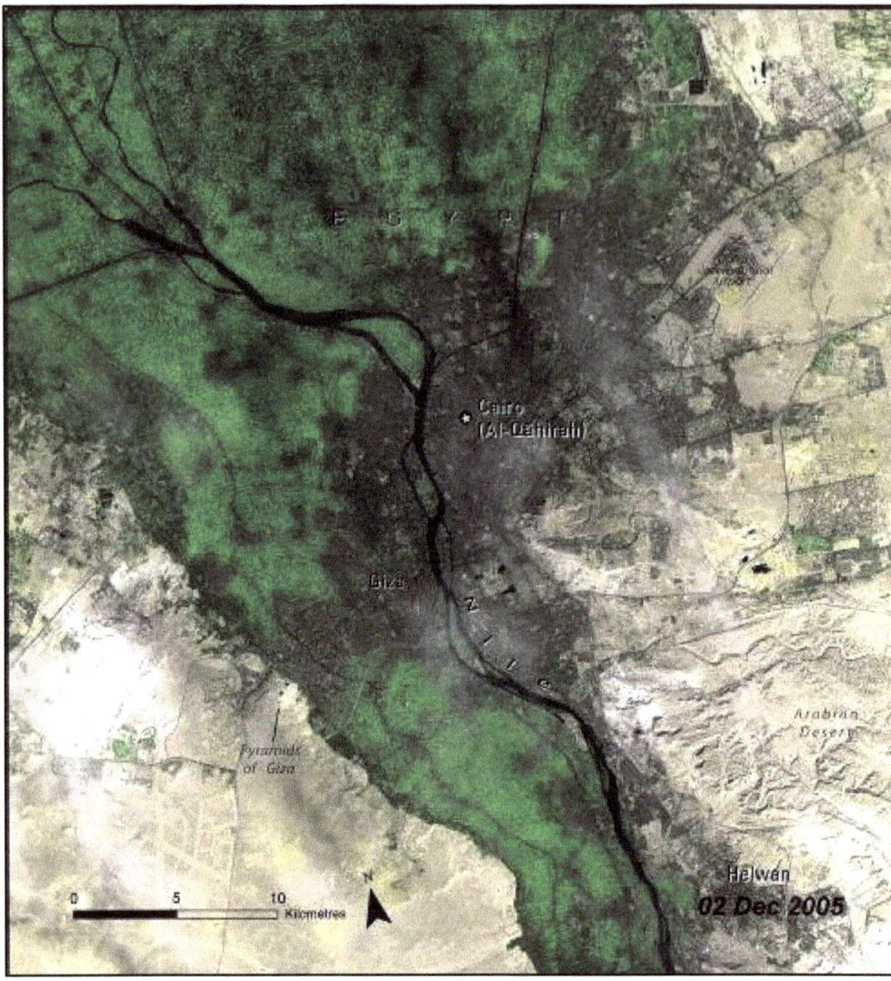
E G Y P T
Cairo
(Al-Qahirah)
Giza
N i l e
Arabian
Desert
Pyramids
of Giza
Helwan
02 Dec 2005
0
5
10
Kilometres
N

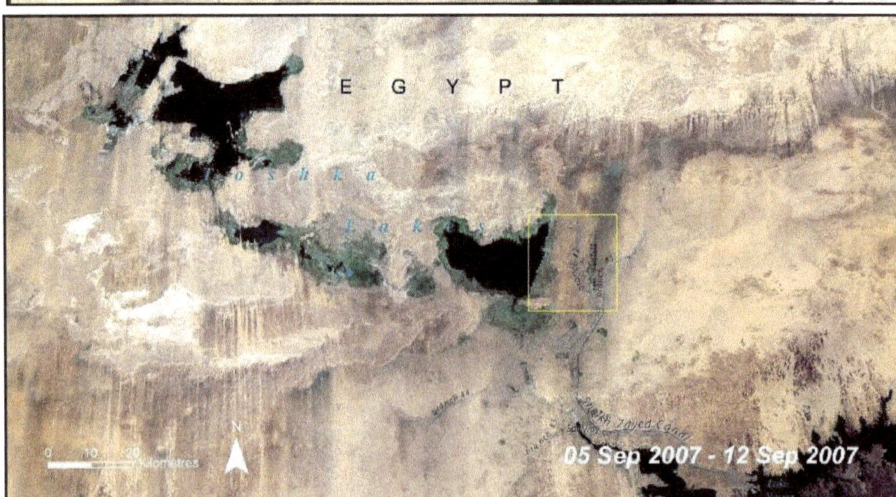
E G Y P T
T o s h k a
L a k e s
0
10
20
Kilometres
N
05 Sep 2007 - 12 Sep 2007

AGREEMENT

Let's all agree that UN nowadays is week and can't give orders or even hardly negotiate some countries for its critical activities that may affect life on our planet. Some examples for this are nuclear power plants which will lead to nuclear pumping. The wars in Iraq, Gulf, Lebanon and many other places have and are damaging the environment and affects will not be erased for decades to come. The big projects causing damage to the surrounding environment, such as the High dam in Egypt, the dam Iceland is building nowadays. Even the UN does not really agree about the black hall experiment in France (Europe), and many other projects which only few of us know but we cannot do anything about all these. A direct refusal to the Global needs, such as Global warming when USA refuses to sign "Kyoto agreement", while Europe on 12 Dec. 2008 just agreed to lower the CO_2 emissions by 20 per cent. The direct damage to the environment for some countries to get more rich, such as cutting trees in Russia or Europe and Brazil or South America in general. The excessive usage of fossil fuel although it is proven to be very harmful for our planet and many other damaging acts for monetary gains. Lack of technical information or technology, such as saving from oil spelling in sees or oceans, and other acts that we don't even have any records about. The new ECO technology, that can solve our problems and save our lives, which people everywhere invent and do not see the light because there is no encouragement in their countries or international encouragements and lack of financial resources. The invention of new ideas that can harm our environment in the long run, this invention people pay lot of money for realisation. Science and Scientists use their way of convincing the community that their way is the best way to follow, the adventure to try what we don't know and discover the unknown. While sometimes the overall view would not clear the minds of scientists. An example of this is the nuclear and atomic power which can be used for good as well as devastating purposes, depending on how and who is using it. Same is the case with cloning and playing with genomes, etc. I know that all UN departments cannot be powerful but at least the new EUN (Environmental United Nation) could prove successful. Therefore, the scientists of the world should come to an agreement in all countries, together with the laymen living with them on this planet and for the sake of our future generations, we should agree on having a powerful EUN, that we all will obey.

Suggested Environmental Information Elements

Water content, water quality, air quality, soil characteristics, land and marines' biodiversity, plant types, earthquake records, wind speed, hurricane expectations, wave characteristics, tsunami characteristics, radiation records, populations, human characteristics, human habits and traditions, human activities, types of production and expected pollutions, rain frequency, clouds, humidity, mines-Metals and diamonds, Oil-natural gases, electromagnetic fields, atmosphere layers' lengths, antiques, historical places, noise pollution, solid wastes and other parameters.

APPLICATIONS ON GOOGLE EARTH

The G-EIS Information can be used for programs such as Google Earth (but according to certain security level).

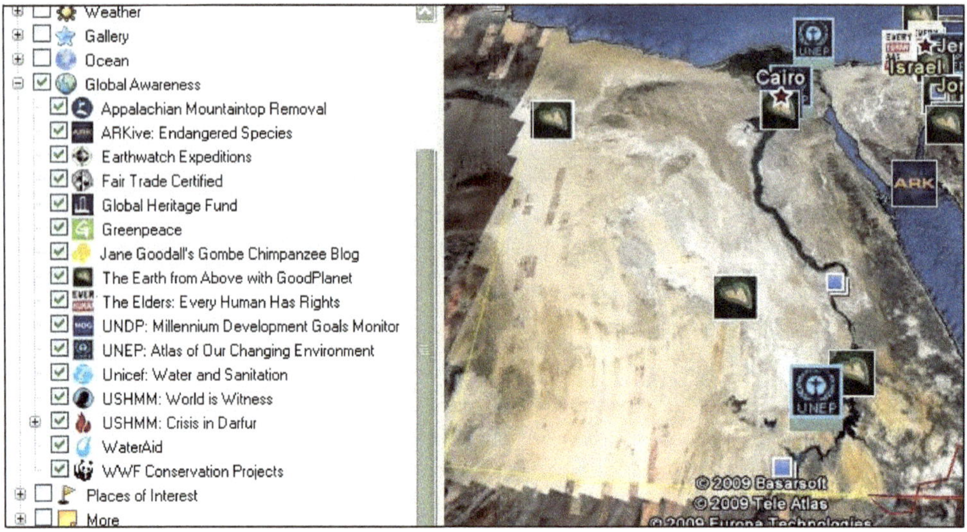

"Lack of data" There is all Environmental Data on Google Earth now!!!!

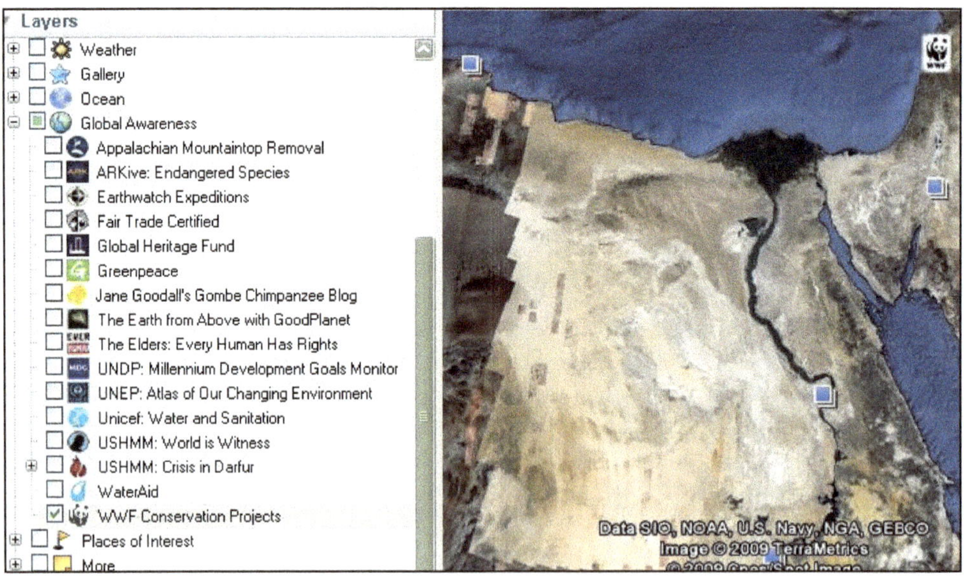

WWF Conservation Projects Are these projects for certain areas only!!!!?

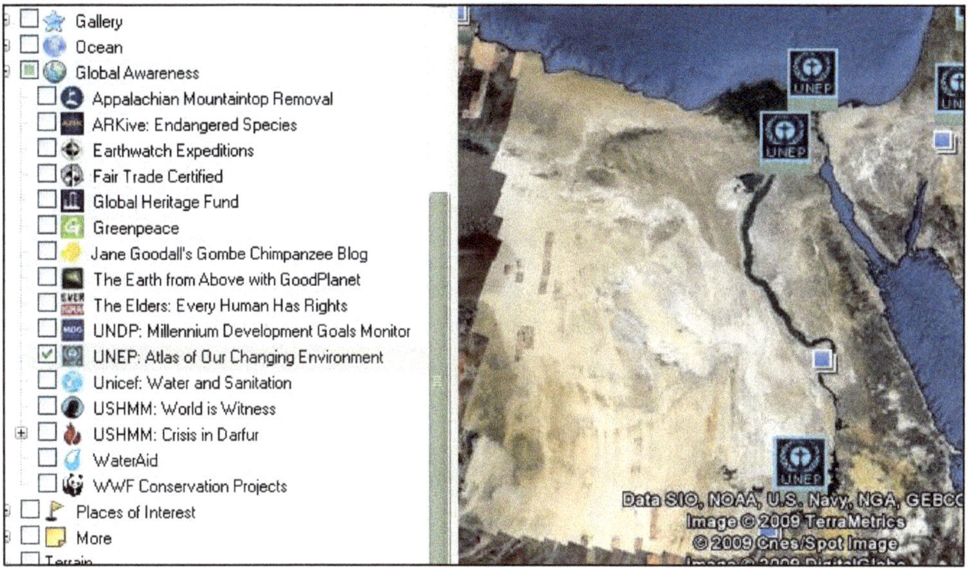

"No data for all the places".

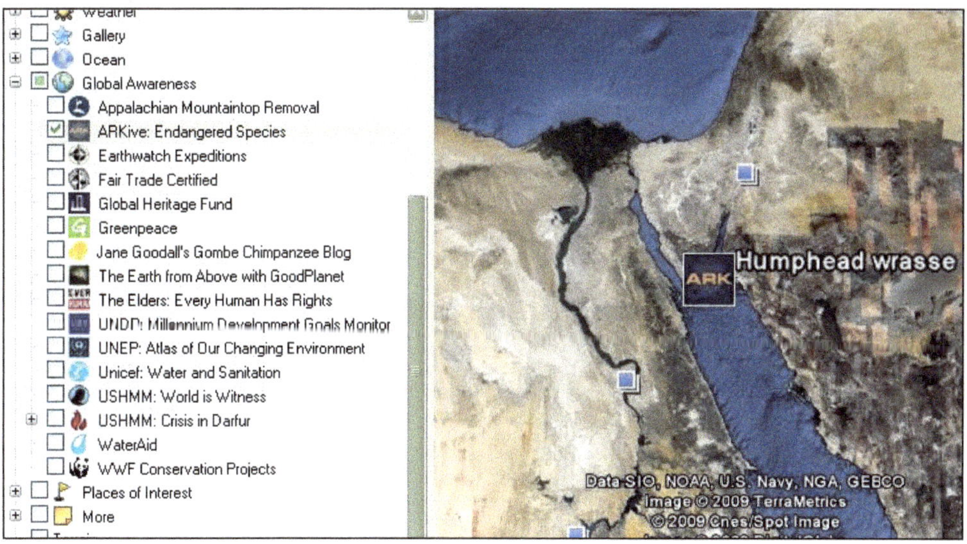

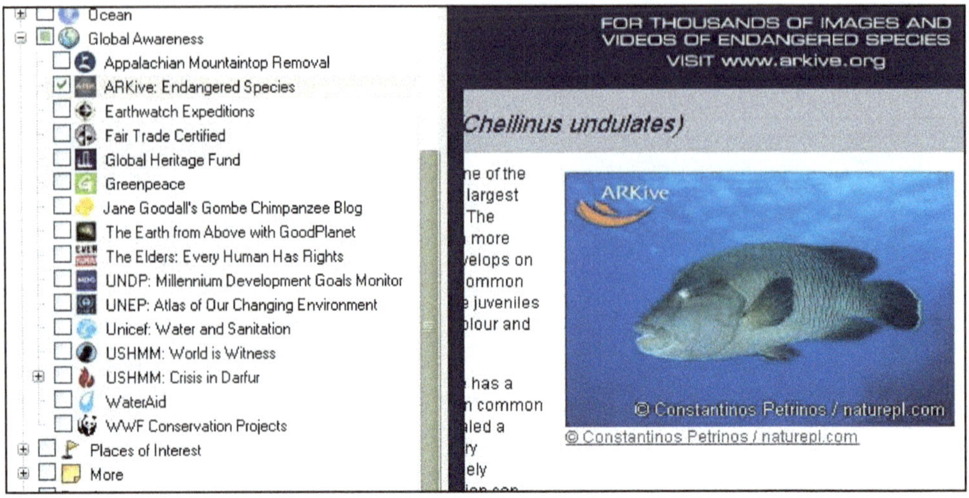

Do we have data bases for all the species on our planet?

CONCLUSION

It is clear that we are in a very huge lack of a Global Information System concerning our Environment. We need to establish a system that we (all countries) will obey for an ultimate control of the pollution that would kill us all or the generations following us; then we and our future generarions will have chance to live a better life.

SO, What do you think? Shall we start to take care of the only planet that we can live on so far towards no pollution for better evolution?

Chapter 13

Effective Criteria and Indices of Desertification and Effect of Agricultural Management on Land Degradation in Iran

Gh. Zehtabian[1] *and H. Khosravi*[2]

[1]*Professor,*

General Director of Academic Affairs, University of Tehran, Iran

[2]*Faculty of Natural Resources, University of Tehran, Karaj, Iran*

ABSTRACT

Desertification is a global problem in the world. Iran, because of susceptible ecosystem and geographic condition, encounters desertification. In this respect, a comprehensive plan entitled" criteria and indices of desertification" is prepared in Iran. Based on global experiences and review of international models, a new method is suggested which includes 9 criteria and 130 indices. Quantitative values assigned to the mentioned factors in geomorphologic work units of arid, semi-arid and humid semi-arid regions lead to establish classes (low, medium. high, very high) and intensity of desertification maps. It's possible to suggest and recommend relevant land use for each region. An application of the model in an arid region of Iran is provided as a case study. Environmental and human factors have different effects on desertification. Soil/water degradation (water/wind erosion), salinity and vegetation removal are of important consequences of desertification. Application of the proposed model in Kashan region, Isfahan province showed that water resources degradation is more important than other factors in desertification.

Beside the adherent potential of Iran for desertification, the population increase and their maltreatment with natural resources has intensified land degradation problems. One of the most significant factors to be verified is the agriculture issue. Therefore, it is essential to do more researches on management systems of agriculture. So, another research was carried out to determine wheather the region is suitable for agriculture or not? If suitable what kind of agriculture management should be applied? To achieve this objective a region in Taleghan, Tehran province was chosen. Some blocks were specified and then different kinds of agriculture were compared in these blocks with Completely Randomized Block Design. Soil sampling in each part of the agricultural land was done and soil factors in depth of 0-30 and 30-60 centimeter and two factors of EC and SAR were measured.

Keywords: *Desertification, Criteria and indices, Soil and water degradation, Land use, Agricultural systems, Agricultural activities, Iran.*

INTRODUCTION

Effective Criteria and Indices of Desertification in Iran

There are vast natural areas in Iran, which are susceptible and fragile ecosystems and have desert condition. According to the new definition of desert, except a narrow strip in north of Iran, other parts of the country encounter desertification problem. There are about 50 million hectares of desert lands in Iran and 120 million hectares are also desert-porene regions (Zehtabian *et al.*, 2008). Wind and water erosion are two major factors of land degradation in these regions, which deliver and deposit about 3 billion m^3 of eroded sediments into reservoirs, rivers and lowlands (Khosravi *et al.*, 2008). Moreover, soil erodibility, improper agriculture and irrigation systems as well as inappropriate application of manure and pesticides increase primary and secondary salinity, which are promoting factors of desertification (Zehtabian *et al.*, 2004 and Zehtabian *et al.*, 2009). It should also be noted that about 50 per cent of 15 million ha of agricultural lands in Iran have salinity problem (Zehtabian *et al.*, 2005 and Zehtabian *et al.*, 2009). As a result of indiscriminate exploitation due to methods applied, land use changes, lack of appropriate pattern of cultivated system (mono cultivated system and multi cultivated system), agricultural inputs, plowing on high slops, excessive consumption of chemical fertilizers, developing deep and semi-deep wells, non promotion of integrated pest, non development of agricultural systems, economic pressure on farmers, resources management and organization problems, unstable process of agricultural management and land degradation have been accelerated in Iran. Groundwater depletion due to increased exploitation and salt/freshwater imbalance has decreased water quality for different uses (Khosravi *et al.*, 2004). These factors in addition to misuse of resources have increased soil degradation but decreased the production and biomass (Zehtabian *et al.*, 2008).

Moreover, deforestation, landuse changes, small size of farmlands, poverty, population and finally illiteracy are among other parameters which affect desertification (Zehtabian *et al.*, 2003). Over last 30 years many activities have been carried out to establish sand dunes and prevent desertification which is threatening towns and villages, by developing biologic measures such as increase of vegetation cover (Raiei Emam 2003). Also, more than 2 million ha were afforested and another program is undertaken for 5 million ha (Khosravi 2003).

Since various factors are responsible for desertification, it is necessary to determine relevant criteria and indices for desired land uses as well as for desertification control. To achieve the above mentioned objective and based on review of international/ national desertification models in the literature, 9 criteria and 130 indices were selected which quantitative and weighted values of these factors determine desertification intensity of each region (Ahmadi *et al.*, 2003, Khosravi *et al.*, 2009).

Consequently, desertification intensity map was classified in four groups (low, medium. high, and very high). It is expected to calibrate the desertification related models for mapping desertification intensity map in arid, semi-arid, and humid semi-arid regions of Iran, which will ease decision making and recommendations for desertification control activities. It is also expected to share these expenses with other country for expand global knowledge about the phenomena.

This study was undertaken to determine if Taleghan region, Tehran province is suitable for agriculture or not? If suitable what kind of agricultural management should be applied prevalently? To achieve this objective a region in Taleghan, Tehran province was chosen. Some blocks were specified and then different kinds of agriculture were compared in these blocks with Completely Randomized Block Design.

Investigation on effects of agriculture management systems on land degradation was carried out in 4 stages: 1) collection of basic information, such as soil data, agriculture productions, area of agriculture lands and producing the slope, aspect by Digital Elevation Model. Changes of factors such as EC, SAR, were compared in two depth of top soil (0-30cm) and sub soil (30-60cm) according to Split Plot Scheme. Six different series of soil have been recognized and they were considered as blocks and, 6 treats with 6 replication including dry farming systems, mono cultivated and multi cultivated systems under irrigation, garden lands, eroded lands and rangelands as control treatments. eRange lands were considered as a part that weren't under influence of human management. 2) the area of sampling was transferred on topography map by using GIS, Arc/Info and IDRISI software. 3) the field studies were carried out and amount of each factor was measured. 4) data collection and measurement analysis with help of MSTATC software.

IMDPA[1] Model

Some international models of desertification such as FAO-UNEP (FAO/UNEP 1977, 1984, 2001 and Bayadgiev 1981), GLASOD (Jie 2002), TAXONOMY, LADA (FAO 2002), MEDALUS (European Commission 1999, Basso *et al.*, 1999, Kosmas *et al.*, 1999, Kosmas 2002, Ladisa 2002, Khosravi 2003) as well as national models including ICD[2] (Ekhtesasi and Mohajeri 1995) and MICD[3] (Ahmadi *et al.*, 2005) were reviewed in this research and 9 criteria were chosen based on previous experiences for desertification intensity mapping.

---

1. Iranian Model of Desertification Potential Assessment
2. Iranian Classification of Desertification
3. Modified Iranian Classification of Desertification

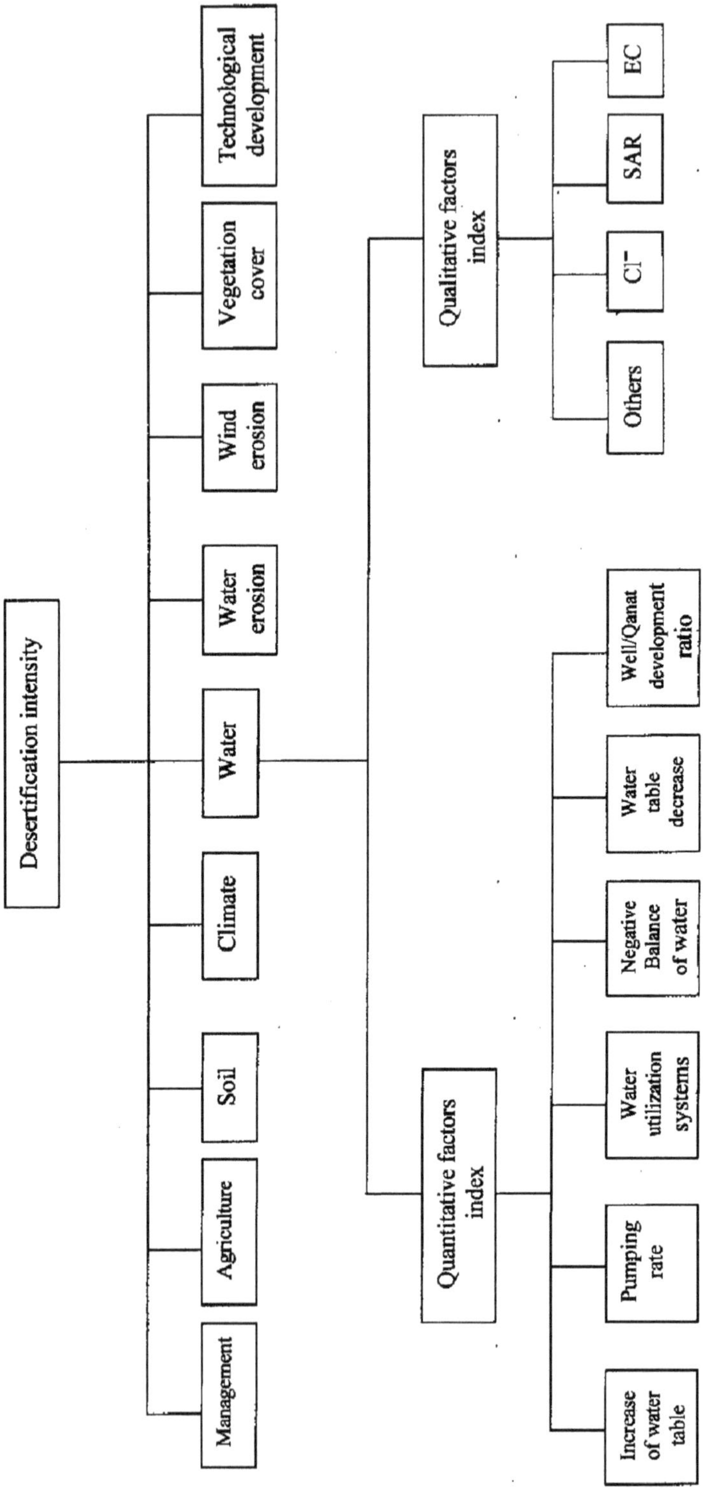

Figure 13.1: IMDPA Criteria as well as the Selected Indices for Water Resources Criteria

All of the criteria as well as the selected indices for water resources criteria are presented in Figure 13.1. A score ranging from 1 to 2 is assigned to each index based on weight of each factor. Finally the value of each criterion was obtained as geometric average of scores of single indices according to the formula:

$$Index - X = [(Layer - 1).(Layer - 2)...(Layer - n)]^{1/n}$$

where:

Index-X: A given criteria

Layer: Index of each criterion

N: Number of indices for each criterion

The role of water resources criteria in desertification is given by the following formula:

Quantitative factors index= (Increase of water table x pumping rate x water utilization systems x Negative balance of water x Decrease of water table x Well/ Qanat development ratio) $^{1/6}$

Qualitative factors index= (CL^{-} x SAR x EC) $^{1/3}$

Water criteria= (Quantitative factors index x Qualitative factors index) $^{1/2}$

The relevant indices determine the values of other criteria like the mentioned example for water criteria. Finally the desertification intensity will be a result of geometric average of 9 criteria as follows:

Desertification intensity = (Water x Soil Water erosion x Wind erosion x Climate x Vegetation cover x Agriculture x Technological development x Management) $^{1/9}$

The geometric average of relevant indices determines values related to other criteria, which ultimately will result in desertification intensity, and class in each geomorphologic work unites of different landuse (Agricultural, rangeland, forest, etc).

Table 13.1: Classification of Desertification Intensity

Low	*Medium*	*High*	*Very High*
1-1.22	1.23-1.32	1.33-1.41	1.42-2

In cases where necessary data are not easily available in the study area, it is better to determine three key indices. In Kashan region, three indices of water criteria including ES, SAR, and groundwater table depletion were selected which are shown in the Figure 13.2.

Assessment of Desertification Intensity Using IMDPA Model (Case Study: Kashan Region)

Kashan is located between 51° 27' E and 33° 51' N and has an arid climate with annual average precipitation of 133.5 mm. This region is included in the Isfahan province in the south of Tehran, with mean annual temperature of 18.8°C.

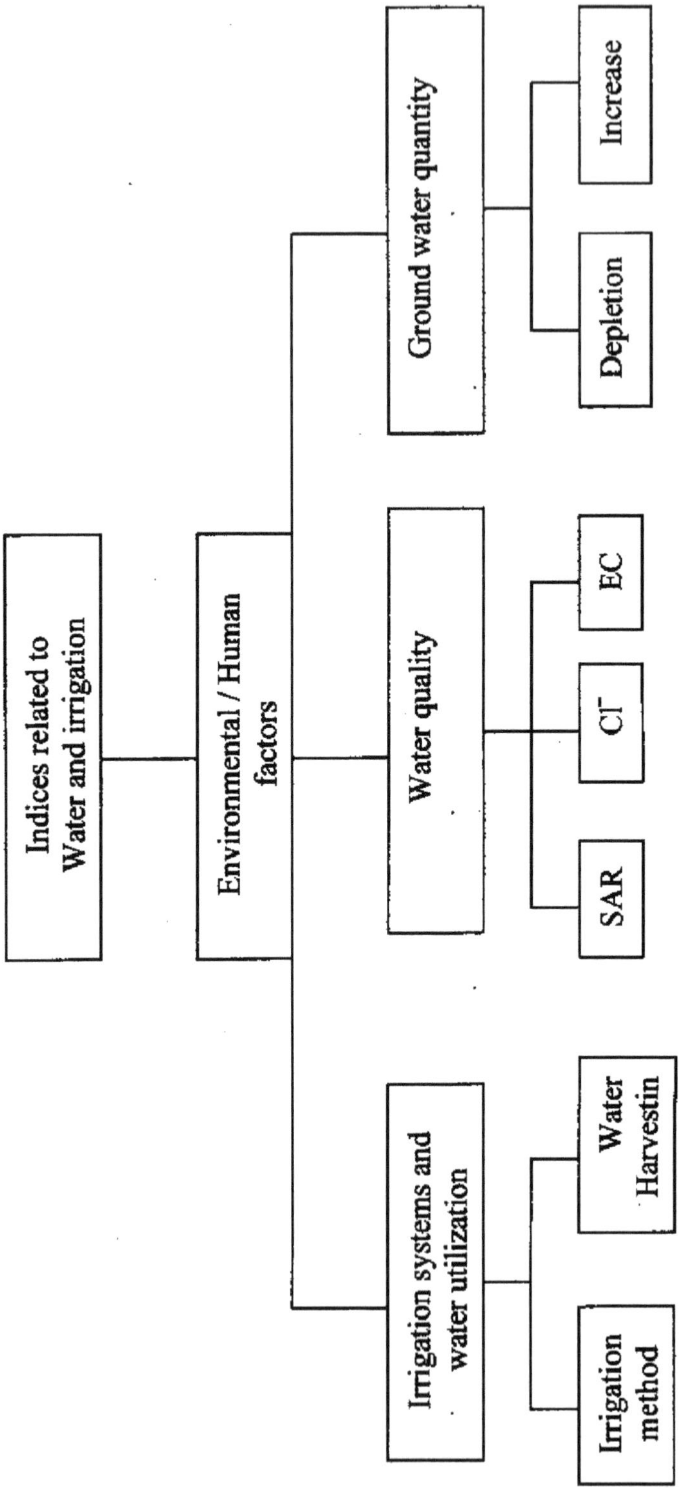

Figure 13.2: Flow Chart Showing Indices Related to Water and Irrigation

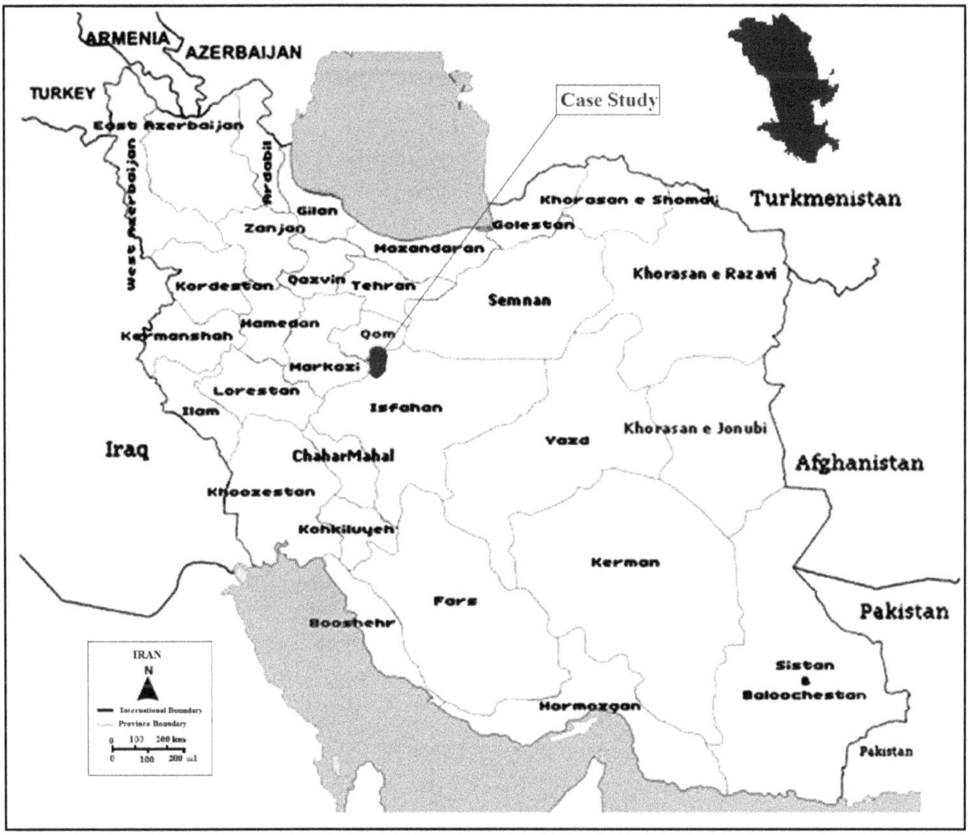

Figure 13.3: Location of the Study Area in Isfahan Province and Iran

The risk of desertification in Kashan region was evaluated on the basis of 7 regional criteria and indices including: Climate, water resources, water erosion, wind erosion, groundwater, soil, vegetation cover and management. Each criterion includes the following indices:

- *Climate Quality Index*: rainfall, aridity, aspect, aridity index
- *Water resources quality*: groundwater table decrease, increase of water table level, Cl^- concentration, water shortage for livestock and wildlife, EC.
- *Wind erosion quality index*: Soil degradation, salinity, alkalinity, wind erosion class, frequency of prevailing winds with speed >6 m/s, wind erosion faces, days with sandy winds with identical in year.
- *Vegetation cover index*: production, regrowth, percentage of plant composition, and percentage of vegetation cover.
- *Soil index*: Texture, depth, slope, drainage and type of geologic formations.
- *Water erosion index*: class of erosion, features of water erosion, damage due to floods and consequent land degradation.
- *Management index*: Policy making and management, type and intensity of land use.

To achieve these criteria, detailed studies of hydrological, geomorphological, erosion, soil and vegetation cover characteristics of the region were conducted and the mentioned indices were individually assessed (Table3 to Table10).

According to the factorial scaling technique, score-ranging form 1(good condition) to 2(deteriorated condition) is assigned to each index. Value "Zero" is assigned to the areas where the measure is not appropriate and/or those, which are not classified.

Finally the value of each criterion was obtained as geometric average of scores of single indices according to the formula:

$$Index - X = \left[(Layer - 1).(Layer - 2)...(Layer - n)\right]^{1/n}$$

where:

Index-X: A given criteria

Layer: Index of each criterion

N: number of indices for each criterion

Seven maps were prepared to study the indices qualities and their influence on desertification trend, and the final map indicating desertification status of this region was prepared using geometric mean of mentioned criteria based on the following formula:

Desertification intensity = $(\text{Groundwater} \times \text{Soil Water erosion} \times \text{Wind erosion} \times \text{soil} \times \text{Climate} \times \text{Vegetation cover} \times \text{Management})^{1/7}$

The risk of desertification in the map was classified into 5 subtypes according to Table 13.2.

Table 13.2: Classification of Desertification Intensity

Qualitative Classification	*Symbol*	*Range of Values*
Urban areas and pools	U	0
Low	I	1-1.22
Medium	II	1.23-1.32
High	III	1.33-1.41
Very high	IV	1.42-2

Table 13.3: Water Resourced Degradation Indices

Index Class	*Low*	*Medium*	*High*	*Very High (a)*	*Very High (b)*
Value	*1.00-1.19*	*1.20-1.39*	*1.40-1.59*	*1.60-1.79*	*1.80-2*
Groundwater table decrease (cm/year)	0-10	10-20	20-30	30-50	>50
Quality of water					
EC (µmhos/cm)	<250	250-750	750-2250	2250-5000	>5000
CL (Mgr/liter)	<250	250-500	500-1500	1500-3000	>3000

Contd...

Table 13.3–Contd...

Index Class	*Low*	*Medium*	*High*	*Very High (a)*	*Very High (b)*
Value	*1.00-1.19*	*1.20-1.39*	*1.40-1.59*	*1.60-1.79*	*1.80-2*
Water resources shortage for people and livestock	Water resources enough for livestock and wildlife requirement	Water resources reasonably enough for livestock and wildlife requirement	Water resources meet livestock and wildlife needs, proportionally	Water resources less than livestock and wildlife requirement	Water resources much less than livestock and wildlife requirement
The Index of Sustainable Development Committee of International Union	10>IUN	10<IUN<20	20<IUN<40	40<IUN<60	60<IUN

Table 13.4: Water Erosion Indices

Index Class	*Low*	*Medium*	*High*	*Very High*
Value	*1.00-1.24*	*1.25-1.49*	*1.50-1.74*	*1.75-2.00*
Class of water erosion	I, II	III	IV	V
WATER EROSION FACES	Lack of water erosion or negligible surface and till erosion, 0 to 5 per cent	Surface erosion associated with medium till erosion 5 per cent to 15 per cent	Intense till erosion, 15 per cent to 30 per cent	Different form of erosion, especially badland, up to 30 per cent of the whole area
Damage due to floods and consequent land degradation	No flooding during year with negligible damage	Flooding occurs only once a year with considerable damage	Flooding occurs 2-3 times a year with considerable damage	Evident flooding during year with considerable damage

Table 13.5: Wind Erosion Indices

Index Class	*Low*	*Medium*	*High*	*Very high*
Value	*1.00-1.24*	*1.25-1.49*	*1.50-1.74*	*1.75-2.00*
Wind erosion class (IRIFR)	I, II	III	IV	V
Wind erosion faces	No wind erosion and sediments accumulate under perennial shrubs like tiny projection	Small Nebka, and medium wind erosion	Inactive Nebka, small dunes with some Yardang and high wind erosion	Active Nebka, sand dunes without vegetation and some Yardang
Percentage of surface gravel sand	> 80 per cent	50 per cent–80 per cent	20 per cent–50 per cent	< 20 per cent

Contd...

Table 13.5–Contd...

Index	*Class*	*Low*	*Medium*	*High*	*Very high*
	Value	*1.00-1.24*	*1.25-1.49*	*1.50-1.74*	*1.75-2.00*
Soil degradation		Very resistant to strong impact or human traffic	Not easily collapsed as a result of human traffic	Powered and collapsed because of human traffic	Powered and collapsed quietly because of human traffic
Days with sandy winds with identical in year		1-2	3-4	5-10	10<
Frequency of prevailing winds with speed >6 m/s		<5	5-10	10-20	20<

Table 13.6: Vegetation Covers Indices

Index	*Class*	*Low*	*Medium*	*High*	*Very high*
	Value	*1.00-1.24*	*1.25-1.49*	*1.50-1.74*	*1.75-2.00*
Percentage of vegetation cover		>50	30-50	10-30	<10
Percentage of plant composition		Invasive species less than 5 per cent of plant composition and annual species less than 25 per cent	Invasive species 5-8 per cent of plant composition and annual species 25-50	Invasive species 25-50 per cent of plant composition and most of the vegetation cover from annual species	Invasive species more than 50 per cent of plant composition and most of the vegetation cover from annual species
Production		>85 per cent of the potential production	65-85 per cent of the potential production	25-65 per cent of the potential production	<25 per cent of the potential production
Regrowth		Regrowth is relatively fit for decreased species	Regrowth of decreased species is low	Regrowth of decreased species is low and negligible	Non-regrowth of decreased species

Table 13.7: Soil Degradation Indices

Index	*Effective Indices on Desertification*			
Texture	SC–SiC	L–SCL–SiCL- SiL	LS–SL	S–C>60 per cent
Value	*1.00-1.24*	*1.25-1.49*	*1.50-1.74*	*1.75-2.00*
Percentage of under ground gravel sand	>50 per cent	20 per cent–50 per cent	< 20 per cent	
Value	*1.00-1.33*	*1.34-1.66*	*1.67-2.00*	
Depth (cm)	>75cm	30-75	15-30	<15cm
Value	*1.00-1.24*	*1.25-1.49*	*1.50-1.74*	*1.75-2.00*

Contd...

Table 13.7–Contd...

Index	*Effective Indices on Desertification*			
Drainage	Fe, Mg in depth of >100cm, water penetration is very fast–soil is moisture during the growth period	Fe, Mg in depth of 30-100cm, water penetration is slowly–soil is relatively moisture		Fe, Mg in depth of <30cm,water penetration is very slowly–soil is moisture for along time
Value	*1.00-1.33*	*1.34-1.66*		*1.67-2.00*
Slope	<6 per cent	6-18	18-35	35 per cent <
Value	*1.00-1.24*	*1.25-1.49*	*1.50-1.74*	*1.75-2.00*
Type of geologic formations	Sedimentary and soil unit, constant in front of destructive factors	Sedimentary and soil unit, relatively constant in front of destructive factors	Sedimentary and soil unit, relatively sensitive in front of destructive factors	Sedimentary and soil unit, really sensitive in front of destructive factors
Value	*1.00-1.24*	*1.25-1.49*	*1.50-1.74*	*1.75-2.00*
Salinity and alkalinity				
EC	<8	8-16	16-32	>32
SAR	> 13	13-30	30-70	>70
ESP	<15	15-30	30-50	>50
Value	*1.00-1.24*	*1.25-1.49*	*1.50-1.74*	*1.75-2.00*

Table 13.8: Policy and Management Indices

Index	*Class*	*Low*	*Medium*	*High*	*Very high*
	Value	*1.00-1.24*	*1.25-1.49*	*1.50-1.74*	*1.75-2.00*
Necessary operation		More than 75 per cent of the area under protection	50-75 per cent of area under protection	25-50 per cent of area under protection	Less than 25 per cent of area under protection
Desertification method		Desertification will occur automatically, following using proper management	Desertification is possible with biological operations and little cost	Desertification is possible with mechanic, biological operation	Very difficult desertification if not impossible, without economic ecological efficiency
Carried out operation		No need to desertification	So far, desertification operations were quiet effective	Relative successful operations were carried out	Operation wasn't successful yet
Paying attention to people cooperation		Good operation between people and engineer	Local people no attention to engineer	To proceed project without paying attention to people	Contrast between local people and engineers

Table 13.9: Management Indices

Original Factors	*Secondary Factors*	*Index* / *Class*	*Low*	*Medium*	*High*	*Very high*
		Value *1.00-1.24*	*1.25-1.49*	*1.50-1.74*	*1.75-2.00*	
Type of land use and intensity of land use	Agricultural product	Land use change	I→→II	III→→II	I, II→→IV	I→→V
		Pay attention to agriculture basics (cultivation, plough, rotation, fallow)	Good attention to agriculture basics and perennial farming and gardening	Proper attention to basics and perennial farming	Improper agriculture basics and unsuitable rotation with fallows less than 6 months	There is no respect to agriculture fundamentals, at all. Such as no rotation, fallow more than 6 months and over cultivation
		Improper use of equipment and chemical fertilizers	Using modern equipment with high efficiency, property. Not using fertilizers and insecticides	Using modern equipment with high efficiency, properly using fertilizers and pesticides	Using equipment with high efficiency very low output, nearly improperly. Using pesticides nearly improperly	Using quiet depreciated equipment with very low output in a wrong way, using pesticides improperly
		Irrigation method	Proper irrigation method	Relative proper irrigation method	Relative proper irrigation method	Wrong and improper irrigation method
		Changing well to subterranean canal	1-3	3-6	6-9	>12

Contd...

Table 13.9–Contd...

Original Factors	*Secondary Factors*	*Index*	*Class*	*Low*	*Medium*	*High*	*Very high*
		Value	*1.00-1.24*	*1.25-1.49*	*1.50-1.74*	*1.75-2.00*	
	Rangeland and shrub land	Overgrazing (Animal unit to potential grazing)		1	1-2	2-3	>3
		Picking shrub		< 10 per cent	10–25 per cent	25–50 per cent	> 50 per cent
		The amount of low efficiency lands which have been left and grassland plough		Low efficiency land not left and there is very little amount of plough (less than 20 per cent)	Land has been left and there is medium amount of plough (20 to 50 per cent)	Land has been left and there is medium amount of plough (50 to 75 per cent)	Low efficiency lands are left and grassland plough is very high (up to 75 per cent)
		Decrease of dominant crop production to potential crop		Low and negligible 85-100	Medium 65-85	High 25-65	Very high <25
	Rural and urban area	Improper rural and urban development		Rural and urban development was nearly fundamentally compatible to the area capacity	Rural and urban development was nearly fundamentally compatible to the area capacity	Rural and urban development wasn't fundamentally compatible to the area capacity	Unlimited rural and urban development, not proportionate to area capacity

Table 13.10: Climate Indices

Index	*Effective Indices on Desertification*				
Rainfall	>650 mm		280-650 mm	<280mm	
Value	*1.00-1.33*		*1.34-1.66*	*1.67-2.00*	
Aspect	E–NE–N–NW		W–SW- S–SE		
Value	*1.00-1.49*		*1.50-2.00*		
Aridity index (P/ETP)	>0.65	0.45-0.65	0.20-0.45	0.05-0.20	<0.05
Value	*1.00-1.19*	*1.20-1.39*	*1.40-1.59*	*1.60-1.79*	*1.79-2.00*

- ✰ Indices Analysis of Desertification Criteria
- ✰ Water Resources Desertification Criterion

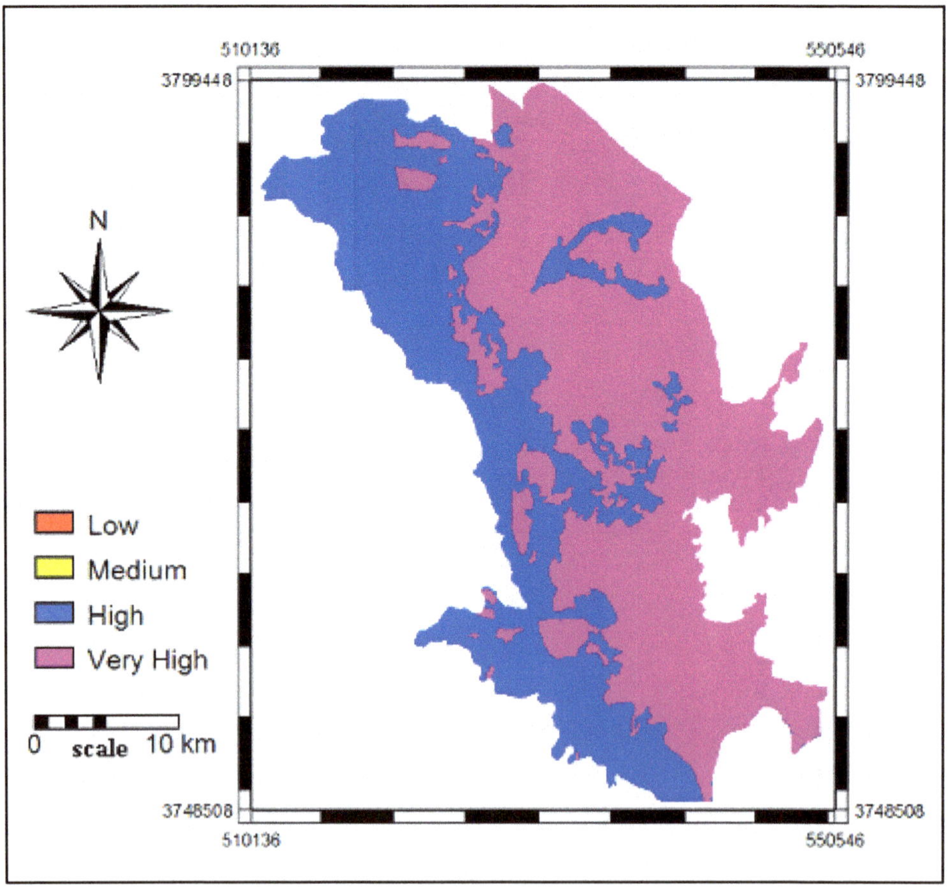

Figure 13.4: Quality of Water Resources Degradation Criterion in Kashan

- ☆ 40.3 per cent of total area is classified as highly deteriorated water resources. This class is more observed in NW, west, SW of Kashan.
- ☆ 59.7 per cent of total area is classified as very highly deteriorated water resources. This class is more observed in NW, E, and SE of Kashan.

After studying mean value of factors involved in water resources deterioration, it's indicated that water crisis index is the most effective factor in increasing desertification intensity of the studied region. Table 13.11 presents all indices influencing water resources deterioration.

Table 13.11: Geometric Average of the Quantitative Values of Water Resources Degradation Index

Order	*Index*	*Value*	*Class*
1	Groundwater table decrease	1.792	Very high
2	EC (µmhos/cm)	1.790	Very high
3	Cl (Mg/l)	1.571	High
4	Water resources shortage	1.647	High
5	The Index of sustainable development committee	1.83	Very high

Water Erosion Criterion

Water erosion map of this region has two classes of low and medium desertification.

After studying mean value of three indices influencing water erosion (Table 13.12).

It's evident that water faces is the most effective factor in desertification among those three indices.

Table 13.12: Geometric Average of the Quantitative Values of Water Erosion Index

Order	*Index*	*Value*	*Class*
1	Class of water erosion	1.087	Low
2	Water erosion class	1.123	Low
3	Damage due to floods and consequent land degradation	1.077	Low

Wind Erosion Criterion

Wind erosion map shows that wind erosion criterion is divided into three classes: low, medium, and high intensity.

After, studying mean value of the indices of wind erosion, it's concluded that surface gravel sand percentage with the value of 1.522 is the most effective factor in increasing desertification intensity in this region (Table 13.13).

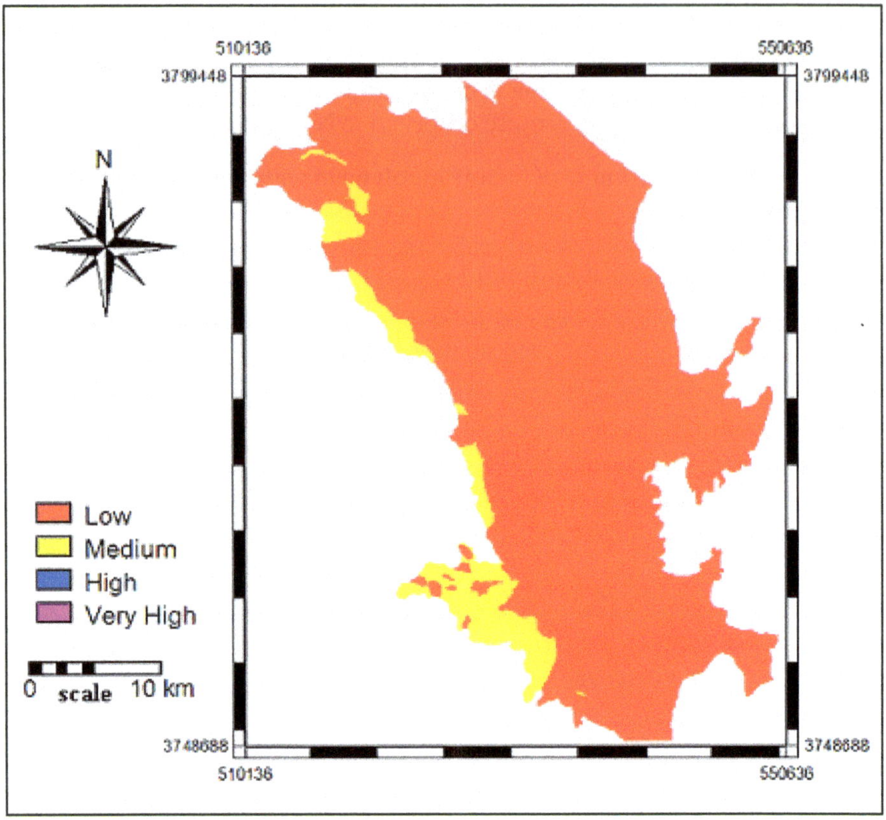

Figure 13.5: Quality of Water Erosion Criteria in Kashan Region, Iran

Table 13.13: Geometric Average of the Quantitative Values of Wind Erosion Index

Order	*Index*	*Value*	*Class*
1	Wind erosion class (IRIFR)	1.378	Medium
2	Wind erosion faces	1.250	Medium
3	Percentage of surface gravel sand	1.522	High
4	Soil degradation	1.307	Medium
5	Days with sandy winds with identical in year	1.49	Medium
6	Frequency of prevailing winds with speed >6 m/s	1.18	Low

Vegetation Cover Criterion

Vegetation cover map represents that vegetation criterion is classified as low, high, and very high intensity desertification.

Results obtained from quantitative mean value of vegetation indices shows that vegetation composition percentage among four studied indices are the most effective one (1.491) (Table 14).

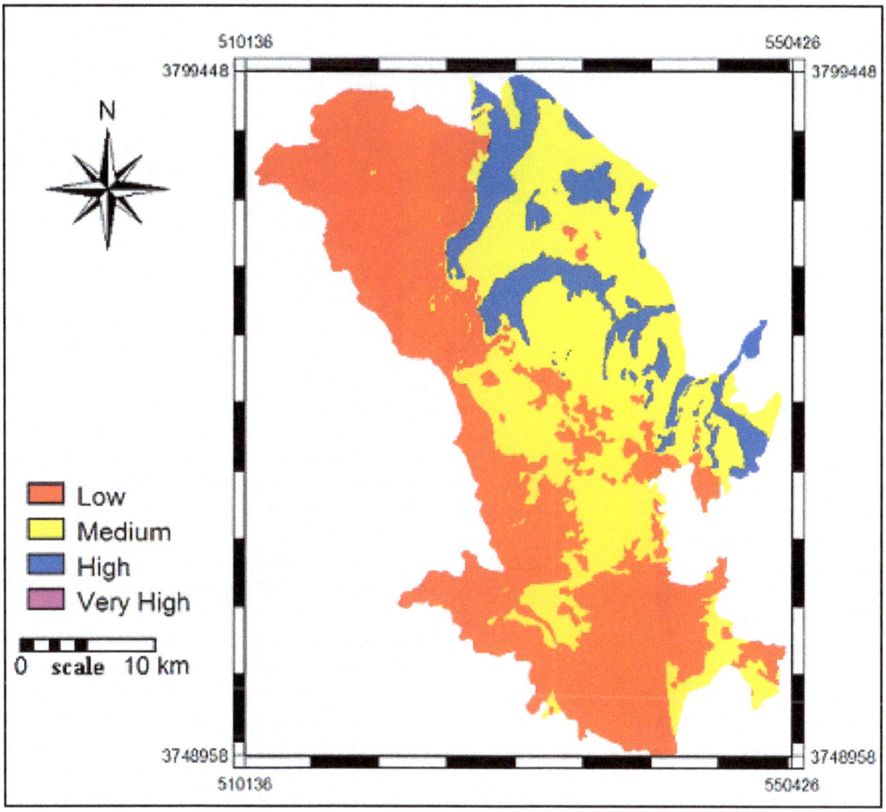

Figure 13.6: Quality of Wind Erosion Criterion in Kashan Region

Table 13.14: Geometric Average of the Quantitative Values of Vegetation Cover Index

Order	*Index*	*Value*	*Class*
1	Percentage of vegetation cover	1.422	Medium
2	Percentage of plant composition	1.491	Medium
3	Production	1.218	Medium
4	Regrowth	1.398	Medium

Soil Criterion

According to an evaluation of chosen indices, as we proceed from the north towards the source and from the west towards the east, soil deterioration becomes more heavy. Soil deterioration map is divided into classes: low and medium.

Using analysis applied to the mean values of effective indices in soil deterioration, percentage is the most important factor in desertification (Table 13.15).

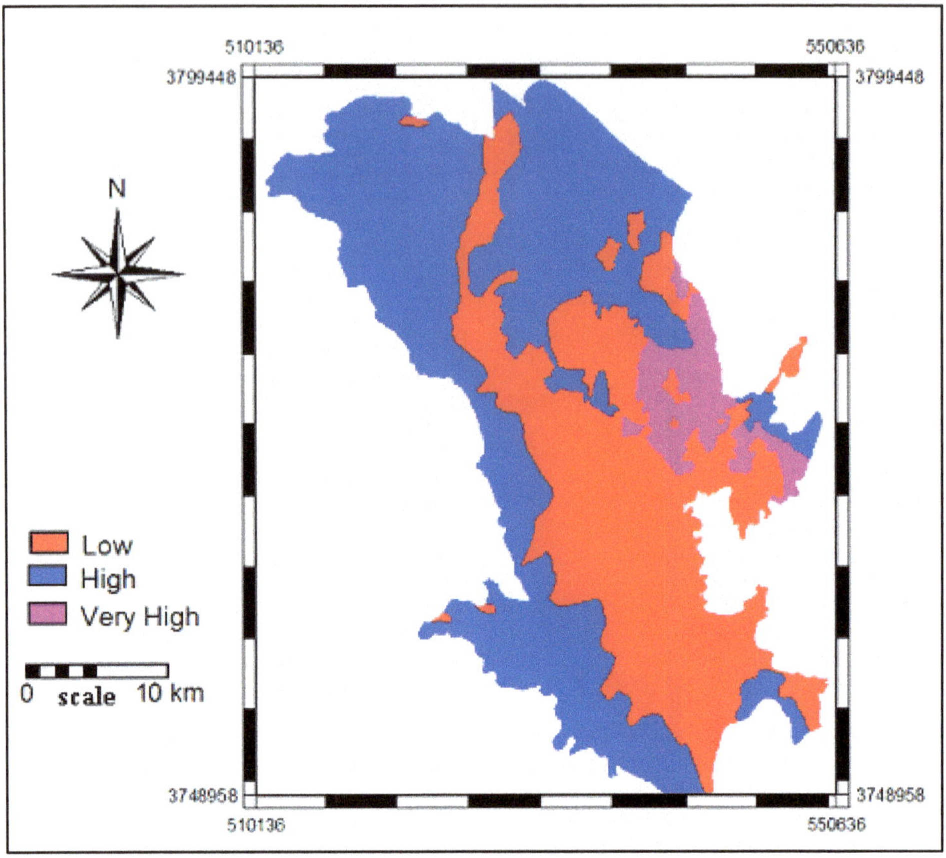

Figure 13.7: Quality of Vegetation Cover Criterion in Kashan Region

Table 13.15: Geometric Average of the Quantitative Values of Soil Degradation Index

Order	*Index*	*Value*	*Class*
1	Texture	1.151	Low
2	Percentage of under ground gravel sand	1.164	Low
3	Depth (cm)	1.007	Low
4	Drainage	1.261	Medium
5	Slope	1.5	High
6	Type of geologic formations	1.481	Medium
7	Salinity	1.266	Medium
8	Alkalinity	1.204	Low

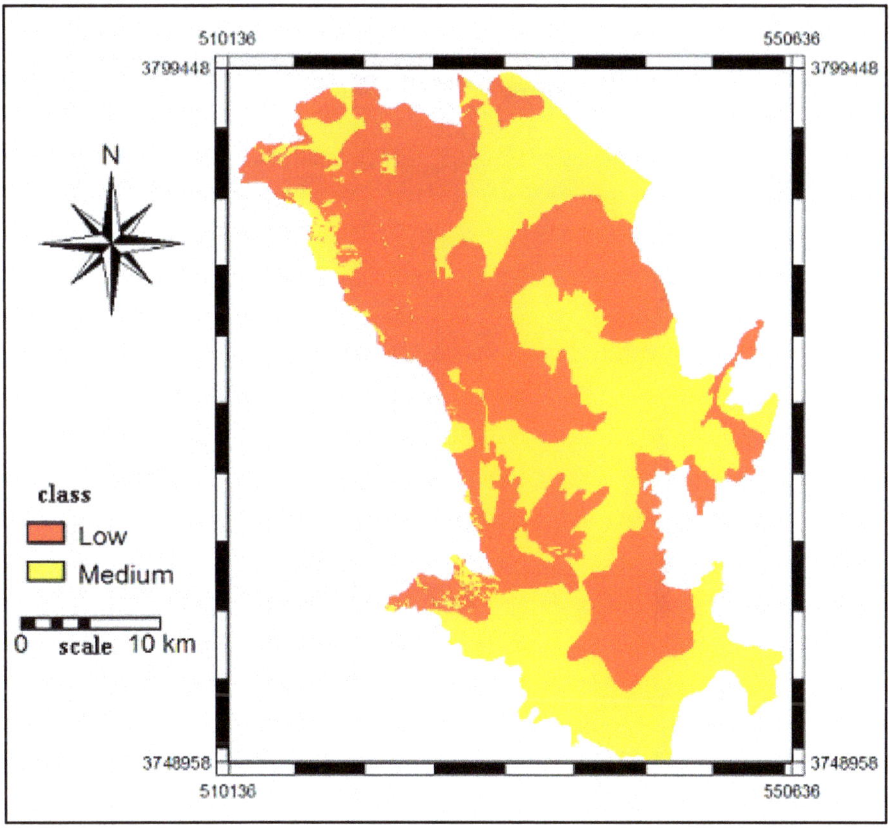

Figure 13.8: Quality of Soil Criterion in Kashan

Climate Criterion

Studies on the climate criterion with respect to chosen indices resulted in two classes including: high and medium desertification intensity.

After estimating mean value of three indices influencing climate, it was concluded that low rainfall had the main role in desertification of Kashan. Table 13.16 represents all indices influencing climatic criterion.

Table 13.16: Geometric Average of the Quantitative Values of Climate Index

Order	*Index*	*Value*	*Class*
1	Rainfall	1.851	Very high
2	Aspect	1.059	Low
3	Aridity index (P/ETP)	1.734	High

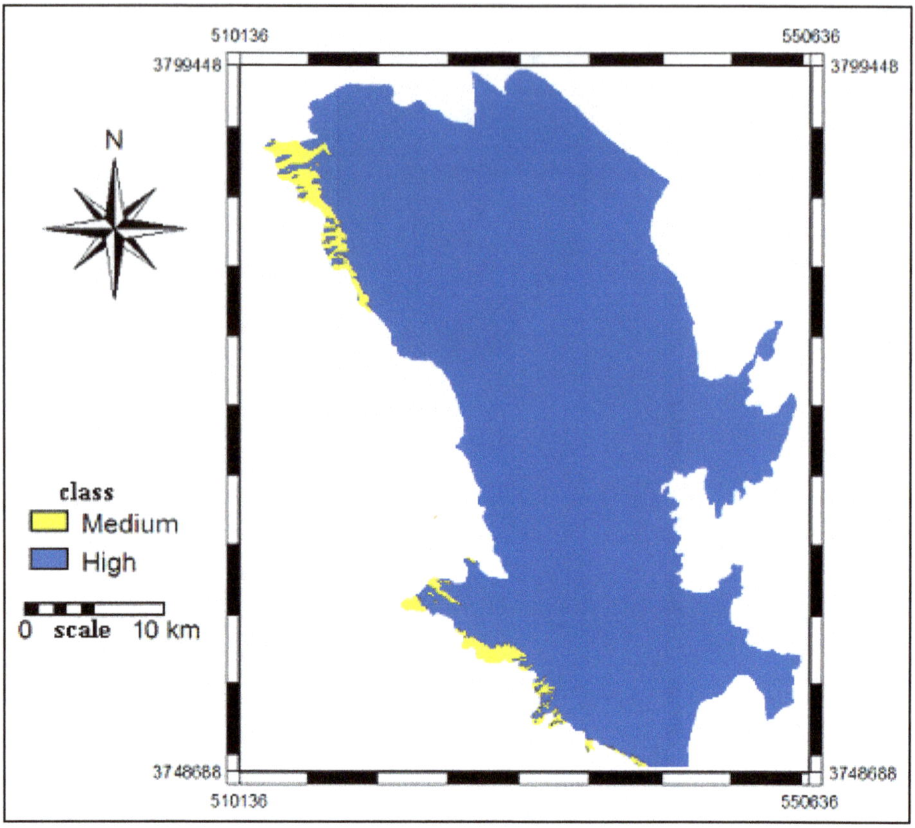

Figure 13.9: Quality of Climate Criterion in Kashan

Management Criterion

The Figure 13.10 shows quality of management criterion intensity for grassland and agricultural use, respectively studying mean value of land use types. It represents that in relation to the former one, anthropogenic factor of cultivation (plough) with numerical value of 1.771 is the most effective factor for desertification intensity and about the latter one, improper irrigation method with numerical value of 1.79 in the main factor of desertification in Kashan area (Tables 13.17 and 13.18).

Table 13.17: Geometric Average of the Quantitative Values of Grassland and Shrub Land Index

Order	*Index*	*Value*	*Class*
1	Overgrazing	1.208	Low
2	Picking shrub	1.181	Low
3	The amount of low efficiency lands which have been left and grass land plough	1.771	Very high
4	Decrease of dominant crop production to potential crop	1.393	Medium

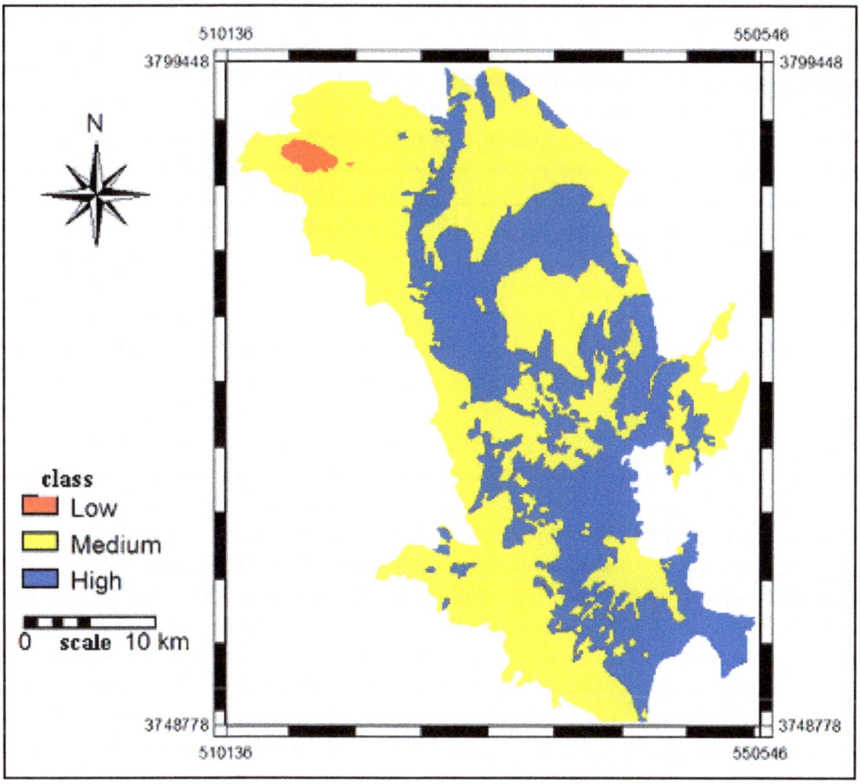

Figure 13.10: Quality of Management Criterion in Kashan Area

Table 13.18: Geometric Average of the Quantitative Values of Agricultural Products Index

Order	*Index*	*Value*	*Class*
1	Land use change	1.114	Low
2	Pay attention to agriculture basics	1.294	Medium
3	Improper use of equipment and chemical fertilizers	1.163	Medium
4	Irrigation method	1.79	Very high
5	Changing well to subterranean canal	1.668	High

In the end, numerical mean value of evaluation index of management and policy-making after field evaluation was equal to 1.297, representing that with regard to management, this region (Kashan)can be classified as medium.

Management criterion map of Kashan shows that with respect to management this region is divided into three classes.

Analysis of Criteria of Desertification in Kashan

Analysis of desertification indices in Kashan region showed that water resources degradation criteria is a major problem in the study area with a geometric average of

1.74 which shows very high class while climate quantity index with a weighted average of 1.56 stands in second order of desertification factor (Figure 13.11).

The results of geometric average of the studied criteria are shown in Table 13.19.

Table 13.19: Geometric Average of the Quantitative Values of Indices

Order	*Index*	*Value*	*Class of Desertification*
1	Water resources degradation	1.74	Very high
2	Climate	1.56	Very high
3	Management	1.41	High
4	Vegetation cover	1.37	High
5	Wind erosion	1.33	High
6	Soil	1.25	Medium
7	Water erosion	1.1	Low

This study based on the geometric average of quantitative indices in Kashan showed that among 43 indices and sub-indices, three indices including depth of rainfall water resources degradation and groundwater table depletion have the most significant effect with the values of 1.851, 1.83 and 1.79 respectively. Also three other

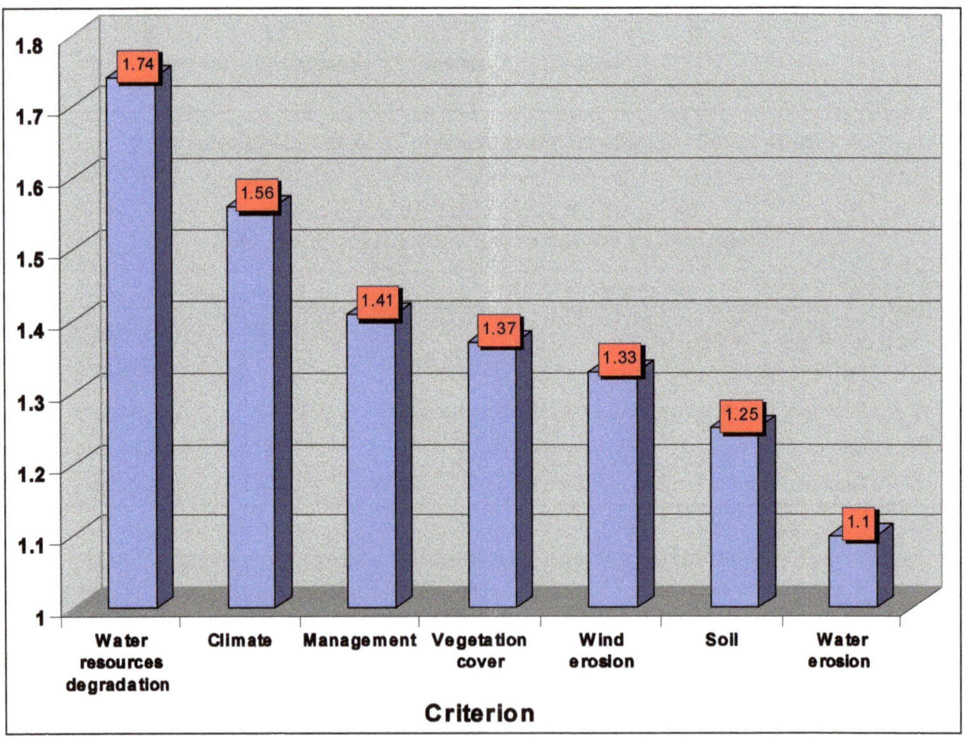

Figure 13.11: Analysis of Weighted Average of Desertification Criteria in Kashan Region

indices including drainage, aspect and damage of flooding had the lowest effect with the values of 1.007, 1.059 and 1.077 respectively. Figure 13.12 shows the sensitivity of the study area to desertification according to the mentioned algorithm.

This study showed that desertification happens in the region with different intensities. Table 13.20 represents the frequency distribution of different classes of desertification in the study area.

Table 13.20: Frequency Distribution of Different Classes of Desertification in the Study Area

Qualitative Classification	*Symbol*	*Range of Values*	*Area (Km²)*	*Relative Area (per cent)*
Urban areas and pools	U	0	8.91	0.970
Low	I	1-1.22	0	0
Medium	II	1.23-1.32	298.67	32.68
High	III	1.33-1.41	360.04	40.07
Very high	IV	1.42-2	240.21	26.28

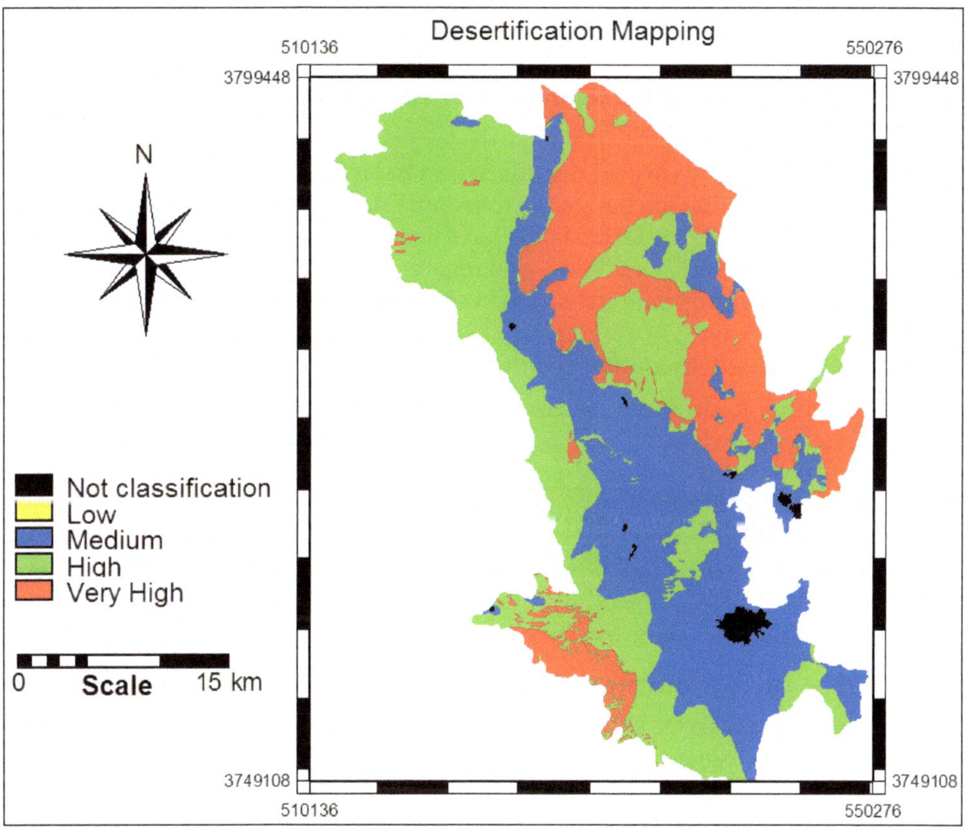

Figure 13.12: Characterization of the Areas Sensitive to Desertification in Kashan

This research showed that geometric average of quantitative factor of desertification intensity (Current condition of desertification) for the region is Ds= 1.372 which shows high levels of desertification.

Effects of Agricultural Management on Land Degradation

Iran is one of the countries located in the desert belt of the world and this has caused major problems. Beside the adherent potential of Iran for desertification, the population increase and their maltreatment with natural resources has intensified these problems (Zehtabian *et al.*, 2005). One of the most significant factors to be verified is the agricultural activity (Zehtabian *et al.*, 2009). Therefor it is essential to do more researches on agricultural lands in different parts of Iran, especially in semi-arid regions where the desert expansion and agricultural land degradation is observed more.

Population growth and food supply is the main problem in Iran. 13 per cent of Iran is arid and 61 per cent is semiarid with a precipitation of less than 100 and between 100-350 mm, respectively (Zehtabian *et al.*, 2005). One of the most significant factors to be verified is the agricultural activities. 18 million hectares of the land is under tillage, 47 per cent of cultivated lands are irrigated rest being dry farming (Zehtabian *et al.*, 2009). But when we use these lands for food production, we should use soil properly. Unfortunately, because of unsuitable land and water management systems the soil loss is more than 15 ton/ha/year. In this case, agricultural systems are very important and we can control the derivative effects by correct management practices (Zehtabian *et al.*, 2009).

There are large areas in arid regions of Iran having low rainfall affected by land degradation caused due to change in environmental condition and anthropogenic activities. Consequently the fertile lands have been changed in to degraded and bare lands. In many parts of arid and semi-arid regions of Iran unwise management of the land as well as inappropriate irrigation systems have seriously led to salinity and soil degradation. A wise management of soil and water resources is inevitable to avoid consequences degradation (Zehtabian *et al.*, 2004, 2005, 2009).

It is essential to do more researches on management systems of agriculture. Some researchers, such as Nahal (1981), Jamshidi (1999) have studied agricultural practices and concluded that incorrect management can affect our land degradation. Nahal (1981) studied the relationship between management protection and expansion of agricultural resources in Syria. He concluded that wrong irrigation practices effect land degradation. Jamshidi (1999) studied the effects of agriculture on desertification and considered climate, soil, and water factors. He concluded that deficit of surface water, salinization of undergroundwater and soil, effect desertification process.

Experience of Agricultural Management System in Taleghan

The study area is about 132500 ha and located in Taleghan Basin, Tehran province (Figure 13.13). It is located in 36° 10´ 58½ to 36° 16´ 58½ N and 50° 34´ 30½ to 50° 44´ 18½ E. The study area has a semi-arid climate with annual average precipitation of 500 mm and mean annual temperature is 18.8°C.

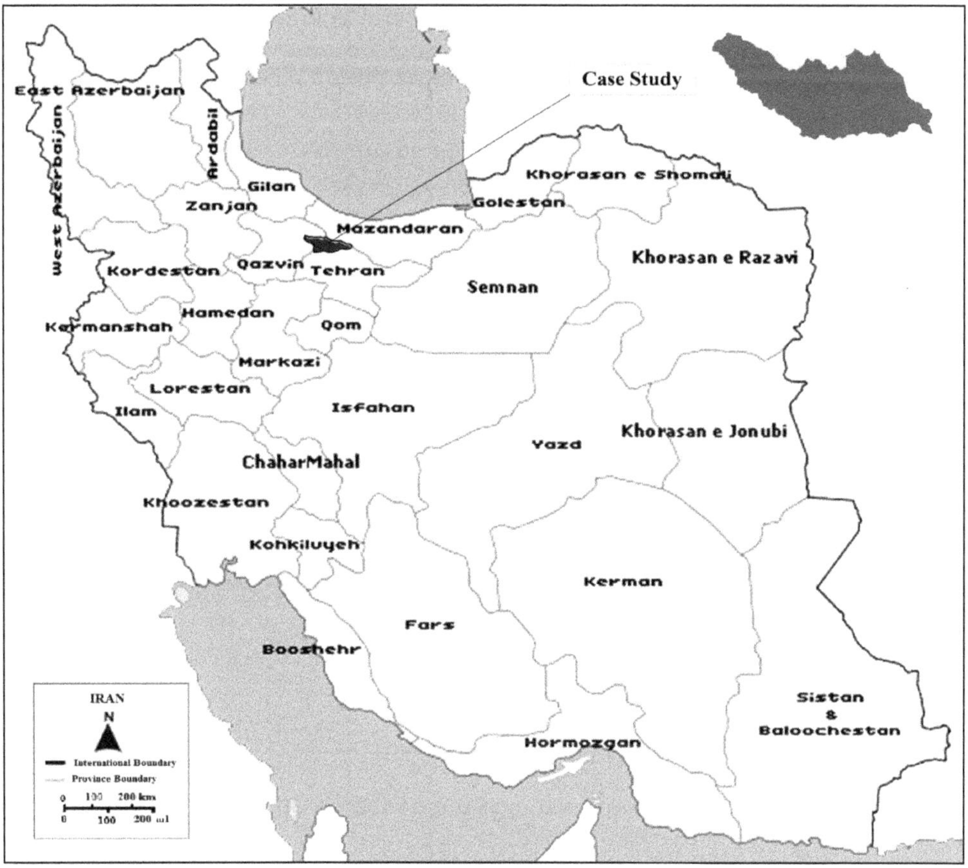

Figure 13.13: Map of Iran Showing the Location of the Study Area

Results of Case Study

Results can be presented in 3 stages. At the first stage the results obtained from testing of data have shown that all data show a normal distribution. A variance analysis was done (Tables 13.22 and 13.23).

Table 13.21: General Information of Factors in Top and Sub Soil

Factor	*Mean*		*Variance*		*Standard Deviation*	
	Top	*Sub*	*Top*	*Sub*	*Top*	*Sub*
EC	0.2583	0.2542	0.000060	0.000084	0.0078	0.00780
SAR	0.1750	0.2492	0.000161	0.000161	0.01270	0.01270

Table 13.22: Variance Analysis of EC According to Split Plot Scheme

Row	*Surface of Error*	*Df*	*Sss*	*Mss*	*F*	*Probability*
1	Treatment	3	0.00292	0.000764	0.48	0.702
2	Rep	5	0.136875	0.027375	17.14	0
3	Error	15	0.023958	0.001597	1.10	0.422
4	Depth Sampling	1	0.000208	0.000208	0.14	0.710
5	Depth × treatment	5	0.008542	0.001708	1.17	0.361
6	Sub Error Factor	18	0.02625	0.001458		
7	Total	47	0.198125			

Table 23: Variance analysis of SAR according to Split Plot Scheme

Row	*Surface of Error*	*Df*	*Sss*	*Mss*	*F*	*Probability*
1	Treatment	3	0.064937	0.02165	1.191	0.171
2	Rep	5	1.172342	0.23447	20.74	0
3	Error	15	0.0169589	0.011306	2.92	0.016
4	Depth Sampling	1	0.066079	0.066079	17.08	0.001
5	Depth × treatment	5	0.064982	0.0129960	3.36	0.026
6	Sub Error Factor	18	0.069630	0.003868		
7	Total	47	1.607558			

A comparison of treatment means was done with the Duncan's mean, in order to show the treatments with significant difference.

The amount of EC in top soil was more than sub soil. Maximum amount of EC in top soil was observed in garden lands. In this layer between mono cultivated systems under irrigation and multi cultivated systems under irrigation there was no significant difference, but between mono cultivated systems under irrigation with other treatments there was significant difference. In sub soil maximum amount of EC belonged to eroded lands. This treatment in sub soil had the most unsuitable treatment (Table 13.24).

Table 13.24: Arranged EC Means in Different Treatments

Row	*Treatment*	*Top Soil*	*Sub Soil*
1	Mono cultivated systems under irrigation	0.3000b	0.3500c
2	Multi cultivated systems under irrigation	0.3000b	0.2750b
3	Dry farming systems	0.2000a	0.2000a
4	Garden lands	0.3250bc	0.3000b
5	Rangelands	0.2000a	0.2000a
6	Eroded lands	0.2250a	0.2000a

The maximum amount of SAR was in mono cultivated systems under irrigation treatment especially in sub soil. In top and sub soil there was significant difference between mono cultivated systems under irrigation treatment with other treatments (Table 13.25). The best suitable treatment is one that has the least amount of EC and SAR. In Table 13.25, suitable and unsuitable treatments for top and sub layers are shown.

Table 13.25: Arranged SAR Means in Different Treatments

Row	*Treatment*	*Top Soil*	*Sub Soil*
1	Mono cultivated systems under irrigation	0.4412b	0.6227f
2	Multi cultivated systems under irrigation	0.2432ab	0.3230c
3	Dry farming systems	0.1028a	0.1245ab
4	Garden lands	0.1207a	0.1082a
5	Rangelands	0.08325a	0.0990a
6	Eroded lands	0.0590a	0.22182a

Table 13.26: Suitable and Unsuitable Factor in Each Treatment in Top Soil

Factor	*Suitable*	*Unsuitable*
EC	Rangelands and Dry farming systems	Garden lands
SAR	Eroded lands, Rangelands and Dry farming systems	Mono cultivated systems under irrigation

Table 13.27: Suitable and Unsuitable Factor in Each Treatment in Sub Soil

Factor	*Suitable*	*Unsuitable*
EC	Eroded lands, Rangelands and Dry farming systems	Mono cultivated systems under irrigation
SAR	Rangelands and Garden lands	Mono cultivated systems under irrigation

CONCLUSION

Initial results provided by this study for determining effective indices and criteria on desertification showed that both environmental and anthropogenic factors play significant role in desertification. In Kashan region the most important factor in desertification is water resources degradation. This is mainly because of better access to groundwater resources, intensive population and numerous wells in plain areas. Other effective factors, which should be mentioned, are surface water harvesting projects to supply water demands for domestic, industrial and agricultural uses. Such managerial factors increase the role of anthropogenic activities in desertification intensity more than environmental parameters.

Accessibility of different criteria and indices on one hand and considering various aspects of land degradation process on the other provide an accurate estimation of desertification condition and intensity as well as determine effective factors in land degradation.

Since the effects of selected criteria and indices varies according to climatic conditions then it is necessary to calibrate them in different climatic regions of Iran (arid, semi-arid, humid semi –arid) which will increase the accuracy and applicability of results.

Current knowledge and lack of sufficient data makes it difficult to introducie a global or even a national model for desertification. Instead it is better to set regional models for different geographical regions.

One of the disadvantageous of the proposed procedure is difficulty of measuring all effective factors because of limiting parameters such as costs, intensive filed work, deficiency of necessary data and information. Meanwhile all indices play major role in desertification process since the number of indices in the current study has been reduced consequently the efficiency of the model has decreased which needs more study to find solutions for overcoming this problem.

Application of desertification models needs continuous updating of data to determine better trend and intensity of desertification for recommending preventive measures.

One of the advantages of this study is establishment of a warning system to record desertification hazard in pilot regions for continuous monitoring and reporting of desertification intensity. If agricultural activities are following a sustainable development, agriculture does not cause land degradation, instead both biological development and production are provided.

REFERENCES

1. Ahmadi, H., Abrisham, E., Ekhtesasi, M.R., 2005. Evaluation and mapping of desertification condition in Fakhr Abad region with the ICD and MICD models, Biaban Journal. Vol. 10, No 1-1, 37-50.
2. Ahmadi, H., Zehtabian, Gh., Azarnivand, H., 2003, Iranian Model of Desertification Potential Assessment, University of Tehran.
3. Babaev, A.G., Kharin, N., Orlovsky G., 1993. Assessment and Mapping of Desertification Processes. A Mythological Guide. Ashghabad.
4. Basso F, Belloti A, Faretta S, Ferara A, Manino G, Pisante M, Quaranta G, Tabemer M. (1999) The Agri Basin In: MEDALUS Project_ Mediterranean Desertification and Land Use. Manual on Key indicators of Desertification and mapping Environmentally Sensitive areas to desertification.
5. Bayadgiev, T. (1981), FAO/UNEP Project of Desertification Assessment and Mapping.
6. Chen Jie, Chen Jing-zhang, Tan Man-zhi and Gong Zi-tong, 2002, Soil Degradation: a Global Problem Endangering Sustainable Development, Science Press, co-published with Springer-Verlag GmbH, V. 12 N. 2, Pages 243-252.

7. Ekhtesasi, M. S, Mohajeri, 1995, Method of Classification for Type and Intensity of Desertification in Iran, second congress on desert regions problems, Kerman, Iran.

8. European Commission, 1999, Mediterranean Desertification and LandUse MEDALUS). MEDALUS Office. Lodon

9. FAO, 2002, Land Degradation Assessment in Dry Land, LADA Project.

10. FAO/UNEP, Desertification Assessment in Dryland (LAND),(2001) United Nations Environment Program, Global Environment Facility (GEF).

11. FAO/UNEP, 1984. Provisional Methodology for Assessment and Mapping of Desertification, Rome.

12. FAO-UNEP/UNESCO/WMO, 1977, Word Map of Desertification at a Scale of 1:25000000

13. Grumblat, J., 1991. Kenya Pilot Study to Evaluate FAO/UNEP Provisional Methodology for Assessment and Mapping of Desertification (Gok, DRSRS).

14. Khosravi H,2004, The Strategies for Prevention of Desert Regions Degradation Using Desertification Models in Kashan, the forth international Iran and Russia conference "Agriculture and Natural Resources" Shahrekord, Iran.

15. Khosravi H, 2003, Application of MEDALUS Model in Kashan Region), M.Sc thesis, National Project, University of Tehran.

16. Khosravi. H, Gh. Zehtabian, M. Ghodsi, 2008, the study of water erosion effect on desertification intensity in Kashan region, The First European Conference of Iranian Scientists in Agriculture and Natural Resources, 8-9 October, Paris, France.

17. Khosravi. H, Gh. Zehtabian, M. Ghodsi, 2009, Effective Factors in Land Degradation in Kashan Region, Iran, International Workshop on Urbanization, Land Use, Land Degradation, and Environment, 28 Sept-01 Oct, Denizli Turkey.

18. Kosmas C, Kirkby M, Geeson N (1999) European Commission the MEDALUS Project Mediterranean Desertification and Land Use.

19. Kosmas, C., Gerontidis, St., Detsis, V., Zafiriou, Th., Marathianou, M., 2002. Application of the MEDALUS methodology for defining ESAs in the Lesvos Islsnd, European Commission. Capri, Italy, June 24-28.

20. Ladisa, G., Todorovic M., Trisorio_liuzzi, G., 2002. Characterization of Area Sensitive to Desertification in Southern Italy. Proc.Of the 2nd Int.Conf. On New Trend in Water and Environmental Engineering for Safety and Life: Eco-compatible solutions for Aquatic Environmental, Capri, Italy.

21. Rafiei Emam A. (2003), Desertification Vulnerability in Varamin Plain. Proc. of the 6th Int. Conf. on Map India 2003, New Delhi, India.

22. National Soil Erosion, Soil Productivity, 1981. A Research Prospective. Journal of Soil and Water Conservation. Vol. 39, P: 82-90.

23. Zehtabian. Gh., H. Ahmadi, H. Khosravi, A. Rafiei Emam, 2005, The Approach of Desertification Mapping Using MEDALUS Methodology in Iran. BIABAN Journal, Vol. 10, No.1.
24. Zehtabian, Gh., H. Ahmadi, H. Khosravi, Caliberation of MEDALUS Model to Present Regional Model For Desertification Intensity(case study: Kashan, Iran). Vol. 60, No. 4, 2008, ISSN: 1025-1316.
25. Zehtabian. Gh., B. Amiri, M. Soori, 2005, The Comparison of Soil Nutrients Among Agriculture land and Rangelands with Emphasis on N, P, K (case study: Kohdasht-Zanjan), Journal of pajohesh and sazandegi, Minisity of Jihad-e-Sazandegi 68, Vol. 18, No. 3.
26. Zehtabian, Gh., R. Hamedi, F. Amiraslani, 2004, the Study of Soil in Northern and Eastern areas of Varamin Plain View of the Role of Minerals in Production and Fertility, Journal of Biaban, University of Tehran, Vol. 10, No. 2.
27. Zehtabian. Gh, H. Khosravi, 2009, Investigation on Effect of Agricultural Management on Land Degradation- a Aase Study in Taleghan Region, Iran. 7th international symposium on plant-soil interaction at low PH, South China University of Technology Press, Guangzhou.
28. Zehtabian. Gh, H. Khosravi, 2009, Optimizing agricultural activities in order to arrive to sustainable development, Center of Excellence for Sustainable Watersheds Management, Faculty of Natural Resources, University of Tehran.
29. Zehtabian Gh. R., A Rafiei (2003). ESAS A New Method and Mapping of Area Sensitive to Desertification. Biaban Journal, Vol. 8, No. 5 120-126

Chapter 14

Urban Ecology of Edirne –Turkey

R. Yilmaz[1], *M. Ozturk*[2], *S. Sakcali*[3], *S. Gucel*[4] *and M. Karakuyu*[5]

[1]*Namık Kemal University, Faculty of Agriculture, Department of Landscape Architecture, Tekirdag, Turkey*

[2]*Ege University, Botany Department, Bornova-Izmir, Turkey*

[3]*Fatih University, Biology Department, Istanbul, Turkey*

[4]*Institute of Environmental Sciences, Near East University, Lefkoşa, Northern Cyprus*

[5]*Fatih University, Geography Department, Istanbul, Turkey*

ABSTRACT

State of Edirne covers an area of 6098 km^2, bordering Greece and Bulgaria. The climate is continental type in general but around Enez city and coastal parts mediterranean climate predominates. In the North of Edirne we come across moist forests, in the middle anthropogenic steppes, in the south dry forests, and alongside the coast macchia and pseudomacchia. These areas have nearly 2500 taxa of plants, 165 birds and 55 mammals. The area abounds in many endemics. There are nearly 56 monumental trees distributed in Edirne with ages varying between 90 (*Taxus baccata*) and 750 (*Quercus robur*) years. Majority belong to the genera *Platanus* and *Quercus*. Total area of grasslands lays around 81.279 ha. However, most of it has been changed into an anthropogenic steppe through different degradative activities. All through its history demographic developments, migrations, industrialization and agricultural activities have resulted in a heavy urbanization in this area resulting in an irrational land use. The population has reached a level of nearly 400 inhabitants, with a major increase in the city proper during the last decade and rural areas have suffered a great loss in the same period. The number of industrial establishments has also increased. More than 1000 ha of prime agricultural land have been used for housing, road construction and other activities. In this study land use in the past and possible future dimensions in relation to urbanisation and industrialisation in the State were investigated in the light of demographic development. The results were evaluated by making observations in different areas and an attempt has been made to put forth measures for the sustainable development.

Keywords: *Urbanization, Land use pattern, Urban ecology, Biodiversity, Human impact.*

INTRODUCTION

The city of Edirne was first established in 5th century B.C and was named differently by different rulers as Odris, Odrisia, Adrianapolis, and Edrinus during the history. The name Edirne was given by the Ottoman King Murat I after conquerring it in 1361. It has served as the second capital of the Ottoman Empire for 92 years until the conquest of Istanbul in 1453 (Inalcik, 1969). All through the history this strategically important Province has been influenced by many factors including historical (wars etc.), migrations from Europe, and rapid industrialisation. All these together with population outburst have lead to an uncontrolled urbanisation. The historical attractiveness of the city is disappearing due to unplanned urbanisation (Figures 14.1a,b). Nearly 80 per cent of the soils of Edirne are arable and include most important productive plains of Turkey Ergene and Merich, which are particularly important for the production of sunflower, rice and wheat. However, most fertile agricultural soils are getting negatively affected both by land degradation as well as environmental pollution. The productive agricultural lands are shrinking every year due to urban sprawl on one hand and built up of industrial establishments on the other hand. Merich, Tunca and Ergene Rivers, irrigating the cultivated fields, receive a heavy transboundry pollutant load, as well as from an indiscriminate discharge of various types of effluents from different industries and domestic sewage locally, which are posing a great threat for the fertile agricultural soils of the area. The agriproducts from these areas produce detrimental effects on water quality, which in

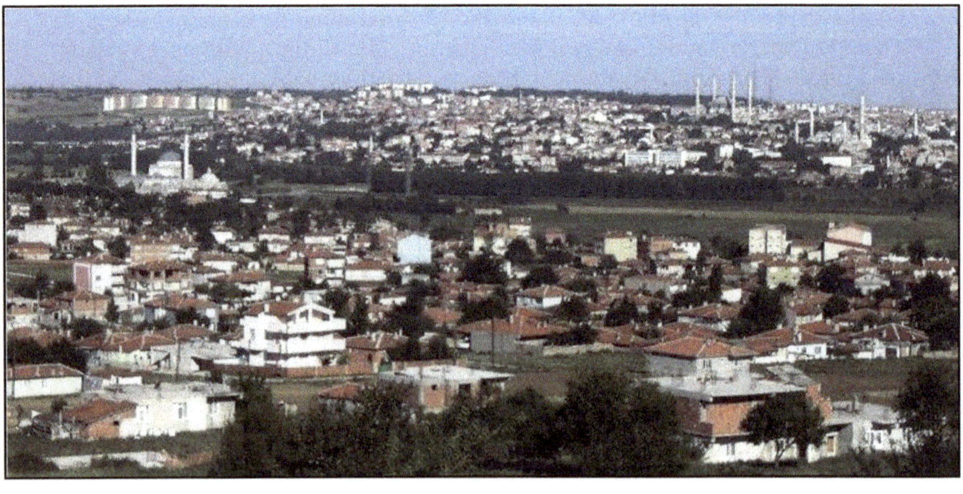

Figure 14.1a: A General View of Edirne City

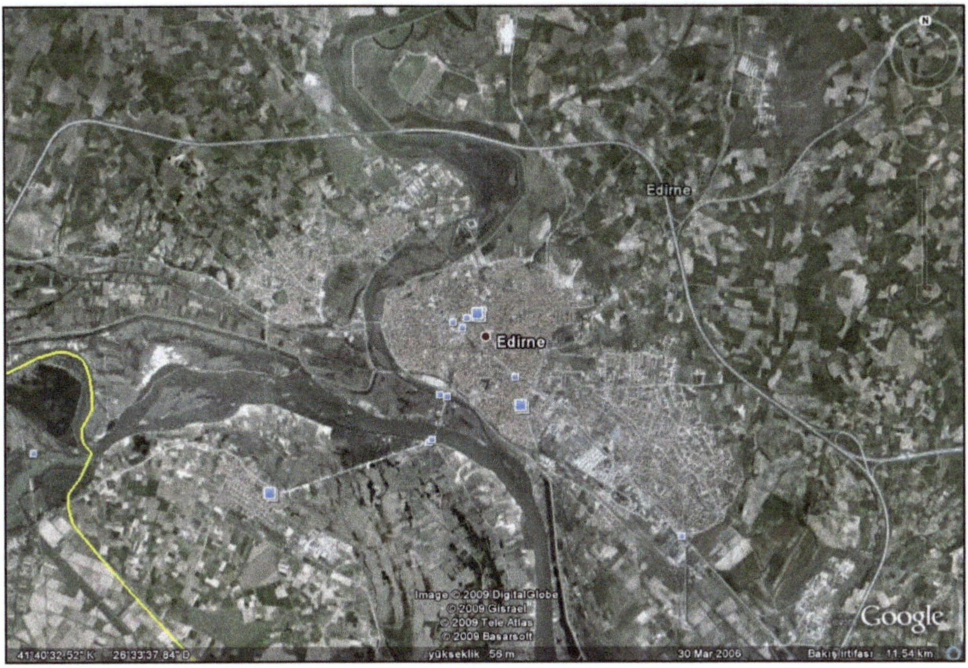

Figure 14.1b: A Satellite Image of Edirne

the long run will produce harmful effects on human beings if control measures are not adopted (Sakçalı *et al.*, 2009). It has been reported that waters from Arda, Tunca, Merich, and Ergene carry II, II, II and III, III and IV classes of pollutants (Sakçalı *et al.*, 2009).

Topography, Geology, Hydrology and Climate

Edirne, with a total area of 6098 km^2 is located in the North-West part of Turkey, serving as the door to Europe through its northern borders with Bulgaria. It is surrounded by Istranca mountains in the North; Merich River and Merich Plain, Ergene Plain in the East; Ergene Basin in the middle section; the plateau area, Koru Mountains and Saros Bay in the South; and Çanakkale and the Aegean Sea in the Southwest and Greece in the West. Southern parts form relatively mountainous area, most important heights being Istranca Mountains, Koru Mountain, Candır Mountain and Uzunkopru Mountains; and most important plains are Tunca, Kazanova, Ipsala and Ergene. Four main rivers namely; Merich, Tunica, Arda and Ergene flow through the State and their waters are used to irrigate the fertile agricultural lands around Ergene basin (Sakçalı *et al.*, 2009). These rivers flow from March to April. The province has 24 municipalities and 248 villages, many natural lakes, dams, reservoirs and ponds. The lakes are Gala, Dalyan, Tasalti, Bucurmene, Sigircik and Pamuklu. Main city of Edirne lies between 41°40′ N and 26°30′ E with an altitude of 51 m asl. Districts of the State are Enez, Havsa, İpsala, Keşan, Lalapaþa, Merich, Süloğlu and Uzunköprü.

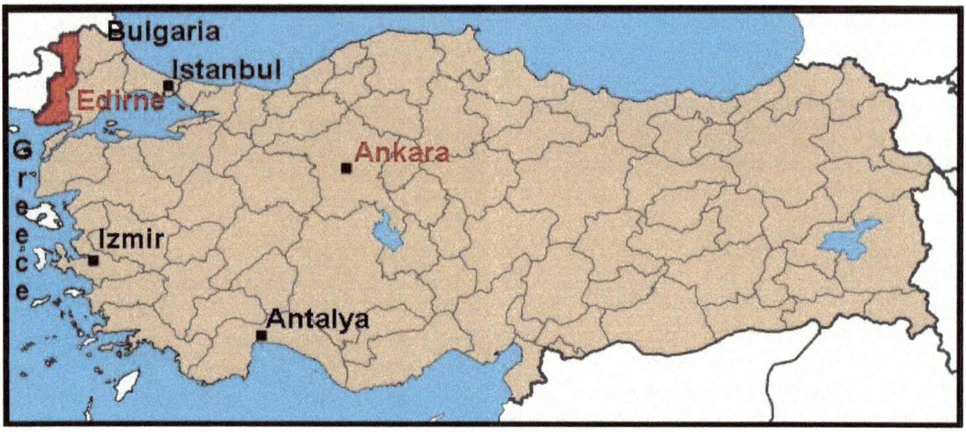

Figure 14.2: Map Showing Location of Edirne Province in Turkey

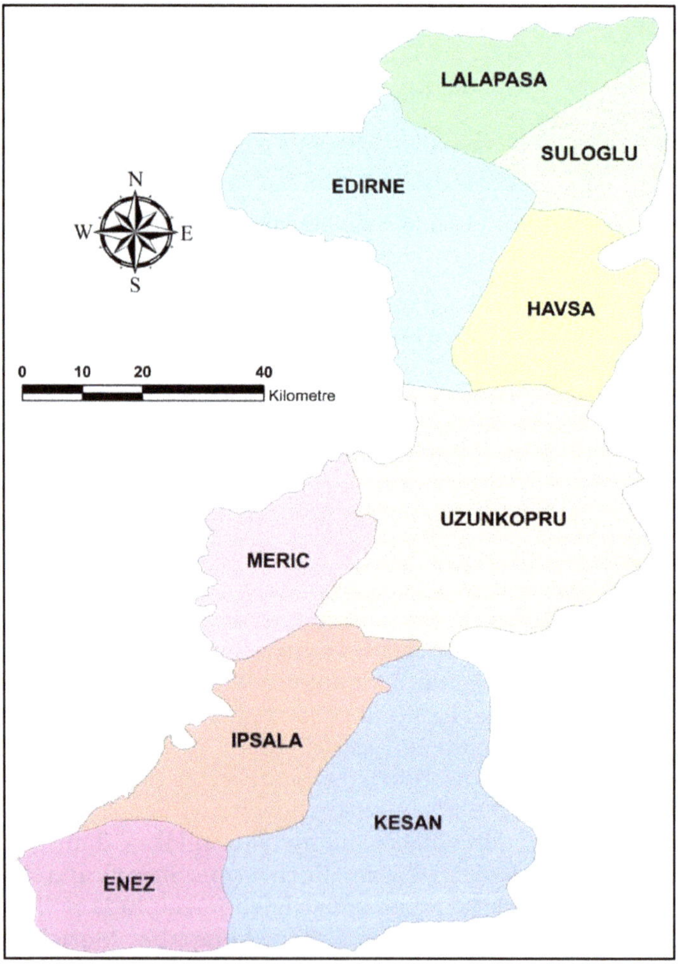

Figure 14.3: Map Showing Different Districts of Edirne Province in Turkey

Topography and geology is not too much rugged. However, the city center in Edirne and area arround Uzunkopru are always getting affected by the floods. The reason being development of city centers towards the stream beds (Figures 14.4a,b).

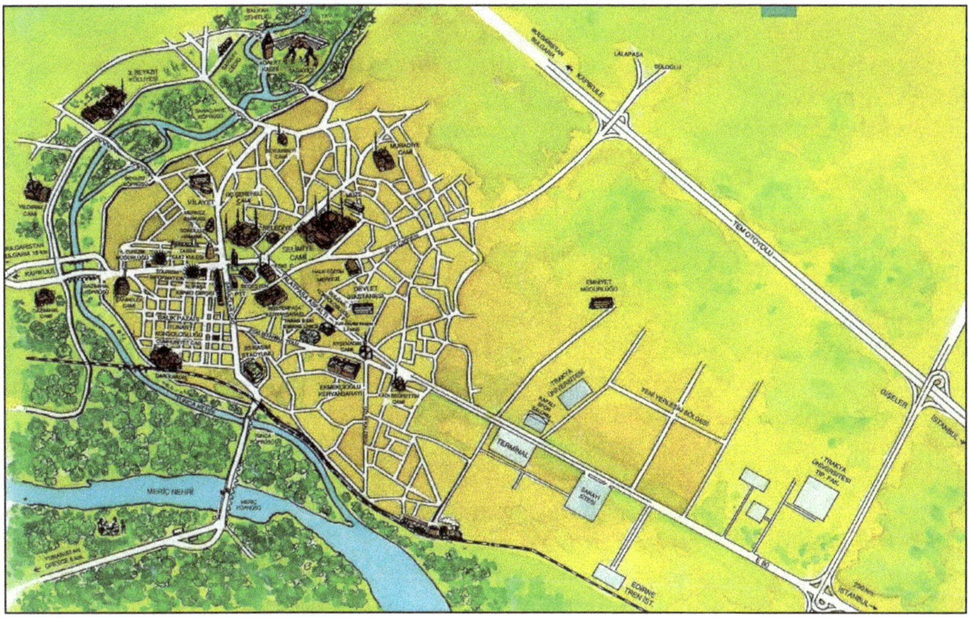

Figure 14.4a: Development of City Around the Stream Beds

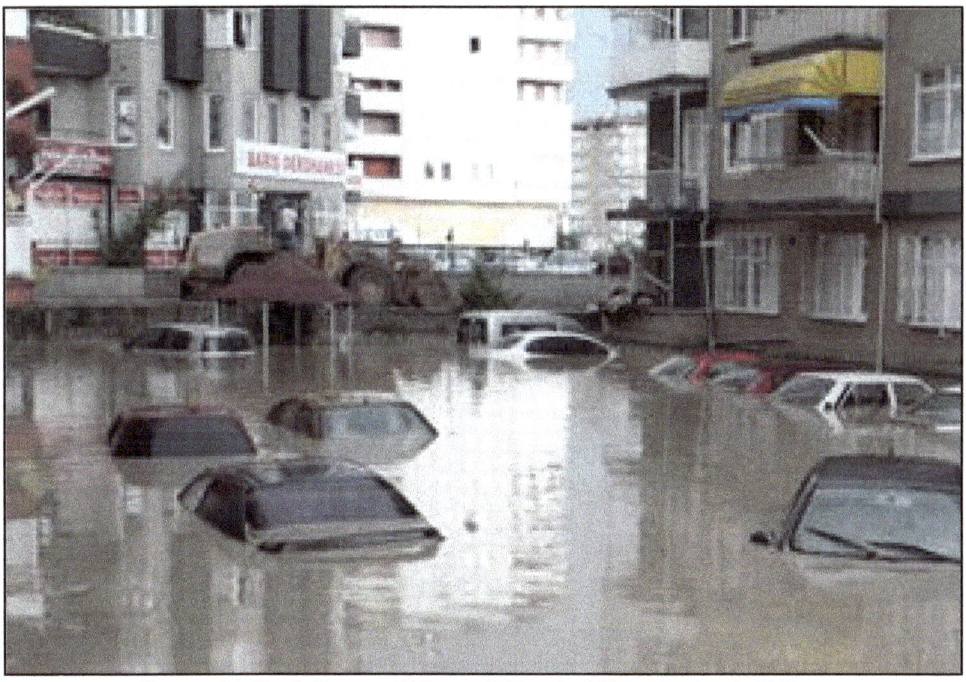

Figure 14.4b: A General View of Flooded City Centre in Edirne

The climate prevailing in Edirne province is continental type with long and cold winters and quite warm summers. The average annual temperature has been recorded as 13.4°C and the highest temperature as 42.2 °C, usually in July. The lowest recorded temperature is –22.2°C generally in January. Average annual rainfall is 585.9 mm and 35 per cent of this falls in summer months, followed by spring with 31 per cent, winter with 24 per cent and autumn with 10 per cent. Along the Gulf of Saros and Enez County coast in the South of the province Mediterranean climate prevails with hot and dry summers, mild and rainy winters (Figure 14.5).

Land Use and Population

Landscape is an area as perceived by people and recognized as a part of natural, historical, cultural and scientific heritage. Lately the landscapes on global scale have undergone several dramatic changes ending up with a loss in biodiversity, agricultural lands and degradation; theirby creating threats to environmental security (Council of Europe, 2000; McGlade, 2004). Many indicators have been developed during the last few years to monitor these changes (Golley and Ballot, 1999; Jackson *et al.*, 2000; Banko *et al.*, 2003;OECD, 2003). These indicators have an important function in the research and decision-making process, and contribute much in the sustainable development (OECD, 2003; US-EPA, 2005). They can demonstrate and provide estimates of actual changes in land use over a certain time, with the help of data about the land cover and land use (Tress and Tress, 2003). These indicators are a particular category of ecological indicators usually based on remotely sensed data or other geographical information, and can be based on a single measure or a combination of measures (EU Com.,1993; Kepner *et al.*, 1995; Nurlu *et al.*, 2008). Land use is one of the most visible effects of humans on the earth and can have effects on both human health and ecological systems.

Edirne is one of the most important industrial and agricultural provinces of Turkey. Major agricultural crops sown in the Province are rice, sugarbeet, sunflower, maize, melon, water melon, and pumpkin. The 75.7, 45, and 7.7 per cent of sunflower,

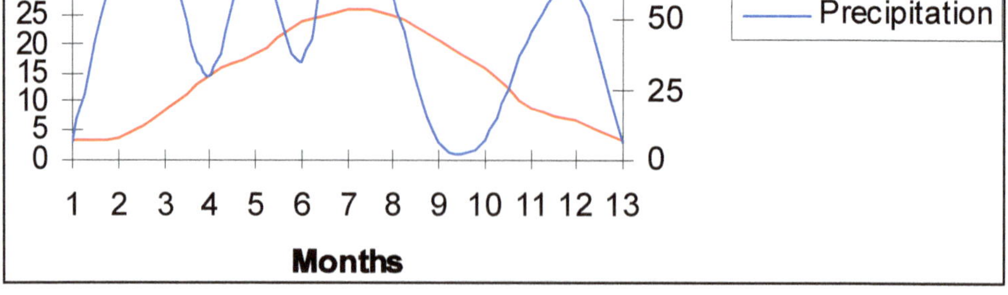

Figure 14.5: Climatic Diagram of Edirne

rice and wheat production in Turkey comes from this region. All these activities are using the water from four rivers namely; Arda, Ergene, Tunca and Merich. On the other hand, all wastes from agricultural activities, pesticides and fertilizers, chemical and textile industries and domestic sewage reach these rivers resulting in pollution. An indiscriminate discharge of different kinds of industrial effluents entering these transboundry rivers leads to a heavy bombardment of pollutants producing phytotoxic chemicals with high degree of organic impurities and toxic inorganic substances which can produce detrimetal effects on the safety of water, agriculture and of course to the humans in the long run (Sakcali *et al.*, 2009).

The distribution of soil types and classes is presented in Table 14.1 and different districts presented in (Figure 14.6). More than 1000 ha of prime agricultural land have been used for housing, road construction and other activities. In 2004 the area of

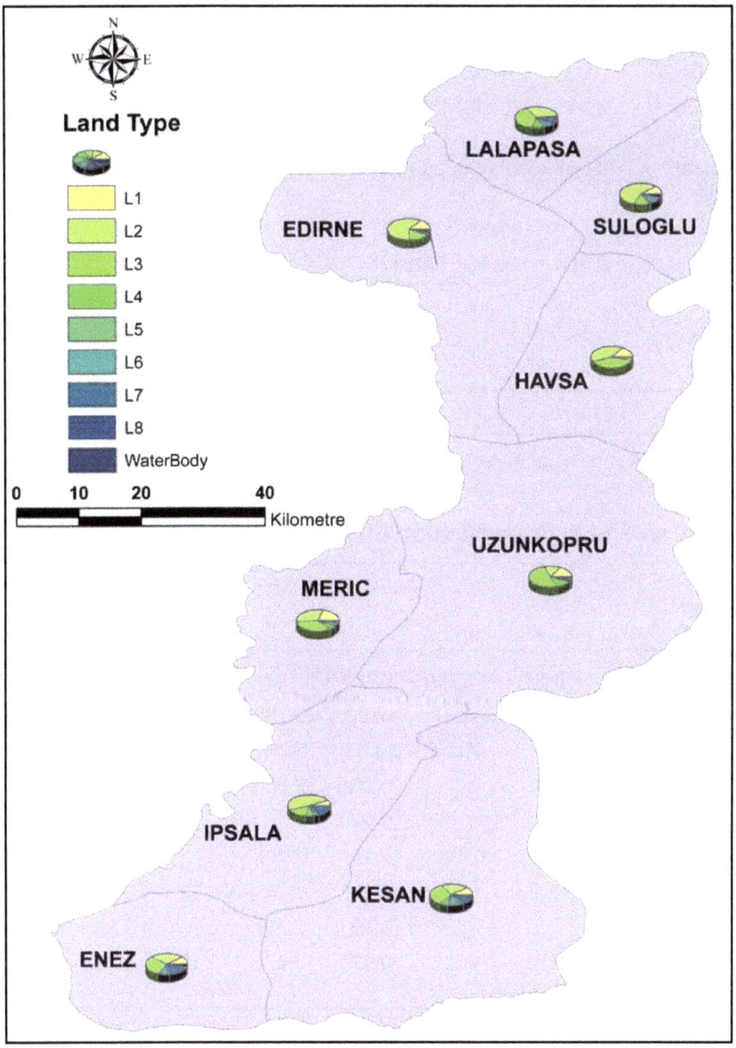

Figure 14.6: Map Showing Land Types in Different Distrcts of Edirne

agricultural land in the city was 65,21 per cent, university area 7,54 per cent, roads 5, 08 per cent, housing areas 6,64 per cent, official buildings 3,89 per cent, woody areas 2,07 per cent, graveyard 0,39 per cent, sport complexes 0,24 per cent, parks 0,2 per cent, historical places 0,10 per cent, gardens for children 0,03 per cent and others 10, 5 per cent. More than 1000 ha of the distribution of soil types and classes are presented in Tables 14.1 and 14.2.

Table 14.1: General Classification of Soils in Edirne Province

Soil Types	*Classes*							
	I (%)	*II* (%)	*III* (%)	*IV* (%)	*V* (%)	*VI* (%)	*VII* (%)	*VIII* (%)
Alluvial soils	58.98	15.70	13.633	6.237	0.18	5.27	–	–
Hydromorphic Alluvial soils	–	–	–	–	4.63	–	95.367	–
Brown Forest Soils	–	11.52	31.505	43.854	–	5.94	7.17	–
Noncalcareous Brown Forest Soils	2.537	26.180	35.792	9.91	–	14.128	11.453	–
Noncalcareous Brown Soils	6.097	42.660	38.28	8.27	–	2.47	2.34	–
Rendzins	–	100	–	–	–	–	–	–
Vertisolls	0.364	69.52	30.116	–	–	–	–	–
Coastal Sand Dunes	–	–	–	–	–	–	–	100
River Flood Zones	–	–	–	–	–	–	–	100
Bare rocks	–	–	–	–	–	–	–	100
Total	**67.978**	**165.58**	**249.336**	**68.271**	**4.81**	**27.808**	**116.33**	**300**

Source: Rural Services Provincial Directorate.

Table 14.2: Classification of Soils in Different Districts

Soil Classification	*Edirne District*	*Süloğlu*	*Enez*	*Havsa*	*İpsala*	*Keşan*	*Lalapaşa*	*Merich*	*Uzunköprü*	*Total*
I	12.71	9.8	8.65	13.92	6.14	9.04	1.7	19.46	15.13	96.55
II	62.12	57.28	31.19	40.87	47.93	25.40	33.60	31.21	16.21	345.81
III	17.63	17.2	25.07	41.24	15.67	28.17	34.53	36.39	50.49	266.39
IV	2.48	1.3	0.6	3.71	9.02	7.78	13.02	5.37	12.87	56.15
V	0.39				0.22			0.14	0.09	0.84
VI	1.06	10	11.99	0.175	5.15	15.88	8.72	5.57	2.27	59.815
VII	1.59		18.33		14.67	12.73	8.22		1.68	57.22
VIII	0.94		0.7		0.16	0.33		0.70	0.07	2.9
Water Level	1.044	4.42	3.31	0.085	0.99	0.67	0.12	1.16	1.15	12.94
Total	100	100	100	100	100	100	100	100	100	900

Source: Rural Services Provincial Directorate.

The population of Edirne in 1927 was 32.2 per cent and it decreased during the following 20 years. From 1950 onwards the population started increasing mainly due to migrations from Bulgaria and other nearby countries (Figures 14.7a,b and 14.8). The urbanization rate started increasing and was recorded as 33.7, 37.9, 42.2, 52, 57.4 and 65.5 per cent in the years 1965, 1970, 1980, 1990, 2000, and 2007 respectively (Yasar, 2002; SIS, 2004). The number of settlements showed a parallel trend with the population increase as 53806, 71914, 102345, 119298 and 141,570 in 1970,1980, 1990, 2000 and 2009 respectively. The most important reason for this increase after 1970's was agricultural mechanization. In 1982 Trakya University

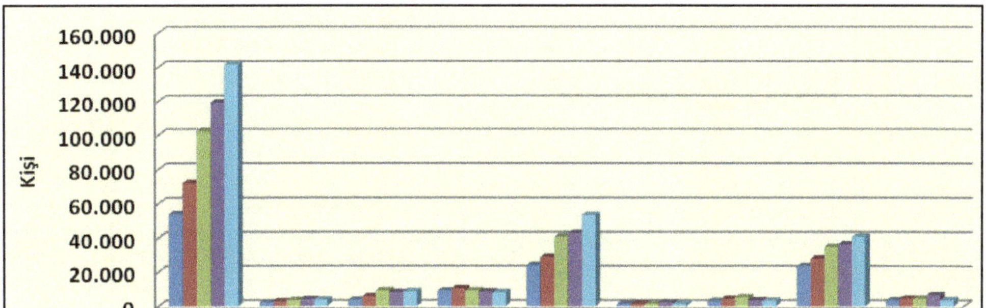

	Edirne	Enez	Havsa	İpsala	Keşan	Lalapaşa	Meriç	Uzunköprü	Süloğlu
1970	53.806	2.239	3.685	8.958	23.801	1.109	3.054	23.326	3.282
1980	71.914	2.878	5.654	10.373	28.884	1.373	4.287	27.873	4.487
1990	102.345	3.509	9.237	9.212	40.656	1.248	5.021	34.741	4.452
2000	119.298	3.941	8.081	8.471	42.755	2.270	3.382	36.162	6.548
2009	141.570	3.820	8.628	8.033	53.391	1.389	3.168	40.587	3.627

Figure 14.7a: Demographic Developments in Different Distrcts of Edirne Province (According to TÝK irrespective of population situation all cities and towns are regarded as cities).

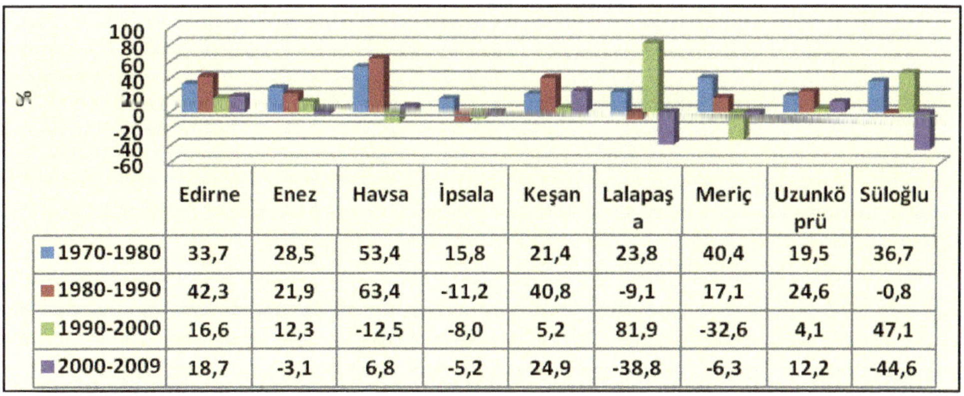

	Edirne	Enez	Havsa	İpsala	Keşan	Lalapaşa	Meriç	Uzunköprü	Süloğlu
1970-1980	33,7	28,5	53,4	15,8	21,4	23,8	40,4	19,5	36,7
1980-1990	42,3	21,9	63,4	-11,2	40,8	-9,1	17,1	24,6	-0,8
1990-2000	16,6	12,3	-12,5	-8,0	5,2	81,9	-32,6	4,1	47,1
2000-2009	18,7	-3,1	6,8	-5,2	24,9	-38,8	-6,3	12,2	-44,6

Figure 14.7b: Changes in the Population of Different Districts of Edirne (Percentage basis)

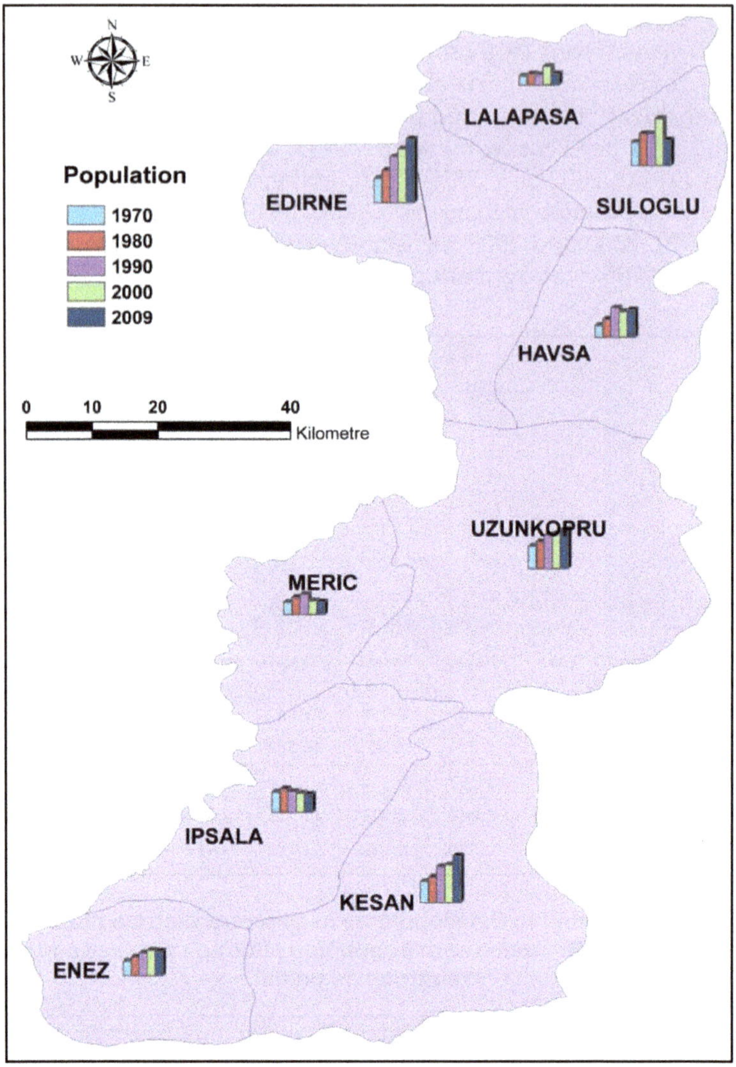

Figure 14.8: Map Showing Population Changes in Different Distrcts of Edirne

was established and this added to the population between 1980-90 together with industrialization and increase in border trade. The highest rate of urbanization in Edirne was recorded in the town of Keshan. It went up from 23,801 inhabitants in 1970 to 53,391 in 2009. The increase was the highest between 1980-1990, rate being 48,8 per cent. This town is located on the main route connecting Istanbul to Canakkale and Greece; and West Anatolia to Edirne and Europe. The town became centre of attraction for industry, trade and development. Other densly populated town is Uzunkopru,which had a population of 23,326 inhabitants in 1970 but went up to 40,587 in 2009. During the last four decades maximum increase in population rate has been recorded (24.6 per cent) between 1980-1990. Outside these towns and Edirne city center, general population has not increased much, infact the population of Ipsala, Meric and Suloglu towns has decreased.

Industry

There are approximately 910 industrial establishments in Thrace, 8 per cent of these are located close to the main city of Edirne. Major 74 industrial organisations include food, textile, stone and soil, paper and package and machinery. Out of these 76 per cent is food industry including oil, flour, rice, forage and dairy products. Second highly distributed industrial sector is textile production, which includes thread weaving, and confections, followed by brick and tile factories. Textile and stone producing or establishments are 11 per cent of the total. Only 2 per cent consist of paper and machinery industry. In The percentage of small scale industries lies around 25 per cent (18), medium scaled 44 per cent (33) and are large scale 14 per cent (10). Most of the small scale industries are food industries dealing with rice, flour and diary products (Table 14.3).

Table 14.3: Distribution of Industrial Organisations in Edirne and its Environs

Districts	*Food*	*Textile*	*Stone and Soil*	*Machinery and Tools*	*Paper*	*Total*
Centre	6	5	3	–	1	15
Havsa	4	1	–	–	–	5
Uzunköprü	24	–	–	1	–	25
Keşan	4	2	3	–	–	9
İpsala	12	–	–	–	–	12
Merich	3	–	–	–	–	3
Lalapaşa	–	–	1	–	–	1
Süloğlu	–	–	–	–	–	–
Enez	3	–	1	–	–	4
Total	56	8	8	1	1	74

Biodiversity

Edirne is covered by a rich humid forest in the north, anthropogen steppe in the middle and maquis and pseudomaquis in the south. A large area of forests has changed into an anthropogenic steppe due to human interventions.

The forests are found at high altitudes, widely distributed around Lalapaþa Muhittin Baba (601 m) and Uzunköprü B. Hacıdağ (359 m). Major dominating species are *Pinus brutia*, *Pinus nigra*, *Pinus pinaster*, *Pinus pinea*. Other species accompanying these are; *Quercus* sp., *Acer* spp., *Tilia silvestris*, *Robinia pseudoacacia*, *Carpinus* spp., *Ailanthus* spp. and *Populus* spp.(Table 14.4). Moist forests are distributed around Edirne city, in the middle anthropogenic steppes, in the south dry forests, and alongside the coast macchia and pseudomacchia.The area of grasslands lies around 81.279 ha.

Table 14.4: Area Covered by Major Forest Tree Species in Edirne

Forest Division	*Pinus brutia*	*Pinus nigra*	*Cedrus*	*Juniper*	*Oak*	*Populur*	*Walnut*	*Horn-beam*	*Total (ha)*
EDİRNE	–	2806	–	–	3879	243	–	–	6928
LALAPAŞA	–	2932	39	–	11587	4	–	13	14575
UZUNKÖPRÜ	192	1672	201	227	18766	401	29	–	21488
Total	192	7410	240	227	34232	648	29	13	42991

Source: Edirne and Kesan Forest Stations.

According to the directorate of Agricultural Services the area of grasslands in different districts of Edirne is 6.211 ha (Edirne centre), 1.980 ha (Enez), 6.300 ha (Havsa), 1.127 ha (Ipsala), 11.519 ha (Keşan), 13.000 ha (Lalapaşa), 13.353 ha (Merich), 7.110 ha (Süloğlu) and 20.679 ha (Uzunköprü). The total area is around 81.279 ha.

The national parks, nature preserved areas and nature parks are found in Enez district, covering an area of 2369 ha. These include Small Gala lake, Pamuklu lake and their environs. In addition to these Merich Delta (Gala Lake) is regarded as an A class reserve and Saros Bay B class marsh (Kilic and Eken, 2004).

We determined 210 plant taxa showing distribution in the province (Davis,1965-1985; Davis *et al.*, 1988). The most prominent ones are; *Abelia grandiflora* Rehd., *Abies bornmülleriana* Mattf., *Acer campestre* L., *A. negundo* L., *A. negundo* L., *A. pseudoplatanus* L., *A. tataricum* L., *Aesculus hippocastanum* L., *A. xcarnea* Hayne, *Ailanthus altissima* (Miller) *Swingle, Albizzia julibrissin Durazz., Alnus glutinosa* (L.) *Gaertn., Amorpha fruticosa* L., *Amygdalus communis* L., *Asparagus acutifolius* L.,*Atraphaxis billardieri* Jaub. Et Spach, *Aucuba japonica* Thunb., *Baccharis halimifolia* L., *Berberis julianae* Schneid, *B. thunbergii* DC., *B. vulgaris* L., *Betula pendula* Roth., *Biota orientalis* Endl., *B. orientalis* Endl., *Buddleia davidii* Franch., *Buxus micropylla* Sieb. and Zucc., *B. sempervirens* L., *Caesalpinia gilliesii* Wall. Ex Hook., *Calycanthus floridus* L., *Camellia japonica* L., *Carpinus betulus* L., *C. orientalis* Miller, *Cassia corymbosa* Lam, *Castanea sativa* Miller, *Catalpa bignonioides* Walt, *Cedrus atlantica* (Endl.) Carr., *C. libani* A. Rich., *Celtis australis* L., *Cerasus mahaleb* (L.) Miller, *Cercis siliquastrum* L., *Chaenomeles speciosa* Nakai, *Chamaecyparis lawsoniana* (A. Murr.) Parl., *Cistus creticus* L., *C. salviifolius* L., *Clematis vitalba* L., *Clematis viticella* L., *Colutea cilicica* L., *Cornus mas* L., *Cortaderia selloana* (Schult.) Aschers., *Corylus avellana* L., *Cotinus coggyria* Scop., *Cotoneaster franchetii* Bois., *C. horizontalis* Decne, *Crataegus oxycantha* L., *C. monogyna* Jag., *C. szovitsii* Pojark, *Cryptomeria japonica* D. Don, *Cupressus arizonica* Grene, *C. sempervirens* L., *Cydonia oblonga* Miller, *Deutzia gracilis* Sieb. and Zucc., *Diospyros kaki* L., *Elaeagnus angustifolia* L., *Eriobotrya Japonica* (Thunb.) Lindl., *Eucalyptus camaldulensis* Dehn, *Euonymus japonica* L., *Fagus orientalis* L., *Ficus carica* L., *Forsythia xintermedia* Zab, *Fraxinus angustifolia* Vahl., *F. exelsior*, *F. ornus* L., *Gleditschia tricanthos* L., *Hedera canariensis* Willd., *H. helix* L., *Hibiscus syriacus* L., *Hydrangea macrophylla* (Thunb.) Ser., *Hypericum androsaemum* L., *Ilex aquifolium*, *Jasminum fruticans* L., *J. oficinale* L., *J. primulinum* Hemsl., *Juglans nigra* L., *J. regia* L., *Juniperus communis* L., *J. horizontalis* L., *J. oxycedrus* L., *J. sabina* L., *Koelreuteria paniculata* Laxm., *Laburnum anagyroides* Med., *Lagerstroemia indica* L., *Laurus nobilis* L., *Laurocerasus officinalis* Roem.,

Ligustrum ovalifolium Hassk., *L. vulgare* L., *Lonicera japonica* Thunb., *Lycium barbatum* L., *Maclura pomifera* (Raf.) Schneider, *Magnolia grandiflora* L., *Mahonia aquifolium* (Pursh.) Nutt., *Malus sylvestris* Miller, *Melia azedarach* L., *Mespilus germanica* L., *Morus alba* L., *M. nigra* L., *M. rubra* L., *Nerium oleander* L., *Olea europaea* L., *O. europaea* L. ssp. *sylvestris* L., *Osyris alba* L., *Paliurus spina-christi* Miller, *Parthenocissus quinquefolia* Graben, *Paulownia tomentosa* (Thunb.) Steud., *Periploca graeca* L., *Philadelphus coronarius* L., *Phillyrea latifolia* L., *Phoenix canariensis* Chabaud., *Picea abies* (L.) Karst., *P. omorika* (Pancic.) Purkyne, *P. orientalis* (L.) Link., *P. pungens* Engelm., *Pinus brutia* Ten., *P. pinaster* Ait, *P. pinea* L., *Pittosporum tobira* (Thunb.) Ait, *Platanus xacerifolia* Willd., *Pistacia terebinthus* L., *Platanus orientalis* L., *Polygonum baldschuanicum* Regel, *Populus alba* L., *P. nigra* L., *P. tremula* L., *P. x canadensis* Moench, *P. x canescens* (Ait) Sm., *Prunus armeniaca* Lam., *P. avium* (L.) Moench, *P. ceracifera* Ehrh., *P. cerasus* L., *P. domestica* L., *P. persica* L. Batsch, *P. serrulata* Lindl., *P. spinosa* L., *P. divaricata* L., *Punica granatum* L., *Pyracantha coccinea* Roem., *Pyrus elaeagnifolia* Pall., *Pyrus communi* L., *Quercus cerris* L., *Quercus coccifera* L., *Quercus frainetto* Ten, *Quercus hartwissiana* Stev., *Quercus infectoria Oliver*, *Quercus ithaburensis* Decne ssp. *Macrolepis* (Kotschy) Hedge et Yaltırık, *Quercus petraea* (Mattuschka) Liebl., *Quercus pubescens* Willd., *Q. robur* L., *Q. trojana* P.B. Webb, *Rhododendron ponticum* L., *Rhus coriaria* L., *Robinia pseudo-acacia* L., *Rosa canina* L., *R. foetida* L., *R. gallica* L., *R. sempervirens* L., *Rosmarinus officinalis* L., *Rubus caesius* L., *R. discolor* Weihe et Nees, *R. sanctus* Schreb., *Ruscus aculeatus* L., *Salix alba* L., *S. babaylonica* L., *S. caprea* L., *S. cinerea* L., *S. matsudana* Koidz, *Sambucus nigra* L., *Sequoia sempervirens* (D.Don.) Endl., *Sophora japonica* L., *Sorbus torminalis* L. Crantz, *Spartium junceum* L., *Spiraea xvanhouettei* L., *Symphoricarpus albus* L. *Blake*, *Symphoricarpus orbiculatus* Mnch., *Syringa vulgaris* L., *Tamarix parviflora* DC., *Taxus baccata* L., *Thuja occidentalis* L., *Tilia argentea* Desf. Ex DC., *T. platyphyllos* Scop.,*T. platyphyllos* Scop., *T. rubra* DC., *Ulmus glabra* Huds., *U. leavis* Pall., *U. minor* Miller, *Viburnum opulus* L., *V. tinus* L., *V. rhytidophyllum* Hemsl., *Vitex agnus-castus* L., *Weigela florida* (Sieb. and Zucc.) A.DD, *Wisteria sinensis* (Sims.) Sweet., *Yucca filamentosa* L., *Zizyphus jujuba* Miller.

There are many endemics like; *Centaurea kilea* (Asteraceae), *Helleborus vesicarius* (Ranunculaceae), *Paronychia argyroloba* (IIlecebraceae), *Fraxinus ornus ssp. cilicia* (Oleaceae), *Anchusa heterophylla ssp. incana* (Boraginaceae), *Asperula Lilaciflora ssp. phrygia* (Rubiaceae), *Crocus biflorus ssp. pulchriclor* (Iridaceae).

The province abounds in monumental trees as well (Dalgıç and Yaltırık, 1999). In all 56 monumental trees have been recorded from Edirne. Major monumental species are; *Celtis australis* (160 years old), *Fraxinus angustifolia* (135-210 years old), *Juniperus oxycedrus* (110 years old), *Morus alba* (130-210 years old), *Platanus xacerifolia* (120- 210 years old), *Platanus orientalis* (135-210 years old), *Quercus cerris* (210 years old), *Quercus infectoria* (210 years old), *Quercus ithaburensis* (210-310 years old), *Quercus pubescens* (130 years old), *Quercus robur* (160-760 years old), *Taxus baccata* cv.'*fastigiata* (100 years old, 10 m tall), *Tilia argentea* (110-135 years old), *Ulmus minor* (160-210 years old) (Figure 14.9).

The State is rich faunistically as well. There are many invertebrates (dragonfly, dung beetle, ants and bees), vertebrates (frogs, birds and mammalians especially rodents) (Table 14.5).

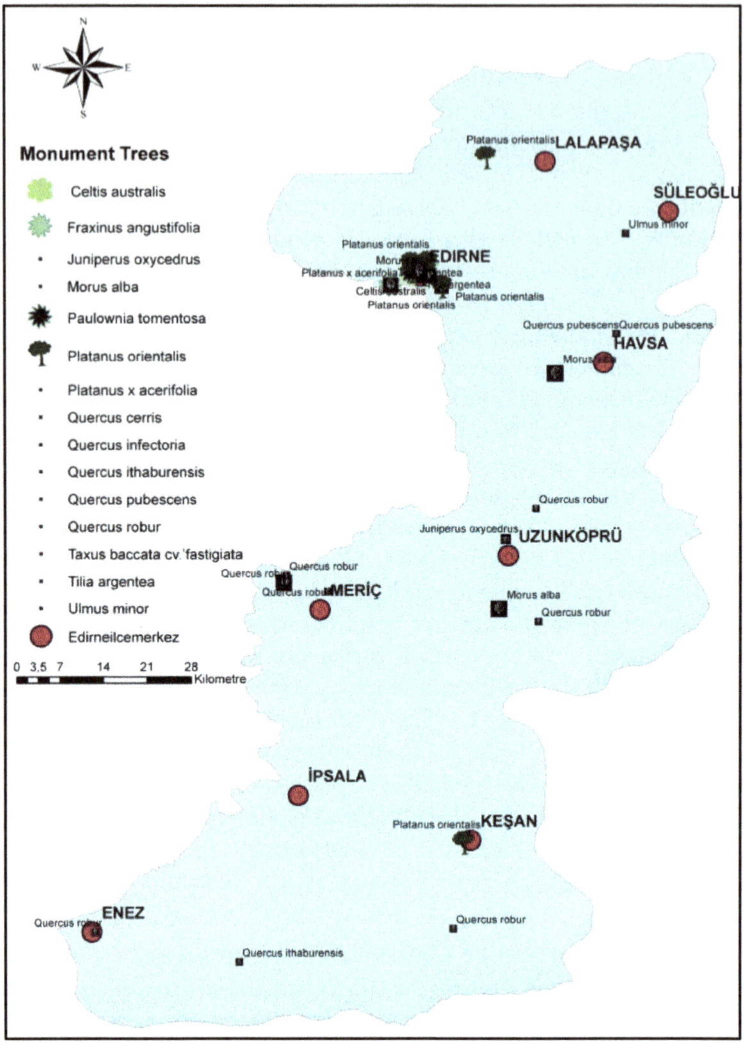

Figure 14.9: Map Showing Distribution of Monumental Trees

Table 14.5: Endemic Species of Thrace (Trakya University)

Groups	*No. of Species in Turkey*	*No. of Species in the State*	*Percentage in the State*
Plants	10	304	0.03
Freshwater Fishes	192	40	21
Salamander, Frogs	18	7	39
Reptiles	79	25	32
Birds	418	165	40
Mammals	138	55	40

CONCLUSIONS

All through its history demographic developments, migrations, industrialization and agricultural activities have resulted in a heavy urbanization in this area resulting in an irrational land use. The population has reached a level of nearly 400.000 inhabitants, with a major increase in the city proper during the last decade and rural areas have suffered a great loss in the same period. The number of industrial establishments has also increased. More than 1000 ha of prime agricultural land have been used for housing, road construction and other activities. Most of the grasslands have changed into an anthropogenic steppe due to different degradative activities. There is a great need for developing sustainable solutions against these degradative forces and environmental pollution (Yilmaz *et al.*, 2006). Riverbeds and floodplains should be cleaned from all unplanned development, in order to reduce the destructions created by floods.

REFERENCES

Banko, G., Zethner, G., Wrbka, T., Schmitzberger, I. 2003. Landscape Types as the Optimal Domain for Developing Landscape Indicators, in: *Agricultural Impacts on Landscapes: Developing Indicators for Policy Analysis*, Proceedings from NIJOS/ OECD Expert Meeting on Agricultural Landscape Indicators, W. Dramstad and C. Sogge, eds., Norwegian Institute of Land Inventory, ISBN 82-7464-308-9, pp. 317-328.

Council of Europe. 2000. European Landscape Convention-Explanatory Report and Convention, Florence.

Dalgic, G., Yaltirik, F.1999. Monumental Trees of Thrace and Protection Measures. Edirne Governors Office Publication No:11, State Cultural Directorate Publ. No:8, Acar Press, Istanbul, Turkey.

Davis, P.H. 1965-1985. *Flora of Turkey and East Aegean Islands*, Vols. I-IX, Edinburgh University Press, UK.

Davis, P.H., Mill, R.R.,Tan, K. 1988. *Flora of Turkey and East Aegean Islands*, Vol. X, Edinburgh University Press,UK.

EU-Com.(European Commission). 1993. CORINE Land Cover: Technical Guide, (Report EUR 12585), Office for Official Publications of the European Communities, Luxembourg, p.111.

Golley, F.B., Bellot, J. (eds). 1999. *Rural Planning from an Environmental Systems Perspective*, ISBN 0 387 98519 0, Springer-Verlag, New York, Inc, p. 377.

Inalcik, H. 1969. Capital Formation in the Ottoman Empire. The Journal of Economic History, Vol. 29, No. 1, 97 pp.

Jackson, L.E., Kurtz, J.C., Fisher, W.S. (eds.). 2000. *Evaluation Guidelines for Ecological Indicators,* U.S. Environmental Protection Agency, Office of Research and Development, Environmental Monitoring and Assessment Program, Washington, D.C., EPA/620/R-99/005.

Kepner, W.G., Jones, K.B., Chaloud, D.J., Wickham, J.D., Riiters, K.H., O'Neill, R.V.1995. *Mid-Atlantic Landscape Indicators Project Plan, Environmental Monitoring and Assessment Program,* U.S.Environmental Protection Agency, Office of Research and Development, National Exposure Research Washington, D.C., EPA/620/R-95/003.

Kilic, D.,Eken, G. 2004. *Turkiye'nin Onemli Kus Alanlari-2004 Güncellemesi,* Doga Dernegi, Ankara, ISBN 975-98901-0-0, p. 232.

McGlade, J. 2004. EU Conference on Changing Land Use in Europe-Getting the Picture, European Environment Agency, 9-12 November, 2004, the Netherlands.

Nurlu E., Erdem U., Ozturk M., Guvensen A., Turk T. 2008. Landscape, Demographic Developments, Biodiversity and Sustainable Land Use Strategy: A Case Study on Karaburun Peninsula, Izmir, Turkey, In: Use of Landscape Sciences for the Assessment of Environmental–Security, Petrosillo, I. *et al.* (eds.), Springer, The Netherlands, pp.357-368.

OECD. 2003. OECD Environmental Indicators–Development, Measurement and Use, Reference Paper, OECD Environmental Performance and Information Division, Paris Cedex, France.

Sakcali,S., Yilmaz, R., Gucel, S., Yarci, C.,Ozturk, M. 2009. Water Pollution Studies in the Rivers of

Edirne State–Turkey. Aquatic Ecosystem Health and Management, 12: 3,313-319.

SIS. 2004. State Institute of Statistics, Census of Population by Administrative Division, Prime Ministry, Republic of Turkey, Ankara, http://www.die.gov.tr.

Tress, B.,Tress, G. 2003. Shaping Future Landscape: The Scenario Approach, in: *Agricultural Impacts on Landscapes: Developing Indicators for Policy Analysis,* Proceedings from NIJOS/OECD Expert Meeting on Agricultural Landscape Indicators, W. Dramstad and C. Sogge, eds., Norwegian Institute of Land Inventory, ISBN 82-7464-308-9, pp. 68-81.

U.S. EPA. 2005. Proposed Indicators for 2007 Report on the Environment (ROE 2007), www.epa.gov/ncea/

Yaşar, O. 2002. Investigation of the demographic developments in the State of Edirne and socio-economic characateristics of migrants coming during the last period. Dogu Cografya Dergisi, S. 21, s. 195-219.

Yilmaz, R.,Sakcali, S., Yarci, C.,Aksoy, A.,Ozturk, M. 2006.Use of *Aesculus hippocastanum* L. as a Biomonitor of Heavy Metal Pollution. Pakistan Journal of Botany, Vol. 38, No. 5, 1519-1527.

Chapter 15

Promoting Sustainable Urban Livelihoods through Sustainable Utilization of Natural Resources in Urban and Peri-Urban areas of the Eastern Cape, South Africa

L.L. Sombalo

Researcher,

Agricultural Research Council,

ARC-ISCW 600 Belverdere Street,

Arcardia, Pretoria, Republic of South Africa

ABSTRACT

South Africa is faced with a major problem of physical, biological and chemical poor natural resource degradation indicators. The population is increasing at an alarming rate because of an influx of people from predominantly rural areas to urban areas. This leads to an increase in the number of urban poor, shortage of food and poor natural resource conservation leading to degradation. An urban agricultural project was initiated in Zwelitsha with the aim to evaluate technologies for their appropriateness, related to composted material and soil and water management for sustainable crop production. The nature of urban agriculture in Zwelitsha offers great opportunities for evaluating the most promising fertility strategies for production in vegetable gardens. Major activities of urban agriculture included cultivation of field and horticultural crops, which are faced with agroclimatic, socio-economic and institutional constraints.The overall goal of the project was to focus on a case study basis on sustainable urban livelihoods and implementation of eco-friendly technologies for sustainable natural resource use. Techniques used are suitable for Zwelitsha because they are cost-effective, conserve resources and require fewer inputs. The idea is to put in place a system that both protects the environment and supplies adequate nutritious food in harmony with nature and uses as little external inputs as possible.

Keywords: *Sustainable agriculture, Land degradation, Urban and peri-urban agriculture, Organic farming.*

INTRODUCTION

South Africa is a middle-income country, but despite this wealth relative to much of Africa, the experience of many South African households is vulnerability to poverty manifested in food insecurity, unemployment, violence and ill health. Access to adequate nutrition is enshrined in the new South African constitution with the State taking measures to ensure that marginalised people meet their food demands (Meadows, 2000). Since 1994 the country has undergone political change, but the population has grown and migration to urban areas including those in the Eastern Cape has been amongst the highest in Africa. As a result, food, land and water resources are under pressure- a factor compounded by uneven distribution of wealth and resources. The heavy urbanization is leading to an increase in the crime rate, community facilities are becoming scarce, land to practice agricultural activities becomes scarce and all these end up in an environmental degradation. The deforestation, overgrazing, salinisation, water logging arising from wrong irrigation, poor quality of public open spaces, soil and water pollution and increased use in agro-chemicals are leading to unsustainable ecosystem health. Because of these problems author visited India in 2004 to learn about their work on sustainable agriculture and the application of eco-friendly technologies through judicious use of the natural resources soil and water, as well as technology generation. Lessons learnt from India on ecological balance and constant regeneration of biodiversity, biological plant-pest control, minimum tillage, companion planting and contour sowing were evaluated for applicability in community food gardens in Zwelitsha. Zwelitsha is a town in the Eastern Cape Province of South Africa. From 1972 to 1981 it served as the provisional capital of the Bantustan of Ciskei, until the capital was moved to Alice, and then to Bhisho. It forms part of the Buffalo City Local Municipality. Urban and peri-urban agriculture is widely practiced in the Zwelitsha township especially vegetable production, which is subsistence orientated. Famers are faced with severe agro-climatic, socio-economic and institutional constraints, including poor availability and high cost of synthetic inputs, in particular chemical fertlizers and plant disease control material. The vision in Zwelitsha is to put in place a system that both protects the environment and supplies adequate nutritious food in harmony with nature, using as little external inputs as possible. The aim of the project was to design and implement appropriate eco-technologies to ensure sustainable resource management and sustainable urban livelihoods through the introduction of organic vegetables production practices. The specific objectives were to promote more efficient soil and water management for sustainable crop production, improving soil fertility and long term sustainability of farming systems through organic farming and implementation of eco-friendly technologies, and finally to improve the level of knowledge and skills for resource poor farmers in Zwelitsha.

METHODOLOGY

A stakeholder interaction platform was developed, which involved forming working committees and communication structures. The structures were already running and the farmers were well mobilized during monthly meetings. These activities provided a basis for forming urban farmers associations or cooperatives for input resources mobilization. PRA tools such as transect walks, plot land use mapping were used as introductory exercise which provided researchers with an initial overview of the area and the farmers to familiarize themselves with the research and to take a lead in identifying which key issues are important to them. Participatory methods, such as semi-structured interviews and visualization, focus group discussions and community mapping were used in triangulation with other information sources.

Participatory research was judged on the involvement and commitment of local people in the process, crosschecking of information from the different sources or collected by different methods, analysis of difference-knowledge from different actors and sectors of the community, and the degree of learning by local people about their own reality and willingness to improve (Pretty *et al.*, 1995, as quoted by ICRA learning materials). An action plan was used to guide the execution of the activities (See appendix 1). A Farmers First approach by Chambers (1997) was adopted as a part of the methodology, the farmers were empowered and principles, methods and choices were given to them on a flexible approach.

These methodologies offered opportunity to reinforce, through a mix of scientific and local knowledge, the four steps in the experiential learning cycles of experiencing (observation), sense making (reflection), drawing conclusions (modification), and action (Figure 15.1).

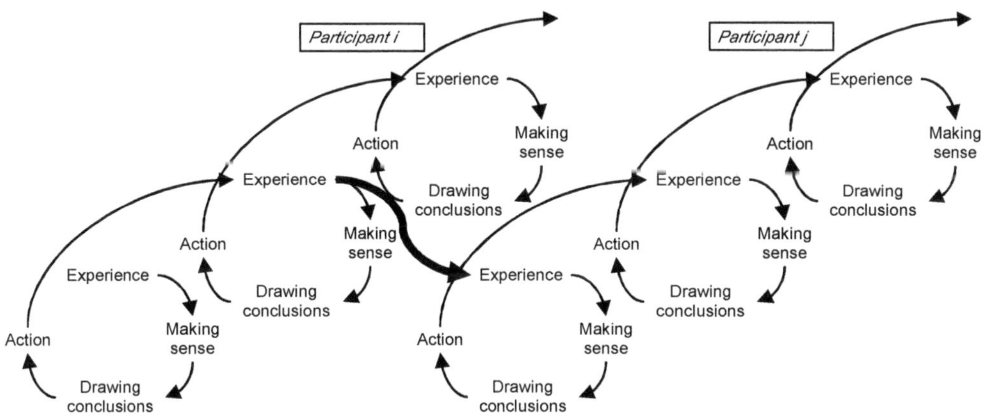

Figure 15.1: Iterative Cycles in an Experiential Learning Process (Douthwaite, 2003)

The activities involved "Training" which covered the need assessment, theoretical training, practical training and demonstrations on production skills; "on-site demonstrations" including establishment of organic gardens in schools, churches, idle state land and community centers as well as demonstrations of Vitamin A rich crops gardens; "after care services" entailing follow-up workshops, refresher courses and entrepreneurial establishment, and finally an "Appraisal" referring to assessment of the project.

FRAMEWORKS AND CONCEPTUAL CHALLENGES

In considering methodologies for situational analysis in urban agriculture the starting point is the identification of a useful conceptual framework. One of these is the "Sustainable Livelihood Framework" (Figure 15.2). It assists in conceptualising the interrelationships between the different dimensions of people's lives and helps to reveal the complexity of urban livelihoods and poverty. It makes the conceptual link between understanding at the household level and at meso and macro level encouraging analysis of how livelihoods as influenced by institutional and policy processes and vice-versa (Sanderson, 2000, Martin *et al.*, 2000, and Carney, 1998). In consolidating the existing knowledge, and identifying knowledge gaps the project was guided by the sustainable livelihoods framework developed by the Institute for Development Studies, University of Sussex, and adapted to accommodate DFID's concerns as described in the book '*Sustainable Rural Livelihoods: What contribution can we make*? (Carney, 1998).

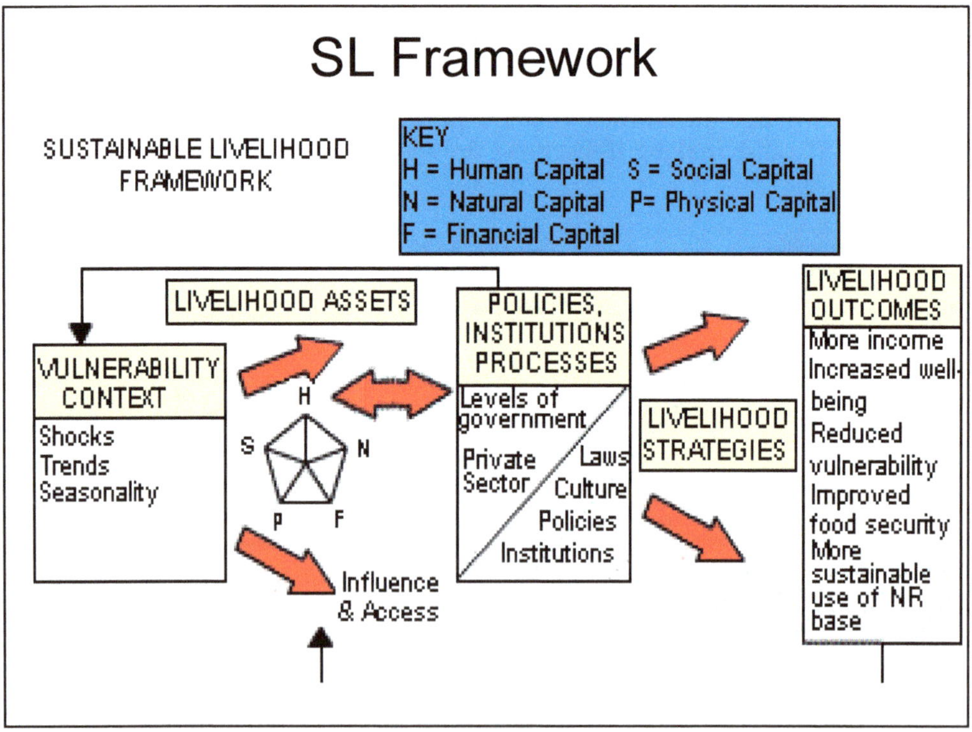

Figure 15.2: Adapted from Carney-1998

PARTICIPATORY MONITORING AND EVALUATION (M&E)

M&E is generally seen as means to assess project efficiency, effectiveness and relevance. It is expected to yield information about project report and accomplishments of targets. It can be viewed as a set of activities and methods to track change in a given situation or system, and/or to assess project progress and impact. It can help us to understand whether;

1. Anticipated changes have actually occurred, and
2. If these are in fact the result of the intervention under review.

As a whole, M& E involves gathering information, analysing data, judging and making decisions. To carry out these processes, M& E should be building upon an initial situation analysis and is likely to use related key parameters/indicators.

The evaluation process assesses how far research or development has achieved its objectives and examines any unplanned outcomes. According to Farrington *et al.* (1997) a monitoring and evaluation (M and E) process can be applied for specific stakeholders of a sustainable UA development (*e.g.* researchers or farmers) or at numerous levels (*e.g.* community or farm level or project and programme level). Each of these levels can have a number of purposes; each requiring different M and E approaches.

Examples of the activities that were monitored in the Zwelitsha project include soil quality, seedling production, vegetable production, income generated by farmers and vegetable consumption. Evaluation measures the change that the project has made to the life and well-being of the target group. It involves assessment before the project was initiated, and again after the project has been running to determine if the objectives have been achieved. The incorporation of mechanisms into this process is to allow for learning, correction and adjustment by all stakeholders posing significant challenges for M and E, especially on community or project level (Allen, 1997). Development of such an M and E process requires clear sets of long and short-term objectives, project inputs, outcomes, outputs, impacts and indicators of success. These must promote accountability and participation, which can be monitored and evaluated by the relevant decision-makers at all levels. For this purpose participatory and systems-based evaluation models are used, which help to facilitate the implementation of the monitoring and evaluation processes (Smith *et al.*, 2001).

FOCUS GROUP DISCUSSIONS

A total of 45 farmers were interviewed during the workshop to assess their motivation to garden (Table 15.1). The table reveals that the motivating factors for Zwelitsha farmers engaged in urban agriculture are predominantly for food security purposes, and for income generation. Food insecurity is the primary concern to the majority of farmers to be engaged in urban agriculture. Secondly, through the intervention of this project the farmers are also able to generate direct income through sales of their produce and indirect income through the reduction of expenditures on food. The income is of great value to these farmers, wheather it comes in the form of cash or "in kind" and moreover they produce organically, consequently food produced organically is healthy, poses less risk to health and environment. The net farm income for these farmers depends on their farming success, availability and cost of basic inputs, yield as determined by technology and access to market. The farmers gain higher yield because the soil is fertile, production inputs are of a high quality, and eco-technologies are being practiced. Marketing is a problem for the farmers because there is a lack of transport to enable them to transport vegetables to King Williams Town for marketing days. Sales occur on the project site with buyers coming to the project. On the otherhand this saves the producer transport costs to the market and also saves on storage costs as the produce is harvested when the consumers are in need of them.

Table 15.1: Ranked Reasons for Engaging in Urban Agriculture

Reason	*Frequency*	*Percentage*
Production for home consumption	25	56
Income generation	8	18
Unemployment	5	11
High prices of food	6	13
Supplementary income	1	2
Total	45	100

Table 15.2 shows that Swiss chard and cabbage is the most common crop grown by the farmers because they can grow it year round. Maize is not grown by the urban farmers at all, since they prefer to plant potatoes for income generation. Farmers are now planting a variety of crops since the intervention of the project. Crop diversification maintains soil fertility and balanced diet.

Table 15.2: Crops Grown by Zwelitsha Farmers

Zwelitsha farmers	Swiss chard (85 per cent); Cabbage (80 per cent); Onions (60 per cent); Beetroot (62) Carrots (70 per cent), Sweet potato (80 per cent) and Potatoes (85 per cent)

ADOPTION RATE OF ORGANIC AGRICULTURE AND CONSERVATION AGRICULTURE

The farmers were asked if they are practicing organic and conservation agriculture principles in their projects and the response was very positive. During the field visits it was observed that farmers are using compost, manures, mulching and biomass decomposition in their projects (Figure 15.3).

About 85 per cent of the farmers interviewed use compost, 80 per cent minimum tillage, 70 per cent kraal manure, 65 per cent crop residues and 60 per cent do mulching. The reason for this high adoption rate is because farmers were capacitated and empowered during the initiation of the project about the concept of sustainability, that the soil fertility depends on the manipulation of organic inputs rather than any

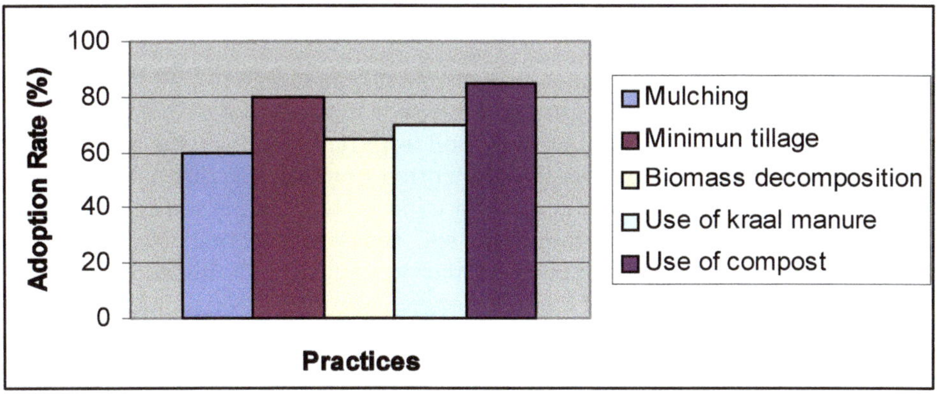

Figure 15.3: Adoption Rate of Eco-technologies

activity and if they managed their natural resources by using organic amendments the future will be bright for the next coming generation.

RESOURCE SUPPORT

The project provided individual gardeners and farmers groups with low cost gardening resources such as manure, seed, seedlings, and organic pest control remedies and demonstration of appropriate organic methodologies. A nursery was also constructed for the multiplication of sweet potato. The farmers were asked if they received inputs (seed and seedlings), did they receive it in time, and did they know how to use them. Fertilizers and pesticides were not supplied to farmers because the project promotes organic farming to enhance agro-ecosystem health, including biodiversity, biological cycles and soil biological activities. Because all the farmers were part of the project they all received adequate inputs as needed at a right time and had know-how of using them. The effectivity and quality of support in resources by the project is depicted in Figure 15.4.

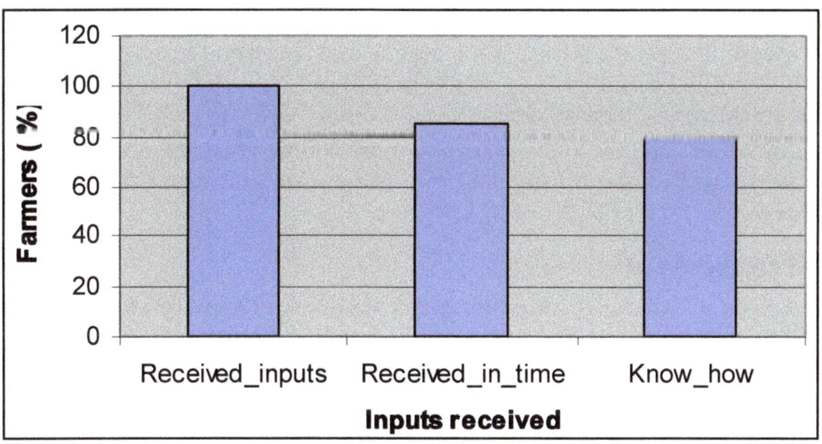

Figure 15.4: Resource Support

TRAINING MODULES

Soil Management

Soil is a medium for plant growth because it supports the plants, stores water and nutrients essential for plant growth. Soil fertility depends on the manipulation of organic inputs to the soil more than any other practice. The farmer's strategy in these projects is to increase the amount of organic material in the soil and to build up a larger permanent humus fraction in the soil. Biomass recycling also plays a chief role in soil fertility management because microorganisms mix the biomass with the soil to improve the conditions of the soil. Constant multiplication of microbes results in an increase in biological activity thereby improving the condition of the soil. Crop residues are left in the field after harvesting because of the good minimal value since they contain appreciable amounts of plant nutrients, controls weeds and check soil erosion. Soil must have sufficient air, water and nutrients for the germination of seeds and the healthy growth and development of small plants. Most soils have been compacted by feet and other traffic and this makes it difficult for the roots to grow down towards the nutrients. During the workshop sessions the farmers were reminded about the environmental factors to be considered when growing vegetables, such as wind, rainfall, moisture and humidity. The wind affects the crops, it was thus recommended to the farmers to use windbreaks or to plant across the direction of the wind. A farmer must not grow crops on clay soil and compacted soil and the planting method that prevents soil erosion are crucial for the success of the project. The importance of crop rotation to avoid the accumulation of pest and depletion of soil fertility was emphasized. Different crops require different nutrients so it is important to rotate the crops to avoid draining of essential elements in the soil.

Water Management

All the project participants had access to tap water provided by local municipality. The schools, churches and clinics are the public institutions that do not have to pay for their water usage. The quality of water was clean and suitable for drinking and irrigation purposes. Farmers do not pay for irrigation water but they adopted water conservation measures such as mulching and *in situ*-moisture conservation, because they are aware that South Africa is a water scarce country. Drum-drip irrigation is going to be installed in one of the projects in the next financial year in anticipation of future water shortages. This system is ideal for these kinds of projects and conditions as it applies water directly to the plants so there is little wastage, reduces water loss through wind and evaporation, is simple to operate and repair and is relatively inexpensive.

Crop Management

Plant pests and diseases can cause a loss of yield and damage vegetables making it difficult to eat and to sell. Regular use of synthetic chemicals such as pesticides is not recommended in this project because they pose a serious health threat to farmers and consumers and they also contaminate the environment and they are expensive. Integrated pest management (IPM) was recommended to farmers because it is a sustainable measure to control pests and diseases. Major steps of IPM include scouting, making rational decisions such as establishment of the density of pest population, severity of damage, applying treatment to reduce pest population below the threshold level. Scouting involves walking through the field early morning and in the evenings to identify problems, evaluate the effectiveness of preventive measures and establish need for treatment. Cultivation control is essential such as crop rotation to prevent build-up of pest and diseases and also the usage of pest resistant cultivars. Crop rotation is the key to the reduction of diseases and pests. Recipes of mixtures used to control pest were discussed with farmers such as the use of liquid soap spray to control aphids, nicotine spray, rhubarb spray for controlling aphids, marigold remedy etc.

Composting Technique

Composting is a way to reduce the volume of organic waste returning to the soil as nutrients which is beneficial for the growth of plants. In this project the farmers were encouraged to make use of compost because it is easy and cheap to make, improves soil fertility, soil texture, aeration and water holding capacity of the soil. Composting increases the activity of earthworms and other natural soil organisms that are beneficial for plant growth. Participants were also made aware about the moisture percentage when doing composting, temperature range, aeration, animal manures, green manuring, plant diversity and natural pest control. Procedures for building of compost pile were explained; starting with brown material, and then adding water, then green material and manure. The maintenance was also explained in detail that the farmer must check water level and air spaces to avoid bad odour and flies (Table 15.3).

Table 15.3: Building a Compost Heap

Layer	*What to be Used*	*How Much*	*Function*
Base	Maize cobs	20 cm deep	Improves aeration
First layer	News papers	1 layer	Keeps moisture
Second layer	Kraal manure	5 cm deep	Facilitate decomposition
Third layer	Greens	15–20 cm deep	Decomposed material
Cover	Cover	5 cm deep	Prevents bad odors

On Site Demonstrations

Demonstrations on using bio-intensive gardening techniques, such as trench bedding, raised beds and no-tillage system were demonstrated on the fields. Organic farming is promoted to avoid the negative health and environmental impacts of agro-chemicals and poisons that causes problems for people with compromised immune systems and save money that many of the target groups do not have. Demonstration of gardening practices is an essential step in any home garden approach because it serves as a training environment where farmers learn how to plant vitamin A rich crops and is also used during awareness of events. Demonstrations were based on seven crops covering the training workshops. Two demonstrations of 8x10 cm were established during practical training with 3 rows of sweet potato, 3 rows of cabbage,

2 rows of spinach, 4 rows of carrots, 6 rows of butternut, 2 rows of beetroot and 3 rows of onion. Practical demonstrating to member groups was done so that each member gains the skills and knowledge to set up their own gardens at home. The hands-on training was a great success because there was full participation from the farmers as well as communities that participated (Figure 15.5).

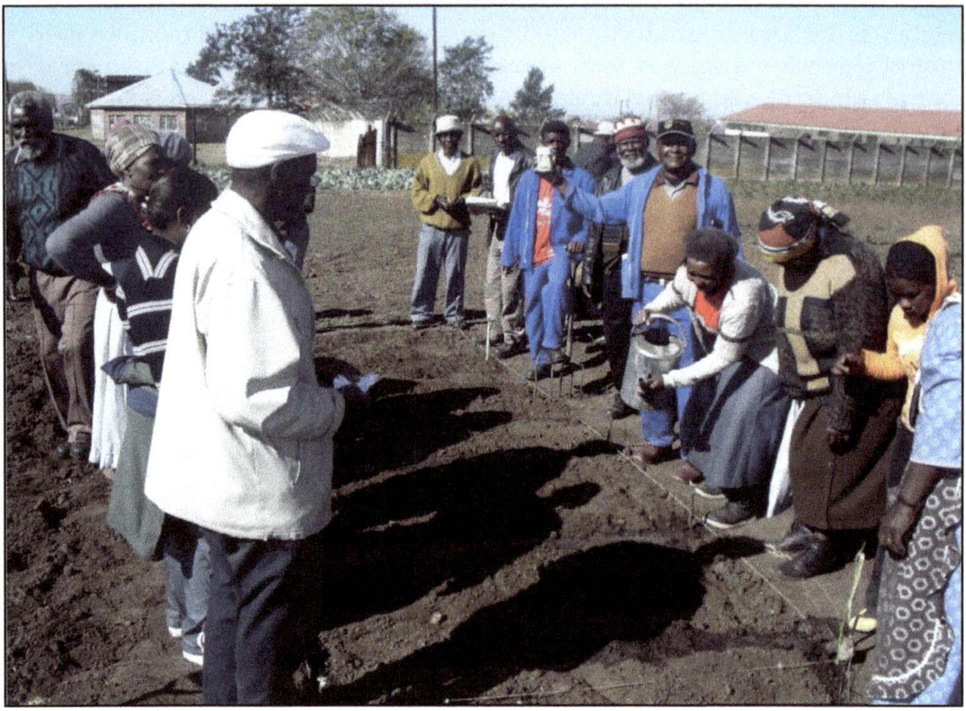

Figure 15.5: Enthusiastic Farmers are Learning by Doing Practicals

On-site Demonstrations in Zwelitsha

Five case studies of community vegetable gardens established in Zwelitsha as a part of this project deliverables were closely monitored to make sure that all the information shared with farmers during training and workshop is implemented. The information included in each case study refers to project establishment and membership, motivation, living strategies, livelihood outcomes and assets, problems experienced and successes and failures. Table 15.4 summarizes and compares these projects in the light of sustainable livelihood framework.

Different motives exist for each project visited, but main one was gardening for home consumption. All these projects are capable of achieving greater yields because of strong social capital (solidarity) and newly gained human capital (knowledge and skills). Good social capital is crucial in gaining access to other forms of capital such as financial, human, natural and physical capital. Financial capital is low for all the members because of high levels of hire purchase and indebtedness and most of them are pensioners and pensions provide the economic underpinning of many households. Marketing poses problems for community food gardens in Zwelitsha possibly because of disadvantaged access such as poor the market.

Table 15.4: Comparison of Five Case Studies in Line with Sustainable Livelihoods Framework

Aspects	*Vukuzenzele Vegetable Project*	*Ikhwezi ILokusa Vegetable Project*	*Loyiso Vegetable Project*	*Empilisweni Vegetable Project*	*Masincedana Vegetable Project*
Location	*School*	*Municipal Land*	*School*	*School*	*School*
Expected livelihood outcomes	Improved food security Income generation Reduced vulnerability	Improved food security Income generation Increased well-being	Secure food Income generation	Subsistence Generate income	Secure food Generate income Reduced vulnerability
Main constraints	Financial capital Lack of extension support	Garden pest	Garden pest	Institutional constraints Social capital Theft	Natural capital
Strength	Dedication Hardworking Independency	Sufficient time for gardening Dedicated Teamwork Show house	Enjoy gardening Sufficient time for gardening Favored by extension officers	Serve as an inspiration for people to start improving their own lives	Dedicated
Vulnerability context	Casual employment/ Insecure jobs			Casual employment replaces agricultural work Resource sink	Earn very little from casual employment

CONCLUSIONS

This project implimentation has shown that promotional strategies should include workshops on organic farming, marketing of organic products, and conservation of the natural environmental. In order to strengthen the long-term promotion, biomass recycling is ideal for our own local conditions. An information day will be a good way to spread the information to the people and to encourage them to do more of what they do presently. Zwelitsha has a very good potential in vegetable production and natural resource management. There has been very little research work on the organic farming system especially for the resource poor farmers. The information on the organic production methods appropriate to the circumstances and needs of the resource poor farmers is limited. Information on organic production methods for resource poor farmers is not well documented so there is a dire need to move organically, to put in place a system that both protects the environment and supplies adequate nutritious food in harmony with nature and uses as little external inputs as possible.

ACKNOWLEDGEMENTS

I wish to express my gratitude to Agricultural Research Council for sponsoring this project and providing me with an opportunity to work on *organic farming and focus on sustainable utilization of natural resources in urban and peri-urban areas of the Eastern Cape.* This is an issue of great relevance to poverty alleviation in urban and peri-urban areas and it is hoped that the project will be refunded in future by any funding agencies for further investigation on improvement of soil fertility by using organic inputs such as cattle and chicken manure, enriched compost, in comparison with (inorganic fertilizer). I am thankful to the Eastern Cape Department of Agriculture and Land Affairs for their networking and support and without their help I would not have been be able to achieve the success described in this report.

RECOMMENDATIONS

Some major preliminary recommendations for the support of Zwelitsha urban farmers and urban food production are:

- ☆ Strengthen the organisation of urban and peri-urban practitioners,
- ☆ Strengthen institutional capacity at local level,
- ☆ Land use regulations that protect urban agriculture uses will encourage investment in UPA and would make credit use viable,
- ☆ Assess land and water tenure/access conditions–develop policy reform proposals and tools,
- ☆ Establish nutrition education/consumer information,
- ☆ Establish linkages with other support services (in particular health),
- ☆ Conduct training in small-scale irrigation techniques and integrated plant protection techniques, and
- ☆ Identify opportunities for small scale enterprise development for *e.g.* recycling of organic materials.

It is recommended that funding agencies should consider providing for researchers, NGO's involved in sustainable urban agriculture, particularly those whose focus is on sustainable utilization of natural resource use.

REFERENCES

Allen, W., 2000. Future directions for research into collaborative learning–helping people maximize the use of technical information within multi-stakeholder environmental management contexts. NRM-change links working paper No. 2.

Eberhard, R. (1989a). *Urban Agriculture: The Potential in Cape Town. The Economics of Small Vegetable Gardens in Cape Town.* Vol.3. Working Paper 89/E3. Town Planning branch, City Planner's Department, Cape Town.

Chambers, R. (1997). "Whose reality counts? Putting the first last", Intermediate Technology Publications.

Farrington, J., Thirtle, C. and Henderson, S., 1997. Methodologies for monitoring and evaluating agricultural and natural resources research. *Agricultural Systems* 55, 273-300.

Hardoy, J.E., Mitlin, D and Satterthwaite, D. (2001). *Environmental problems in an urbanizing world: Finding solutions for cities in Africa, Asia and Latin America.* Earthscan, London.

Harris, P.J.C., Lloyd, H.D., Hofny-Collins, A.H., Barrett, AR and Browne, A.W. (1998). *Organic agriculture in sub-Saharan Africa: farmer demand and potential for development.* Kenilworth: Emmerson Press.

Lhopitallier4, L. (2000). A participatory zoning process at region level: the amatola district zoning project. Eastern Cape Province. South Africa.

Meadows, K. (2000). The Social and Institutional Aspects of Urban Agriculture in the Cape Flats, Western Cape, South Africa. NRI Report no 2526.Natural Resource Institute, University of Greenwich, UK.

Phnuyal, B., Archer, D. and Cottingham, S. (1998) Reflections on REFLECT, PLA Notes, Number 32: 27–30.

Raviv, M, Chen, Y. and Inbar, Y. (1986). The use of peat and compost as growth media for container-grown plants. In: Chen, Y. and Avnimelech, Y. (eds). *The role of Organic Matter in Morden Agriculture.* Dordrecht: Martinus Nijhoff Publ. 257-287

Smith, H.J., Steyn, C.E. and Jansen van Rensburg, H.G., 2001. An approach to promote agricultural development of urban and rural farming communities. Report GW/A/2001/49. ARC-ISCW, Pretoria.

United Nations Development Programme (1996). Sustainable Livelihoods (SL).

www. livelihood connect.org

Appendix 1
Revised Action Plant for Zwelitsha Urban Agriculture Project September to 2008/2009

What	*How*	*With whom*	*When*	*Where*	*Who is Responsible*	*Expected Outputs*
Compile business plan for farmers	By stating their focus, strengths, opportunities, weaknesses and threats and spell out their marketing plan.	Farmers	August September	ARC-ISCW	Lulama	Business plans
Purchase inputs for the farmers to practice conservation agriculture practices	Purchased from Lucia, local inputs supplier	Farmers, researcher and extension officers	September	King Williams	Lulama and Farmers	Purchased inputs were potato seedlings, cabbage, spinach, beetroot, butternut, green pepper and onions
Monitoring production	Regular field visit to experimental site. Physical observations and discussions with farmers.	Extension officers and farmers	October	Zwelitsha urban agriculture projects	Researcher, extension officers and farmers	Improved production methods

Contd...

Appendix 1–Contd...

What	*How*	*With whom*	*When*	*Where*	*Who is Responsible*	*Expected Outputs*
Activities below were not executed because of Insufficient funds						
Experimental design and treatment selection	Randomized Complete Block Design	ARC-ISCW, Biometrics and Extension officers	October November	ARC-ISCW and ILI	Researchers and technicians	Proper planning resulting to good results
Trial layout and site selection	Field planning	Researchers and farmers	November	Zwelitsha	Researchers Extension officer for site identification and farmers	Trial layout
Data collection	Soil and plant samples	Researchers, Farmers and Extension officers	December	Trial site in Zwelitsha	Researchers and farmers	Exchange of information between researchers, extension officers and farmers over the growing season. Focus for discussion
Maintenance of collaborative trial	Duration of the trial	Farmers, researcher and extension officers	November December	Trial site in Zwelitsha	Researchers and farmers	Well maintained trials
Data recording, storage and analysis	After harvesting	Farmers, researcher and extension officers	February March	Trial site in Zwelitsha	Researchers, extension officers and farmers	Compiling a final report
Reporting and feedback	Open day or field day	Researcher and leader farmers	February March	Trial site in Zwelitsha	Researchers, extension officers and farmers	Field day at trial site

Chapter 16

Computation of Urban Heat Island Created Due to Changes in Land Cover and Land Use: Case Study of Twin Cities of Pakistan

Saroop Shahid, M.A. Baig*, U. Khattak and Adnan Zahoor

**Faculty of the Institute of Geographic Information System (IGIS), National University of Sciences and Technology (NUST), Islamabad, Pakistan*

ABSTRACT

Changes in land use-land cover (LULC) pattern in relation to urban growth have made city environments congested, mainly due to lack of proper planning. This noticeably has an impact on the local temperature of the area, creating Urban Heat Island (UHI) effect which abundantly prevails in almost all densely populated cities around the globe. The main metropolitan region of Islamabad-Rawalpindi (twin cities) in Pakistan has experienced rapid urbanization leading to increased urban expansion. This study relates the changes in the LULC types in the twin cities and the land surface temperature (LST) during the summer season in order to estimate the UHI effect. Advanced Spaceborne Thermal Emission and Reflection Radiometer (ASTER) imageries of 2001 and 2006 were utilized for the retrieval of LST and LULC types. Normalized Difference Vegetation Index (NDVI), Normalized Difference Bare Index (NDBaI) and Normalized Difference Built-up Index (NDBI) were used to extract LULC attributes for quantitative and qualitative image analysis. Results indicated that variation in LST for both the cities was affected fundamentally by the spatial distribution and area extent of LULC type. Strong positive correlation between NDBI and NDBaI index with LST illustrated significant contribution of built-up and bare land types towards the rise in urban temperatures. Flaws in LULC planning were clearly evident through high urban temperatures in the city of Rawalpindi as compared to Islamabad.

Keywords: *Land surface temperature; ASTER; LULC; UHI; Remote sensing; GIS.*

INTRODUCTION

Several physical features of landscape combine to influence local climate variations, main among these being topography, proximity to a water body and urbanization. Urbanization involves urban growth in terms of both area and population; the latter being directly proportional to the former. These developments lead to an impervious built-up/urban land which drastically changes the surface physical characteristics, most important one being surface temperature variation [11]. Therefore, during the urban climate investigations, the factor of land surface temperature (LST) becomes very important [9] [8]. LST alters the air temperature and stands as a major player in the surface energy balance by efficiently partitioning latent heat fluxes and consequently the surface radiant temperature responses as a function of varying surface soil water content and vegetation cover, thus significantly affecting the local weather and climate [4].

The shifting trends in temperature can be clearly witnessed while travelling from urban downtown area to suburban or rural surroundings, or vice-versa. This is due to distinct landscapes of both areas; the urban largely comprising of buildings, roads, parking lots, and other paved surfaces, while the rural consisting of more natural forest, pastures and other flora. This variability in land cover and land use (LULC) causes urban areas to suffer from higher solar radiation absorption and a greater thermal conductivity and capacity [11] [12]. On hot summer days, one can feel the waves of blistering heat emanating from roads and dark buildings, which keep urban areas hot, even long after the sunset, while the rural areas have already begun to cool rapidly. This phenomenon of customized climate is referred to as the Urban Heat Island (UHI) effect, and has been documented for over 150 years [9].

The concept of UHI was first proposed by Manley (1958), who stated that a 'thermal isolated island' is generated when a city expands to such an extent that it changes the properties of its underlying surface and eventually suffers from severe air pollution, releasing substantial waste heat producing temperature considerably higher as compared to the surrounding rural regime. The U.S. Environmental Protection Agency (EPA) defines the phenomenon of UHI through the same reasoning that '*Heat islands form as vegetation is replaced by asphalt and concrete for roads, buildings, and other structures necessary to accommodate growing populations. These surfaces absorb, rather than reflect the sun's heat, causing surface temperatures and overall ambient temperatures to rise*' [3]. UHIs are found to have the potential of directly influencing the health and wellbeing of urban dwellers. Due to increased temperature, they also raise the degree and extent of heat waves within the cities. Research has shown that mortality rate during a heat wave increases exponentially with the maximum temperature [7].

The primary purpose of this study was to evaluate the phenomenon of UHI effect in the twin cities of Islamabad and Rawalpindi due to rapid urbanization in the region, and determine the relationship between surface temperature and LULC type, followed by exploring the impact of such changes on the surface temperature and the resulting UHI effect. This relationship was invrestigated by making use of RS and GIS technology, through the utilization of ASTER L1B imageries for the retrieval of LST and map LULC types in the twin cities during the summer months of May 2001 and 2006 respectively.

STUDY AREA

Study area for the present research encompasses the main metropolitan region of Islamabad-Rawalpindi which lies between 72°452 and 73°302 East longitude and 33°302 and 33°502 North latitude. The area encompasses portions of Islamabad district, the Rawalpindi district, and the Abbottabad district (Figure 16.1).

Islamabad and Rawalpindi cities are commonly referred to as twin cities due to their proximity to one another, but nevertheless boast contrasting way of life and LULC patterns. Islamabad, having a population of 0.805 million (1998 census) and a master-planned rectangular grid pattern nestled against the Margalla Hills, is the national capital and the center of all governmental and corporate activities. On the other hand, Rawalpindi with a population of 1.406 million (1998 census), is an older and much larger city lying to the south on Swan River and is a center of industrial and commercial activities.

The population of the twin cities is rapidly increasing leading to increased urban expansion. Both have experienced increase in population due to rural-urban migration and drastic developmental projects. Islamabad has grown rapidly since construction

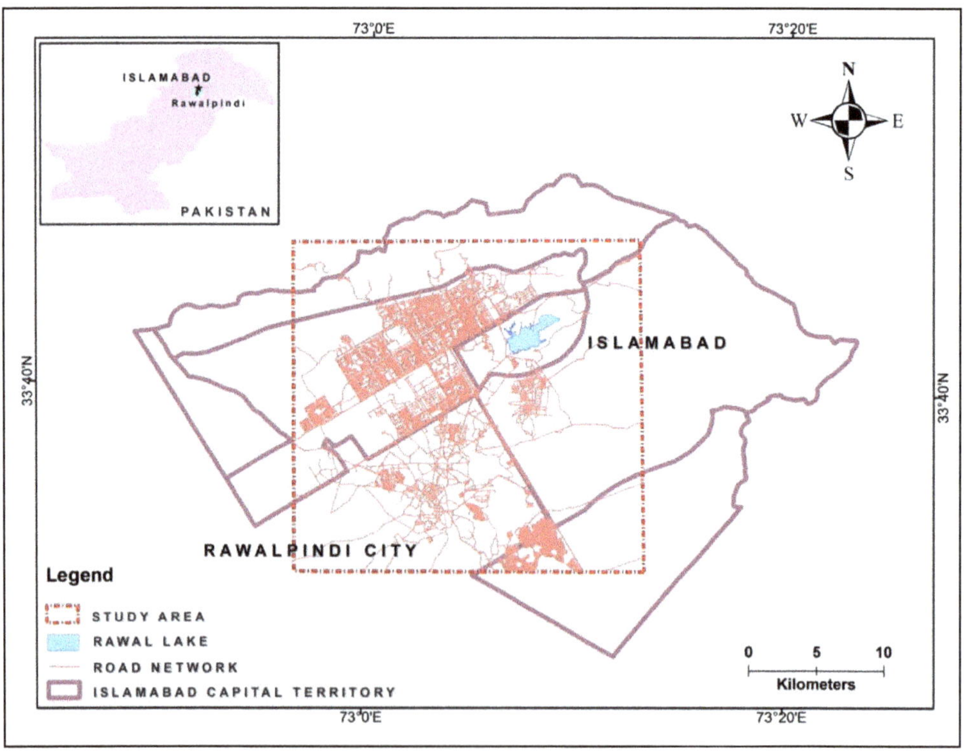

Figure 16.1: Map of Study Area Encompassing the Main Metropolitan Areas of Islamabad and Rawalpindi

began in 1961. In Rawalpindi alone, where land use is divided between commercial and residential, unplanned urban growth has been rampant, particularly in areas where basic infrastructure is available.

While Islamabad's growth is monitored through rapid development of square sectors (following its grid like master plan), Rawalpindi's urban growth is comparatively less controlled with city blocks small and constricted according to space than its neighbour, hence it is spreading alarmingly and encroaching into adjacent other non-urban land. Inadequate urban services, especially sewerage, drainage and solid waste management, have worsened quality of life and environmental conditions. The cities are therefore ideally suitable for the analysis of UHI phenomenon due to its mixture of LULC types and the rapid urbanization.

METHODOLGY

Satellite Data

Two cloudless ASTER L1B images acquired on May 16, 2001 and May 30, 2006 were used respectively, purchased from LP DAAC (NASA). The satellite images were already atmospherically corrected. The images were then geo-referenced and projected to a common Universal Transverse Mercator (UTM) coordinate system for analysis.

Meteorological Data

Two meteorological parameters, temperature and rainfall, were utilized for this study. The data for these parameters was collected from Pakistan Meteorological Department (PMD) Islamabad. As satellites have their own respective image acquisition time, it was essential to obtain hourly meteorological data apart from the average data for the acquired image dates.

A step by step methodology was exercised to analyze the temporal land use-land cover changes and UHI effect in the region (Figure 16.2).The climate data was utilized for relating the ambient air temperature with the near surface temperature derived from the respective imageries. The processed imageries were used for LULC classification, retrieval of vegetation, bareness and built-up indices, and calculation of brightness temperature from radiance in order to finally retrieve the required LST in the study area.

LULC Type Classification

Supervised Classification was done by using the maximum likelihood method. Seven (07) LULC classes were assigned, namely; built-up, forest, woodland, grassland, cropland, transitional area and water.

Brightness Temperature

Brightness Temperature (BT) is basically the blackbody temperature, which is a measure of the photons received by the respective sensor's wavelength presented in units of temperature. Given that the photons are also emitted by the water vapour present in the atmosphere apart from those emitted by the underlying surface, it is therefore the combination of both [5]. BT can be derived from the radiances computed

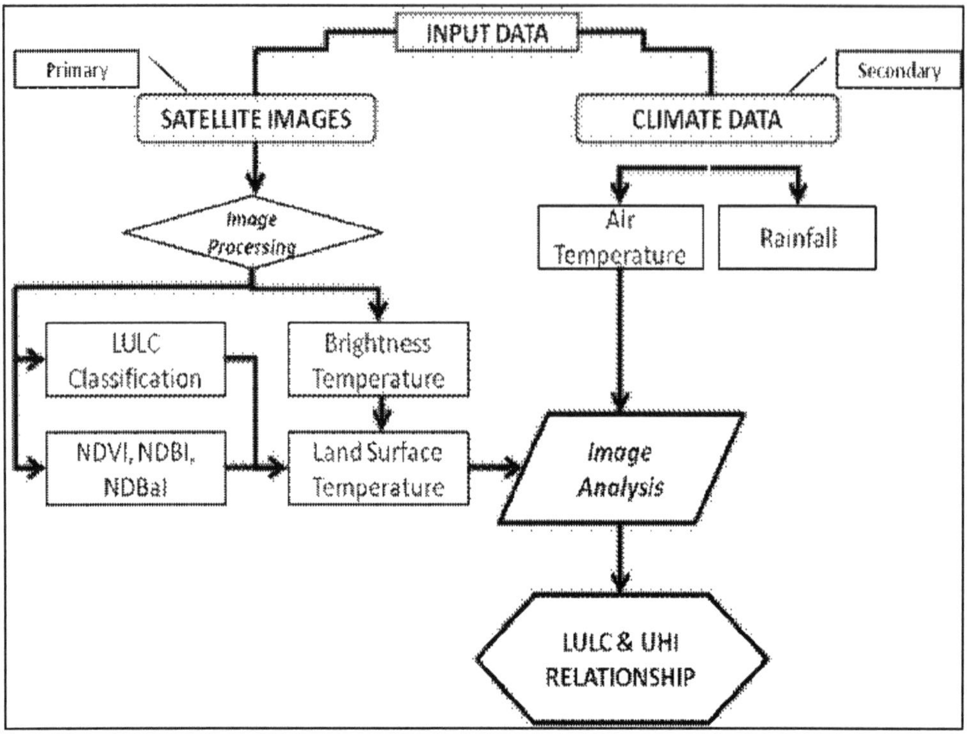

Figure 16.2: Methodology Flow Chart Summarizing the Main Events Involved in the Processing of Data for Determining LULC and UHI Relationship

from thermal infrared sensors at the top of the atmosphere (TOA) by using the Plank's law [2].

ASTER L1B imagery has its radiometric and geometric coefficients applied. Hence, when the thermal bands of the images were imported into ERDAS Imagine 9.1, their DN values were automatically converted into radiance values. ASTER consists of five thermal bands. For computation of BT, ASTER band 13 (10.25–10.95 μm) was selected because the spectral width of this band is close to the peak radiation of the black-body spectrum given off by the urban surface' [10]. The spectral radiance of band 13 was converted to at-sensor brightness temperature by replacing the sensor response function with a delta function at the sensor's central wavelength, as shown in eq 1.

$$BT = \frac{c_2}{\lambda_c \mathrm{Ln}\left(\frac{c_1}{\lambda_c^5 \pi L_\lambda} + 1\right)} \quad \text{................(1)}$$

where,

c_1: First radiation constant = 3.74151 x 10^{-16} $mW * cm^{-2} * sr^{-1} \mu m^{-1}$

c_2: Second radiation constant = 0.0143879 $m*K$

λ_c: Central wavelength for ASTER band 13

L_λ: Radiance observed by sensor

For evaluation of LST in association with the major land cover types, three basic indices were used namely; NDVI, NDBaI and NDBI, which target the three major land cover types, *i.e.* vegetation, bare land and built-up.

Computation of NDVI

NDVI is the most widely used vegetation index developed in 1998 [6]. It tells about the presence, vigour and density of vegetation in an area through the band ratio of infrared and near infrared band wavelengths. The values generated range from -1 to +1, with positive values indicating vegetated areas and negative values signifying non-vegetated surface features.

For ASTER, the equation is given by:

$$\text{NDVI} = \frac{(\text{Band 3N} - \text{Band 2})}{(\text{Band 3N} + \text{Band 2})} \qquad \text{................(2)}$$

Computation of NDBI

NDBI was used for extracting the built-up area from the image [13]. For ASTER imagery, there wasn't any index developed that was sensitive to the built-up area. Hence this research proposed this index for ASTER imagery, through its derivation from the equations given for Land sat 5 imagery by matching the wavelengths of the respective bands (eq 3).

$$\text{NDVI} = \frac{(\text{Band 4} - \text{Band 3N})}{(\text{Band 4} + \text{Band 3N})} \qquad \text{................(3)}$$

Computation of NDBaI

For retrieval of bare land from the image, NDBaI was used [14]. This index clearly differentiates the bare land pixels from those of built-up, as both these LC types have almost the same spectral characteristics. However, this index was again not present for ASTER in literature; therefore it was proposed by the authors through the same method as described earlier for NDBI (eq 4).

$$\text{NDBaI} = \frac{(\text{Band 4} - \text{Band}\,[13+14])}{(\text{Band 4} + \text{Band}\,[13+14])} \qquad \text{................(4)}$$

Retrieval of LST

Since the brightness temperature gives the temperature of both; atmosphere and surface, it wasn't sufficient in accurately relating the UHI and LULC type changes in the study area. Hence it was essential to derive the LST in order to evaluate the relationship with precision. For this, the retrieval of emissivity [$\varepsilon_\downarrow(o)$] was required first, which was accomplished by adopting the method given in eq 5 [1]. This method uses the values of NDVI as the means for calculating emissivity of an image.

$$\varepsilon_0 = 1.009 + 0.047.\mathrm{Ln(NDVI)} \qquad \text{................(5)}$$

After the calculation of emissivity, the final step for the retrieval of LST was accomplished by taking into account the brightness temperature previously calculated, through eq 6, which is:

$$\mathrm{LST} = \frac{\mathrm{BT}}{\varepsilon_0^0.25} \qquad \text{................(6)}$$

Results and Discussion

LULC Changes

Islamabad and Rawalpindi experienced drastic changes in the spatio-temporal LULC patterns between 2001 and 2006. Major developments took place in Rawalpindi where haphazard settlement gave rise to a more contiguous pattern of built-up area as compared to Islamabad where mostly sectors emerged according to a plan. Figure 16.3 shows the LULC classified images and changes in the study area for the years 2001 and 2006.

Through percentage-wise analysis of LULC changes, built-up area was found to be the most rapidly growing land cover type; increasing from 24.15 per cent in 2001 to 34.2 per cent in 2006, *i.e.* almost a 10 per cent increase in area (Table 16.1). As the built-up area increased, the percentage of transitional area declined from 20.17 per cent in 2001 to 5.65 per cent in 2006, due to constant development. Forest area almost remained the same while the woodland area declined by 10 per cent. Grassland area which basically consists of parks, greenbelts along roads and playgrounds increased from 6.98 per cent to 21.95 per cent. This class mainly fell in Islamabad city, because its master plan allots considerable green area to the city.

Table 16.1: Percentage Area of LULC Classification Scheme for the Years 2001 and 2006

Class Name	*2001*		*2006*	
	Area(Km²)	*Percentage*	*Area(Km²)*	*Percentage*
Built-up	152.22	24.15	215.53	34.20
Cropland	118.21	18.76	116.76	18.53
Forest	34.11	5.41	33.60	5.33
Woodland	151.28	24.00	88.51	14.04
Grassland	44.02	6.98	138.33	21.95
Transitional Areas	127.12	20.17	35.63	5.65
Water	3.29	0.52	1.89	0.30
Total	630	100	630	100

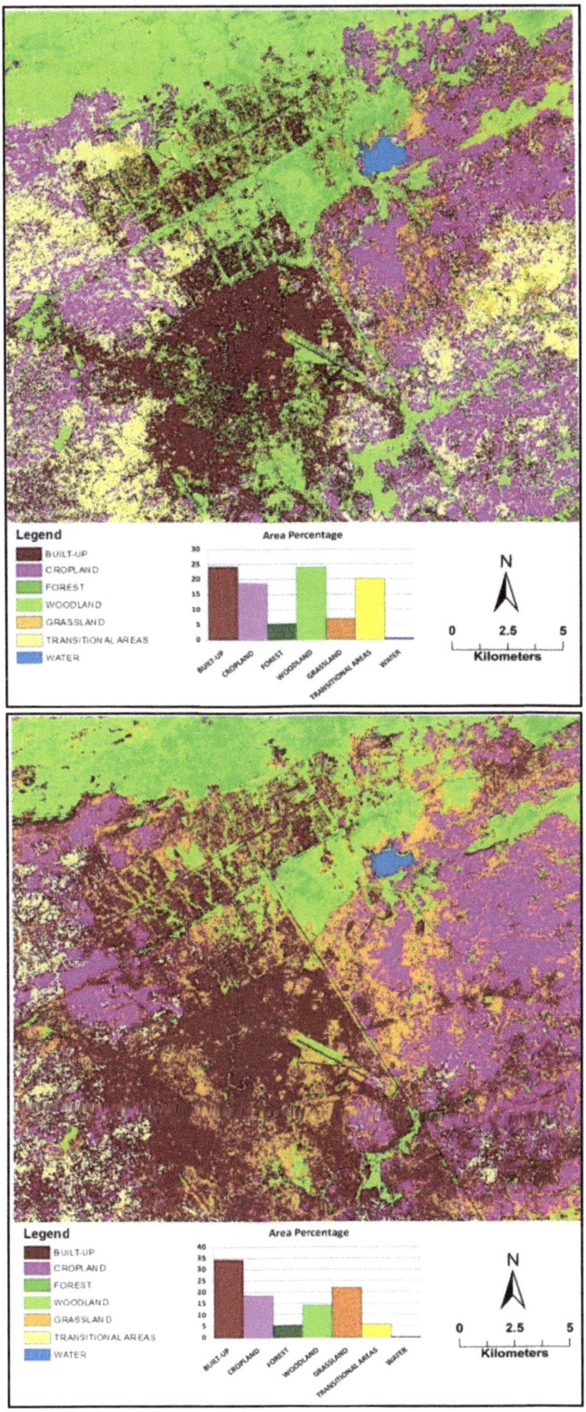

(a)

(b)

Figure 16.3: LULC Classification of Islamabad-Rawalpindi Derived from ASTER, Showing the Respective Classes and Area Percentage for 2001 (a) and 2006 (b)

Spatial Distribution of LST over LULC Types

It was seen that high LST resulted in May 2006, with temperatures soaring up to approximately 43°C (Figure 16.4b). Forest area and water showed the lowest temperature while built-up and barren land had the highest. An increasing trend was witnessed in the mean LST of all the LULC types during 2001 and 2006 (Table 16.2). Islamabad showed a dispersed form of variation in LST while Rawalpindi had major hot spots in the congested urban area. The data on the agricultural activities in the study area plays a major role in the interpretation of the LST images. The month of May fell in the late part of the crop harvesting season when the area of cropland was in a transitional stage. The *kharif* season crops (*jowar, bajra,* etc.) in the area are harvested between June and October, while rabbi season crops (wheat, vegetables, etc.) are harvested between November and April. Hence, high LST was observed in cropland areas in both the years due to low moisture content and barren areas. Figure

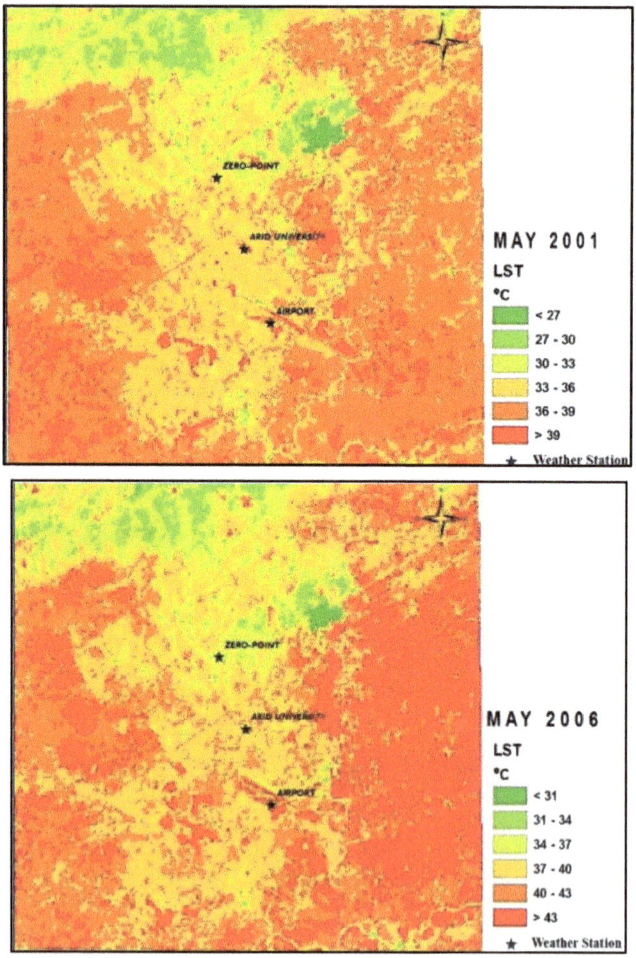

(a)

(b)

Figure 16.4: LST of Islamabad-Rawalpindi Derived from ASTER, Showing the Respective Temperature Ranges and Weather Stations for 2001 (a) and 2006 (b)

16.5 demonstrates linear trend of LST with the expansion of built-up areas from 2001 to 2006, therefore portraying a significant positive correlation between the two (R^2= 0.8508).

Table 16.2: Mean LST of LULC Types of 2001 and 2006 from ASTER Imagery

LC Type	*Mean LST (°C)*	
	2001	*2006*
Built-up	36.71	41.48
Cropland	39.94	43.57
Forest	29.44	33.34
Woodland	32.43	35.37
Grassland	33.84	37.12
Under Development	38.45	42.8
Water	26.23	29.99

Ambient air temperature data was obtained from three weather stations having specific time of image acquisition in order to compare it with LST retrieved through image analysis. The comparison showed the variation in temperature near the surface and in the air, depicting the intensity of surface UHIs during daytime (Table 16.3).

LST and Various Indices

The indices NDVI, NDBI and NDBaI were obtained from ASTER imageries of 2001 and 2006 respectively. Index value ranges were found to be different for different images due to variation in seasonal, atmospheric and ground conditions of the respective images. NDVI value range for May 2001 showed the lowest value of –0.06 to 0.55 as compared to May 2006 (Figures 16.6a and b). Positive values showed high vegetation cover, mostly the forest area, while negative values showed built-up, barren

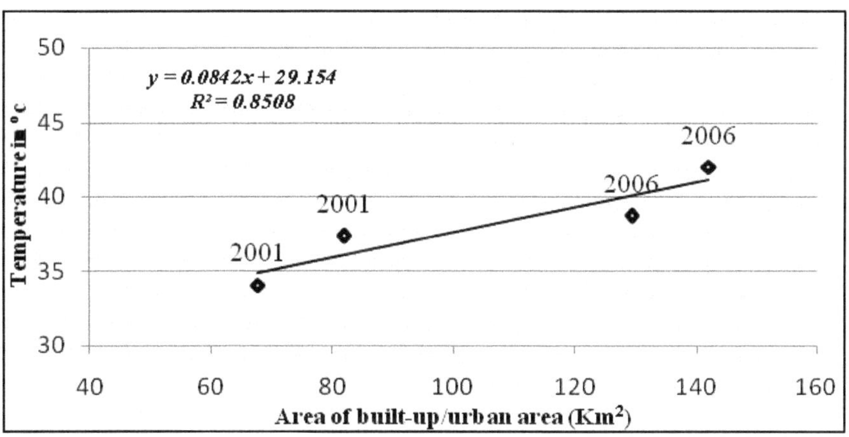

Figure 16.5: Scatterplot Showing Positive Correlation Between LST and Area of Built-up/Urban Land in the Study Area for 2001 and 2006

land and water. NDBI values for all the images collectively fell in the range of 0.09 to 0.63 (Figures 16.6c and d), the highest showing built-up and some barren land while the lowest showing vegetation and water. NDBaI showed the most barren land in 2001 having the value range of 0.01-0.18 and the least barren land in 2006 (0.09–0.17) as shown in Figures 16.6e and 16.6f.

Table 16.3: Difference Between Ambient Air Temperature Recorded from MET Stations and Surface Temperature Calculated on the Same Location in the Study Area through Image Analysis

Date	*Met Stations*	*Surface Temp (from image) °C*	*Air Temp (from station) °C*	*Difference °C*
16/05/2001	Zero Point	35.85	33.5	2.35
	Arid University	36.38	33.7	2.68
	Airport	37.33	33.6	3.73
30/05/2006	Zero Point	38.81	36	2.81
	Arid University	40.31	38.8	1.51
	Airport	39.93	37.2	2.73

The land surface or near land surface temperature can be affected by the nature of land surface cover, ranging from the bare ground to built-up areas to vegetation cover types of variable density. This relationship between the land cover and resulting temperature was quantitatively analyzed for the present study through statistical analysis.

NDVI can be used as a surrogate measure for the density and vigour of vegetation for the evaluation of variation in temperature. Through the association with LULC types retrieved for the period of 2001 and 2006, the results for regression analysis of LST and NDVI showed that a strong negative relationship exists between the two. This negative relationship is also being reported by many studies of thermal remote sensing for urban and rural environments [4].

Although the correlation results were substantial for the selected years, the ambiguity faced by the differentiation of built-up area from barren land through NDBI made the mean NDBI values of the latter more than the former. The mean NDBI for cropland and transitional areas was the same and collectively more than that of built-up area in May 2001 ($R^2 = 0.9586$), whereas for May 2006 the mean NDBI of built-up area and cropland area was same and collectively more than for the transitional areas ($R^2 = 0.9064$). The mean values of NDBaI for the corresponding LULC types aptly analyzed the relationship between the bareness and its corresponding LST, indicating a statistically significant correlation between the variables in May 2001 ($R^2 = 0.9331$). Table 16.4 shows the overall regression parameters for all the above mentioned relationships between indices and mean LST of LULC types respectively.

Figure 16.6: Indices NDVI (a and b), NDBI (c and d) and NDBal (e and f) Derived from ASTER, Showing their Respective Value Ranges for the Years 2001 and 2006 Respectively

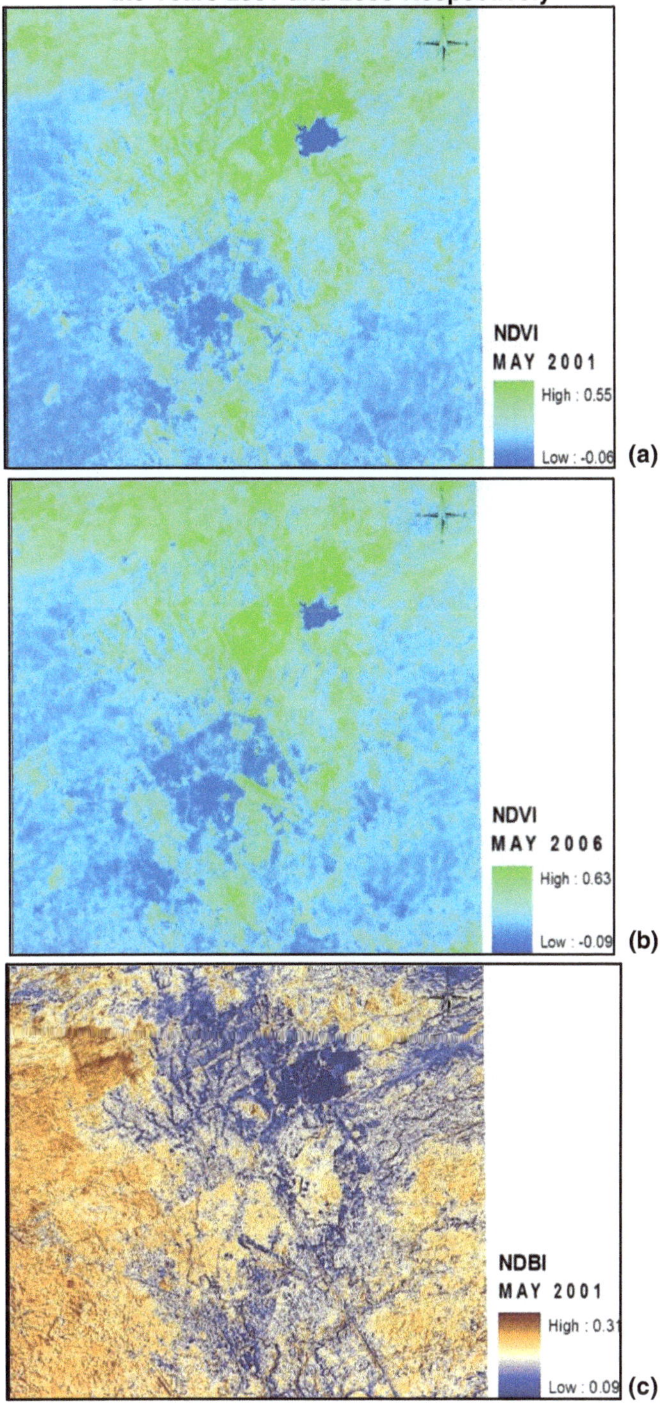

Figure 16.6–Contd...

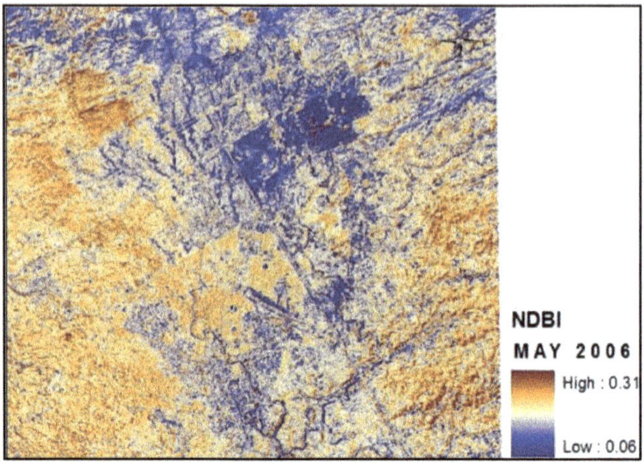

(d)

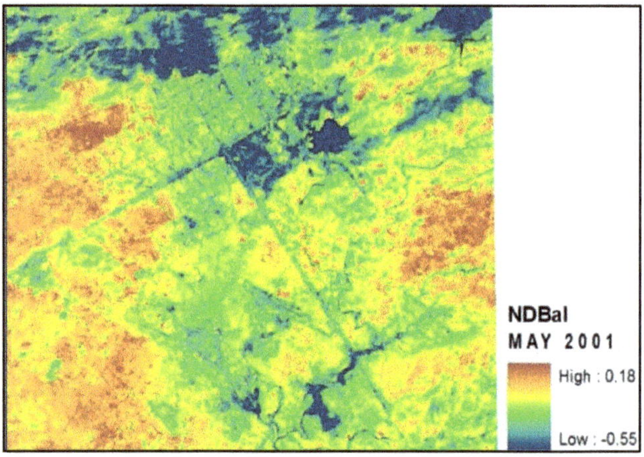

(e)

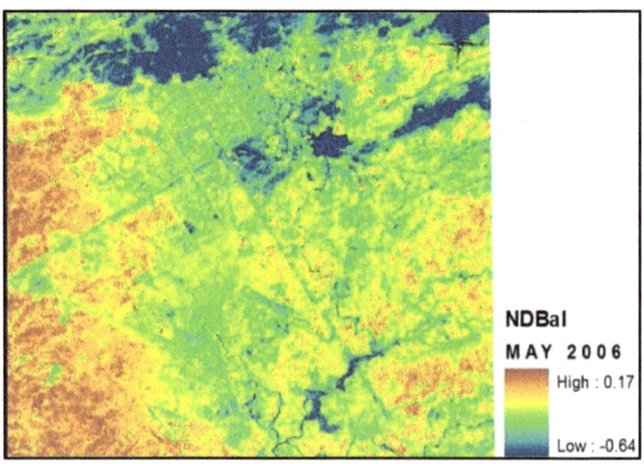

(f)

Table 16.4: Regression Parameters of Linear Relationship Between the LST and Respective Indices of NDVI, NDBI and NDBal

	NDVI-LST		*NDBI-LST*		*NDBal-LST*	
	2001	*2006*	*2001*	*2006*	*2001*	*2006*
r	–0.9168	–0.8416	0.9741	0.9520	0.9659	0.9169
R^2	0.8406	0.7084	0.9586	0.9064	0.9331	0.8408

r: Correlation coefficient; R^2: Regression coefficient.

CONCLUSION

All the analyses in this study were based on the interpretation of remote sensing data, thus suggesting that remote sensing data are helpful for analysing regional UHI effects. Results indicate that the built-up area expanded in the twin cities, while natural vegetation cover decreased. The variation in LST, which determines urban surface thermal spatial patterns and intensities, was affected fundamentally by the spatial distribution and areal extent of LULC type in both the cities. Changes in LULC types modified the reflectance of the land surface and thus affected the heat and moisture fluctuations. Quantitative analysis between LST and indices showed that great differences in temperature existed within a LC type, apart from variations amongst different LC types in the study area. Strong positive correlation between NDBI and NDBaI index with LST illustrated the significant contribution of built-up and bare land types towards the rise in urban temperatures. Islamabad being a master planned city had its LULC type well defined and marked, having a balanced LST in its urban area, but still the temperatures were high. However, contrary to its neighbour, Rawalpindi's haphazard development and no urban planning background made it a victim of UHI effect which was more pronounced in the year 2006 due to more impervious surface and less vegetation. Hence, the observed changes in LULC were mainly attributed to poor land use planning and inconsistent government policies, which is clearly evident through high urban temperatures in the city of Rawalpindi as compared to Islamabad.

REFERENCES

Bastiaanssen, W. G. M., Menenti, M., Feedes, R. A., Holtslag, A. M. (1998a). A Remote Sensing Surface Energy Algorithm for Land (SEBAL): Formulation, Part 1 formulation, Journal of Hydrology 212-213, pp 198-212.

Buechley R. W., J. Van Bruggen, and Trippi L. E. (1972). Heat island = death island? Environmental Research 5: 85–92.

Dash, P., Gottsche, F. -M., Olesen, F. -S., and Fischer, H. (2002). Land surface temperature and emissivity estimation from passive sensor data: Theory and practice—current trends. International Journal of Remote Sensing, 23(13): 2563"2594.

D. Ward (2008-04-15). Urban Heat Islands. Retrieved on 18-02-2009. *http://www.windows.ucar.edu/tour/link=/earth/climate/urban_heat_islands.html*.

Dousset, B. and Gourmelon, F. (2003). Satellite multi-sensor data analysis of urban surface temperatures and land cover. Photogrammetry and Remote Sensing Sensing of Environment, 106, 375–386.

Image Science and Analysis Laboratory, NASA-Johnson Space Centre. The Gateway to Astronaut Photography of Earth. Retrieved 07-03-2009. http://earth.jsc.nasa.gov/debrief/Iss007/topFiles/ISS007-E-13090.htm.

Purevdorj, T.S., Tateishi, R., Ishiyama., T. and Honda, Y. (1998). Relationships between per cent vegetation cover and vegetation indices. International Journal of Remote Sensing 19 (18): 3519–3535.

Small, C. (2006). Comparative analysis of urban reflectance and surface temperature. Remote Sensing of Environment 104, 168–189.

Voogt, J.A. and Oke, T.R. (2003). Thermal remote sensing of urban climates. Remote Sensing of Environment 86, 370–384.

Weng, Q., Lu, D. and Schubring, J. (2004). Estimation of land surface temperature–vegetation abundance relationship for urban heat island studies. Rem. Sens. Env. 89: 467–483.

Xian, G. and Crane, M. (2006). An analysis of urban thermal characteristics and associated land cover in Tampa Bay and Las Vegas using Land sat satellite data. Remote Sensing of Environment 104 147–156

Yuan, F. and Baver, E.M. (2007). Comparison of impervious surface area and normalized difference vegetation index as indicators of surface urban heat island effects in Land sat imagery. Rem. Sens. Env. 106: 375–386.

Zha, Y., Gao, J., Ni, S. (2003). Use of normalized difference built-up index in automatically mapping urban areas from TM imagery. International Journal of Remote Sensing 24 (3), 583–659.

Zhao, H. M., and Chen, X. L. (2005). Use of normalized difference bareness index in quickly mapping bare areas from TM/ETM+. Geoscience and Remote Sensing Symposium, 3(25–29), 1666"1668.

Chapter 17

Environmental Effects and Mitigation Measures of Urban Road TAG (Tarsus-Adana-Gaziantep) Highway in Adana City

Muzaffer Yücel, Nurgül Konaklı, Aysel Güzelmansur and Deniz Çolakkadıoğlu

Cukurova University, Agricultural Faculty, Department of Landscape Architecture

ABSTRACT

It is commonly known that highways cause comprehensive environmental effects during construction and operating stage. Many countries have started to use and practice new concepts such as; sustainable transportation, highway management to decrease natural and social negative effects of transportation on environment.The research area has a total surface of 6 210,26 ha which includes 500 m wide away from North and South side of the highway and 60 km long route within the Adana urban boundary which is lying between " Adana West side" (Tarsus paying Office) and " Adana East side" (Gaziantep paying office) at TAG (Tarsus Adana Gaziantep) highway. To simplify the traffic on that side of the highway entrance and exit ramps from the 7 different points (reciprocal 14 entrances and exits) were added for urban traffic access in recent years. Thus, intensity of usage has been increased. The environmental problems such as; noise, soil, air, and visual pollution have been increased by the intensity of traffic.

The aims of the study were to 1) investigate the natural and socio cultural structures of TAG highway which take place in Adana urban boundary and 2) determine the environmental problems such as noise, soil and visual pollution on the natural environment and human health, and then improve new suggestions according to sustainable aspect to reduce these problems on the city. We studied Pb in the soil and it has been found under the limits according to the Soil Problem

Control Regulation dated 10.12.2001. Besides, we determined noise measurement at 126 points to establish the noise problem and then results have been compared with the regulation of Environmental Noise Assessment and Management Regulation on Environmental Noise Limit Values dated in 07/03/2008. It has been determined that while maximum immediate measurement has been exceeded the limitation at 70 noise measurement points, according to average values have been exceeded at 35 noise measurement points.

Planning and design principles oriented tangible suggestions have been improved to decrease negative effects of the highway on the natural, physical and socio-cultural structures and integration with urban fabric.

Keywords: *TAG highway, Noise pollution, Soil pollution, Visual pollution.*

INTRODUCTION

One of the significant elements of landscape is the highway that has been experienced intensive interaction between human and environment. Besides, its impact on natural resources and scenic quality of landscape, it affects comfort of social life. Many countries have started to use and practices new concept such as sustainable transportation, highway management to decrease negative effects of transportation on natural and social environment Highways are formed from a network of roads in the world. Although it divided the landscape in segments, it should be an integral part of the landscapes [1].

Comparing with other transportation means, highway transport has been developing very rapidly in Turkey in the last thirty years and highway transportation takes place more than 90 per cent of the total transport [2]. Some environmental problems such as traffic safety, noise, air, soil, and visual pollution has been significantly increased during the development of highways.

It is important to ensure convenient, health and safety living environment for residents in big cities. It is possible to protect city's natural resources to improve environmental problems of daily living comfort and to increase social welfare. One of the main objectives of landscape planning studies is repairing damaged natural areas as much as possible and increasing the resistance to environmental conditions. In this perspective, determining route compatible with environmental factors and redacts or eliminates of the environmental problems, after it started to traffic uses.

TAG (Tarsus-Adana-Gaziantep) highway is one of the most important investment with reaching total 262 kilometers and 54 kilometers long highway connecting roads in Turkey. TAG highway is one of the main artery connections among Europe and Middle East countries and it has an intensive traffic load.

This research includes the section of TAG highway within Adana city's urban limits (between two paying offices) (Figure 17.1). In recent years, entrance and exit ramps from 7 different points (reciprocal 14 entrances and exits) from that section of the highway were given to urban traffic access to simplify the urban traffic to increase the road use intensity. Appreciable increase in environmental problems such as noise, soil and air pollution have been observed with increasing traffic intensity.

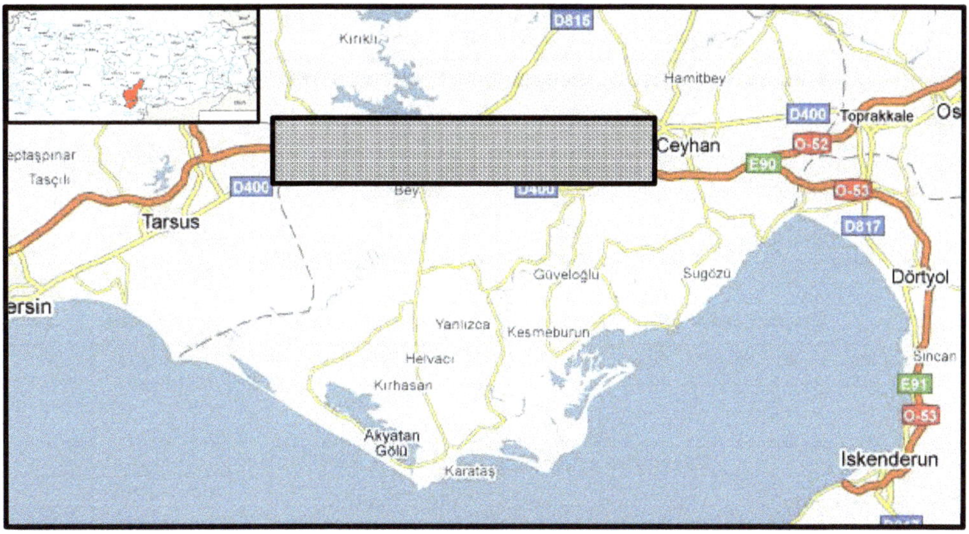

Figure 17.1: Location of Study Area

Urban residents living near a roadway experience higher exposure to motor vehicle emissions compared to those living further away.

Sometimes the junctions that connect the highway to the urban road network increase the accident risks in terms of traffic safety. In this context, the natural and socio-cultural structure of the section of the highway within the city limits was analyzed. Negative effects to the environment and human health were determined and improved tangible proposals about minimizing these effects.

MATERIALS AND METHODS

Research area comprises 60 km long route within the Adana urban boundary which is lying between "The west side of Adana" (Tarsus paying Office) and "The east side of Adana" (Gaziantep paying office) at TAG (Tarsus-Adana-Gaziantep) highway (Figure 17.1). It has a total surface area of 6 210, 26 ha, which includes 500 m wide away from North and South side of the highway. Also there are 7 junctions connecting city roads to the highway and they are located between the two above mentioned paying offices.

1/5 000 scale municipal plan, 1/25 000 scale topographic maps, soil maps, implementation projects of highway and decibel measuring device were used as study materials. Moreover, face to face questionnaire was applied to residents living near highway and literatures about this subject were used.

The research procedure is consisted of 4 steps (Figure 17.2). In the first step, natural and socioeconomic structure of the research area and nearby was analyzed and potential land use was determined.

In the second step, potential environmental negative impacts, the relations and contradictions between the other uses were identified. Junctions provide connections

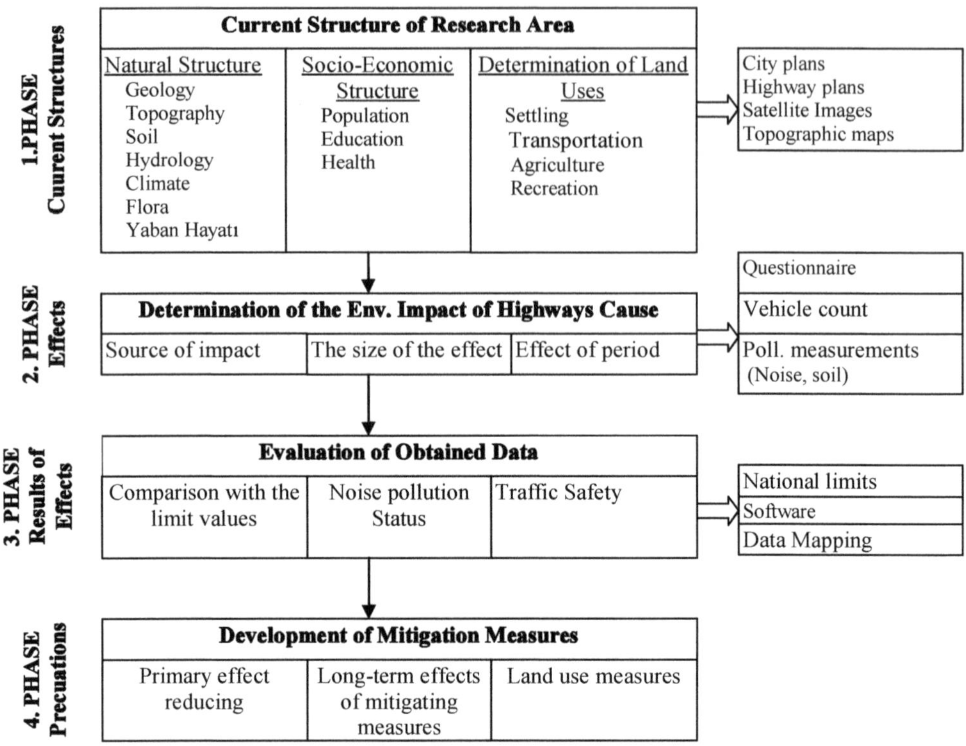

Figure 17.2: Flowchart of the Methodology

between highway and cities, which are opened for traffic use by 2006, were evaluated from the view point of planning and design principles and traffic safety. At this stage, the effect of the highway as the noise was the first; soil and visual pollution were analyzed with different measurements and evaluations. The sentence is not clear??? A questionnaire study was conducted in urban residential settings with high road-traffic noise exposure and influenced degree of residents was determined.

Third step was the evaluation phase and it included the analysis of collected data. Impacts were compared with the designed limit values.

In the last phase of the study proposals that minimize negative effects and supplement scenic and functional unity were developed. Proposals were determined according to effects and affected elements at the priority regions and uses of natural and man-made planning elements on nearby and along the highway within the context of urban plans.

RESULTS AND DISCUSSION

The research was done to determine the negative effects of the highway on residents living nearby of the highway and to improve tangible suggestions based on a planning and design principles to decrease influence of the highway on the natural, physical and socio-cultural structures and to contribute integrating with urban fabric.

Socio-cultural Structure

A questionnaire was made for residents who live around the highway in the research area to provide how and at which level they were affected from the highway. The dataset utilized in the present study were based on questionnaire data obtained from 493 residents. Survey sheet was composed of 15 questions as 14 of them were optioned and 1 of them was open–exit question. 8 questions are determined demographical structure and socio-economic conditions of residents and 7 questions are determine how and at which level they were influenced from the impacts of the highway.

In the first stage, individuals participating in the survey were asked to be in 2 options as "positive and negative" effects of highway. If its effects were negative they were asked options to be in a total of 5 as "noise pollution, air pollution, soil pollution, visual pollution and other" and if positive options to be in a total of 2 "providing ease of transportation and other"

More than half of the residents (56 per cent) have been effected positively from the highway, while 38.9 per cent of them have been effected negatively. 3.7 per cent of the residents have been effected both positively and negatively from the highway (Table 17.1).

Table 17.1: Negative or Positive Effect of Highways According to Age and Gender Distribution (Per cent)

Age	*Gender*	*Effects from Highway*				
		Not Answered	*Negative*	*Positive-Negative*	*Positive*	*Total*
7-18 age	Male	0.0	29.2	0.0	70.8	100.0
	Female	0.0	33.3	0.0	66.7	100.0
19-30 age	Male	4.3	29.8	2.1	63.8	100.0
	Female	0.0	37.5	6.3	56.3	100.0
31-45 age	Male	0.0	33.3	0.0	66.7	100.0
	Female	0.0	47.1	7.1	45.7	100.0
46-60 age	Male	0.0	54.2	2.1	43.8	100.0
	Female	0.0	50.0	2.1	47.9	100.0
61-	Male	5.3	36.8	0.0	57.9	100.0
	Female	0.0	42.9	7.1	50.0	100.0
Exposure Rates from highway		1.4	38.9	3.7	56.0	100.0

According to Table 17.2, 27.2 per cent of residents have been affected negatively from highway through the noise pollution, 2 per cent of them from air pollution, 0.8 per cent from soil pollution, 1.8 per cent visual pollution, the rest (2 per cent) from other factors. 55.8 per cent of residents have been affected negatively, didn't give any reason about the negative effects of highway (Table 17.2). On the other hand, 10, 4 per cent gave more than one reason about the negative effects of highway.

Table 17.2: Negative Effects of Highways According to Age and Gender Distribution (Per cent)

Age	Gender	Negative Effects of Highway							
		Not Answered	Noise Pollution	Air Pollution	Soil Pollution	Visual Pollution	Other	More than One Pollution	Total
7-18 age	Male	70.8	20.8	0.0	4.2	0.0	0.0	4.2	100.0
	Female	66.7	27.8	0.0	0.0	5.5	0.0	.0	100.0
19-30 age	Male	61.7	22.3	3.2	.0	2.1	2.1	8.6	100.0
	Female	56.3	22.5	1.3	1.3	5.0	3.8	9.8	100.0
31-45 age	Male	59.7	20.9	3.0	1.5	0	1.5	13.4	100.0
	Female	45.7	37.1	1.4	1.4	1.4	2.9	10.1	100.0
46-60 age	Male	41.7	39.6	4.2	0.0	2.1	2.1	10.3	100.0
	Female	47.9	29.2	2.1	0.0	0.0	2.1	18.7	100.0
>61	Male	63.2	26.3	0.0	0.0	0.0	0.0	10.5	100.0
	Female	57.1	28.6	0.0	0.0	0.0	0.0	14.3	100.0
Exposure Rate of residents affected negatively from highway		55.8	27.2	2.0	0.8	1.8	2.0	10.4	100.0

About 57.4 per cent of residents have been indicated simplify of transportation as a positive effect of highway, while 3.9 per cent have been indicated other factors. In addition, 2.4 per cent indicated that not only transportation but also other effects are positively effects the highway (Table 17.3). Because 38.9 per cent of residents have been effected negatively (Table 17.1), 36.9 per cent of residents didn't explain any reason about positive effect of the highway (Table 17.3).

Table 17.3: Positive Effects of Highways According to Age and Gender Distribution (per cent)

Age	*Gender*	*Positive Effects of Highway*				
		Not Answered	*Ease of Transpor-tation*	*Simplify of Transportation Other*	*Other*	*Total*
7-18 age	Male	25.0	70.8	0.0	4.2	100.0
	Female	27.8	61.1	0.0	11.1	100.0
19-30 age	Male	31.9	66.0	1.1	1.1	100.0
	Female	32.5	57.5	3.8	6.3	100.0
31-45 age	Male	32.8	61.2	4.5	1.5	100.0
	Female	41.4	55.7	1.4	1.4	100.0
46-60 age	Male	47.9	39.6	6.3	6.3	100.0
	Female	45.8	45.8	2.1	6.3	100.0
61-	Male	26.3	73.7	0.0	0.0	100.0
	Female	42.9	50.0	0.0	7.1	100.0
Exposure Rate of residents affected positively from highway		36.3	57.4	2.4	3.9	100.0

Much than half of the residents (54.6 per cent) have been effected negatively from highway noise while 43.4 per cent of them have not been effected from this (Table 17.4). About 66.7 per cent of the woman whose age between 46–60 have been constituted of most negative effects from highway. On the contrary the biggest group among which have not been affect noise pollution of highway consist of man whose age between 7-18 ages (66.7 per cent) (Table 17.4).

One of the results of questionnaire was determining connection between education level and negative effects of highway for the residents. The most negatively affected group (66.7 per cent) have not been educated residents, even do not literacy. On the other hand, the group which is the most positively effected from the highway is educated by high school (62.1 per cent) (Table 17.5).

Soil Pollution

The most significant impact of motor vehicle is the lead accumulation. Therefore, 4 localities on 2 transect were chosen to determine the soil pollution and soil samples were taken from 50m 100m, 150m and 500m away from the roadside. Findings (upper limit 85.95 mg/kg) were compared with the limit values (300 mg/kg) of Soil Pollution

Control Regulation dated 10.12.2001 [3]. The lead content of roadside soils was measured under the limit values. Soils within 50 m of the motorway had higher amounts of lead (70.39 mg/kg) than the background level (54.21 mg/kg on 500 m) (Table 17.6). Our results indicated that lead pollution was high nearby highway and was decreased with increasing distance from the road.

Table 17.4: Disturbance from Highway Noise According to Age and Gender Distribution (per cent)

Age	*Gender*	*Effects from highway noise negatively*			
		Not Answered	*Yes*	*No*	*Total*
7-18 age	Male	0.0	33.3	66.7	100.0
	Female	0.0	44.4	55.6	100.0
19-30 age	Male	2.1	46.8	51.1	100.0
	Female	1.3	58.8	40.0	100.0
31-45 age	Male	4.5	47.8	47.8	100.0
	Female	1.4	65.7	32.9	100.0
46-60 age	Male	2.1	64.6	33.3	100.0
	Female	0.0	66.7	33.3	100.0
>61	Male	5.3	47.4	47.4	100.0
	Female	0.0	64.3	35.7	100.0
Exposure Rate of residents affected negatively from highway noise		2.0	54.6	43.4	100.0

Table 17.5: Education Level According to Highway Effects (Per cent)

Education	*Effects From Highway*				
	Not Answered	*Negative*	*Positive-Negative*	*Positive*	*Total*
Illiterate	0.0	66.7	0.0	33.3	100.0
Primary	1.6	51.6	4.1	42.6	100.0
Secondary school	1.6	37.1	1.6	59.7	100.0
High	1.1	32.2	4.5	62.1	100.0
Undergraduate	0.0	36.1	2.5	61.5	100.0
Exposure rate of residents about affect from highway	1.4	38.9	3.7	56.0	100.0

It is observed that, there was 70.39 mg/kg lead per one kilogram soil on 1, transect soil samples from 50 m away from roadside. For 2nd location, 100 m away from roadside, was 60.60 mg/kg, 3rd location from 150 m was 57.32 mg/kg, and the 4th location on 1st transect was 500 m away from roadside was 54.21 mg/kg (Table 17.6). When we compare lead accumulation differences between first and last location on first transect, we can see easily there has been 16 mg/kg differences. This data shows that the highway recives considerable lead pollution. The 2nd transect showed similar amounts of lead accumulation. But the last location has the highest value of the lead accumulation and this is because; it is very close to secondary road.

Table 17.6: Comparison Limits Value of Lead Accumulation in the Soil Survey with "Regulation of Soil Pollution Control"

Heavy Metal	*Location*	*Lead Content in 1 kg Soil (mg/kg) (dry weight)*	*Heavy Metal Limit Value According to "Regulation of Soil Pollution Control"*
			pH>6mg/kg Oven Dry Soil
Lead	1. Transect		300
	1	70.39	
	2	60.60	
	3	57.32	
	4	54.21	
	2. Transect		
	1	43.77	
	2	41.67	
	3	32.60	
	4	85.95	

The other important soil pollution is loss of agricultural land. Research area consist of 6 210.26 ha area which include 500 m wide away from North and South side of the highway and 60 km long route within the Adana. About 266.45 ha of this area is occupied by the highway, the other 5.943,81 ha area is uses by this highway. The second part of the highway is still opened to any use, while 162, 66 ha from occupied part it completely in use and it consist of I and II class of the soil. This part of area can no longer uses any more agricultural land.

There is a regulation which is called "Regulation of Agricultural Land Protection and Use" dated in 25.03.2005, Official Gazette with number of 25766 in 8th theme, identifies prime agricultural areas. These kinds of areas have some specialties such as; suitable for vegetable production because of physical, chemical and biological properties of soil combination, suitable topographic condition, etc. Furthermore, these kinds of lands have local, national and international importance. This type of land has to be used as agricultural area and not other type of uses. So, such areas are accepted as loss of agricultural land.

Noise Pollution: Noise level was measured at different length and high on different months of the year and at different times of the day during daytime in order to determine highway noise pollution. Measures were made by "2010 model Deltaohm noise meter" at 126 points with instantaneous noise measurements and average levels were taken into account. The study area divided into 8 regions according to intersection of road. Some points were measured more than one times. The highest level in measurements was taken into account. 70 points in the total level of metrics were compared the regulation of Environmental Noise Assessment and Management Regulation on Environmental Noise Limit Levels dated in 07/03/2008 [4]. When compared the effects of highway, 3 health institutions and 22 educational institutions which located in the study areas were considered. The regulation in the field of "education, culture, and health fields (65 dB (A))", "dwellings found to be concentrated area (68 dB (A)) "and" industrial areas (72 dB (A)) "specified separately and limit levels have been taken into consideration. As a result of this comparison, it was found that while maximum immediate measurement has been exceeded the limitation at 70 points (the highest 116.6 dB (A)), 35 points more or less have been exceeded the average level (the highest 92.7 dB (A)).

Noise was measured very detailed to define noise pollution in the TAG highway. Measurement results were compared with the limit level in legislation. This comparison, made by taking into account the highway junctions on the highway, as a result, research area is divided into 8 regions. 126 points measured during the project, but not all of them taken into account. In the region at the same point more than one measurement were considered at the highest average measured level.

Results of 70 measurements are given in Table 17.7. Each region's noise limit levels were compared with legislation of "Regulations of Environmental Noise Assessment and Management" which dated in 07.03.2008 considering the education, culture, health, settlement and industry in the area.

As can be seen in Table 17.7, the first region is around Karaisalý junction. This region usually consists of agricultural land; there is not any educational institution, hospital and industrial area. However, a village named Kuyumcular is located about 500 m away from the highway.

The second region occurs around the airport junction. This region is under the construction pressure, due to the intense residential development north of the city remained especially in recent years. So, this region consists of very dense houses. Because of this reason, the area within the context of regulation for the areas where have intensively houses. Limit level for this kind of area is 68 dB (A). The last two sentences are not clear. Table 17.7 shows that measurement for noise pollution through west-east direction in 18.06.2008 between 13.00 and 13.09 hours and through east-west direction in the same day between 10.05, 12.08 and 12.16 hours are higher than limit level which is indicated in legislation.

The third region is around Kasým Ener, Alparslan Türke° Boulevard. There is neither any educational institution nor hospital. On the other hand, there are intensive house density. Especially even after road build up area started on north and west direction. For this reason limit level is 68 dB (A). 4 of 10 noise measurement of the area are higher than 70 dB (A). Although the other measurement are under 68 dB (A), they are very close to limit level.

The fourth region is around Mavi Boulevard Tellidere junction. Residential areas very common and it is very dense here. South and north parts of the highway have very dense residential areas. In addition, Adana Baraj High School which is one of the biggest in Adana is located in the region. According to legislation, limit level of the region is 68 and 65 dB (A). It is noted that most of points where noise measurement was done, were higher than the limit levels. In addition, of instantaneous measurement results are very close to the limit levels.

Table 17.7: Comparison Limiting Level of Instantaneously Measurement Points Level with "Regulation of Environmental Noise Assessment and Management"

Region	*Direction*	*Date*	*Time*	*Average Level (dB(A))*	*Limiting Level (dB(A))*		
					Educational, Cultural and Hospital	*Residential Area*	*Industrial Area*
Around	West–East	18.06.2008	12:40	78,2	–	68	–
Adana (West)		18.06.2008	12:45	80,4			
Karaisalı	East–West	18.06.2008	12:22	74,1			
Junction		18.06.2008	12:30	74,4			
Adana (North)	West–East	18.06.2008	13:00	81,2	–	68	–
Airport		18.06.2008	13:09	78,5			
Junction		01.09.2008	18:13	57,4			
		01.09.2008	18:15	57,1			
		01.09.2008	18:17	55,2			
		01.09.2008	18:28	58,1			
		01.09.2008	18:35	55,5			
		01.09.2008	18:40	55,4			
	East–West	18.06.2008	10:05	75,4			
		18.06.2008	12:08	80			
		18.06.2008	12:16	79,4			
		21.07.2008	11:55	67,5			
		21.07.2008	12:05	63,9			
		21.07.2008	12:20	46,3			
		21.07.2008	12:30	50,3			
		21.07.2008	12:40	63			
Kasım	West–East	18.06.2008	13:15	76,8	–	68	–
Ener,		18.06.2008	13:20	78,5			
Alparslan	East–West	18.06.2008	11:55	71,6			
Türkeş		18.06.2008	11:57	71,2			
Boulevard		21.07.2008	10:30	65,3			
		21.07.2008	10:45	60,9			
		21.07.2008	10:50	67,2			
		21.07.2008	10:58	59,9			
		21.07.2008	11:08	63,4			
		21.07.2008	11:15	57,5			

Contd...

Table 17.7–Contd...

Region	Direction	Date	Time	Average Level (dB(A))	Limiting Level (dB(A))		
					Educational, Cultural and Hospital	Residential Area	Industrial Area
Mavi	West–East	31.08.2008	11:20	68,8	65	68	–
Boulevard		02.09.2008	18:37	68,5			
Tellidere		02.09.2008	18:43	74,2			
Junction		02.09.2008	18:45	75,3			
		02.09.2008	18:52	63,9			
		02.09.2008	18:55	66,3			
	East–West	21.07.2008	10:00	68,9			
		21.07.2008	10:10	60,6			
		02.09.2008	18:00	73,9			
		02.09.2008	18:03	72,5			
		02.09.2008	18:12	56,9			
		02.09.2008	18:20	56			
Beyazevler,	West–East	18.06.2008	13:37	92,7	–	68	–
Adnan		18.06.2008	13:41	88,6			
Menderes	East–West	18.06.2008	11:08	81,7			
Boulevard		18.06.2008	11:18	80			
Junction		21.07.2008	9:15	54,2			
		21.07.2008	9:20	56,2			
		21.07.2008	9:30	53,6			
Around	West–East	21.07.2008	14:05	65,2	–	68	–
Çukurova		21.07.2008	14:20	60,3			
University		21.07.2008	14:35	59,6			
Junction	East–West	18.06.2008	11:04	78,5			
		18.06.2008	16:00	77,1			
Adana	West–East	18.06.2008	13:49	79	–	68	–
East Junction,		18.06.2008	13:53	89,2			
Aroun Kozan		21.07.2008	15:40	62,7			
Exit		21.07.2008	15:50	62,4			
		21.07.2008	16:00	64,2			
		21.07.2008	16:20	55,9			
	East–West	18.06.2008	15:40	78,7			
		18.06.2008	15:47	77,3			
		01.09.2008	16:30	61,8			
		01.09.2008	16:33	56,9			
		01.09.2008	16:41	63,3			
		01.09.2008	16:44	61,2			
Yakapınar,	West–East	18.06.2008	13:59	79,4	–	–	72
Misis,		18.06.2008	14:11	79,9			
Organized	East–West	18.06.2008	15:24	73,8			
Industrial Area		18.06.2008	15:33	73,8			

The sixth and seventh regions are around Beyazevler, Adnan Menderes Boulevard. Duygu Café and around places also includes the region. Although there is not any educational institute and hospital, residents take place intensively. The region is the most crowded and it has the highest noise levels. It was noticed that especially morning and evening measurements exceed 90 dB(A) and the highest level is 116, 6 dB (A).

The last region is around Yakapýnar, Misis, Organized Industry area. There is not any educational institute and hospitalin this region. The noise limit level is 72 dB(A) for such industrial areas. Noise limits are much more than 72 dB(A) in all noise measurement points.

According to Table 17.7 which has record of instantaneous noise measurements in research area, the noisiest region is around Duygu Cafe Mavi Bulevard, Tellidere. After this region Beyazevler (it includes hospital junction), Adnan Menderes Boulevard and around and Adana (West) respectively. Besides this region, Yakapýnar, Misis, Organized industry and surroundings in Organized Industry Region are the noise level is very high.

When all of these measurements compared with Table 17.8, it can be seen that all average measurement points are more than 45 dB (A), which cause problem of insomnia. Especially it is very effective on "sleeping trouble" problem when windows are open. As also in Table 17.7, average level in 39 points is higher than 65 dB (A) and it showed that this level causes some health problems, such as rising of blood pressure, heart rate deceleration in the kidneys in hormonal changes, and changes breathing.

Table 17.8: Noise impacts on people [5]

Noise Level dB(A)	*Noise Source*	*Physical and Psychological Effects on Human*
20–30	Moving leaves, whisper, working clock	Psychological negative effect
45–50	Noise coming from outside to the house that has closed window	Difficulty in fallen asleep
65–70	Road that has heavy traffic, vac, typewriter	The rise of blood pressure, heart rate deceleration, hormonal changes, breathing changes
90–120	Moped, saw, disco, airport	Difficulty hearing in a short time
>120	Jet aircraft, siren whistle, cool handle knobs far away 2 meter, air compressor	Hearing difficulties and severe pain

It is known that, highway effect not only on human health, but also on animal and all of biological life. But studies about noise effect on animals are very restricted [6] made a study to associated loss of biotope quality in an area where it has continuous noise such as highway.

According to the study, one point of noise measurements which took place on Beyazevler, Adnan Menderes Boulevard, it measured in 18.06.2008 at 13.37 o'clock as 92.7 dB (A). It shows that loss of habitat quality of birds 100 per cent on the point

(Table 17.9). Moreover, 31 of all points in the study area loss of habitat quality of birds is about 70-100 per cent. However, it can be expected that, all the points measured in study area birds' habitat quality loss will be between 10-40 per cent.

Table 17.9: Comparison Between Loss of Habitat Quality of Birds in Noisy Places with Number of Measurement on TAG Highway

Noise Level dB(A)	*Loss of Habitat Quality Per cent (Reck* et al., *2001)*		*The Number of Measurement Points that Habitat Loss is Expected*
	much than 90 dB(A)	100	1
90-70	(approximately per cent 70-100)	85	31
70-59	(approximately per cent 40-70)	55	54
59-54	(approximately per cent 30-50)	40	67
54-47	(approximately per cent 10-40)	25	70

Visual Pollution

People can face with visual pollution types in all fields. Generally some types of urban visual pollution are generally; air pollution, lack of green areas, monotone and very crowded structuring, shortness of sight, the exterior image the chaos, debris, irregular traffic, haphazard waste, irrationality and casualness in sheet sign [7] [8].

Types of visual pollution create different impacts according to individual differences. Some of these impacts; disorder as a result of distraction, eyes fatigue, alert reduction, unwillingness, emotional transformation and a decrease in diversity of thought, difficulty in adaptation because of confusion, irritability, anger, behavior disorders, pessimism as a result of visual pollution, a decrease in the joy of living and increase in psychosomatic illnesses, mental trauma because of unwanted images, nausea, dizziness [7] [8].

There is not any regulation to determine visual pollution. The most important visual pollution is dividing into two part of Seyhan River, where has the best visual impact in landscape between agricultural areas and settlement. In addition, the retaining wall in the entrance of Fehmi Özeltürkey Tunnel is created the other visual pollution.

There have been different visual pollution observed around the highway. Not only people who use the highway, but also people living around the highway create visual pollution by their waste. In addition, there are some huge retiring walls, quarry materials and waste storage area around Ceyhan Viaduct which creates another source of visual pollution.

Traffic Safety

TAG highway has created a very positive impact in terms of the traffic safety and it has ensured to reduce the urban traffic congestion. TAG highway constructed 7 (mutual 14 entrances and exits) new entrance to city center to reduce urban traffic density. However, the junction of Çukurova University Balcalý on the direction of Gaziantep, has been evaluated under "risk" in terms of traffic safety both having very short access road and a building which is located in the middle of view and preventing the vision of road for driver. Moreover, Kasým Ener-Alparslan Türkeþ entrance doesn't have any signboard to give information about this junction.

CONCLUSION

All of the environmental problems could be reduced or eliminated. Some principles based on planning and designs are described in detail in the following. It consists of primary and long term effects preventive/mitigation, reducing negative effects on natural, physical and socio-cultural structure and make conductivity with the urban fabric.

There are some principles to reduce noise pollution around highway. One of them is covering the highway about 150 meters before and after Fehmi Özeltürkay overpass. In addition, on this covering highway can be used as a recreational area like sports, and other activities. It is required a construction that a "noise prevention wall" between Fehmi Özeltürkay tunnel and after 2 km to Adana (North) Airport junction. This "noise prevention wall" should be in the north and south direction of the highway. *"Noise Prevention Wall"* should be constructed opposite of the Fehmi Özeltürkay overpass and Kozan direction, where there are intensive buildings. But it should be finished even after TOKI houses.

As the presence of noise prevention walls, forming the walls of building materials is also very important. These walls need to be constructed with materials which are compatible with nature and can be used with plants instead of concrete materials. Thus, we can increase visual effect and reduce the noise pollution.

A large section of highway is within the borders of Adana city. So, urban area around the borders of highway is under potential structuring risk. There should be a buffer zone created about 150–250 m according to topographic structure around highway to reduce the noise and the effects of the pollution. Thus, the problem of housing-noise can be solved easily without *"Noise Prevention"*.

Not only lack of maintenance but also incomplete planting activities, there are some open abandoned areas on the highway slopes. It can be seen easily, especially on the slope. Because filling slopes cannot keep enough water for plant growing, so plants died in a short time. This kind of areas needs additional maintenance. Some areas especially intersections between Adana Airport–Adana West side (Tarsus paying office) and Adana (East) side Adana East side (Osmaniye paying office) need additional planting activities.

The area in the centre strip along the research area is requires planting in addition essential planting work on the slope along the highway. The plants in refuge have very important function especially to keep against the light from the headlamps of vehicles. But plants at the centre strip along the research area were not seen, or a planting operation that will perform this function and composition was not detected. It is tried to be prevented with artificial materials against the light from the headlights of vehicles. Both the long-term effectiveness can be lost in this case and a visually unpleasant appearance occurs. Refuges are required to be planted with species in terms of both functional and visual quality, appropriate to design principles and belonging to the natural vegetation.

Degraded areas nearby highway should be repaired according to biological principals to reduce visual pollution, and increase visual effect. Materials quarry around the Ceyhan Viaduct should be rehabilitated according to "Mining Activities and Corruption of the Land to Nature to Recovery Regulations" dated 14/12/2007 and announced in the Official Gazette No 26 730 [9]. Additionally, waste storage areas should be rehabilitated according to "Solid Waste Control Regulation" which dated 14.03.1991 and announced in the Official Gazette No 20 814 [10].

Çukurova University Balcalý junction should be rearranged because of high traffic safety risk. The most important thing is the elimination of buildings which blocks the view at the entrance of the road. Secondly, the access road should be extended in a secure way out to the boulevard.

ACKNOWLEDGEMENTS

This paper is based on a study performed within the project number ZF 2007 BAP 27 supported by Scientific Research Project Unit of Çukurova University.

REFERENCES

1. B., Öztürk, Kent İçi ve Kent Dıþı Karayolu Ulaşım Sisteminde Bitkilendirmenin Trafik Tekniği Yönünden İşlevleri, Uluslararası Trafik ve Yol Güvenliği Kongresi, Gazi Üniversitesi, Ankara, (2002).
2. ÖİKR, 9. Kalkınma Planı Karayolu Ulaşımı Özel İhtisas Komisyonu Raporu. Devlet Planlama Teşkilatı. Ankara, (2006).
3. Toprak Kirliliğinin Kontrolü Yönetmeliği. 10.12.2001 tarihli ve 24609 sayılı Resmî Gazete (2001).
4. Çevresel Gürültünün Değerlendirilmesi ve Yönetimi Yönetmeliği, 2008. 07.03.2008 tarih ve 26809 sayılı Resmi Gazete, (2008).
5. A. Mavruk, "Yüreğir ve Seyhan (Adana) İlçelerinde Ana Arterlerdeki Toz ve Gürültü Dağılım Haritalarının Hazırlanması", Yüksek Lisans Tezi.Çukurova Üniversitesi. Adana, (2005).
6. H. Reck, C. Herden, J. Rassmus, R. Walter. Die Beurteilung von Lärmwirkungen auf freilebende Tierarten und die Qualität ihrer Lebensräume–Grundlagen und Konventionsvorschläge für die REGELUNG VON Eingriffen nach § 8 NatSchG. Angewandte Landschaftsökologie, Bundesamt für Naturschutz (Hrsg.), Heft 44, s. 125-151, Bonn, (2001).
7. S. Bodur, R. Kucur, Görüntü Kirliliği Üzerine. Ekoloji Dergisi, Sayı:12, Temmuz-Ağustos-Eylül 1994, (1994).
8. Gülüm, K. Çevre Kirlenmesi Sorunu, Çevre Eðitimi ve Yerel Yönetimlerin Rolü. www.yerelsiyaset.com, (2008).
9. Madencilik Faaliyetleri ile Bozulan Arazilerin Doğaya Yeniden Kazandırılması Yönetmeliði, 14.12.2007 tarih ve 26 730 sayılı Resmi Gazete, (2007)
10. Katı Atıkların Kontrolü Yönetmeliği, 14.03.1991 tarih ve 20 814 sayılı Resmi Gazete, (1991).

Chapter 18

An Investigation on Pedestrians in Denizli City, Turkey within the Framework of Urbanization

Yetiş Şazi Murat and Sabit Kutluhan

Pamukkale University, Faculty of Engineering,

Department of Civil Engineering, 20070, Kınıklı/Denizli, Turkey

ABSTRACT

Most of the urban areas in Turkey have been developing without proper planning. This situation creates problems for residents and public users. Pedestrians in the cities are one of such problems. An analysis on pedestrians and pedestrian areas of Denizli city has been considered in this study. Pedestrian traffic flows have been observed and speed-flow-density relationships have been investigated. Analysis of the speed-flow-density diagrams showed that the logarithmic expression can be used to determine pedestrian flows. The levels of service of sidewalks for three intersections are used in this study. Recommendations for increasing level of service have been presented. It is expected that an increase in level of service can meet user's expectations and can also contribute well-planned urbanization. Pedestrian crossings have also been investigated. The analysis has been realized using video camera records. In the analysis, use of pedestrian crossing, crossing in green signal, crossing in red signal, non-usage of pedestrian crossing, crossing at right hand side and crossing at left hand side are analysed. The rate of using pedestrian crossing is determined as about 80 per cent in the analysis. A survey is applied to pedestrians to get information about traffic perception levels and availability of pedestrian areas. In questionnaire, the aim of trip, satisfaction level and problems encountered, perception of safety level, level of obeying traffic rules, use of facilities for pedestrians, use of sidewalks, use of pedestrian signals etc. are asked to participants. Results of the survey are analyzed and traffic perception levels of pedestrians are evaluated.

INTRODUCTION

The mode of urbanization is an indicator of civilization. Urban areas are developed considering needs of residents and users. However, demands of users are not met in every time because of dynamic structure of the development needs. Pedestrians are one of the important components of the traffic. Although pedestrians are ignored in design of related facilities, they should be taken into account in planning and design of urban areas and detailed analysis should be made for this purpose. Signalized intersections, Railway stations, bus stops, shopping centers, metro stations etc. are some areas that have heavy pedestrian traffic and are required for special emphasis. In these areas, pedestrian flows should be observed and speed, volume, density and level of services should be determined (Kutlu, 1963; Özdirim, 1994). In the analysis, environmental effects (restrictions) should also be dealt with in addition to general conditions and the results obtained should be evaluated using this point of view.

Pedestrian flows and behaviors have been considered in many researches. Speeds of pedestrians have been investigated by Tarawneh (2001) for pedestrian signal timing design. Pedestrians were classified by ages, and gender. Effects of these classifications on pedestrian speeds were investigated in the study. Pedestrian signal timings were adjusted using these speed and age classifications and their behaviors on crosswalks have also been observed. Dandan *et al.* (2007) assessed pedestrian level of services on sidewalks considering the relationship between the pedestrians' subjective perceptions and the quality of the road, physical facilities as well as the traffic flow operation. Keegan and O'Mahony (2003) investigated the effect of countdown timer on pedestrian behaviour.at signalised pedestrian crossings in Dublin. It was seen that after installing countdown timer green man signal crossing rose from 65 per cent to 76 per cent. Hamed (2001) considered pedestrian behavior at pedestrian crossings on divided and undivided streets. Results showed that pedestrians behave differently or have different waiting times for two sides of divided streets. Yang *et al.* (2006) dealt with pedestrian behavior at signalized crossings using micro-simulation approach. They classified pedestrians as law-obeying and opportunistic ones. They developed a simulation model and tested by observations. It is understood that the micro-simulation results were agreed with the field results. Hongfei *et al.* (2009) searched pedestrian flow characteristics analysis and models in China. They focused on speed-flow-density (space) relations of pedestrian flows at different parts of a transport terminal and found that quadratic and logarithmic equations could be used in modelling.

In the current study, we have taken into account and pedestrian flows in Denizli, Turkey and facilities for pedestrians are dealt with. A survey was carried out including some inquiries about traffic perception and availability of pedestrian areas. Pedestrian traffic flows have been observed and speed-flow-density relationships were investigated.

PEDESTRIAN FLOW ANALYSIS

Pedestrian flow analysis is one of the planning issues when considering urbanization. Analytical procedures have been discussed and developed by researchers. The procedure includes capacity and level of service analysis of the facilities serving pedestrians. Details about the procedure can be found in literature. The Highway Capacity Manual (HCM) is one of the main references published by Transportation Research Board of National Academy of Sciences in the USA (Transportation Research Board, 1985). Procedures are defined in HCM for the following types of pedestrian facilities:

- *Walkways and sidewalks*: facilities such as terminals, sidewalks, stairs, and paths designated exclusively for pedestrians.
- *Pedestrian queuing areas*: areas where pedestrians stand temporarily, while waiting to be served. Queuing areas are found at elevators, transit platforms, and street crossings.
- *Shared off-street paths*: paths physically separated from highway traffic for the use of pedestrians, bicycles, skateboards, and other nonmotorized traffic.
- *Pedestrian crosswalks*: pedestrian crossings at signalized and unsignalized intersections.
- *Pedestrian facilities along urban streets*: designated pedestrian sidewalks on urban streets, incurring the impacts of both uninterrupted flow and fixed interruptions.

The procedures can be applied with respect to goal of the research and the corresponding facilities. In pedestrian flow analysis, some familiar traffic terms are used. Before details of the procedure, these terms are defined in the following list:

1. Pedestrian speed is the average pedestrian walking speed, generally expressed in units of feet per second. In this study, metric units system are used.
2. Pedestrian flow rate is the number of pedestrian passing a point per unit time, expressed as pedestrians per 15 minutes or pedestrians per minute; "point" refers to a perpendicular line of sight across the width of a walkway.
3. Unit width flow is the average flow of pedestrians per unit of effective walkway width, expressed as pedestrians per minute per foot.
4. Platoon refers to a number of pedestrians walking together in a group, usually involuntarily, because of signal control and other factors.
5. Pedestrian density is the average number of pedestrians per unit of area within a walkway or queuing area, expressed as pedestrians per square foot.
6. Pedestrian space is the average area provided for each pedestrian in a walkway or queuing area, expressed in terms of square feet per pedestrian; this is the inverse of density, but is a more practical unit for the analysis of pedestrian facilities.

Pedestrian flows are similar to vehicular flow in terms of speed and maneuver. Other measures more specially related to pedestrian flow include the ability to cross a pedestrian traffic stream, to walk in the reverse direction of a major pedestrian flow, and to generally maneuver without conflicts and changes in walking speed.

In this study, besides the level of service analysis, the main factors related to pedestrian flows such as speed and space are also taken into consideration and relationships between these factors are investigated. The fundamental relationship between speed, density and volume for pedestrian flow is parallel to vehicle flow. Pedestrian speed decreases with an increase in volume and density. On the other hand, pedestrian space decreases as density increases. The relationship between density, speed and flow for pedestrians is the same as vehicular traffic streams, which is:

$$V = S \times D$$

where,

V is the pedestrian flow, expressed as pedestrians per minute per foot; S is speed, expressed as feet per minute; D is density, expressed as pedestrians per square foot.

The flow variable used in this equation is the unit width flow. An alternative and more useful expression can be developed using the reciprocal density, or space as follows:

$$V = S/M$$

where,

V is pedestrian flow, S is speed, and M is space. In this study, sidewalks are analyzed using the fundamental relationships defined above. The speed- flow- density (space) relations of pedestrians are investigated.

STUDY AREA AND DATA COLLECTION

Denizli is a medium size city with a population of 400000 people. Pedestrian crossings and sidewalks near intersections are the main study areas. Pedestrians are observed at the pedestrian crossings and usage rates of pedestrian crossing have been determined. Sidewalks are also observed and speed-flow-occupancy relations are considered at different segments.

The pedestrian flow data are collected at urban signalized intersections of the city. First, the sidewalks around the signalized intersections are classified into three segments. Then the pedestrians are observed and recorded by a camera. The video recordings are used in analysis. Each pedestrian is considered separately and speed and occupancy figures are calculated. The observations are realized at different three periods in the day. Duration of each observation is selected as one hour. Sample pedestrian flow data obtained from sidewalks are given in Table 18.1.

ANALYSIS AND SURVEY

Two types of analyses were carried out for pedestrians: (1) crossing and (2) sidewalk analysis. Three urban sections (signalized intersections) that have high pedestrian traffic volumes were recorded by a video camera. The analyses were made using the recorded data.

Table 18.1: Pedestrian Flow Data Obtained in a Selected Intersection of Denizli City

Number of People	Area(m^2)	Occupancy (peds/m^2)	Average Speed (m/min)
1	61.07	0.016	23.476
1	61.07	0.016	33.296
6	61.07	0.098	36.971
9	61.07	0.147	41.739
14	61.07	0.229	48.103
30	61.07	0.491	53.093
18	61.07	0.295	57.415
27	61.07	0.442	62.829
22	61.07	0.360	67.088
8	61.07	0.131	72.976
7	61.07	0.115	77.787
1	61.07	0.016	82.865
1	61.07	0.016	95.323
1	61.07	0.016	103.684

Pedestrian Crossing Analysis

The pedestrian crossings of three signalized intersections are recorded and the analysis has been achieved using video recordings. The crossing in green signal, crossing in red signal, non-usage of pedestrian crossing, crossing at right hand side and crossing at left hand side were investigated (Anatürk and Çetin, 2008). Intersection layout and sample results of analysis are given in Figure 18.1, Tables 18.2 and 18.3.

Table 18.2: Pedestrian Crossing Data

Direction	Use of Peds. Crossing	Non use of Peds. Crossing	Use of Green Signal and Peds. Crossing	Use of Red Signal and Peds. Crossing	Use of Right Side in Peds. Crossing	Use of Left Side in Peds. Crossing	Total Pedestrian
1	20	4	10	11	17	13	30
2	26	6	20	12	16	16	32
3	31	8	27	12	22	17	39
4	29	6	22	13	18	17	35
5	171	16	119	68	121	66	187
6	84	9	54	39	58	35	93
7	39	4	23	20	30	13	43
8	36	4	22	18	23	17	40

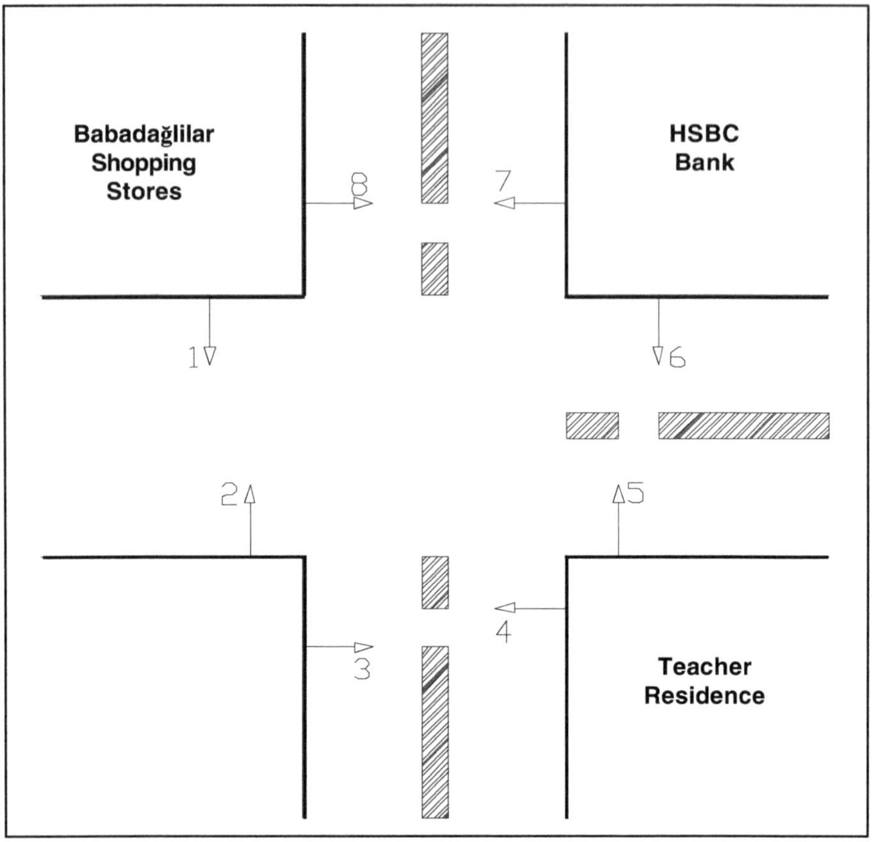

Figure 18.1: The Babadağlilar Intersection Layout

Table 18.3: Rates of Pedestrian Crossing Usage

Direction	*Rate of Peds. Crossing*	*Rate of Green Signal*	*Rate of Right Side Usage*
1	0.87	0.63	0.57
2	0.81	0.63	0.50
3	0.79	0.69	0.56
4	0.83	0.63	0.51
5	0.91	0.64	0.65
6	0.91	0.53	0.55
7	0.91	0.53	0.70
8	0.90	0.55	0.58

As can be seen in Tables 18.2 and 18.3, the rates of pedestrian crossing are acceptable. But the use of green signal and right side usage in crossing are very poor. In this analyses, the 75 per cent is considered as critical rate.

Sidewalk Analysis

Sidewalks are also considered in the analysis; pedestrian flows at sidewalks are observed. The speed- flow and occupancy (density) values of pedestrian flows have been investigated. Sidewalks of signalized intersections' arms are carried out and the levels of service of sidewalks have been determined regarding the speed, pedestrian space, and pedestrian flow rate. The Highway Capacity Manual (HCM) procedure is used in the analysis (TRB, 1985). Table 18.4 shows the recommended level of service criteria for sidewalks in the HCM.

Table 18.4: Pedestrian Level of Service Criteria on Sidewalks (TRB, 1985)

Level of Service	*Space (sqft/ped)*	*Average Speed (ft/min)*	*Flow Rate (Ped/min/ft)*	*Volume/Capacity Ratio*
A	≥ 130	≥ 260	≤ 2	≤ 0.08
B	≥ 40	≥ 250	≤ 7	≤ 0.28
C	≥ 24	≥ 240	≤ 10	≤ 0.40
D	≥ 15	≥ 225	≤ 15	≤ 0.60
E	≥ 6	≥ 150	≤ 25	≤ 1
F	< 6	< 150		

The sidewalks are divided into segments based on the pedestrian density and usage. 100m total distance from corners is taken into consideration for each arm. Pedestrians are recorded by video and are analyzed. The levels of service are determined for each segment considering the speed, pedestrian space and pedestrian flow rate values. The levels of services determined using the criteria given in Table 4 for the Babadağlılar Intersection are given in Figure 18.2. As can be seen from this figure, the levels of services are varied from A to E. It is based on pedestrian density on

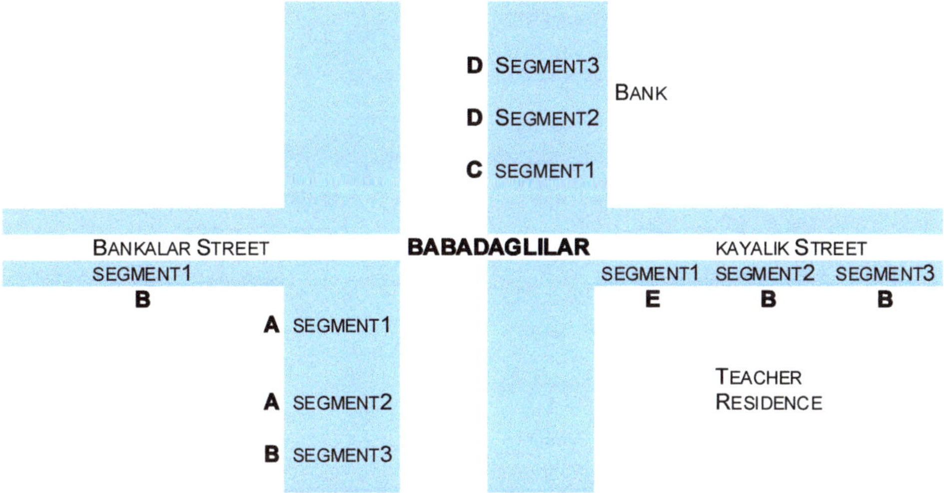

Figure 18.2: Sidewalk Analysis for the Babadağlılar Intersection

the segments, for instance, the density of pedestrian on the Kayalık Street Segment 1 is higher than the Segment 2. Therefore average speed and pedestrian spaces on Segment 1(E) are lower than that of Segment 2 (B).

The levels of service of the Babadağlılar intersection arms are different from each other. For example, while levels of service for the Kayalık Street are being determined as E, B, and B for each segment respectively, for the Istasyon Street; these levels are C, D, and E. These differences are related to physical and environmental conditions of these streets. On the Istasyon Street, there are shopping stores and a bank office that affects the pedestrian flows negatively. But there is no obstruction that interrupts the pedestrian flows on 2nd and 3rd segments of the Kayalık Street.

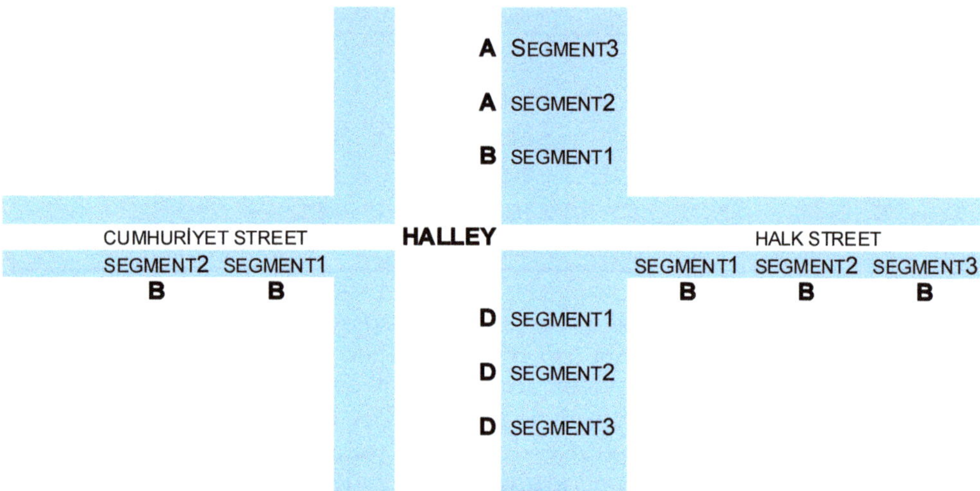

Figure 18.3: Sidewalk Analysis for the Halley Intersection

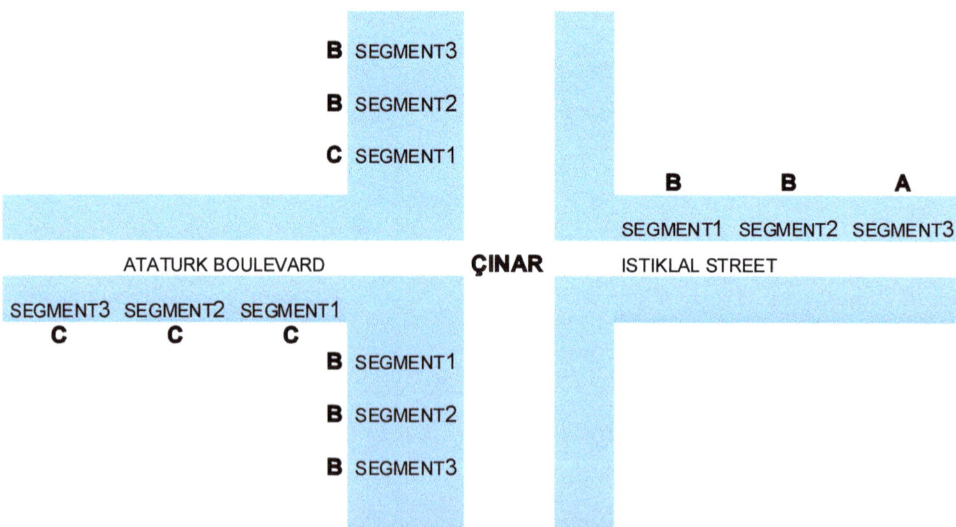

Figure 18.4: Sidewalk Analysis for the Çınar Intersection

Similar observations and analyses were also made for the Halley and Çınar intersections. Results of the analysis for the Halley and the Çınar intersections are given in Figures 18.3 and 18.4. Because of shopping stores and bus terminal located around the Halley intersection, pedestrian traffic flows on arms of the intersection varies. The levels of service of the arms located on the Istasyon Street are determined as D. This finding supports the result obtained for previous intersection (Babadağlılar).

The Speed-Flow-Density relations are also investigated on the sidewalks. The data obtained are grouped and the following relations are acquired. The sample relations obtained for the Babadağlılar Intersection Segment 1 are shown in Figures 18.5 and 18.8. Figure 18.5 shows that exponential or linear expressions can be used

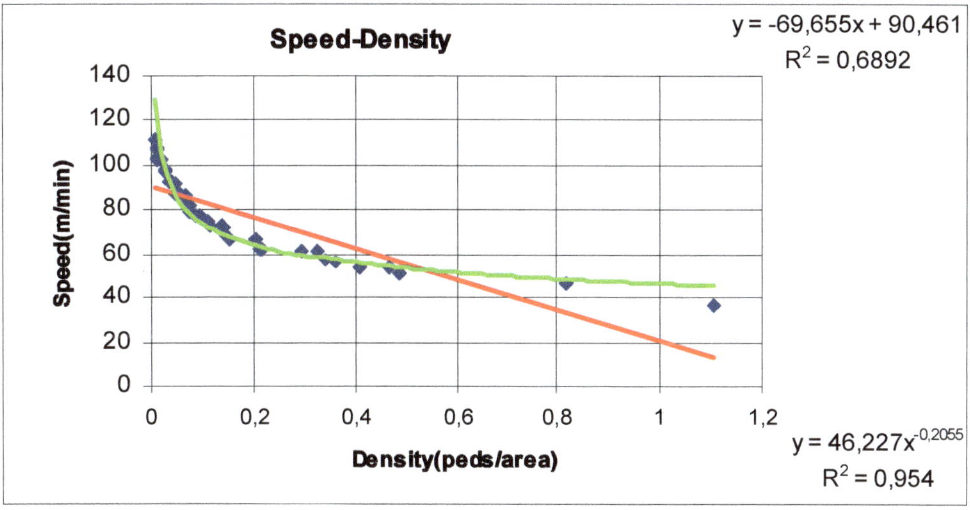

Figure 18.5: Speed-density Relations for the Babadağlılar Intersection Segment 1

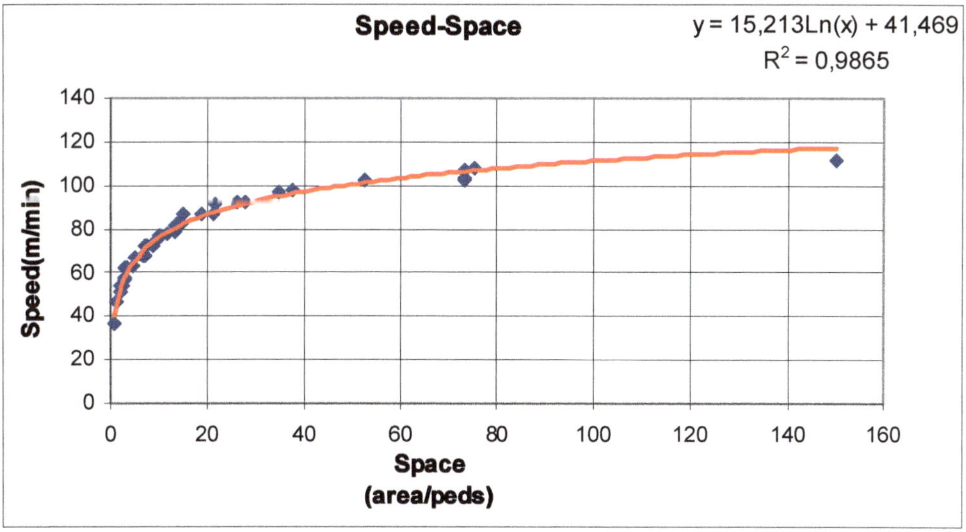

Figure 18.6: Speed-Space Relations for the Babadağlılar Intersection Segment 1

for speed-density relations. As can be seen in Figure 18.6 logarithmic expression can be used for representation of speed-space relation.

Figure 18.7 depicts that polynomial expression can be reasonable for volume-speed relations. On the other hand, the exponential expression can be used for volume-space relations, given in Figure 18.8.

Similar relations and expressions are also obtained for the Halley and the Çınar intersections. Speed-volume-density relations samples for other segments of the intersections are shown in Figures 18.9–18.13.

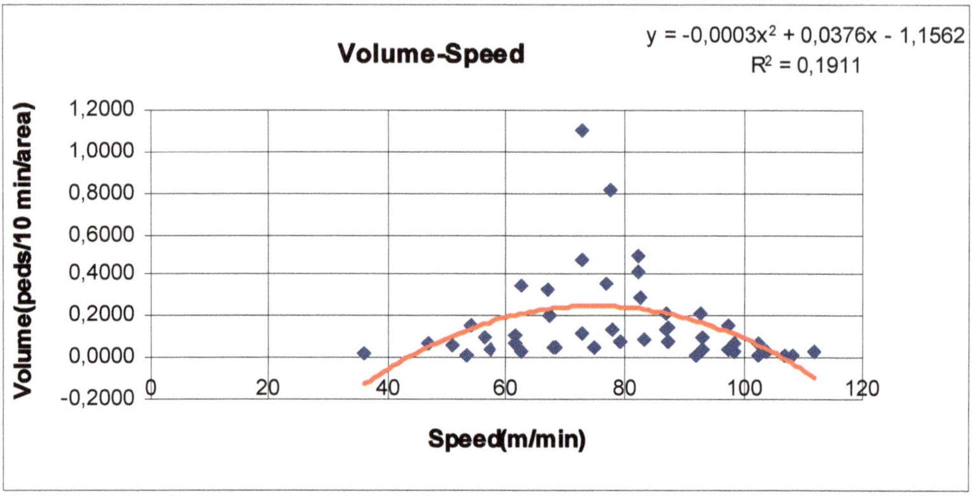

Figure 18.7: Volume-Speed Relations for the Babadağlılar Intersection Segment 1

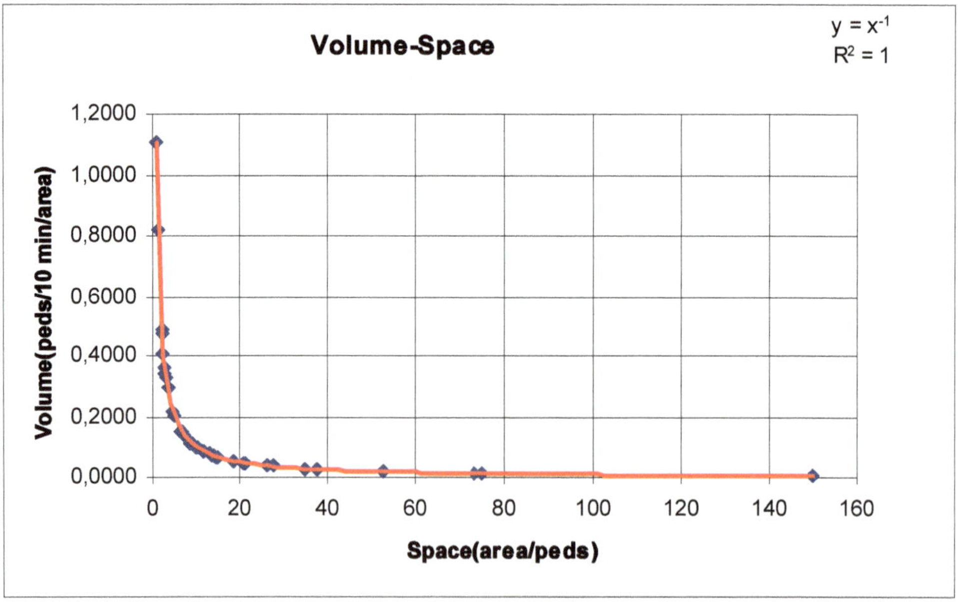

Figure 18.8: Volume-Space Relations for the Babadağlılar Intersection Segment 1

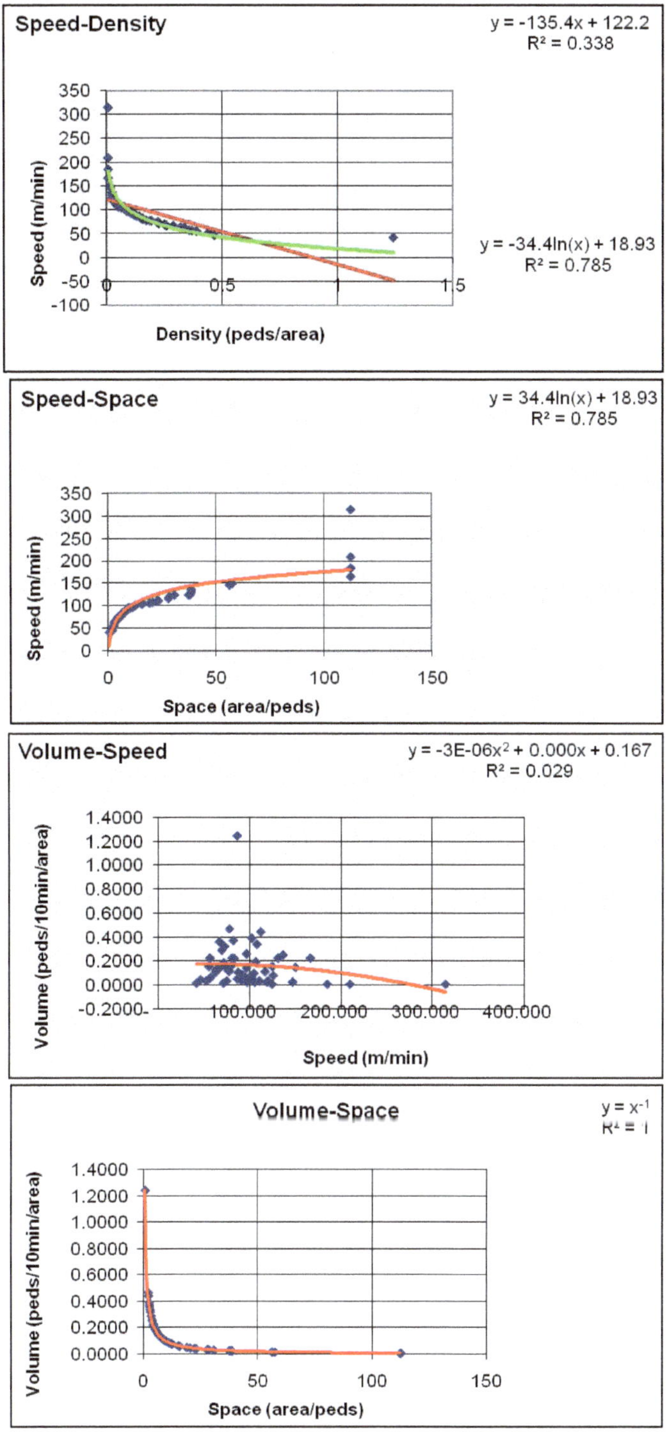

Figure 18.9: Fundamental Relations of Pedestrian Flows for the Çınar Intersection Segment 1

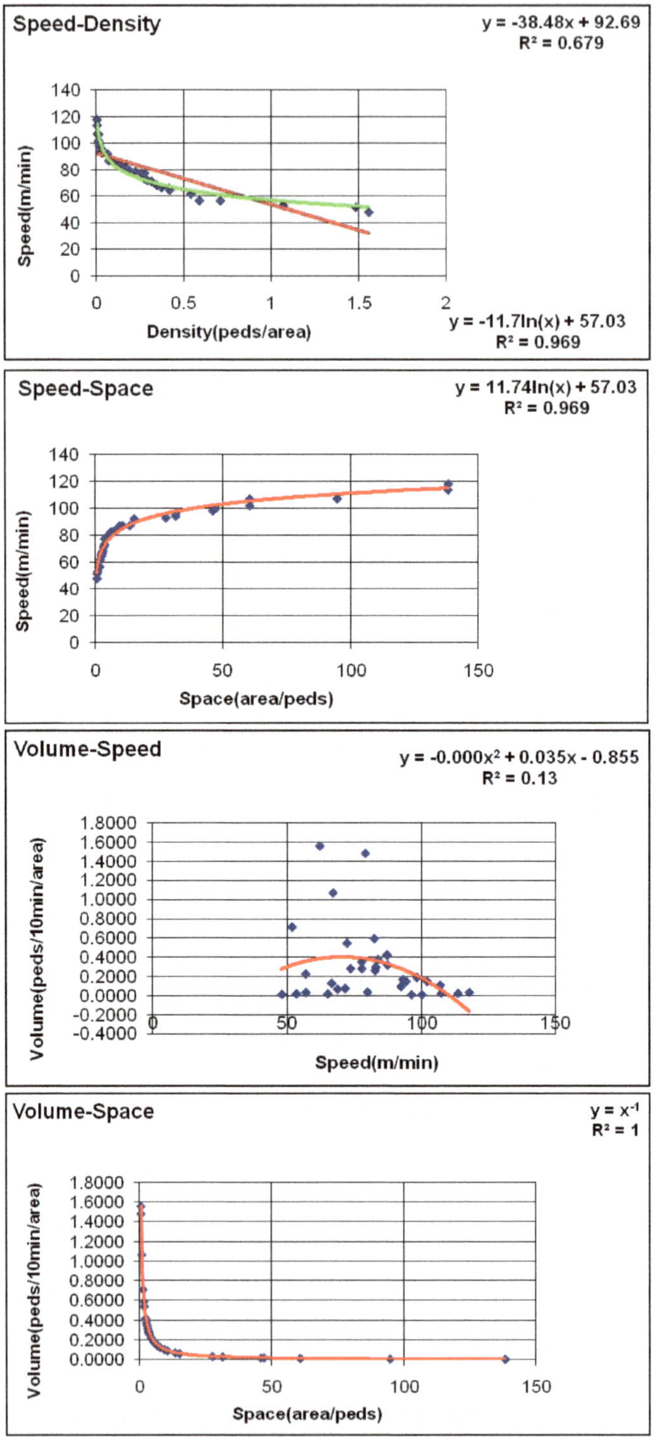

Figure 18.10: Fundamental Relations of Pedestrian Flows for the Halley Intersection Segment

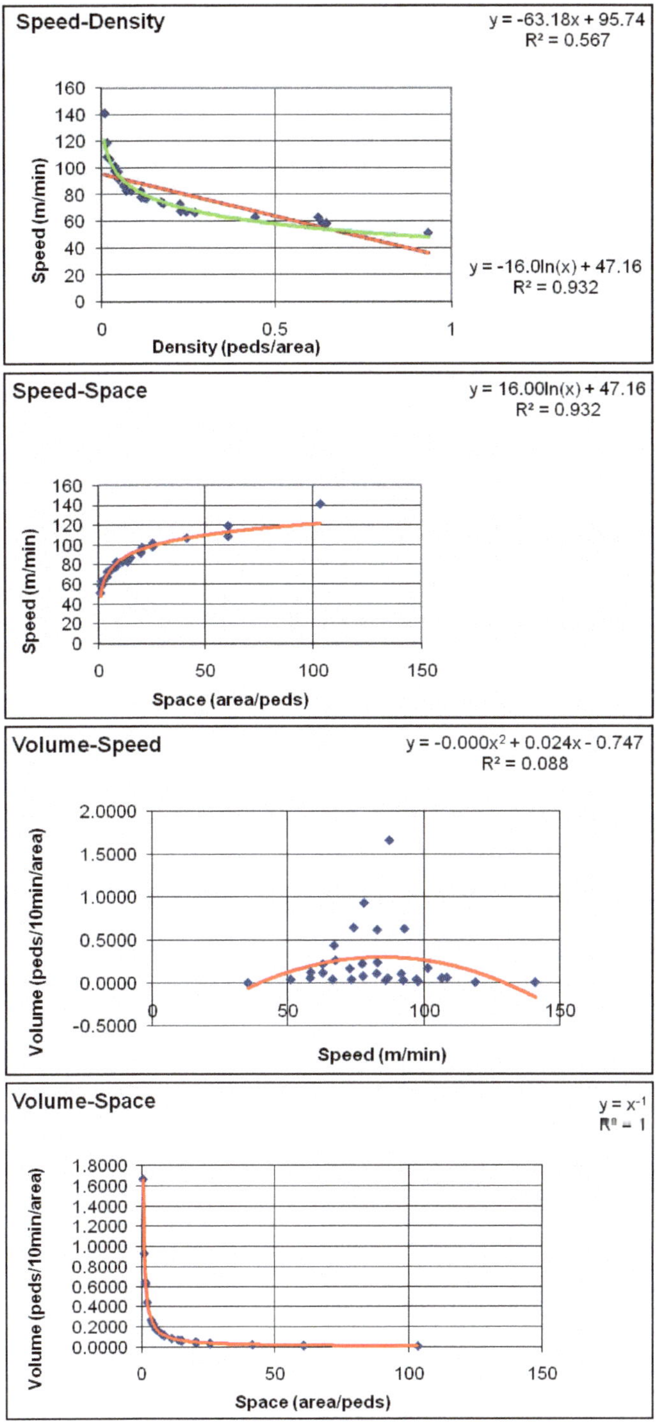

Figure 18.11: Fundamental Relations of Pedestrian Flows for the Babadağlılar Intersection Segment 3

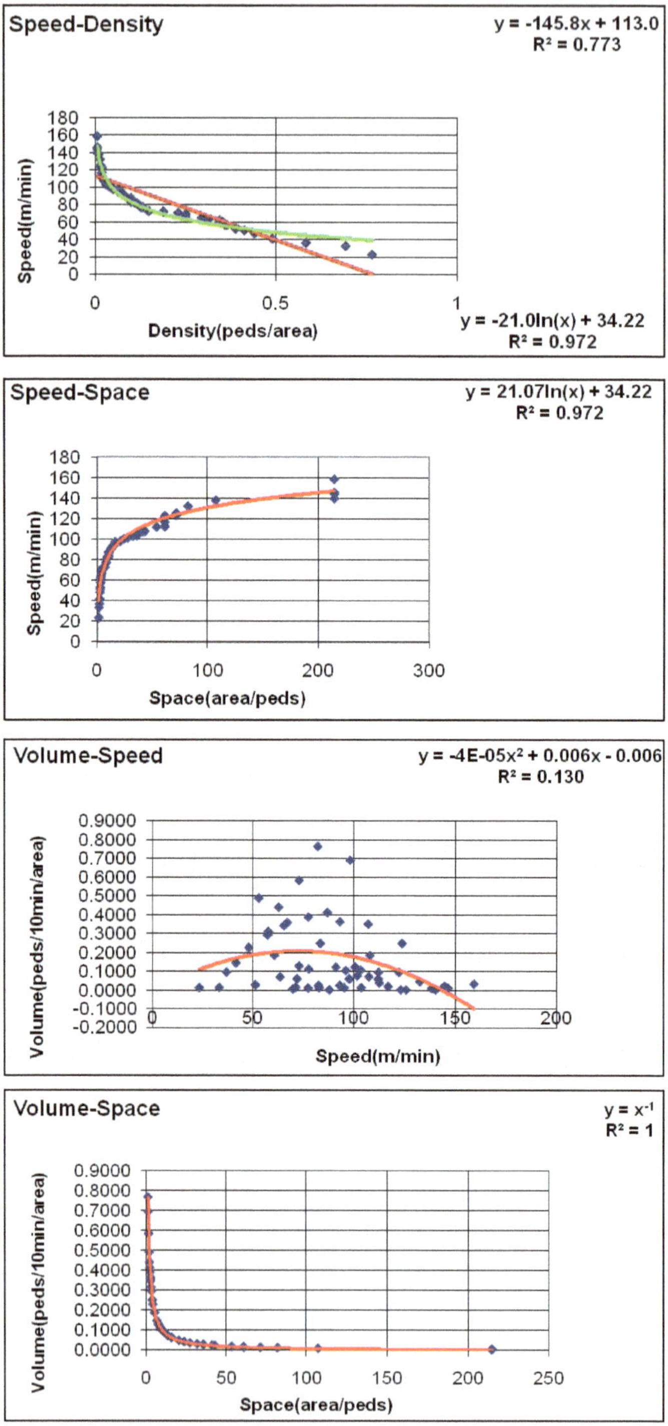

Figure 18.12: Fundamental Relations of Pedestrian Flows for the Çınar Intersection Segment 3

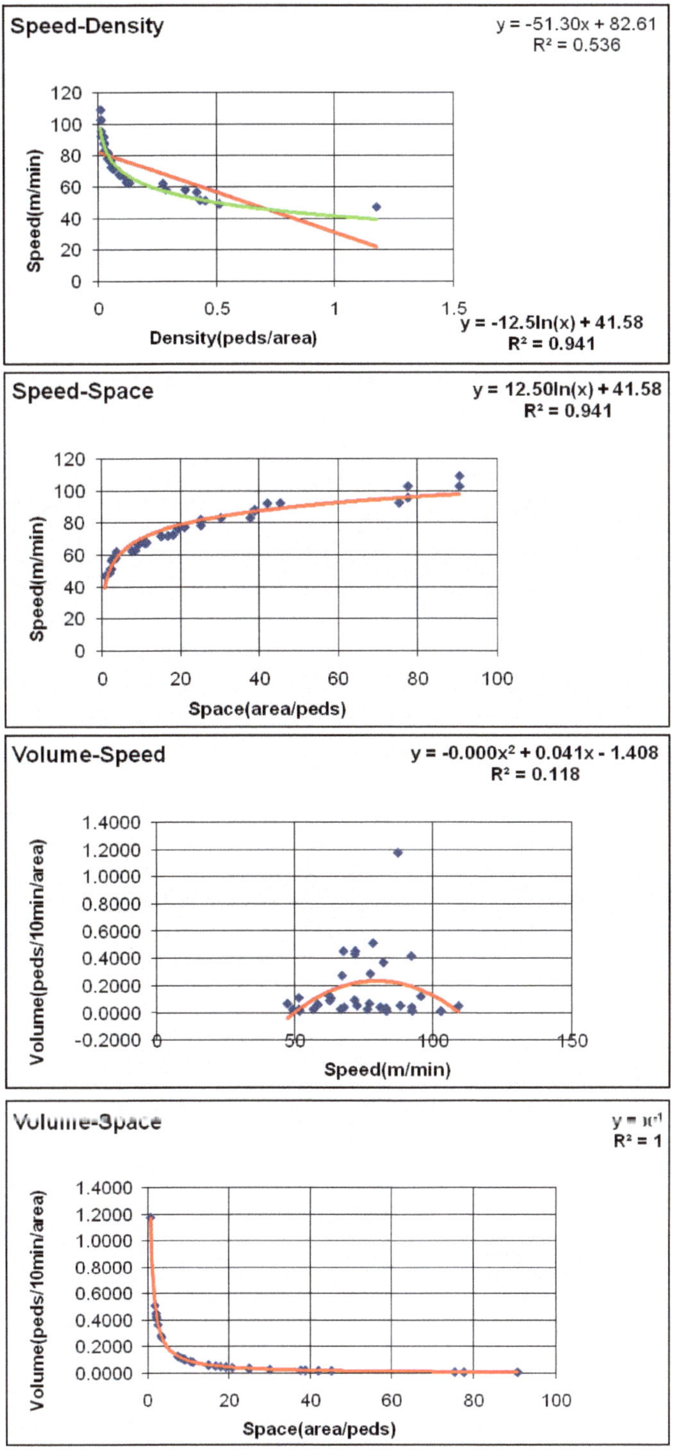

Figure 18.13: Fundamental Relations of Pedestrian Flows for the Halley Intersection Segment 3

Some characteristics of pedestrian traffic flows are calculated based on these relations. The free flow speed and jam (congestion) density values are obtained for all segments of intersections. Table 18.5 shows free flow speed and jam density values for the Babadağlılar, Çınar and Halley intersections.

Table 18.5: Free Flow Speed and Jam Density Values for Segments for the Babadağlılar, Çınar and Halley intersections

Intersection	*Segment 1*		*Segment 2*		*Segment 3*	
	Free Flow Speed (m/min)	*Jam Density (peds/ m^2)*	*Free Flow Speed (m/min)*	*Jam Density (peds/ m^2)*	*Free Flow Speed (m/min)*	*Jam Density (peds/ m^2)*
Babadağlılar	125	1.1	120	1.45	118	1.7
Cinar	185.2	1.25	130.8	0.84	144.5	0.77
Halley	113.5	1.55	110.3	1.33	97.7	1.17

All segments of the Çınar intersection are better than the Halley and the Babadağlılar intersections (Table 18.5). Especially the values for Segment 2 and Segment 3 of the Çınar intersection are quite well compared to other intersections. The pedestrian traffic volumes of the Halley and the Babadağlılar intersections are higher than the Çınar intersection. On the other hand, dimensions of the sidewalks of those intersections are different and narrow than the Çınar intersection's have.

Pedestrians Survey

A survey is applied for pedestrians to get information about traffic perception levels and availability of pedestrian areas. 20 questions were asked to participants in the survey. Total numbers of participants were 1400. In questionnaire, the aim of trip, satisfaction level, problems encountered, perception of safety level, level of obeying traffic rules, use of facilities for pedestrians, use of sidewalks, use of pedestrian signals etc. were investigated. The survey results are given in Figures 18.14–18.16 shows the number of participants classified for gender, age and graduation levels respectively.

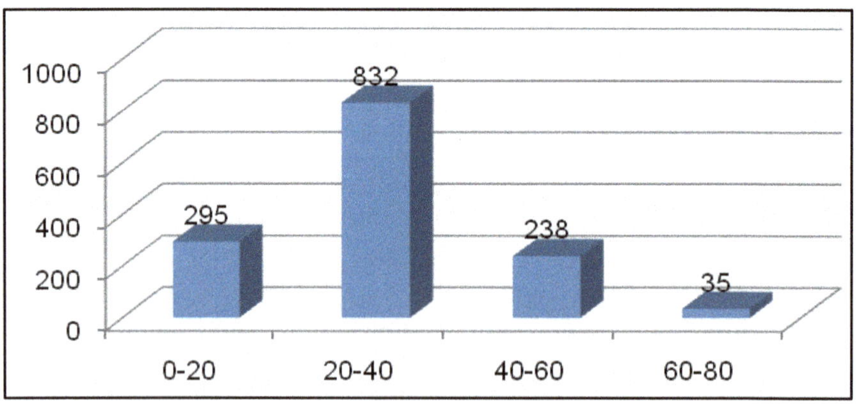

Figure 18.14: Distribution of Participants for Age in the Surveyed Area

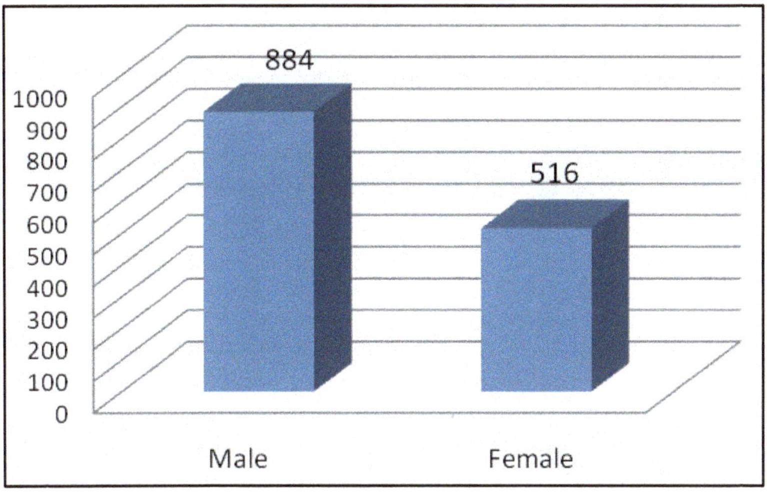

Figure 18.15: Gender Distribution of Participants for Denizli City

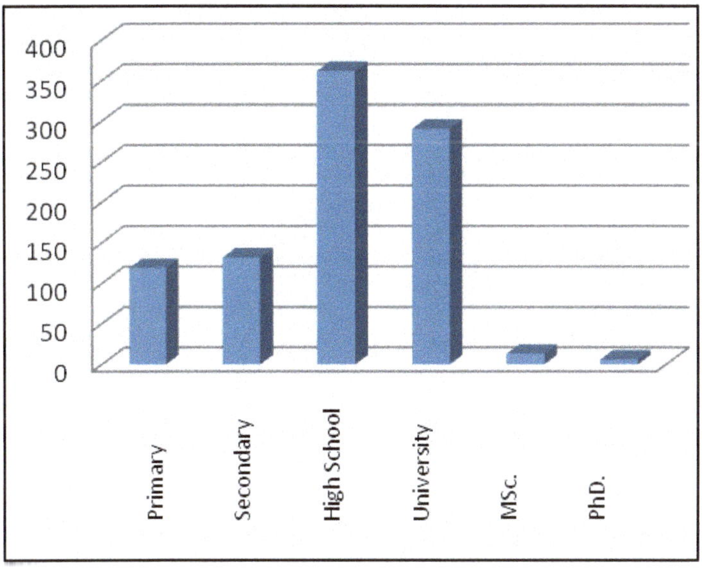

Figure 18.16: Distribution of Graduation Levels In the Surveyed Area

In pedestrian survey, walking durations of participants were also investigated. Figure 18.17 shows that, most of the participants walk at least 30 minutes.

The most important problem faced in the traffic is also asked to participants in the questionnaire. Figure 18.18 depicts the problems encountered by pedestrians. Results showed that the most important problem is the conflict with vehicle traffic at uncontrolled and signalized intersections. The second problem that has high percentage is insufficient sidewalk width. Other problems are defined as insufficient crosswalks and interactions between pedestrians.

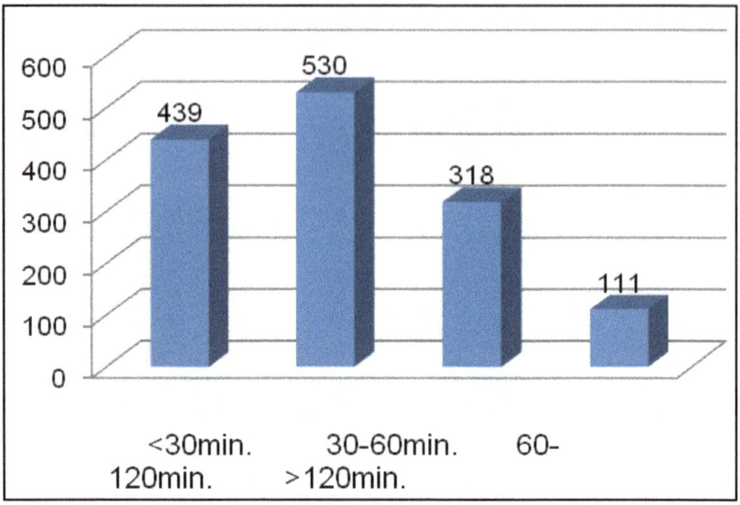

Figure 18.17: Distribution of Walking Times

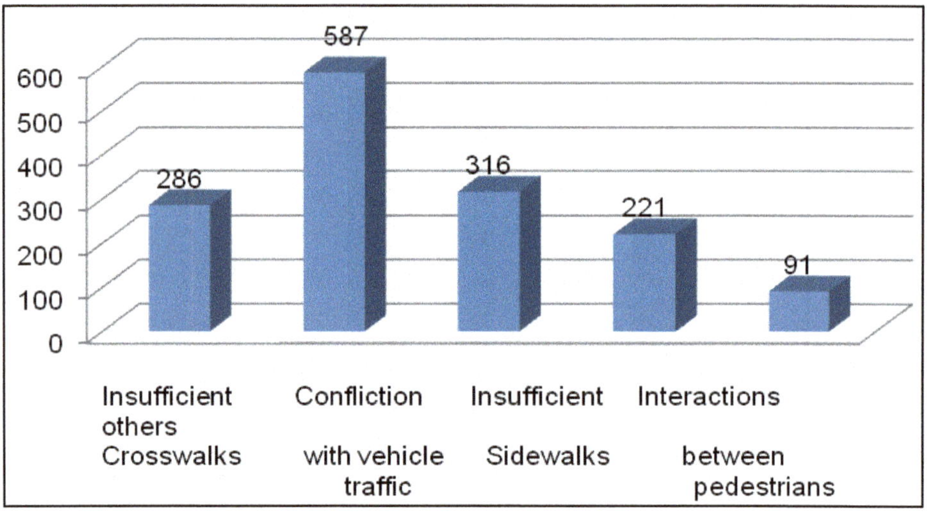

Figure 18.18: Distribution of Problems Encountered by Pedestrians

One of the important issues for pedestrian traffic flows is related to rules and regulations. This phenomenon was asked to participants in the survey (Figure 18.19). It is observed that most of the pedestrians have certain information about traffic rules and regulations. On the other hand, significant parts of them have insufficient information about the subject.

Sidewalks usage is also dealt with in the questionnaire. In the survey, walking sides (*i.e.* right hand, center and left hand sides) are asked to participants. Distribution of answers is shown in Figure 18.20. Results show that, the use of right hand side has

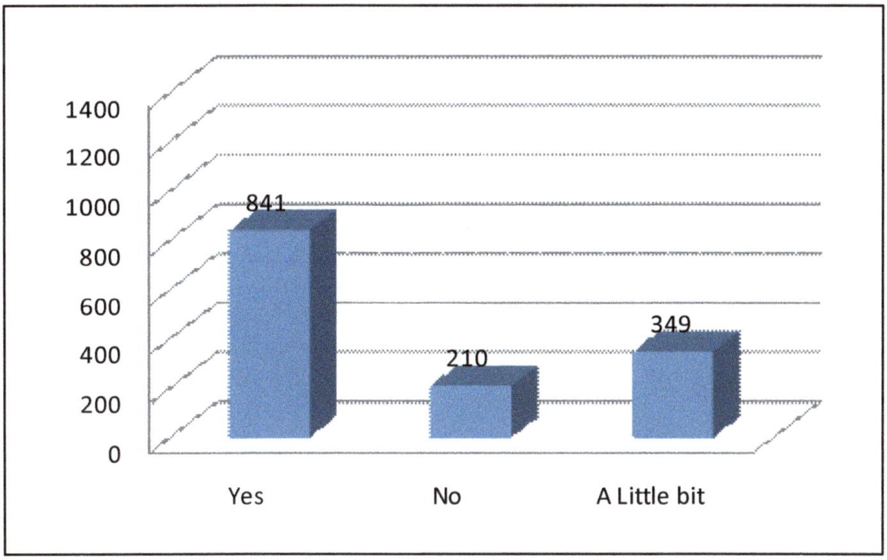

Figure 18.19: Distribution of Pedestrian Information Levels About the Traffic Rules and Regulations

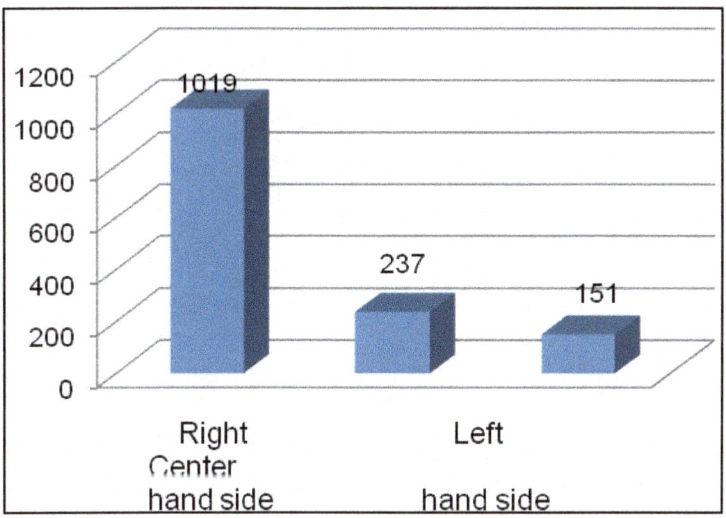

Figure 18.20: Distribution of the Sidewalks Usage

the highest value (72 per cent). However, use of left hand side and the center part have relatively low percentages.

Use of pedestrian crossing is also researched. The usage frequency is asked to participants. Figure 18.21 depicts that the usage frequency of pedestrian crossing. It was found that significant part of pedestrians frequently use the crosswalks. Pedestrian signal usage is one of the research subject that are taken into consideration in the survey. Results are shown in Figure 18.22.

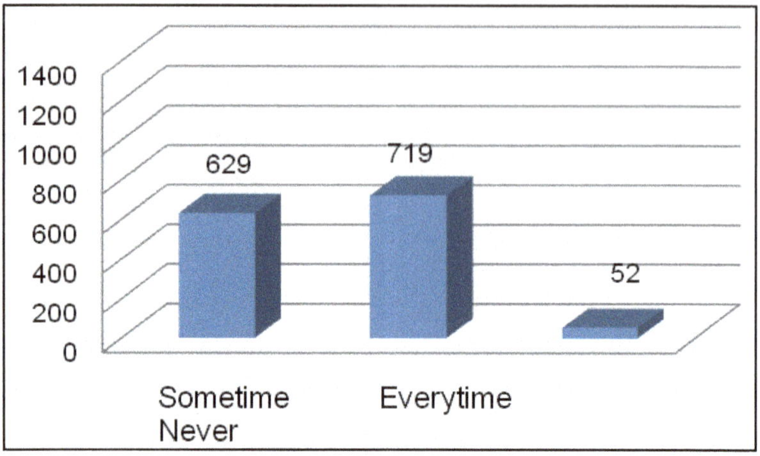

Figure 18.21: Distribution of Pedestrian Crossing Usage

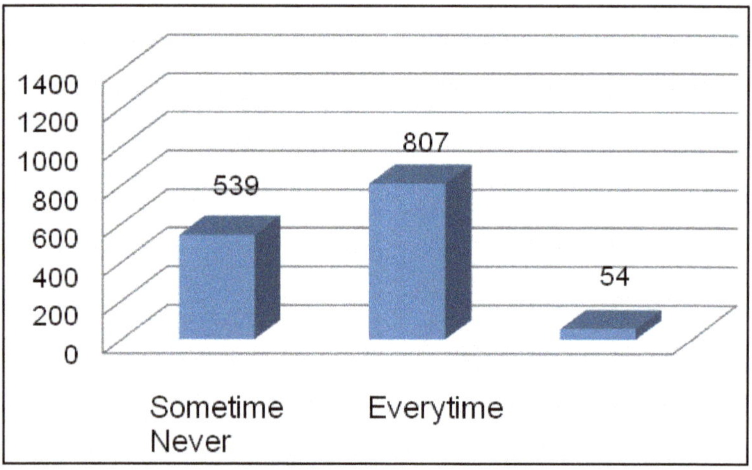

Figure 18.22: Distribution of Obeying Pedestrian Signal

RESULTS AND DISCUSSION

Based on the analyses, the following results are derived and summarized.

- ☆ The pedestrian crossing analysis shows that use of pedestrian crossing in urban areas is reasonable. But use of green signal and right side usage in crossings are poor. This is very interesting and can be related to educational and cultural level. On the other hand, the pedestrian survey results are different from the pedestrian crossing analysis. This finding shows that perceived use of pedestrian crossing is lower than the realized one.
- ☆ The pedestrians use green signal especially in the peak hours. In off-peak periods pedestrians prefer to seek headways of vehicles. This result shows

that both the crossing analysis and the pedestrian survey results are compatible about the subject.

- ✰ The drivers have problems about obeying traffic rules and regulations particularly related to pedestrians such as right of way rules in right turns.
- ✰ The level of service values of segments varies based on the sidewalk dimensions and pedestrian traffic volumes. The net width of sidewalks affects level of services of segments. In Babadağlılar and Halley intersection, the net width is affected by some stores and therefore bottlenecks can be occurred on sidewalks.
- ✰ The lowest level of service values is obtained as D for three segments of Halley intersection. This situation is occurred, because of high pedestrian volumes and net width of sidewalks. This arm of the intersection has the highest volume among all of the arms at all periods of daytime. Besides many stores located along the arms have some obstructions on the sidewalks. Therefore, the levels of service values of three segments are obtained as the lowest. The level of service can be increased with prohibition of sidewalks' occupation by stores.
- ✰ The logarithmic expressions are attained as significant for speed and density relations. This result can be used for pedestrian flows instead of linear relations. The exponential and polynomial expressions are also significant for other relations. These findings can be used for pedestrian flow modelling. The observations should be increased to generalize these findings.

REFERENCES

Anatürk B., Çetin Z. (2008) "Investigation of Pedestrian Traffic in Denizli Province", Graduation Project, 58p., Pamukkale University, Faculty of Eng., Dept. of Civil Eng.

Dandan T, Wei W., Jian L, Yang B. (2007) Research on Methods of Assessing Pedestrian Level of Service for Sidewalk J Transpn Sys Eng and IT, 7(5), 74"79.

Hamed, M. (2001). Analysis of pedestrians' behavior a pedestrian crossings. Safety Science 38, 63-82.

Hongfei J., Lili Y., Ming T. (2009). "Pedestrian Flow Characteristics Analysis and Model Parameter Calibration İn Comprehensive Transport Terminal", Journal of Transportatıon Systems Engıneerıng and Informatıon Technology, 9(5), 117-123.

Keegan, O, Mahony, O. (2003) Modifying pedestrian behaviour Transportation research Part A 37, 889–901

Kutlu, K. (1963). Traffic Surveys, İTÜ Press, 110p.,İstanbul, Turkey

Özdirim, M. (1994). Traffic Engineering Volume1-2, KGM Press, Ankara, Turkey.

Tarawneh, M.S.(2001). Evaluation of pedestrian speed in Jordan with investigation of some contributing factors Journal of Safety Research 32, 229– 236.

Transportation Research Board (1985) Highway Capacity Manual, Washington D.C USA

Yang J., Deng W., Wang J., Li Q, Wang Z, (2006) Modeling pedestrians road crossing behavior in traffic system micro-simulation in China Transportation Research Part A 40, 280–290.

Chapter 19

Paradox of Halophyte Utilization as Biofuel Resources and Land Sustainability

Mohammad Kafi[1] *and Munir Ozturk*[2]

[1]*Centre of Excellence for Especial Crops, Ferdowsi University of Mashhad, Iran*

[2]*Ege University, Botany Department, Bornova-Izmir, Turkey*

ABSTRACT

Economic growth through heavy industrialization has been based on utilization of energy locked in fossil fuels. Many countries are facing serious pollution problems from these sources. Alternative in this connection could be use of renewable energy like biomass. Biomass utilisation is advantageous because it includes agricultural residues, is readily available, reduces need for fossil fuels, foreign dependency, and provides bio-fuel. However, there are some disadvantages as well such as; energy comes mainly from plants which must be harvested thus leading to a depletion of soil nutrients, and land use to grow biomass materials may be used for other purposes. However, the vast areas under natural salt-affected degraded lands could fill this gap. These lands are characterized by a cover of saline plant diversity called "halophytes" which can be used for biofuel production. Halophytes are an untrapped source of bioenergy. It is possible to explore the extent of seasonal salt accumulation in these plants and the possible uses of saline biomass to generate bioenergy. The advantage of developing biofuel from sea water growing halophytes compare to other types of biomass is that saltwater plants are not dependent on freshwater, which is in increasingly short supply, and can instead be irrigated using plentiful sea water supplies. Halophyte agriculture could free up freshwater supply and arable lands for food production and also they will produce a cooler, wetter land surface, which could lead to rainfall in areas of the world where rainwater is in short supply. Major arid or semi-arid halophyte agriculture problems include pumping and draining the required high volumes of irrigation water from sea or ocean sources. Halophyte agriculture ecosystems are very vulnerable; therefore all environmental issues such as biodiversity loss, soil and water depletion and salinization should be concerned. On the other hand halophytes mostly benefits from C4 photosynthesis pathway, so they are more productive compared with C3 conventional crops and they can produce economically enough biomass for changing to the ethanol. The traditional halophyte bioenergy should be firstly consumed in the local community to help the strengthens of desert ecosystems and benefit of any industrial biofuel from inland and coastal resources should also spend for environmental improvement.

INTRODUCTION

Agriculture is life, food, fiber, forage, and lately a source of fuel. Later is termed as biofuel or bioenergy because of its relation to the biomass from crops/plants, organic waste products, sewage sludge and household garbage. Biomass (solid) is converted to produce electricity, heat, liquid fuel (ethanol, methanol, butanol, biodiesel) or gas (methane, H_2, CO, gasification, pyrolysis). In fact bioenergy is not far from petroleum due to the fact that it also comes from the plants just living million years back.

It is important to remember that humans have faced many challenges through the ages but what we need today is a holistic, comprehensive approach to energy, including its social and cultural dimensions. Education will play a great role in the use of biofuels in sustainable development, capacity building, and better communication with the general public in developing countries. The agricultural residues, animal wastes, biomass from seawater irrigated halophytic species seem to be interesting as an energy source. The research and development in energy efficiency and renewable energy sources should be encouraged all over the globe, halophytic biomass used in the biofuel production should be increased by employing fast growing species, biofuel firms should take an active part in this production system, detailed realistic economic and cost-benefit comparisons with other alternatives will prove helpul, levels of incentives that could be introduced by the government for the construction of digesters should be determined, and finally if governments, regulators, energy companies and consumers work together, energy development will serve all people in the world and can be a real catalyst for peace (Ozturk *et al.*, 2003).

An achievement in the amount of emission reductions through the use of biofuels will vary much. It depends on choices made in feedstock selection and production through final fuel use. There are many technologies available today to produce biofuels from different feedstocks. The research is going on the lowering biofuel costs, GHG emissions, and land and water resource needs, and on improving compatibility with fuel distribution systems and vehicle engines (Ozturk *et al.*, 2003, 2007).

There are large uncertainties in estimates of the amount of energy that could be supplied by biomass. Estimates of the amount of global energy that could be supplied by biomass in 2050 range from 40 to more than 1,000 exajoules (EJ). To put these numbers in perspective, 40 EJ is close to current global biomass-derived energy (10 per cent of global energy demand). On the other hand, 1,000 EJ conceivably could be enough to meet total 2050 energy demand (UNDP, 2000; GTZ, 2006). Factors that contribute to this wide range of estimates include uncertainty about future crop yields and prices, water availability, and competition for land. Just as the cost of feedstocks plays a large role in costs of current ethanol and biodiesel products, feedstock costs are expected to play a similarly significant role in cellulose-based fuel costs, and therefore in amounts that can be economically produced. It has been estimated, for example, that if cellulosic conversion costs decrease by half and their efficiencies improve considerably, cellulose-based biofuels could be economically competitive at a farm-gate feedstock price of $30 per dry ton or less (again, this depends on petroleum prices and other factors). At this price, however, it is estimated that only 500 million dry tons of biomass would be available by 2025 from a combination of woody crops, wood and agricultural wastes, and perennial grasses grown as energy crops (Sheehan 2007; Pena, 2008). The fact that cellulosic conversion costs likely need to decrease by half to be competitive is illustrative of the cost challenges facing emerging biofuel options. It will take significant, consistent expenditures in biofuel R&D for the emerging pathways to become cost-competitive with petroleum fuels.

There are abundant resources in every region of the world to meet growing global energy demand, but countries need to diversify their energy sources by keeping all options in this connection open. World proven reserves of petroleum lie around 1.033.2 billion barrels, natural gas 5.141.6 trillion cubic feet and coal 1.087.2 billion tons (Ozturk *et al.*, 2003). Out of this 26.6 billion barrels, 83 trillion cubic feet and 5.04 billion tons were used in 1998 (Wright and Nebel, 2002). Oil production in OPEC countries lied around 34.8 million barrels per day in the year 2000 and is estimated to lie around 45 million barrels at present (Ozturk *et al.*, 2003,).The percentage of energy sources used for generating electrical power in the world is 17.7 (nuclear), 14.8 (natural gas), 38.4 (coal), 9.3 (oil), 18.4 (hydro), and 1.4 (wood+refuse+renewables) (Wright and Nebel, 2002). A movement has started where countries are now trying hard for moving away from the petroleum gradually. During the coming decades we may see biofuels providing a full 25 per cent of the world's energy needs.

Although there are some dangers in this field but "Bioenergy" has an enormous opportunity for farmers, especially in the industrialising countries. Although, some people believe that conversion of productive agricultural lands into biofuel fields is not good solution, but in our opinion discrediting biomass energy sources would not be fair, because these can be marketed as a solution to economic and energy security, agricultural production, rural development, solution for environmental issues, and release the developing world from dependence and insecurity.

Current biofuel feedstocks and human existence are highly dependent on the familiar glycophytes rice, corn, wheat, potatoes, soy beans, palm oil, and nut plants, which cannot tolerate salt (Hendricks and Bushnell, 2008). Scientists should consider those sources of biomass production that not disturb current agricultural resources. Halophytes are highly specialized plants with a great tolerance to salt. They can germinate, grow, and reproduce in areas of high-saline solutions, coastal shorelines, marshes, inland lakes (coastal-halophytes), huge potential irrigation areas such as desert regions with subterranean brackish water aquifers (desert-halophytes), and directly in oceans or seas (Hendricks and Bushnell, 2008). There is some probability that plants started as halophytes, moving from the sea to the shores and marshes (Hendricks and Bushnell, 2008). These plants provide shoreline erosion protection and feeding areas for birds, fish, and animals. They are coming to the foreground as the salinity of freshwater-irrigated systems continues to rise as does the population demand for food and energy biofuel feedstocks (Hendricks and Bushnell, 2008).

Selectivity and nurturing of both micro and macro halophytes to provide proteins, oils, and biomass while enduring a wide range of environmental conditions may provide human needs as limits to freshwater food and energy are reached (Khan and Weber, 2006). Harvesting ocean halophytes such as seaweed, algae, and kelp becomes more difficult because of storms, conversion, and transport, yet the oceans are vast reservoirs of both CO_2 and nutrients (nearly 80 per cent of required plant nutrients exist in sea water) (Bushnell, 2007). Halophyte agriculture is not that well understood, yet it has potential to provide yields similar to glycophytes (alfalfa) (Glenn *et al.*, 1998)

LIMITATIONS

Similar to glycophyte crop production that not all arable land are suitable for all crops, all nonarable or desert soils also are not suitable for irrigated halophyte crops; or high-yielding algae ponding or bioreactors a viable option in various climates and environments (Hendricks and Bushnell, 2008). For all fuel-food biomass, suitable water resources, quantitatively and qualitatively, are a necessity. Some 43 per cent of the Earth's land is arid or semi-arid and 97 per cent of the Earth's water is in its oceans (Glenn *et al.*, 1998). Worldwide estimates of coastal lands and inland salt deserts suitable for growing halophytes are set at 130 million hectares, of which perhaps 10 per cent resides in the U.S. West and Southwest. Assuming that $13{\times}10^6$ ha desert soils can produce Salicornia bigelovii (glasswort species) oil at the rates given by Glenn *et al.* (1998) of 0.2 kg/m^2 (2000 kg/ha), the zeroloss conversion would supply 2 per cent of 2005 U.S. (~20.3 Mbbl/day) oil usage. At the projected production rate of 60 bbl/acre of oil (148 bbl/ha or 2.17kg/m^2) this percentage jumps to nearly 22 per cent, providing. Of course, halophyte cultivation needs to plan supply the necessary irrigation water, soil, weather, nutrients, and disease-free environment. Irrigating large tracks of desert land with salt water can have one unintended anthropomorphic effect, namely turning adjacent desert into farmland. The evaporating water from these plants will change the local weather pattern, resulting in condensation, and change dry climates to wetter climates. This is the exact reverse effect that is happening in large tracks of land next to the Sahara desert where the desert soaks up the moisture and expands into the adjacent farmlands. Such an event has been a critical issue in Beijing for the last few decades. Due to rampant deforestation, the Gobi desert has been creeping towards Beijing for the last few decades and persons returning after several decades have noted it is now much drier and sandier than it used to be. Putting large amount of water into the desert for the water to vaporize can potentially make places like Beijing a little wetter and promote vegetation growth, which will in turn put more water in the atmosphere. Ultimately, it has the possibility of halting or outright reversing the desert growth. (Hendricks and Bushnell, 2008)

FUEL VERSUS FOOD: ADVERSARY

The major issue of biomass production is choosing between producing food or oil because there is a limited amount of arable land. That is why saline agriculture as an undeveloped source of both food and fuel is so attracting. Saline agriculture, when combined with aqua culture provides food, fuel feedstocks, habitat, and redistribution of freshwater resources. Seawater irrigation requires a lot of water, some 12000 m^3/year for each hectare. Seawater irrigation also provides some 80 per cent of the nutrients needed for plant growth and requires only additions of phosphorous, iron and nitrogen, suggesting the need for genetically modified nitrogen-fixing halophytes (Bushnell, 2007).

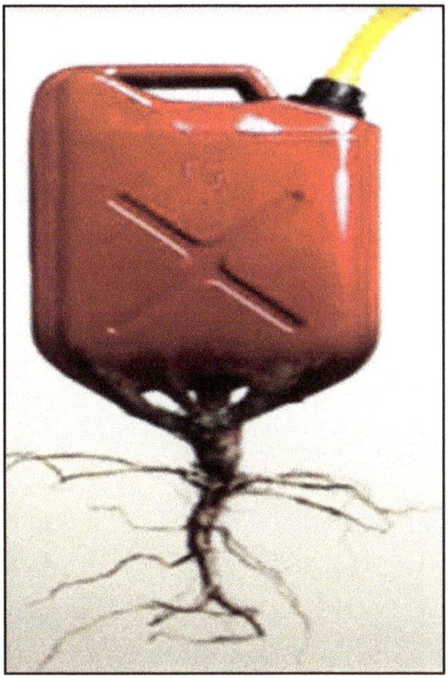

Figure 19.1: Bioenergy from Plants

SOURCES OF BIOFUELS

According to Wang *et al.* (2007) biofuel production pathways with current research and development efforts can be classified as sugar crops like sugarcane, sugarbeet, sweet sorghum (for ethanol), starch crops like corn, wheat, cassava, sweet potato (for ethanol), cellulosic biomass like crop residues (corn stover, rice straw, wheat straw etc.), forest residues, municipal solid wastes, energy crops (e.g., switchgrass, fast growing trees, etc.), black liquor from paper mills (for ethanol), corn, sugar beet (for butanol), soybeans, rapeseeds, palm oil, jatropha, waste cooking oil, animal fat (for biodiesel/renewable diesel), cellulosic biomass via gasification (for Fitscher-Tropsch diesel, hydrogen, methanol) and finally algae (for hydrogen, ethanol) (Armagan and Batýgý, 2007). The renewable energy sources are growing fast and on the way to become a cornerstone of the energy and environmental policies. Annual turnover has reached 30 billion Euros or about 50 per cent of the world market, and more than 200 000 jobs have is an old concept (Wright and Nebel 1998).

The traditional, often non-commercial sources of fuel provide nearly 20 per cent of the world's energy today, but are not reported in most statistics about global energy (Owen and Chiras, 1995). In many developing countries these sources of fuel are a large proportion of the energy available. Biomass provides upto 90 per cent of the total energy demand of these nations (Enger and Smith, 2000). Since 43 to 44 per cent of the Earth landmass is arid or semi-arid, there is the potential for developing a multiplicity of seawater-irrigated regions to produce halophyte crops and combine them with aquaculture. It is the potential of halophytes to provide both sources of food and fuel feedstocks (without affecting current sources) that excites our interest.

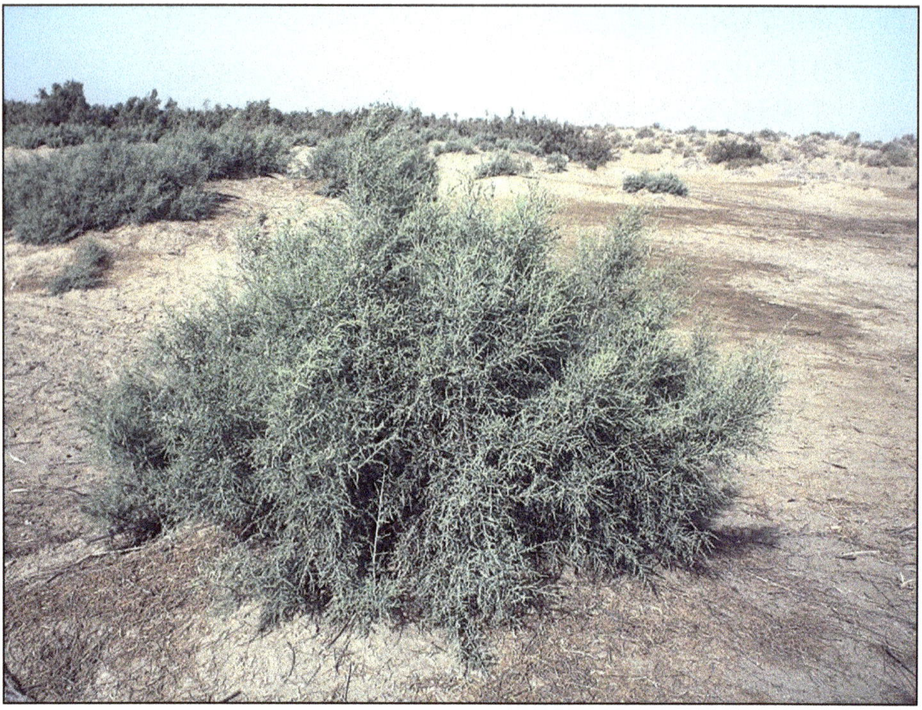

Figure 19.2a: *Tamarix dioica* Used for Energy

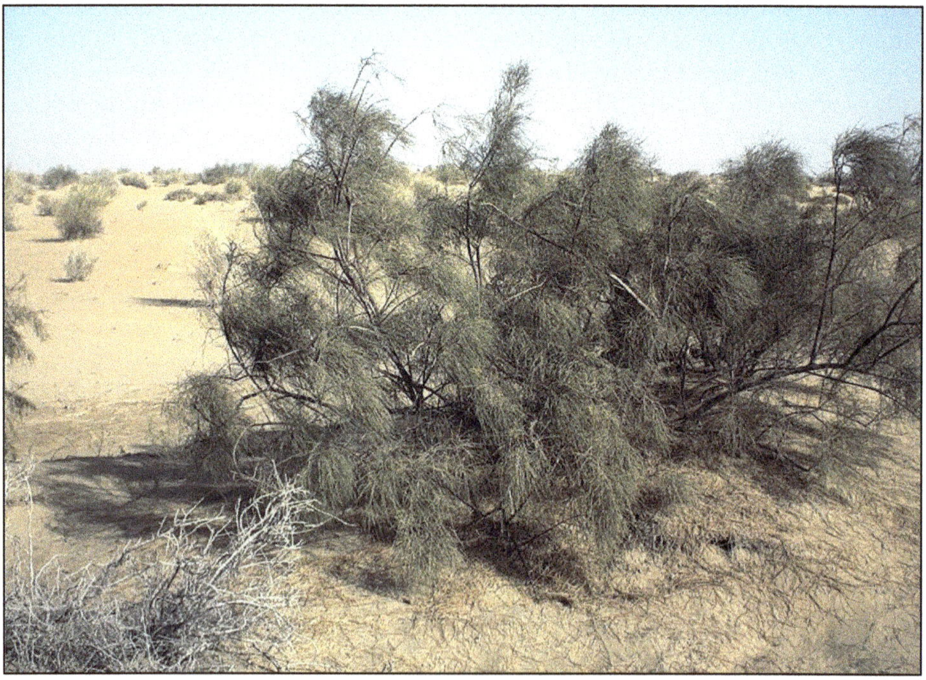

Figure 19.2b: *Tamarix aphylla* Used for Energy

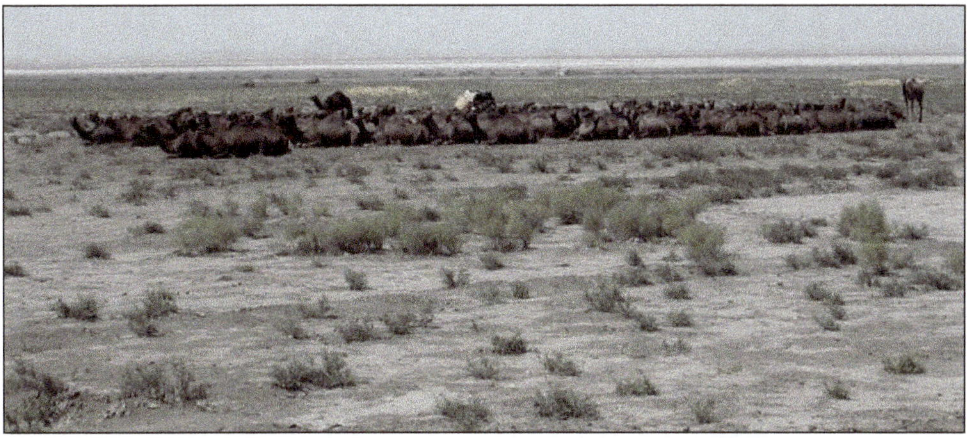

Figure 19.3: Camel Feeding in the Arid Zone

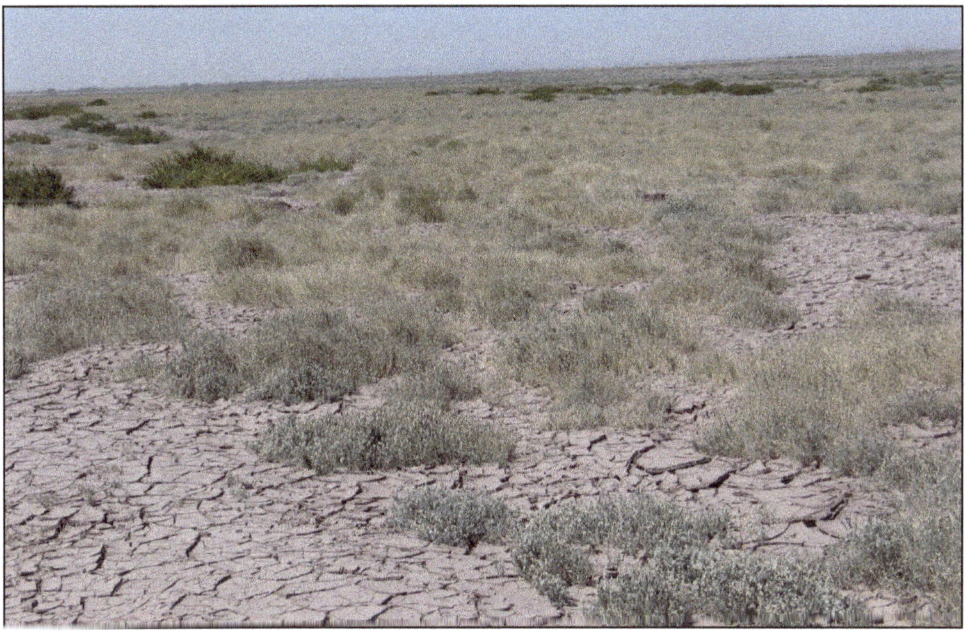

Figure 19.4: Saline Lands

In addition to sea water irrigation for halophyte production there are many other sources of inland water (surface or ground) that due to high salinity are not suitable. Examples of capable halophyte for biofuel growing include salicornia, an annual salt-marsh plant with an oil content of 15 to 35 per cent, and seashore mallow, a perennial which grows on coastal marshlands or inland brackish lakes and has an oil content of 18 per cent. The oil from salicornia is similar to safflower oil and

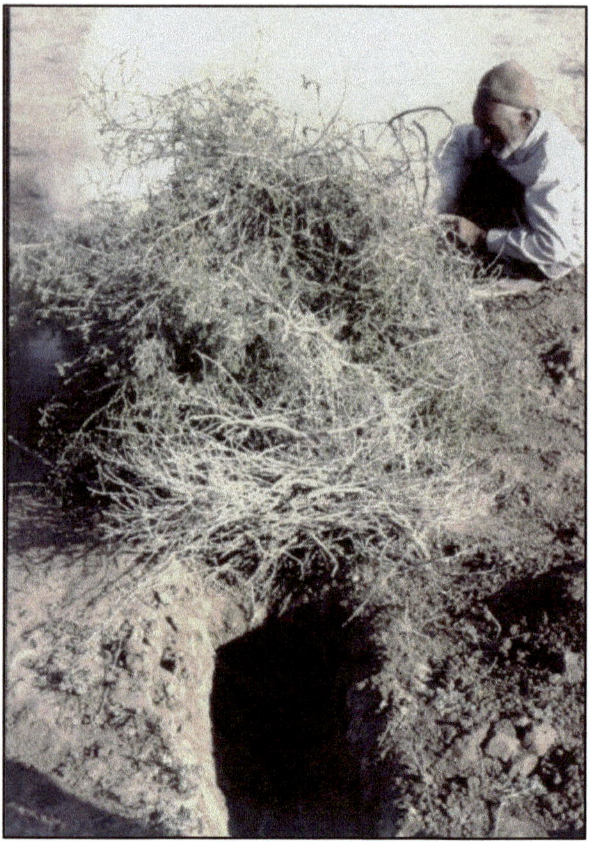

Figure 18.5: Halophytes Used as Energy Source

seashore mallow to that of cottonseed oil. Halophytes represent a non-edible feedstock source that would be grown on hectares not currently being utilized for edible production.

Salicornia is reported to be tolerant of salt levels up to twice that of seawater, it has more than six years of field trials in Mexico, and could generate more than 80 gallons of oil per acre for biodiesel production (Weber, 2009). *Salicornia bigelovii*, a leafless annual salt-marsh plant with green jointed and succulent stems, is indigenous to the Arabian Sea coasts of Pakistan and India on the margin of salt lakes and Celon (Sri Lanka), the U.S. Southwest, and other parts of the world.

Seashore mallow (*Kosteletzkya virginica*) is a novel salt-tolerant perennial crop derived from a salt marsh plant with an oil content of approximately 18 per cent and residual meal that contains 30 per cent protein. This plant can be grown on saline land and produce vegetable oil on underutilized or non-arable land. As reported by researchers at the University of Delaware, seashore mallow has a productive life of about a decade and the oil is very similar to cottonseed oil in fatty acid composition

(Weber, 2009). Seashore mallow has been evaluated in more than four years of field trials in the Delaware Coastal Plain and could generate more than 30 gallons per acre of oil for biodiesel production. Williams *et al.* (2008) Studied use of saline land and wastewater for growing a potential biofuel crop *Arundo donax* L. in Autralia and reported that Arundo can be used for carbon credits. Arundo has high potential to be the basis for a new biofuel industry with effective weed risk management and proper tools and land management.

Distichlis spicata is the only C4 species most suited to the high temperatures and high-radiation regimes during the warm summer months of southern Australia. This species not only has theability to grow and spread in saline waterlogged soils, but it also produces valuable green feed in moist saline discharge soils during the summer period. Such forage has real value for a mixed farming system.

BIOMASS PRODUCTION

The biomass energy which is just deriving energy from photosynthesis is not a new name it is a simple process of obtaining energy from the chemical energy stored in biomass and has been used for thousands of years (Nebel and Wright, 1998). Humans began to use different sources of energy to make their lives more comfortable. Biomass as a renewable source of energy can be used as a solid fuel or converted into liquid or gaseous forms for the production of electric power, heat, or chemicals, or for use in vehicles. Recently Malaysia has developed an invigorative progrmme of biomass based power generation which includes wastes from palm oil industry and other sources (Ibrahim *et al.*, 2002; Pimental 2002; Perlack, 2005; Schnepf, 2006; Gustavsson *et al.*, 2007).

Nearly 2.5 billion people in the developing countries depend on wood for cooking and heating, and 1.3 billion people are meeting their needs through deforestation but reforestation is very slow. This might result in creating problems for approximately 3 billion people in near future (Ozturk *et al.*, 2003). Many different kinds of fuels are produced from biomass. The most common of these are biodiesel and bioethanol (Wu *et al.*, 2006; Caesar *et al.*, 2007; Edwards *et al.*, 2007; RFA, 2007). A big potential exists in the field of bio-fuels (Cakir, 2006; Hunt and Stair, 2006). World biodiesel production has gone up from about 1 billion litres in 1995 to around 10 billion litres in 2007 and ethanol production from 18 billion litres in 1995 to 58 billion litres in 2007 (WWI, 2006; JCTR, 2007; RFA, 2007; Pena, 2008).

The facts which make biomass energy attractive is that it has great potentials like; readily available significant renewable and recycle resource, reduces foreign dependency, adds secondary value to agricultural crops, creates habitats, increases biodiversity, helps cut down on wastes, is inexpensive, decreases greenhouse gas emissions, reduces need for fossil fuels, stimulates rural development, and environmental considerations, decreases dependency on foreign fossil fuels, and liberalisation of energy market. However, there are also some disadvantages like plants must be harvested which can deplete soil nutrients, which could add to the food shortage, land used to grow biomass is needed for other purposes, it is labor intensive, cost is higher, needs storage rooms, burning plants releases carbon dioxide, particulates and other pollutants, more than 75 organic compounds are released from wood burning, some of which are suspected to be carcinogens.The potentials and obstacles in the biomass energy are outlined below.

Economical and Energy Security

Potentials: alternative to petroleum-based fuels, diversification of fuel supplies, reduction of oil dependence.

Obstacles: low net energy, low energy efficiency, competition with fossil fuels, global population growth, urbanization trends, rising energy consumption, growing human demand for food, food vs. Fuel, climate change uncertainties and shrinking global resources.

Agricultural Production and Rural Development

Potentials: transformation of agriculture, increase of farm incomes, strengthening rural economies, a solution to eliminate/reduce agricultural subsidies, a way to utilize the surplus production of food.

Concerns: crop monocultures, GM crop, extensive use of fertilizers/pesticides, mechanization, agricultural intensification, competition for land, ownership concentration, evictions, exploitation of farm workers.

Environmental Issues

Potentials: reduction of GHG emissions, lowering urban pollution levels, waste reduction.

Risks: extensive use of water, water contamination, soil depletion, soil contamination, deforestation, land-use changes, biodiversity loss due to monoculture, invasive and non-native species, emissions of GHGs and other pollutants, GM crops.

Reasons Behind Current Biofuel Expansion

Peak oil': shortage of fossil fuels, climate change, energy security, high oil prices, crisis in rural areas of OECD countries.

SALINE LANDS AND HALOPHYTES AGAINST THE OBSTACLES

In the arid regions freshwater sources are unable to meet the demand of increasing population and saline inland basins are increasing due to the rising level of groundwaters. There are huge salt-affected areas distributed in the world (1 billion ha) and another billion ha overlies the saline aquifers. The saline habitats are inhabited by more than 3500 species of halophytes (Lieth, 2003). They occupy a wide range of habitats from wet coastal marshes to dry, inland saline deserts. A number of them could survive and produce enough biomass by seawater irrigation.

The patents have been issued for halophyte crops. Nearly 10,000 "natural" halophyte taxa are distributed globally; out of these 250 are potential "staple" crops. Presently research is ongoing on halophilic halophytes related to the topic more salt faster the growth. More than 100 halophyte taxa are now under "trial" for "commercial" utilization. The percentage of total number of halophyte species used as fuel are eight (Chenopodiaceae), seven (Tamaricaceae), four (Polygonaceae), four (Solanaceae) and one (other species). Out of these the energy value of the wood of *Tamarix aphylla* on burning is 17.4 KJ per g dry matter which is only 8 per cent lower than the average value for conventional firewood from temperate forests. Under conditions of 150-200 mm rainfall a minimum net profit of 230 $ per hectare per year has been reported from *Tamarix* plantations (*BOSTID, 1990*). The halophytes have a potential to provide both food and fuel feedstock. Since nearly 45 per cent on our planet is arid or semi-arid and 97 per cent of the water is saline, there is a potential for developing a multiplicity of seawater-irrigated regions to produce halophyte crops and combine them with aquaculture.

The arid or semi-arid halophyte agriculture includes pumping and draining the required high volumes of irrigation water from sea or ocean sources. The biggest advantage of using seawater for agriculture is that 97 per cent of all water is seawater and has meager chances to "Run Out". The seawater contains a wide variety of important minerals and approximately 80 per cent of nutrients required for agriculture in the proximity of a number of dry/desert areas is achievable through irrigation water. This type of agriculture could free up freshwater supply and arable lands for food production.

Basically based on the definition of biofuels, halophytes have been used as biofuels for centuries, and today most of the fuel needs of people who live in the deserts come from halophyte burning. The only difference between the old and modern biofuel is that in the old system, plants were burnt directly whereas in the modern system these are changed into ethanol. In the old system halophytes were consumed by local societies in a eco-friendly manner but in the modern system we should wait and see how humans will manage the sustainability of the ecosystem. Some scientists are confident that within coming decade commercial aircraft could be powered using a type of biofuel derived from saltwater plants, or halophytes, grown in desert areas and irrigated with sea water (Boeing, 2007). While the concept may sound far-fetched, some manufacturers are following developments in this area "with interest", and an oil company is calculating on the benefits in this connection. Approximately 22 countries are carrying out small experimental activities in the cultivation of halophytes for use in food production.

TRAITS OF AN IDEAL BIOMASS/BIOENERGY CROP

The best plants for biomass energy production are the ones with C4 photosynthesis, with wide adaptation, must possess high yield potential, should be resistant to adverse abiotic and biotic factors, can be used as rotation in cropping system, should be noncompetitive with food and feed, must use existing agriculture infrastructure, be non invasive, winter standing, with desirable composition, characteristics for recycling of nutrients and can be a perennial/annual (Dorado, 2006; Kim and Dale, 2005; Greene, 2006; Doornbosch and Steenblik, 2007).

However, halophyte agricultural systems are vulnerable as such; all environmental issues such as biodiversity loss, soil and water depletion and salinization should be fully investigated. These systems can be divided into coastal and inland systems. In the inland systems water should be pumped up from underground or diverted from saline rivers which need a high amount of energy, and more importantly, if the basin is a closed one, excess water cannot be drained out to the sea, salt will accumulate in the soil and after a period soil might be so much saline and useless for even halophyte growing. Halophyte communities in inland ecosystems are in use traditionally in many areas, therefore, changing the system should be done with precaution, and otherwise it might cause more pressure on the rest of vegetation and accelerate soil erosion. The halophyte bioenergy extraction should be designed according to the requirements of local societies such as biogas production consumed in the local communities to help in the strengthening of desert ecosystems. Benefits of any industrial biofuel extraction from inland and coastal halophyte resources need to spend for environmental improvement.

CONCLUSIONS

The opportunity to support the agricultural sector is an extremely important and powerful motivation for pursuing biofuels worldwide. However, any benefits to the agricultural sector must be weighed against impacts on food prices and land use, both of which are also major international concerns. In addition, existing companies with expertise in specific steps are exploring partnerships. United States, Brazil and China are the world's largest ethanol producers, producing 4.8, 4.5, and 1 billion gallons respectively in 2006 (Davis and Eter, 2007; Faguendes *et al.*, 2007; Wang *et al.*, 2007). In Europe, more biodiesel is produced than ethanol, with rapeseed used as the primary feedstock. A very wide variety of feedstocks are used currently to produce biofuels. They differ in the yields, types of lands on which they can be grown, and fuels into which they are processed (Ramachandra *et al.*, 2000; Ozturk *et al.*, 2007; Johansson and Azar, 2007).

Halophyte agricultural ecosystems are vulnerable; therefore all environmental issues such as biodiversity loss, soil and water depletion and salinization should be investigated before giving a start in this connection. These can be divided into coastal and inland systems. In inland systems water should be pumped up from underground or diverted from saline rivers. However, diversion will require a high economic input, and more importantly, if the basin is a closed one, excess water can prove dangerous in the long run because soils may become much saline and useless for growing even halophytes. Halophyte communities in inland ecosystems are in use traditionally in many areas, therefore change in the system should be done with precaution, and otherwise it might cause more pressure on the rest of vegetation and accelerate soil erosion. The halophyte bioenergy extraction should be firstly designed based on the requirement of the local societies such as biogas production consumed in the local communities to help the strengthening of desert ecosystems. Benefits of any industrial biofuel extraction from inland and coastal halophyte resources should also spend for environmental improvement.

Finally there is a need for new generation of biofuels which are economically acceptable and possess significantly lower green house gases (GHG) profiles. These should come from feedstocks that put less pressure on agricultural lands, water, and other habitats. Policy makers should try to balance the competition among transportation fuel users and feedstock producers as well as biofuel manufacturers, agricultural producers and consumers of agricultural products, and land use for biofuels against land use for other purposes (Sheehan, 2002; Koplow, 2006; Pena and Sheehan, 2007; Rotman, 2008). They should also try to assess interactions between biofuel policies, the wide array of existing policies that affect land use-particularly crop supports and trade policies-and land-use components of climate agreements. All these developments should be in agreement with the achievement of climate change goals, as such; support for biofuels should be linked to their GHG profiles. Majoroty of biofuel production are dependent to fesh water consumption, while in many part of the world there are other priotities for the limited water resources. Therefore, halophyte agriculture could free up freshwater supply and arable lands for food production and also they will produce a cooler, wetter land surface, which could lead to rainfall in areas of the world where rainwater is in short supply. Major arid or semi-arid halophyte agriculture problems include pumping and draining the required high volumes of irrigation water from sea or ocean sources. Inland halophyte cultivation in the deserts are vulnerable, and special care should be paid to prevent soil degradation, and also consider the needs of rural societies that they are dependent to the natural flora of the desrts.

REFERENCES

Armagan, B., Batgı, S. 2007. The potential biogas production from livestock manures: A Case Study for Gap Region-Sanliurfa. International Conference on Environment: Survival and Sustainability, 19-24 February *2007*, Near East University, Nicosia-Northern Cyprus.

Boeing. 2007. Virgin pushing biofuel-run planes. United Press International. http://www.upi.com/Energy/Briefing/2007/04/24/boeing_virgin_pushing_biofuelrun_planes/4016/. Virgin 747 flight testing.

BOSTID (Board on Science and Technology for International Development).*1990.* Saline Agriculture: Salt-Tolerant Plants for Developing Countries. National Academy Press, Washington, DC, 130 pp.

Bushnell, D.M. 2007. Seawater/Saline Agriculture for Energy, Water, Land, Food and Warming. Large Scale Disasters: Prediction, Control and Mitigation, Mohamed Gad-eLHak (Ed.), Cambridge University Press.

Caesar, W., Riese, J., Seitz, T. 2007. Betting on Biofuels. The McKinsey Quarterly, May 2007.

Çakir, M. 2006. Biofuel production from sugar beet in EU and politics. Pankobirlik, Ankara, 17 (87).

Davis, B., Eter, L.2007. Brazil's Sugar-Cane Ethanol Gets a Boost from IMF Report. *The Wall Street Journal*.October 18.

Doornbosch, R., Steenblik, R. 2007. Biofuels: is the cure worse than the disease? OECD. Paris. September 11-12.

Dorado, M.P. 2006. Approach to the economics of two vegetable oil based biofuels in Spain. *Renewable Energy*, 31, 1231-37.

Edwards, R., Larive, J-F., Mahieu, V., Rouveirolles, P. 2007. Well-to-wheels analysis of future automotive fuels and powertrains in the European context. Version 2c, http://ies.jrc.ec.europa.eu/WTW.

Enger, E.D., Smith, B.F. 2000. *Environmental Science.* McGraw Hill, pp.434.

Faguendes de Almeida, E., Bomtempo, J.V., Souze e Silva. 2007. The Performance of Brazilian biofuels: An economic, environmental and social analysis. Energy Economics Group, Institute of Economics, Federal University of Rio de Janeiro.

Glenn E.P., Brown J.J., O'Leary J.W. 1998. Irrigating crops with seawater. *Sci Am.* 279:76-81

Greene, N. 2006. Biofuels and the environment: Promises and challenges. September 14.

GTZ (Deutsche Gesellschaft fur Technische Zusammenarbeit). 2006. Conference Handout. Biofuels for transportation: Global potential and implications for sustainable agriculture and energy in the 21st century. Bonn, May 16-17. www.gtz.de/biofuels-conference.

Gustavsson, L., Holmberg, J., Dornburg, V., Sathre, R., Eggers, T., Mahapatra, K., Marland, G. 2007. Using biomass for climate change mitigation and oil use reduction. Energy Policy. 35.

Hendricks, R.C., Bushnell, D.M. 2008. Halophytes energy feedstocks: back to our roots. The 12th International Symposium on Transport Phenomena and Dynamics of Rotating Machinery, Honolulu, Hawaii, February 17–22, 2008.

Hunt, S., Stair, P. 2006. *Biofuels Hit a Gusher,* Vital Signs 2006-2007; World watch Institute: Washington, DC, 40-41.

Ibrahim, H. 2002. Small renewable energy power programme for the promotion of renewable energy power generation. 1st Meeting of ASEM Gren IPPs Network, Bangkok.

JCTR (Jefferies Clean Tech Review). 2007. Biofuels in 2007: A U.S. Perspective, 2, 1.

Johansson, D.J.A., Azar, C. 2007. A scenario based analysis of land competition between food and bioenergy production in the US. *Climatic Change.* 82:267-291.

Khan, M. A. and D. J. Weber. 2006. Ecophysiology of High Salinity Tolerant in Plants. Springer, Netherlands.

Kim S., Dale, B.E. 2005. Life cycle assessment of various cropping systems utilized for producing biofuels: Bioethanol and biodiesel. *Biomass and Bioenergy* 29:426–439.

Koplow, D. 2006. Biofuels—At What Cost? Global subsidies initiative, International Institute for Sustainable Development. Geneva.

Licht, F.O. 2006. *Ethanol: World Production by Country.* World Ethanol and Biofuels Report 4(17), 395.

Lieth, H., Mochtchenko, M. 2003. Cash Crop Halophytes, Recent Studies: Tasks for Vegetation Science—38, Kluwer Academic Publishers, The Netherlands. pp:240.

Owen, S.O., Chiras, D.D. 1995. Natural resource conservation-management for a sustainable future. Prentice Hall, New Jersey, USA, pp.586.

Ozturk, M., Ergin, M., Kucuk, M. 2003. Sustainable use of biomass energy in Turkey.13th Science Conference of the IAS, Kuching, Sarawak, Malaysia.

Ozturk, M., Gucel, S., Armagan, B. 2007. Agricultural residues in Turkey and their rRole in the bioenergy production. 2nd FAO Regional Expert Consultation on the *"Utilization of Agricultural Residues with Special Emphasis on Utilization of Agricultural Residues as Biofuel"* Cairo-Egypt.

Pena, N. and J. Sheehan. 2007. Biofuels for transportation. CDM Investment Newsletter. Number 3/2007.

Pena, N. 2008. Biofuels for transportation: A Climate Perspective Prepared for the Pew Center on Global Climate Change, pp:32.

Perlack, R. D., Wright, L. L., Turhollow, A., Graham, R. L., Stokes, B., Urbach, D. C. 2005. Biomass as feedstock for bioenergy and bioproducts industry: the technical feasibility of a billion-ton annual supply, US Dept. of Energy and the US Dept. of Agric. by Oak Ridge Nat. Lab.

Pimentel, D. 2002. Limits of biomass utilization. Encyclopedia of physical science and technology. New York: Academic Pres, 159–171.

RFA (Renewable Fuels Association) 2007. *Ethanol Industry Outlook: Building New Horizons,* Washington, DC.

Ramachandra, T.V., Joshi, N.V., Subramanian, D.K. 2000. Present and prospective role of bioenergy in regional energy system. *Renewable Sustain Energy Rev,* 24 (4), 375–430.

RFA (Renewable Fuel Association). 2007. Ethanol Industry Outlook 2008: Changing the Climate. http://www.ethanolrfa/objects/pdf/outlook/RFA_Outlook_2008.pdf./

Rotman, D. 2008. The Price of biofuels. MIT Technology Review. https://www. technologyreview. com/Energy/19842.

Schnepf, R. 2006. *European Union Biofuels Policy and Agriculture: An Overview,* CRS Report for Congress, (http://www.usembassy.it/pdf/other/RS22404.pdf).

Sheehan J, Aden A, Riley C, Paustian K, Brenner J, Lightle D, Nelson R, Walsh M, Cushman J. 2002. Is ethanol from corn stover sustainable? Adventures in cyber-farming: a life cycle assessment of the production of ethanol from corn stover for use in a flex fuel vehicle. Draft report Colorado, National Renewable Energy Laboratory.

Sheehan, J. 2007. Biofuels and climate change, a sustainable opportunity? Presented at the Colorado Agricultural Outlook Forum, Denver, CO, February 21.

Thornley, P. 2006. Increasing biomass based power Generation in The UK., *Energy Policy,* 34, 5, 20087 -89.

UNDP (United Nations Development Program), United Nations Department of Economic and Social Affairs, and World Energy Council. 2000. World Energy Assessment: energy and the challenge of sustainability.UNDP. New York, New York.

Wang, M., Wu, M., Huo, H. 2007. Life-cycle energy and greenhouse gas emission impacts of different corn ethanol plant types. *Environ. Res. Lett.* 2,024001 (13pp).

Weber J. A. 2009. Feedstock Supplies for U. S. Biodiesel Production. MARC-IV Consulting, Inc.

Williams, C., Biswas, T., G. Schrale, S. Heading J., Virtue. 2008. Use of saline land and wastewater for growing a potential biofuel crop (*Arundo donax* L.). Irrigation Australia 2008 Conference May 20, Melbourne.

Wright, R.T., Nebel, B.J. 2002. Environmental Science. New Jersey, USA, pp.681.

Wu, M., Wang, M., Huo, H. 2006. Fuel-cycle assessment of selected bioethanol production pathways in the United States *ANL/ESD/06-7* Center for Transportation Research, Argonne.

WWI (Worldwatch Institute). 2006. Biofuels for transportation: Global Potential and Implications for Sustainable Agriculture and Energy in the 21st Century. Washington DC.

Chapter 20

Restoring Saline Habitats: Identification and Name Changes in Halophytes of the Arabian Peninsula

S.A. Ghazanfar

Royal Botanic Gardens Kew, Richmond, Surrey TW9 3AB, UK

ABSTRACT

Much of the coastal areas of the Arabian Peninsula, especially in the north-eastern region are damaged, in some cases beyond recovery. A relatively long history of subsistence farming and date plantations along coasts, and the use of water from aquifers for irrigation, has resulted in increasing salinity of coastal soils. Along with that, the ever growing building of beach hotels and resorts, private homes, and development of beaches for amenity, all have damaged the halophytic plant-life enormously. In several countries of the Arabian Peninsula there is now a growing concern for the restoration of damaged landscapes. In order to restore the coastal areas, correct identification of halophytes and knowledge of their biology is necessary. Recent molecular studies of the families of the halophytic flora have resulted in several name changes and affiliations of taxa that are potentially important for any coastal restoration work. This paper gives the nomenclatural changes of halophytes of the Arabian Peninsula and recommends the use of vegetation maps and vegetation profiles as models for restoration programmes.

Keywords: *Arabian Peninsula, Biogeography, Distribution, Halophytes, Nomenclature, Restoration.*

INTRODUCTION

A relatively long history of subsistence farming and date plantations along much of the coasts of the Arabian Peninsula, and the use of water from aquifers for irrigation and domestic use, have resulted in increasing salinity of coastal soils. Coupled with that, the recent expansion of 'good value for money' tourist destinations into the eastern parts of the Arabian Peninsula has led to the fast and unsustainable development of beach resorts and coastal amenities there. This is especially true for the United Arab Emirates and the Sultanate of Oman, where beach hotels and resorts dot more or less the entire accessible coastline.

Coastal developments have had a detrimental impact on the terrestrial and marine biodiversity over a period of time which is now becoming apparent. For example, an extensive dieback of mangroves in an area of the Batinah region of Oman is a direct cause of the construction of a coastal road. Similarly, several date plantations and parks on the Batinah are showing dieback due to the increased salinity of the soil which has been caused by excessive removal of potable water from aquifers and the incursion of seawater.

In the last two decades there is a growing concern in several countries of the Arabian Peninsula for the loss and degradation of plant and animal life. There are now attempts to conserve and preserve and restore damaged landscapes, and to bring back the plant and animal diversity and richness that was once there not too long ago. Dams have been made in several locations in the Hajar mountains to recharge the aquifers, livestock farms are being fenced so as to reduce grazing pressures on landscapes, and projects to restore degraded land are in place (Pers. Obs. 2009).

Restoration projects are especially useful as they aim to run sustainably. Albeit they are long term, but within a few years, restored landscapes begin to attract the local fauna and habitats are created for many plants, especially the annual species to return. Whereas all these efforts are a welcome sign, it is important for anyone involved in any restoration project to have a full knowledge of plants that have to be used for restoration purposes.

In this paper I will review the halophytic vegetation of coastal areas native to the Arabian Peninsula, their nomenclatural changes, distribution and recommend suitable species that may be used for the restoration of coastal areas.

THE HALOPHYTIC FLORA OF THE ARABIAN PENINSULA

This paper is based on a survey of mostly published literature (references given as appropriate) pertaining to halophytic vegetation of the Arabian Peninsula. Nomenclature follows the accepted names as cited.

About 140 taxa are recorded halophytes in the Arabian Peninsula. This constitutes about 4 per cent of the total flora. Not surprisingly the majority of halophytes belong to the families Chenopodiaceae, Poaceae, Zygophyllaceae, Fabaceae and Plumbaginaceae. Table 20.1 shows the distribution of halophytic taxa and their families in the Arabian Peninsula. These data are in accordance with that found for halophytes of the world (1, and references therein) (Table 20.2).

Key species in the saline habitats of Arabia are nearly always perennial. The predominant life-forms are succulent, semi-woody dwarf shrubs belonging to the

Table 20.1: Number of Taxa in Different Plant Families in the Arabian Peninsula. See Appendix A for the list of taxa.

Family	*Number of Halophytic Species*	*Family*	*Number of Halophytic Species*
AcanthaceaeAizoaceae	14	JuncaceaeLiliaceae	21
Apocynaceae	2	Mimosaceae	2
Arecaceae	1	Najadaceae	1
Asteraceae	2	Orobanchaceae	1
Boraginaceae	2	Plumbaginaceae	6
Chenopodiaceae s.str.	41	Poaceae	17
Caryophyllacae	8	Portulacaceae	1
Ceratophyllaceae	1	Potamogetonaceae	1
Convolvulaceae	2	Rhizophoraceae	2
Cymodoceaceae	3	Ruppiaceae	1
Cynomoriaceae	1	Salvadoraceae	1
Cyperaceae	4	Tamaricaceae	3
Fabaceae	5	Typhaceae	1
Frankeniaceae	1	Zygophyllaceae	11
Hydrocharitaceae	2		

families Chenopodiaceae s. str., Zygophyllaceae and Plumbaginaceae, and hemicryptophytes with runners and spiny leaves belonging to the families Poaceae and Juncaceae. Annual succulents such as *Bienertia cycloptera* and *Tetraena simplex* are exceptions. Coastal species are either obligate halophytes like the representatives of the families Chenopodiaceae s. str., Frankeniaceae and Plumbaginaceae, or salt tolerant genera from unspecialized families, such as *Sporobolus* and *Aeluropus* (Poaceae), or salt secreting species such as *Avicennia* (Acantahceae) and *Limonium* (Plumbaginaceae). The most common coastal and salt tolerant species are *Arthrocnemum macrostachyum, Halocnemum strobilaceum, Halopeplis perfoliata, Salsola* spp., *Suaeda* spp. (Chenopodiaceae s. str.), *Aeluropus lagopoides, Odyssea mucronata, Sporobolus spicatus, S. consimilis* (Poaceae), *Juncus rigidus* (Juncaceae), *Tetraena* spp. (Zygophyllaceae), *Limonium* spp. (Plumbaginaceae) and *Avicennia marina* (Acanthaceae).

Table 20.2: Families of Flowering Plants in which Halophytic Genera Occur Most Frequently (Adapted from 1)

Family	*Percentage of Halophytic Genera*
Chenopodiaceae	44
Aizoaceae	15
Cyperaceae	14
Caryophyllaceae	11
Poaceae	7
Arecaceae	6
Asteraceae	3
Fabaceae	3

The saline and sabkha ecosytems where halophytes grow are found throughout the Arabian Peninsula. Large parts of coastal areas and several inland sabkhas are present. Several halophytic communities are recognised based on their distribution on a saline gradient. Details of the vegetation of coastal and sabkha vegetation can found in Deil (2 and references therein), and are briefly mentioned here.

INTERTIDAL VEGETATION

Seagrasses

Sea grasses are the only group of angiosperms that are able to live permanently submerged in the sea. Eleven species of sea grasses are recorded from the Arabian Peninsula grasses (of which four or five are more common) (3); their distribution controlled by a complex of soil, temperature, salinity and light characters. As far as I know there have been no attempts in the restoration of shallow coastal bays which have been polluted or degraded in some way or another and where the sea grass flora has been damaged.

Mangroves

Avicennia marina has been used to restore damaged or degraded mangrove stands in the Arabian Peninsula with some success. Of the three recorded species (Appendix 1), *Avicennia marina* is the most common and abundant being tolerant of low temperatures (12–35°C) and high salinities (40–50 per cent) (3, 4). The distribution of mangroves indicates that cold winter temperatures rather than salinity limit their northernmost extent, and that the mangroves may formerly have been more abundant in the Gulf than they are at present (3). The seeds of *Avicennia marina* are recalcitrant, and therefore cannot be stored for a long period, with seed viability declining after 10 days (5). The highest germination is recorded in 21 per ml salinity in terms of per cent germination, leaf size (cm^2), internode length (cm) and plant height (cm). Attempts to introduce this species in the intertidal zones, with growing the propogules in greenhouses has met with success in Kuwait (6).

TERRESTRIAL VEGETATION

Coastal vegetation is one of the best studied vegetation types in Arabia, with the salt marsh vegetation along the western and the eastern coasts of the Peninsula having received particular attention (2 and references therein). Figures 20.1–20.5, give examples of the vegetation profiles and are representative of the coastal vegetation in the Arabian Peninsula and would be useful in any restoration work on coastal ecosystems.

NOMENCLATURE

Recent molecular work on angiosperms has resulted in a new classification in the APG system (7, 8) which is now accepted by majority of systematic botanists. Amongst the halophytic families found in the Arabian Peninsula, classification is as follows:

Angiosperms

Magnoliids: Rhizophoraceae

Monocots: Cymodoceaceae, Hydrocharitaceae, Liliaceae, Najadaceae, Ruppiaceae.

Commelinids: Arecaceae, Cyperaceae, Juncaceae, Poaceae, Typhaceae.

Eudicots: Core eudicots: Aizoaceae, Amaranthaceae (includes subfamilies Amaranthoideae, Chenopodioideae, Gomphrenoideae, Salicornioideae, Salsoloideae), Caryphyllaceae, Cynomoriaceae, Frankeniaceae,, Plumbaginaceae, Polygonaceae, Portulacacaceae, Tamaricaceae.

Rosids: Eurosids I: Zygophyllaceae; Fabaceae (includes subfamilies Caesalpinioideae, Mimosoideae and Papilionoideae).

Eurosids II: Salvadoraceae.

Asterids: Euasterids I: Apocynaceae (includes Asclepiadaceae); Acanthaceae (includes *Avicennia*), Boraginaceae; Convolvulaceae, Orobanchaceae.

Euasterids II: Asteraceae.

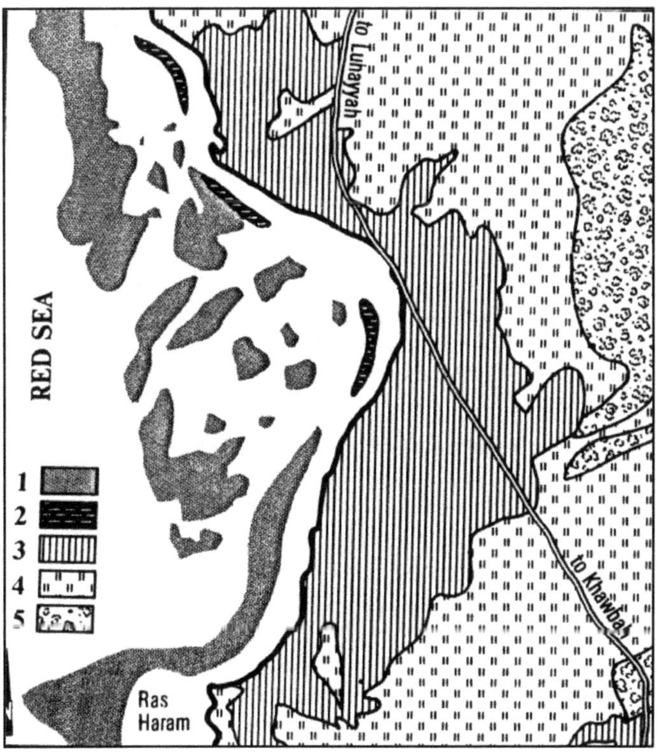

Figure 20.1: Coastal Vegetation Between Luhayyah and Al Hudayda in Yemen, Typical of the Tihama Coast

1. Coral reef; *Avicennia marina*; 3. Mosaic of *Suaeda fruticosa, Aleuropus-sporpbolus* grass community on small dunes and Zygophyllum-Suaeda community in sandy depressions; 4. *Panicum turgidum-Odyssea mucronata* community on mobile dunes; 5. *Caroxylon spinescens-Schweinfurthia pterosperma* community on consolidated dunes. Adapted from (2).

In the APG II system, of 2003 (unchanged from the APG system of 1998), Chenopodiaceae is included in the family Amaranthaceae (Order Caryophyllales). The monophyly of this new, broadly defined Amaranthaceae has been strongly supported by both morphological and phylogenetic analyses. The main differences between Amaranthaceae (s. str.) and Chenopodiaceae (s. str.) are the membraneous petals and stamens often united in a ring structure in Amaranthaceae.

Cronquist (9) includes *Avicennia* in Verbenaceae, but more recently it has been placed in the mono-generic family Avicenniaceae by several taxonomists. Recent phylogenetic studies have suggested that *Avicennia* is derived from within Acanthaceae, and is included in that family in the APG II System.

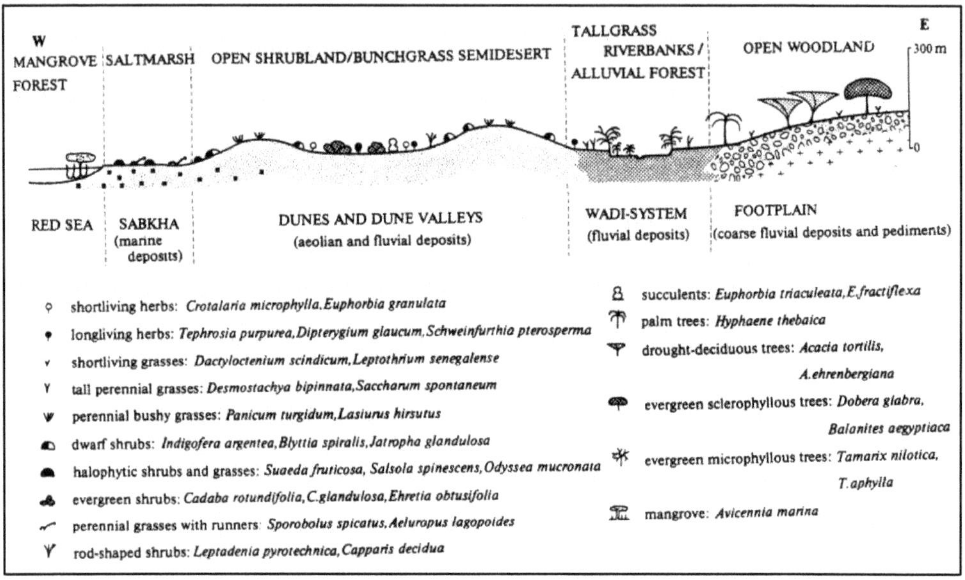

Figure 20.2: Natural Vegetation of the Tihama in Yemen. Adapted from (13).

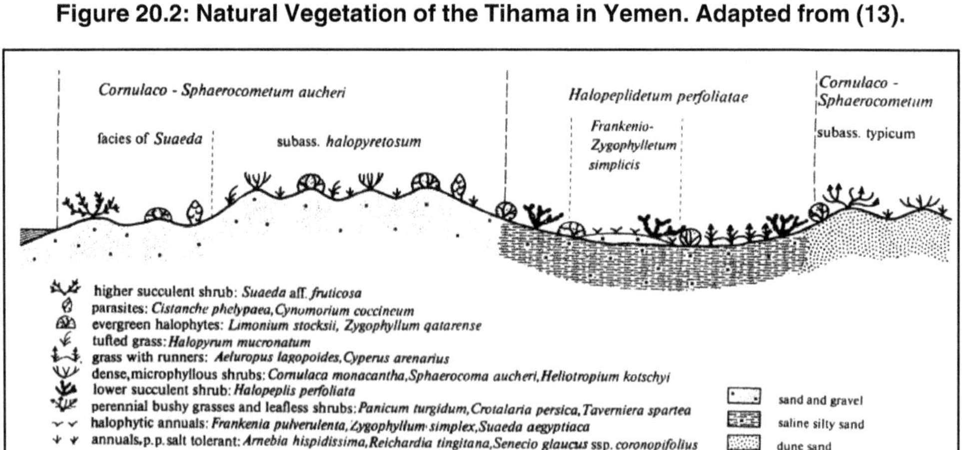

Figure 20.3: Coastal Dune and Sabkha Vegetation Near Dubai, UAE. Adapted from (14).

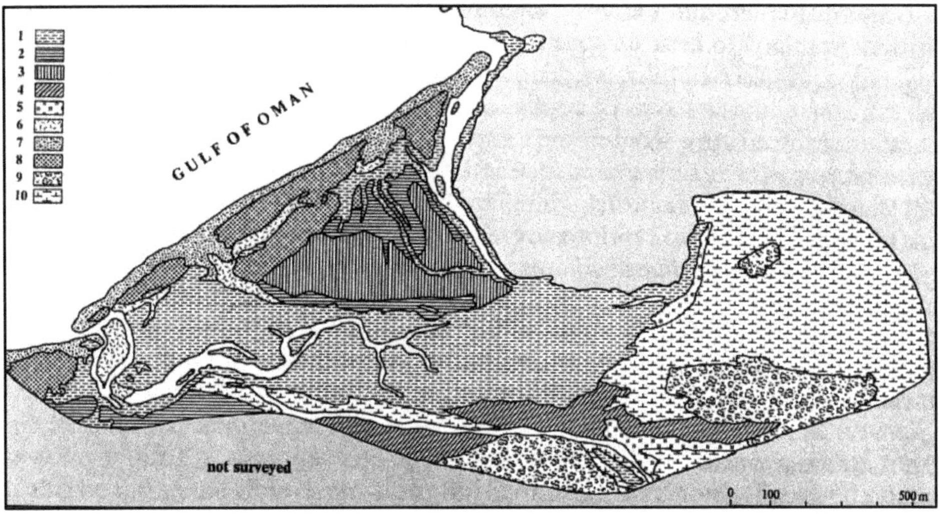

Figure 20.4: Vegetation of the Qurm Nature Reserve Near Muscat, Oman in 1986. Adapted from (15).

This habitat and the vegetation there no longer exists as it was 20 years ago due to developmental projects in this area. This would serve a good reference for any restoration in this area. 1. Mangroves; 2. Halopeplis community; 3. Suaeda community; 4. Mixed halophytic and forbs; 5. *Aleuropus lagopoides* community; 6&7. Saline beach zone flooded at high tide; 8. *Halopyrum mucronatum* community; 9. Open Acacia tortilis and Prosopis cineraria woodland with *Euphorbia larica*; 10. Cutivated land.

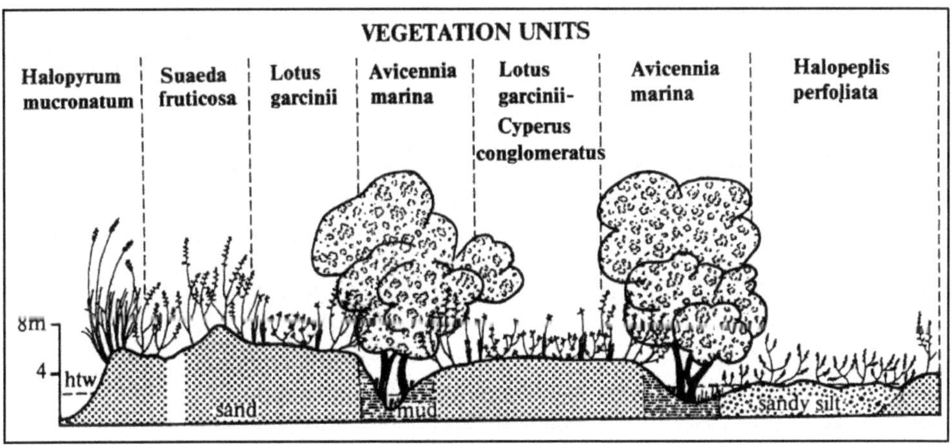

Figure 20.5: Zonation of Vegetation Types in Salt Marsh Near Muscat, Oman. Adapted from (16).

Recent molecular work on the Old world Salsoleae s.l. (Chenopodiaceae s.str.) (10) and the revision of Salicorniaceae (11,12) has changed the classification of several genera in the tribe Salsoleae which includes several species in the Arabian Peninsula.

RESTORING COASTAL ENVIRONMENTS

Before any restoration has to be taken it is important to know the plant communities and their composition which had once inhabited there. A full vegetation survey of the area together with the geomorphology, soil and climatic data is mandatory for any restoration work. A vegetation survey would list the dominant and associated species and the dominant plant communities in an area. In many instances if the land is very degraded only the "hardy" or "non palatable" species remain. If imagery prior to degradation is available (*e.g* . before construction of a building or road, etc.) or a similar area is available that is not degraded or less so, a survey should be carried out there so that a reference for restoration (in terms of species richness and alpha diversity) can be established. A model for restoration should then be prepared for the site. Detailed vegetation maps and vegetation profiles are very useful references in restoration programmes as the key species and their distribution is well recorded and vegetation described in some detail (Figures 20.1–20.5).

Species and plant communities should be planted to keep the species composition as natural as possible. Whole plants can be salvaged from development sites and used for restoration, or species propagated from seed or cuttings. It must be kept in mind that for any plant restoration project irrigation may be necessary at first and plants eventually weaned off artificial watering. It is best therefore, as a start, to plant in natural depressions, runnels and wadis where there may be some subsurface water or where water accumulates from dew or rain.

CONCLUSION

For any restoration project it is important to have the correct identity of the species. Vegetation maps and profiles are very useful in constructing a model for restoration as are images taken prior to the modification or degradation of site.

REFERENCES

1. Flowers, T.J., Hajibagheri, M.A. and Clipson, N.J.W. (1986). Halophytes. The Quaterly Review of Biology 61(3): 313–337.
2. Deil, U. (1988). Coastal and sabkha Vegetation. In: Ghazanfar, S.A. and Fisher, M. (eds), *Vegetation of the Arabian Peninsula*. Kluwer Academic, Netherlands 209–228.
3. Sheppard, C.R.C., Price, A.R.G. and Roberts, C. (1992) *Marine Ecology of the Arabian Region: Patterns and Processes in Extreme Tropical Environments*. Academic Press, London.
4. Böer, B. (1996c) Plants as soil indicators along the Saudi coast of the Arabian Gulf. *Journal of Arid Environments* 33: 417–423.
5. Pammenter N. W., Farrant, J. M. and Berjak, P. (1984). Recalcitrant Seeds: Short-term Storage Effects in *Avicennia marina* (Forsk.) Vierh. Annals of Botany 54: 843–846.

6. Abo El.Nil, Mostafa (2001). Growth and establishment of mangrove, *Avicennia marina* on the coastline of Kuwait., Wetland ecology and managemnet 9(5) 421–428.
7. Angiosperm Phylogeny Group (APG) (1998)
8. Angiosperm Phylogeny Group (APG) (2003). An update of the Angiosperm Phylogeny Group classification for the orders and families of flowering plants: APG II. *Botanical Journal of the Linnean Society* 141: 399–436.
9. Cronquist, A. (1981). An integrated system for the classification of Angiosperms.
10. Akani, H., Edwards, G. and Roalson, E.H. (2007). Diversification of the Old World Salsoleae s.l. (Chenopodiaceae): molecular Phylogenetic analysis of nuclear and chloroplast data sets and a revised classification. Int. J. Plant Sc. 168(6): 931–956.
11. Scott, A.J. (1977). Reinstatement and revbision of Salicorniaceae J. Aghardh (Caryophyllales). Bot. J. Linn./Soc. 75: 357–374.
12. Kadereit, G., Ball, P., Beer, S. Mucina, L., Sokoloff, D., Teege, P., Yaprak, A.E., Freitag, H. (2007). A taxonomic nightmare come true: phylogeny and biogeography of glassworts (*Salicornia* L., Chenopodiaceae). *Taxon* 56(4): 1143–1170.
13. Al-Hubaishi, A. and Müller-Hohenstein, K. (1984) *An Introduction to the Vegetation of Yemen. Ecological Basis, Floristic Composition, Human Influence.* Deutsche Gesellschaft für Technische Zusammenarbeit, Eschborn.
14. Deil, U. and Müller-Hohenstein, K. (1996) An outline of the vegetation of Dubai (UAE). *Verhandlungen der Gesellschaft für Ökologie* 25: 77–95.
15. Kürschner, H. (1986c) A study of the vegetation of the Qurm Nature Reserve, Muscat area, Oman. *Arab Gulf Journal of Scientific Research* 4: 23–52.
16. Frey, W. and Kürschner, H. (1986) Masqat Area (Oman), remnants of vegetation in an urban habitat. Pp. 201-221 in *Contributions to the Vegetation of Southwest Asia.* H. Kürschner, ed. Beihefte Tübinger Atlas Vorderen Orients, Reihe A (Naturwissenschaften) Nr. 24. Dr Ludwig Reichert Verlag, Wiesbaden.

Appendix 1: Accepted Names of Halophytes of the Arabian Peninsula

Aizoaceae

Mesembryanthemum nodiflorum L., Sp. Pl. 480 (1753); Syn.: *Chlorophytum nodiflorum* L. (1753).

Sesuvium sesuvioides (Fenzl) Verdc., Kew Bull. 1957, 349 (1957); Basionym: *Diplochonium sesuvioides*Sesuvium portulacastrum Syst. Nat., ed. 10. 2: 1058 (1759).

Trianthema triquetra Willd., Ges. Naturf. Fr. Berlin Neue Schriften 4: 181 (1803).

Zaleya pentandra (L.) Jeffery, Kew Bull. 14 (2): 238 (1960); Syn.: *Trianthema pentandra* L. (1767).

Apocynaceae: Asclepiadoidae

Calotropis procera (Aiton) W.T.Aiton, Hortus Kewensis 2: (1811). Combination indicated but not made by R.Brown, Mem. Wernerian Soc. 1: 39 (1811).

Pentatropis nivalis (J.F.Gmel.) D.V.Field and J.R.I.Wood, Kew Bull. 38(2): 215 (1983). Basionym: Asclepias nivalis J.F.Gmel. (1791).

Arecacae

Phoenix X dactylifera

Asteraceae

Pluchea dioscorides DC., Prodr. 5: 450 (1836).

Pulicaria hadramautica Edinb. J. Bot. 50(1): 79 (1993).

Acanthaceae

Avicennia marina Vierh., Denkschr. Kaiserl. Akad. Wiss. Wien. Math.-Naturwiss. Kl. lxxi. 435 (1907).

Boraginaceae

Moltkiopsis ciliata (Forssk.) I.M.Johnst., J. Arn. Arbor. 34: 3 (1953); Basionym: *Lithospermum ciliatum* Forssk.(1775).

Heliotropium bacciferum Forssk., Fl. Aegypt.-Arab. 38 (1775) sensu lato. Syn.: *H. ramosissimum* (Lehm.) DC. (1845); *H. undulatum* Vahl var. *ramosissimum* Lehm. (1831); *H. kotschyi* Bunge (1869) nomen nudum; *H. tuberculosum* (Boiss.) Boiss. (1879); *H. persicum* auct.: Boiss. (1879), non Lam. (1789); *H. lignosum* Bornm. (1937) nomen nudum; *H. fartakense* O.Schwartz (1939); *H. bacciferum* Forssk. subsp. *lignosum* (Vatke) Kazmi var. *fartakense* (O.Schwartz) Kazmi (1970).

Amaranthaceae: Chenopodioideae

Anabasis ehrenbergii Schweinf. ex Boiss., Fl. Orient. 4(2): 970, in nota. 1879 [Apr-May 1879].

Arthrocnemum macrostachyum (Moric.) K.Koch, Hort. Dendrol. 96, no. 3 (1853).

Basionym: *Salicornia macrostachya* Moric (1820); *Salicornia glauca* Delile (1813) non Stocks (1812);

Arthrocnemem glaucum (Delile) Ung.-Sternb. (1876).

Atriplex farinosa Forssk., Fl. Aegypt.-Arab. 123 (1775). *A. hastata* Forssk. (1775) non Linn.

Atriplex stocksii Boiss., Diagn. Pl. Or. Nov. ser. 2(4): 73 (1859).

Atriplex griffithii Moq. var. *stocksii* (Boiss.) Boiss., Fl. Or. (1879). Syn.: *A. sokotranum* Vierh. (1903); *A griffithii* Moq. subsp. *stocksii* (Boiss.) Boulos (1991).

Atriplex leucoclada Boiss., Diagn. Pl. Or. Nov. ser. 2 (12): 95 (1853) var. *inamoena* (Allen) Zohary, Fl. Palest. 1: 147 (1966). Syn.: *A. inamoena* Allen (1939).

Bassia muricata (L.) Asch., Beitr. Fl. Aethiop. 1: 289 (1867). Syn.: *Salsola muricata* L. (1767); *Kochia muricata* (L.) Schrad. (1809).

Bassia eriophora (Schrad.) Asch., Beitr.. Fl. Aethiop. 187 (1867). Syn.: *Kochia eriophora* Scghrad. (1909).

Bienertia cycloptera Bunge, Trudy Imp. S.-Petersb. Bot. Sada vi, ii, 425 (1879) and Boiss., Fl. Or. 4: 945 (1879).

Caroxylon cyclophyllum (Baker) Akhani and E.H.Roalson, Int. J. Pl. Sci. 168(6): 947 (2007).

Basionym: *Salsola cyclophylla* Baker, Bull. Misc. Inform. (Kew Bull.) 1894: 340 (1894).

Caroxylon imbricatum (Forssk.) Moq., Prodr. (DC.) 13(2): 177. (1849). Basionym: *Salsola imbricata* Forssk., Fl. Eygpt.-Arab. 57 (1775); *Chenopodium baryosmon* Schult. ex Roem. and Schult. (1820);); *Salsola baryosma* (Roem. and Schult.) Dandy (1950); *Caroxylon imbricatum* (Forssk.) Akhani and E.H.Roalson (2007) nom.superfl. later homonym.

Caroxylon spinescens (Moq.) Akhani and E.H.Roalson, Int. J. Pl. Sci. 168(6): 948 (2007).

Basionym: *Salsola spinescens* Moq., in DC., Prodr. 13(2): 179 (1849).

Caroxylon villosum (Schult.) Akhani and E.H.Roalson, Int. J. Pl. Sci. 168(6): 948 (2007). Basionym: *Salsola villosa* Schult. (1820).

Cornulaca aucheri Moq., Chenopodium Monogr. Enum. 163 (1840). Syn.*Cornulaca leucacantha* Charif and Aellen (1950).

Cornulaca monacantha Delile, Fl. Aegypt., Ill. 206, t 22, f.3 (1814).

Halocnemum strobilaceum (Pallas) M.Bieb., Fl. Taur.-Caucas. 3: 3 (1819). Basionym: *Salicornia strobilacea* Pallas (1771)

Halogeton alopecuroides (Delile) Moq., Chenop. Monogr. Enum. 161 (1840).

Halopeplis perfoliata (Forssk.) Bunge ex Schweinf. and Aschers, Fl. Aethiop. 289, nomen; et ex Ung.-Sternb. in Atti. Congr. Bot. Firenze, 874 (1876) 329. Basionym: *Salicornia perfoliata* Forssk. (1775).

Halothanmus bottae Jaub. and Spach, Ill. Pl. Orient. 2: 50, t. 136 (1845). Syn.: *Caroxylon bottae* (Jaub. and Spach) Moq.(1849); *Salsola bottae* (Jaub. and Spach) Boiss. (1879).

Haloxylon persicum Bunge ex Boiss. and Bushe, Nouv. Mém. Soc. Imp. Naturalistes Moscou 12: 189 (1860).

Haloxylon salicornicum (Moq.) Bunge ex Boiss., Fl. Orient. 4(2): 949 (1879). Basionym: *Caroxylon salicornicum* Moq. (1849); *Hammada elegans* (Bunge) Botsch (1964); *Hammada salicornica* (Moq.) Iljin (1948).

Kaviria rubescens (Franch.) Akhani, Int. J. Pl. Sci. 168(6): 948 (2007). Basionym: *Salsola rubescens* Franch., Sert. Somal. 60 (1882); *Salola hadramautica* Baker (1894); *Salsola leucophyla* Baker (1894).

Salicornia europea L., Sp. Pl. 1: 3 (1753).

Salsola setifera (Moq.) Akhani, Int. J. Pl. Sci. 168(6): 946 (2007). Basionym: *Anabasis setifera* Moq., Chenop. Monogr. Enum. 164 (1840).

Salsola rosmarinus (Ehrenb. ex Boiss.) Akhani, Int. J. Pl. Sci. 168(6): 946 (2007). Basionym: *Seidlitzia rosmarinus* Ehrenb. ex Boiss., Fl. Or. 4: 951 (1879).

Salsola drummondii Ulbr., Nat. Pflanzenfam. 2, 16C: 256 (1934). *Salsola obpyrifolia* Botsch and Akhani (1989).

Salsola schweinfurthii Solms-Laub., Bot. Zeit. 59: 173 (1901).

Suaeda aegyptiaca (Hasselq.) Zohary, J. Linn. Soc. Bot. 55: 635 (1957). Basionym: *Chenopodium aegyptiacum* Hasselq. (1757); *Suaeda hortensis* Forssk. ex J.F.Gmel. (1791); *Suaeda baccata* Forss. Ex J.F. Gmelin (1791); *Schanginia hortensis Suaeda hortensis* (Forssk. ex Gmelin) Moq. (1840); *Schanginia baccata* (Forssk. ex J.F.Gmel.) Moq.; *Schanginia aegyptiaca* (Hasselq.) Aellen (1964).

Suaeda moschata A.J.Scott, Kew Bull. 36(3) 558 (1981).

Suaeda monoica Forssk. ex J.F.Gmel., Syst. Nat. ed. 1791: 2, 503 (1791).

Sevada schimperi Moq. in DC., Prodr. 13(2): 154 (1849).

Ceratophyllaceae

Ceratophyllum demersum L., Sp. Pl. 992 (1753).

Caryophyllaceae

Herniaria maskatensis Bornm., Mitth. Thuring. Bot. Vereins 6: 51 (1894).

Polycarpaea spicata Wight and Arn. in Ann. Nat. Hist. ser. 1 (3): 91 (1831).

Polycarpaea jazirensis R.A. Clement, Edinb. J. Bot. 51(1): 53-54 (1994).

Sphaerocoma aucheri Boiss., Fl. Or. 1: 739 (1867).

Xerotia arabica Oliver in Hk., Icon. Pl. 24, t. 2359 (1895).

Polycarpon succulentum J. Gay in Rev. Bot. Bull. Mens. 2: 372 (1846).

Spergularia diandra (Guss.) Heldr. et Sart. in Heldr., Herb. Graec. Norm. no. 492 (1855); Syn.: *Arenaria diandra* Guss. (1827).

Spergularia marina (L.) Gris., Spic. 1: 213 (1843).Syn.:*Arenaria rubra* L. var. *marina* L. (1753).

Convolvulaceae

Cressa cretica L. Sp. Pl. 223 (1753).

Ipomoea pes-caprae (L.) R.Br., Narr. Exped. R. Zaire 477 (1818). Basionym: *Convolvulus pes-caprae* L. (1753).

Cymodoceaceae

Halodule uninervis Boiss., Fl. Or. 5: 24 (1882).

Syringodium isoetifolium (Asch.) Dandy, J. Bot. 77: 116 (1939).

Thalassodendron ciliatum (Forssk.) Hartog, Verh. Kon. Ned. Akad. Wet., Afd. Nat. Sect. 2, 59(1): 88 (1970).

Cynomoriaceae

Cynomorium coccineum L., Sp. Pl. 2: 970. 1753

Cyperaceae

Cyperus arenarius Salzm. Ex Steud., Syn. Pl. Glumac. 2(7): 46 (1854) publ. (1855).

Cyperus conglomerates Vahl, Enum. Pl. 2, 334 (1805).

Cyperus laevigatus L., Mant. 179 (1771).

Schoenoplectus littoralis Palla, Sitz. Zool.-Bot.Ges.Wien. 38: 49 (1888).

Fabaceae

Alhagi graecorum Boiss., Diagn. Pl. Or. Ser. 1, 9: 114 (1848). Syn.: *A. maurorum* DC., 1825) non Medik.

Lotus garcinii DC., Prodr. 2: 212 (1825).

Taverniera lappacea (Forssk.) DC., Prodr. 2: 339 (1852). Basionym: *Hedysarum lappaceum* Forssk. (1775).

Taverniera spartea (Burm.f.) DC., Prodr. 2: 339 (1852). Basionym: *Hedysarum spartium* Burm.f. (1768).

Crotalaria saltiaina T.Anders., Bot. Rep. T. 648 (1812).

Frankeniaceae

Frankenia pulverulenta L., Sp. Pl. 332 (1753).

Hydrocharitaceae

Halophila ovalis (R.Br.) Hook.f., Bot. Antart. Voy. III, 2: 45 (1858).

Halophila stipulacea Asch., Sitz. Ges. Naturf. Freunde Berlin 3(1867).

Juncaceae

Juncus rigidus Desf., Fl. Atlant. 1: 312 (1798).

Juncus acutus L., Sp. Pl. 1: 325 (1753).

Liliaceae

Dipcadi biflorum Ghaz., Kew Bull. 51(4): 805 (1996).

Mimosaceae

Acacia tortilis (Forssk.) Hayne, Arzneigew. 10: I, t. 31 (1827). Basionym: *Mimosa tortilis* Forssk. (1775).

Prosopis cineraria (L.) Druce, Rep. Bot. Soc. Exch. Cl. Brit. Isles 1913, 3: 422 (1914). Basionym: *Mimosa cineraria* L. (1753).

Najadaceae

Najas marina L., Sp. Pl. 2: 1015 (1753).

Scrophulariaceae

Cistanche tubulosa (L.) Cout., Fl. Portugal: 571 (1913). Basionym*: Lathraea phelypaea* L. (1753). Syn.: *Orobanche tinctoria* Forssk. (1775); *Cistanche tinctoria* (Forssk.) Beck (1904); *Phelypaea tubulosa* Schrenk (1840); *Cistanche tubulosa* (Schrenk) Hook.f. (1884).

Plumbaginaceae

Limonium axillare (Forssk.) Kuntze, Rev. Gen. Pl. 2: 395 (1891). Syn.: *Statice axillaris* Forssk. (1775).

Limonium carnosum (Boiss.) O. Kuntze, Rev. Gen. Pl. 2: 395 (1891).Syn.: *Statice carnosum* Boiss. (1879).

Limonium cylindrifolium Verdc. ex Cufod., Bull. Jard. Bot. Natl. Belg. 30 (Suppl.) 661 (1960).

Limonium milleri Ghaz. and J.R.Edm., Edinb.J. Bot. 60(1): 15 (2003).

Limonium sarcophyllum Ghaz. and J.R.Edm., Edinb.J. Bot. 60(1): 13 (2003).

Limonium stocksii (Boiss.) Kuntze, Rev. Gen. Pl. 2: 396 (1891). Syn.: *Statice arabicum* Jaub. and Spach (1844);

Statice stocksii Boiss. in DC. (1848); Boiss. (1879).

Poaceae

Aeluropus lagopoides (L.) Trin. Ex Thwaites, Enum. Pl. Zeyl.: 374 (1864). Syn.: *Aeluropus littoralis* auct. non (Gouan) Parl.

Aristida abnormis Chiov., Pirotta, Fl. Eritrea 48 (1903).

Arundo donax L., Sp. Pl., 1: 81 (1753).

Echinochloa crusgalli (L.) P.Beauv., Ess.Agrostogr.: 53: 161 (1812). Basionym: *Panicum crusgalli* L. (1753).

Halopyrum mucronatum (L.) Stapf, Hook.f., Icon. Pl. 25: t. 2448 (1896). Basionym: *Uniola mucronatum* L.

Panicum antidotale Retz., Observ. Bot. (Retzius) iv. 17 (1786).

Panicum turgidum Forssk., Fl. Aegypt.-Arab. 18 (1775).

Paspalidum desertorum (A.Rich.) Stapf, Fl. Trop. Afr. 9(4): 585 (1920). Basionym: *Panicum desertorum* A. Rich. (1850).

Paspalum distichum L., Syst. Nat. Ed. 10, 2: 855 (1759).

Paspalum vaginatum Sw., Prodr.: 21 (1788).

Phragmites australis (Cav.) Trin. ex Steud., Nomencl. Bot. Ed. 2, 2: 324 (1841). Syn.: *P. communis* Trin. (1820); *Arundo donax* Forrsk. (1775) non L.

Sporobolus consimilis Fresen., Mus. Senckenberg. 2: 140 (1837).

Sporobolus helvolus (Trin.) T.Durand and Schinz., Consp. Fl. Afric. 5: 820 (1895). Basionym: *Vilfa helvola* Trin. (1837).

Sporobolus ioclades (Nees ex Trin.) Nees, Fl. Afric. Austr. Ill. 1: 161 (1841). Syn.: *S. arabicus* Boiss. (1853); *S. jemenicus* Pilg. Ex Schwartz (1939); *S. kentrophyllus* (K.Schum.) Calyton (1997).

Sporobolus spicatus (Vahl.) Kunth, Revis. Gramin. 1: 67 (1829). Syn.: *Agrostis virginica* Forssk. (1775).

Sporobolus virginicus (L.) Kunth, Revis. Gramin. 1: 67 (1829).

Urochondra setulosa (Trin.) C.E.Hubb., Hook., Icon. Pl. 35: t. 3457 (1947). Basionym: *Vilfa setulosa* Trin. (1840).

Portulaccaceae

Portulaca oleracea L., Sp. Pl. 445 (1753).

Rhizophoraceae

Rhizophora mucronata Lam., Encycl. 6(1): 189 (1804).

Bruguiera gymnorrhiza (L.) Sav., Ecycl. 4: 696 (1798).

Ruppiaceae

Ruppia maritima L., Sp. Pl. 1: 127 (1753).

Salvadoraceae

Salvadora persica L., Sp. Pl. 1: 122 (1753).

Tamaricaceae

Tamarix mascatensis Bunge, Tentamen 60 (1852).

Tamarix aphylla (L.) G. Karsten, Deutsch. Fl.: 641 (1882). Synomyms: *Thuja aphylla* L. (1755) p.p.; *Tamarix orientalis* Forssk. (1775); *T. articulata* Vahl (1791), *nom. illegit.*

Tamarix aucheriana (Decne.) Baum, Monogr. Rev. Tamarix: 148 (1978). Synomyms: *Trichaurus aucherianus* Decne. ex Walpers (1843); *T. passerinoides*, Boiss. (1867), non Del. ex Desv.

Typhaceae

Typha domingensis Pers., Syn. Pl. 2(2) : 532 (1807).

Zygophyllaceae

Fagonia indica Burm.f., Fl. Indica 102, t. 34, f.1, (1768).

Fagonia luntii Bak., Kew Bull. 1894: 330 (1894).

Fagonia mahrana Beier, Nord. J. Bot. 21(5): 451, 1 (2002).

Fagonia ovalifolia Hadidi, Fl. Iran. 98: 2, t.1, (1972).

Fagonia schweinfurthii (Hadidi) Hadidi, Oester. Bot. Z. 121: 272 (1973). Syn.: *Fagonia arabica* Edgeworth and Hook. F. (1874) non L.; *F. indica* Burm.f. var. *schweinfurthii* Hadidi (1972).

Nitraria retusa Asch., Verh. Biot. Prov. Barndenberg 18: 94 (1876).
Tetraena alba (L.f.) Beier and Thulin, Pl. Syst. Evol. 240: 35 (2003). Basionym: *Zygophyllum album* L.f. (1762).
Tetraena hamiensis (Schweinf.) Beier and Thulin, Pl. Syst. Evol. 240: 35 (2003). Basionym: *Zygophyllum hamiensis* Schweinf. (1899).
Tetraeana qatarensis (Hadidi) Beier and Thulin, Pl. Syst. Evol. 240(1-4): 36 (2003). Basionym: *Zygophyllum qatarense* Hadidi (1978); *Z. Coccineum* auct., non Linn.; *Z. Smithii* Hadidi, nom. nud.; *Z. hamiensis* var. *qatarense* (Hadidi) Jac. Thomas and Chaudhary (2001).
Tetraena simplex (L.) Beier and Thulin, Pl. Syst. Evol. 240: 36 (2003). Basionym: *Zygophyllum simplex* L. (1767).
Tribulus arabicus Hosni, Bot. Not. 130: 261 (1977). Syn.: *T. omanensis* Hosni (1978).

Chapter 21
Halophytes as Medicinal Plants

Muhammad Qasim, Salman Gulzar, M. Ajmal Khan*

**Institute of Sustainable Halophyte Utilization, University of Karachi, Karachi – 75270, Pakistan*

ABSTRACT

Halophytes are plants of significant economic potential which can contribute tremendously toward the environmental restoration besides a potential source of medicine. This paper reviews available literature about the status of 45 coastal and near-coastal species traditionally used as medicine for 7 different types of disease conditions. Local inhabitants have a long history of traditional plant use as medicine to cure almost all types of ailments. Many of these plants are perennial halophytic herbs and shrubs including a few herbaceous annuals. The aim is to highlight the importance of halophytes as medicinal plants and indicate their potential utilization at industrial scale.

INTRODUCTION

Review of literature indicates that an increase in research interest on halophytes which reflect recognition of their immense potential as a valuable resource and cash crop (Rozema and Flowers, 2008; Khan *et al.*, 2009). Use of saline and brackish water resources has been recommended for growing cash crop as food, fuel, fiber, fodder and medicine for the ever increasing human population (Rozema and Flowers, 2008). Recently, a salt tolerant plant *Panicum turgidum* Forssk. has been shown to be useful as alternative cattle feed for maize when irrigated with brackish water on saline soils (Khan *et al.*, 2009). The *Panicum–Suaeda* intercropping system helped maintain soil salinity at low levels and *Panicum turgidum* was reported to produces more than 50 tons/ha biomass annually as fodder for cattle (Khan *et al.*, 2009).

Fatty acids profile of some halophytic species holds great promise for production of commercial vegetable oil with high poly-unsaturation equivalent to conventional oil seed crops such as canola and sunflower (Khan and Qaiser, 2006). Seeds of halophytes, such as *Suaeda fruticosa* Forssk. ex J.F. Gmelin, *Arthrocnemum macrostachyum* (Moric.) C. Koch, *Salicornia bigelovii* Torr., *S. brachiata* Roxb., *Halogeton glomeratus* (M. Bieb.) C. A. Mey., *Kochia scoparia* (L.) Schrad. and *Haloxylon stocksii* (Boiss.) Benth. and Hook. posses a sufficient quantity of high quality edible oil with un-saturation ranging from 70-80 per cent (Weber *et al.*, 2001). Shoots of *Salicornia bigellovi, Sesuvium portulacastrum* (L.) L., *Chenopodium album* L., *Portulaca oleracea* L. and *Suaeda maritima* (L.) Dumortier are used for vegetables, salads and pickles in various parts of the country. Fuel wood is also traditionally obtained from salt tolerant perennials such as *Prosopis, Tamarix, Salsola, Acacia, Suaeda, Kochia, Capparis* and *Salvadora* (Dagar, 1995).

Khan and Qaiser (2006) reported more than 400 species of halophytes in Pakistan with almost 100 species or 1/4 of the total number from coastal areas. Most of the halophytes of arid areas are perennial shrubs or perennial herbs which can provide a sustainable supply of plant material from multiple harvests using non-conventional saline irrigation system on poor quality soil and brackish water irrigation. With appropriate management and monitoring, saline irrigation and cropping systems hold great promise for tapping the highly productive saline systems for meeting the basic needs of local communities and also for ensuring high economic returns in salt prone areas.

This paper is a brief review of the medicinal importance of coastal halophytic species along the Arabian Sea coast which are reported for their use in treating a number of ailments in local communities (Table 21.1). The information will help us to find patterns of plant use and their relationship with other human community attribute that binds man and nature together.

CURRENT SALINITY SCENARIO

Good quality land and freshwater need to be utilized primarily for growing food for the ever growing population. However, immense potential lies not only in reclamation of salinized lands by improving soil fertility but also in better utilization of these untapped resources for cultivating non-conventional salt tolerant cash crop plants–the Halophytes. About 6.5 million hectares of land are salt affected out of which approximately 2 million hectares are slightly to moderately affected but these can successfully be utilized to produce medicinal plants besides other uses.

The consumption of medicinal herbs in Pakistan was estimated at 631.5 tons worth about Rs. 1.43 billion annually (Anon., 1983). However, 90 per cent of these herbs are imported from countries like Sri Lanka, China and India which grow a variety of medicinal plants for local use as well as for export to neighboring countries and earn considerable foreign exchange. Pakistan lacks the basic infrastructure to efficiently exploit the full potential of naturally growing herbs even for domestic usages while there is a constant decrease in indigenous plant species biodiversity due to both urbanization as well as global climate changes (Jonathan *et al.*, 2004). Concerted research efforts are required to domesticate indigenous medicinal plants to be used as cash crops on an urgent basis.

Table 21.1: Traditional Uses of Halophytes Reported as Medicine

Sl.No.	Name of Species (Family)	Habit	Life Form	Part Used	Preparation	Medicinal Use	Reference
1.	*Acacia nilotica* (L.) Delile (Mimosaceae)	T	P	L,B,G	Decoction	Asthma, diarrhea, demulcent	Ilahi, 2008
2.	*Achyranthes aspera* L. (Amaranthaceae)	H	P	R	Powder	Asthma, cough, pneumonia, joint pain, labor discomfort	Qureshi and Bhatti, 2008; Savithramma *et al.*, 2007
3.	*Aerva javanica* var. *bovei* Webb (Amranthaceae)	Sh	P	L,F,WP	Paste/ Decoction	Wounds, jaundice, diabetes, cough, headaches	Qureshi and Bhatti, 2008; Hammiche and Maiza, 2006
4.	*Alhagi maurorum* Medic. (Fabaceae)	Sh	P	WP	Decoction	Laxative, diuretic, eye infections	Ilahi, 2008; Goodman and Ghafoor, 1992
5.	*Amaranthus viridis* L. (Amaranthaceae)	H	A	WP	Potherb	Constipation, gall bladder/ kidney stones	Qureshi and Bhatti, 2008
6.	*Arthrocnemum indicum* (Willd.) Moq (Amranthaceae/Chenopodiaceae)	Sh	P	WP	Ashes	Alexipharmic	Agoramoorthy *et al.*, 2008
7.	*Atriplex stocksii* Boiss (Amranthaceae/ Chenopodiaceae)	Sh	P	L, WP	Infusion	Fever, jaundice, dropsy, liver disease	Qureshi *et al.*, 2009
8.	*Calotropis procera* (Ait.) Ait. f. (Asclepidaceae)	Sh	P	WP	Paste	Tooth and stomach aches	Ilahi, 2008
9.	*Capparis decidua* (Forssk.) Edgew. (Capparidaceae)	Sh	P	F, B	Decoction	Carminative, aphrodisiac, ulcer, cough, asthma, stomach aches	Ilahi, 2008; Goodman and Ghafoor, 1992
10.	*Chenopodium album* L. (Amranthaceae/ Chenopodiaceae)	H	A	WP	Potherb	Constipation	Qureshi and Bhatti, 2008
11.	*Citrullus colocynthis* (L.) Schrad. (Cucurbitaceae)	H	P	F, R	Toothache, Pills	Toothache, Constipation, bleeding, piles, diabetes, leucorrhea, Asthma	Goodman and Ghafoor, 1992; Qureshi and Bhatti, 2008; Savithramma *et al.*, 2007

Contd...

Table 21.1–Contd...

Sl.No.	Name of Species (Family)	Habit	Life Form	Part Used	Preparation	Medicinal Use	Reference
12.	*Cleome brachycarpa* Vahl ex DC. (Capparidaceae)	H	A	WP	With sesame oil	Itching, Joint pain, skin diseases, inflammation	Qureshi and Bhatti 2008; Shafi *et al*., 2001
13.	*Cleome viscosa* L. (Capparidaceae)	H	A	WP	Juice	Ear infection, pain, deafness	Qureshi and Bhatti, 2008
14.	*Corchorus depressus* (L.) Stocks (Tiliaceae)	Sh	P	WP	Infusion	Dysuria	Goodman and Ghafoor, 1992
15.	*Cressa cretica* L. (Convolvulaceae)	H	p	L	Paste	Sores	Qureshi and Bhatti, 2008
16.	*Cymbopogon jwarancusa* (Jones) Schult. (Poaceae)	G	P	WP	Decoction	Fever, catarrh, joint pain, inflammation	Qureshi and Bhatti, 2008; Goodman and Ghafoor, 1992
17.	*Cynodon dactylon* (L.) Pers. Poaceae)	G	P	WP	Paste	Cuts, wounds	Qureshi and Bhatti, 2008
18.	*Cyperus rotundus* L. (Cyeraceae)	Se	P	R	Poultice	Acne, Carbuncle	Qureshi and Bhatti, 2008
19.	*Desmostachya bipinnata* (L.) Stapf (Poaceae)	G	P	R	Potherb	Carbuncle	Qureshi and Bhatti, 2008
20.	*Digera muricata* (L.) Mart (Amranthaceae)	H	A	WP	Powder	Constipation	Qureshi and Bhatti, 2008
21.	*Euphorbia caducifolia* Haines (Euphorbiaceae)	Sh	P	La	Raw	Increase male sexual vigor	Goodman and Ghafoor, 1992
22.	*Fagonia indica* Burm. f. (Zygophyllaceae)	Sh	P	WP	Decoction	Skin eruption, malarial fever	Qureshi and Bhatti, 2008; Goodman and Ghafoor, 1992
23.	*Glinus lotoides* L.	H	A	WP	Powder	Blood purifier	Qureshi and Bhatti, 2008
24.	*Gynandropsis gynandra* (L.) Briq. (Capparidaceae)	H	A	R, L	Decoction	Stomach ace and skin inflammation	Shafi *et al*., 2001
25.	*Haloxylon stocksii* (Boiss.) Benth. and Hook. (Amranthaceae/Chenopodiaceae)	Sh	P	WP	Decoction	Tooth and stomach aches	Ilahi, 2008

Contd...

Table 21.1–Contd...

Sl.No.	Name of Species (Family)	Habit	Life Form	Part Used	Preparation	Medicinal Use	Reference
26.	*Heliotropium curassavicum* L. (Boraginaceae)	H	P	R	Paste	Boils	Qureshi and Bhatti, 2008
27.	*Heliotropium strigossum* Willd. (Boraginaceae)	Sh	P	WP	Paste	Boils, Ulcer, wounds	Shafi *et al.*, 2001
28.	*Heliotropium europaeum* var. *lasiocarpum* (F. and M.) Kazmi (Boraginaceae)	Sh	A	L	Raw	Earache	Shafi *et al.*, 2001
29.	*Ipomoea pescaprae* (L.) Swt. (Convolvulaceae)	H	P	WP	Decoction	Diarrhea, pains, vomiting, inflammation of legs, piles	Agoramoorthy *et al.*, 2008
30.	*Leucas urticifolia* (Vahl) Rr. B. (Labiatae)	H	A	WP	Decoction problems	Cold, fever, gastrointestinal	Shafi *et al.*, 2001
31.	*Peganum harmala* L. (Zygophyllaceae)	H	P	WP,S	Decoction	Anti-microbial, colic, lumbago/ Rheumatisms, back pains	Ilahi 2008; Hammiche and Maiza, 2006
32.	*Portulaca oleracea* L. (Portulacaceae)	H	A	L,F,S	Infusion	Stomach ache, gastric trouble	Husain *et al.*, 2008
33.	*Rhizophora mangle* = *Rhizophora mucronata* Poir. (Rhizophoraceae)	Sh	P	F,R,W	Maceration, tea	Diarrhea, diabetes	Ferreira, 2009
34.	*Suaeda monoica* Forssk. ex J. F. Gmelin (Chenopodiaceae)	Sh	P	WP	Paste	Ointment for wounds	Agoramoorthy *et al.*, 2008
35.	*Salvadora oleoides* Decne. (Salvadoraceae)	Sh	P	L,F,S	Decoction	Purgative, aphrodisiac, diarrhea	Shafi *et al.*, 2001; Goodman and Ghafoor, 1992
36.	*Salvadora persica* L. (Salvadoraceae)	Sh	P	L,F	Decoction	Asthma, cough	Shafi *et al.*, 2001; Savithramma *et al.*, 2007
37.	*Salsola imbricata* Forssk. (Amranthaceae/Chenopodiaceae)	Sh	P	WP	Infusion	Insecticidal, Vascular hypertension	Hassan *et al.*, 2005, Hammiche and Maiza, 2006

Contd...

Table 21.1–Contd...

Sl.No.	Name of Species (Family)	Habit	Life Form	Part Used	Preparation	Medicinal Use	Reference
38.	*Solanum surattense* Burm.f. (Solanaceae)	Sh	P	F, WP	Pills	Bronchial, asthma, headache, cough, joint pain and chest pain, vomiting, burning feet	Qureshi and Bhatti 2008; Savithramma *et al.*, 2007; Ilahi, 2008; Goodman and Ghafoor, 1992
39.	*Suaeda fruticosa* Forssk. ex J.F. Gmelin (Amranthaceae/Chenopodiaceae)	Sh	P	L	Water and Alcoholic extracts	Antibacterial	Rashid *et al.*, 2000
40.	*Trianthema portulacastrum* L. (Aizoaceae)	H	A	L	Decoction	Asthma	Qureshi *et al.*, 2009; Savithramma *et al.*, 2007
41.	*Tribulus terrestris* L. (Zygophyllaceae)	H	A	R	Infusion	Urinary calculi, sperma-torrhoea, general debility, Asthma, Kidney stones	Qureshi and Bhatti, 2008; Savithramma *et al.*, 2007; Hammiche and Maiza 2006; Husain *et al.*, 2008
42.	*Withania somnifera* (L.) Dunal (Solanaceae)	Sh	P	F, L	Decoction	Anthelmintic, leucorrhoea, tuberculosis	Husain *et al.*, 2008
43.	*Zaleya pentandra* (L.) Jeffrey (Aizoaceae)	H	P	R	Powder	Influenza, phlegmatic cough	Qureshi and Bhatti, 2008
44.	*Ziziphus nummularia* (Burm. f.) Wight and Arn. (Rhamnaceae)	Sh	P	L, F, B	Decoction	Digestive, blood purifier, sores, skin diseases	Ilahi, 2008
45.	*Zygophyllum simplex* L. (Zygophyllaceae)	H	A	L	Decoction	Helminthiases	Hammiche and Maiza, 2006

Sh: Shrub; H: Herb; T: Tree; G: Grass Se: Sedge; L: Leaves; WP: Whole plant; F: Fruit; R: Root; S: Seed; B: Bark; La: Latex; A: Annual; P: Perennial.

DISTRIBUTION

Halophytes are found in a range of environmental conditions depending on their ability to tolerate various combinations of salts in the substrate (Song *et al.*, 2008) and may be grouped in various ways such as into coastal, near coastal and inland species depending on the proximity to the open sea or according to the type of habitat they occupy such as salt desert, salt flat or salt marsh plants.

RESULTS AND DISCUSSION

Halophytes are adapted to tolerate harsh saline and arid conditions by synthesizing a number of active chemicals to maximize their fitness, many of which are a source of potent medicine against a number of chronic ailments. Coastal environments were reported to be more stressful than inland due to higher soil salinity, greater light intensity and more frequent diurnal and seasonal climatic conditions (Gulzar and Khan, 1998). Numerous reports have documented the traditional uses of medicinal plants in rural and tribal areas all over the world as a successful home remedy against different ailments (Hamburger and Hostettmann, 1991; Hammiche and Maiza, 2006; Husain *et al.*, 2008).

TRADITIONAL USES

The ethnobotanical data collected from different sources covers 7 health related conditions *i.e.*, digestive, skin, respiratory, pain, fever and toothache, genito-urinary and others. A summary of the different ethnobotanical uses of 45 documented species (Table 21.1) and their per cent distribution with respect to type of disease condition (Table 21.2) is presented here. In almost all cases whole plants or plant parts are used as a cure for various illnesses which indicates the presence of beneficial chemicals.

DIGESTIVE CONDITIONS

Twenty four per cent of the species reported here are used as purgatives, carminatives and as anti-helmintics in the treatment of digestive disorders like stomach pain, vomiting, diarrhea (Figure 21.1).

Acacia nilotica L. Delile, *Citrullus colocynthis* L. Schrad. and *Ipomoea pes-caprae* (L.) Swt. are the most frequently used species. *Chenopodium album* and *Digera muricata* (L.) Mart are used to treat constipation and bowel movement. *Haloxylon stocksii* and *Portulaca oleracea* are used for the treatment of stomach ache (Ilahi, 2008; Husain *et al.*, 2008), for ulcer (Ilahi, 2008; Shafi *et al.*, 2001) and for vomiting and gastric trouble (Agoramoorthy *et al.*, 2008; Shafi *et al.*, 2001; Qureshi and Bhatti, 2008; Savithramma *et al.*, 2007; Ilahi, 2008). *Capparis decidua* (Forssk.) Edgew. is a popular carminative (Ilahi 2008) while *Withania somnifera* (L.) Dunal and *Zygophyllum simplex* L. are used to cure helminthiases (Husain *et al.*, 2008; Hammiche and Maiza, 2006). *Salvadora oleoides* Decne. is also used as a purgative (Shafi *et al.*, 2001) while *Rhizophora mangle=Rhizophora mucronata Poir.*is used to treat diarrhoea (Coelho-Ferreira, 2009).

Skin Conditions

Skin conditions account for 15 per cent of plant species listed (Table 21.1; Figure 21.1). *Cleome brachycarpa* Vahl. ex DC. and *Cynodon dactylon* (L.) Pers. are frequently cited for the treatment of wounds, itching and inflammations. *Fagonia indica* Burm. f. and *Heliotropium strigossum* Willd. are also used in the treatment of wounds and burns (Shafi *et al.*, 2001) and leaves of *Suaeda monoica* Forsk. ex Gmel. to make an antiseptic ointment for skin eruptions (Agoramoorthy *et al.*, 2008). *Heliotropium curassavicum* L. is used for healing boils (Qureshi and Bhatti, 2008); *Ziziphus nummularia* (Burm. f.) Wight and Arn. for skin infections (Ilahi, 2008) while paste from *Aerva javanica* (Burm. f.) Juss. ex J.A. Schultes leaves for treating skin diseases (Qureshi and Bhatti, 2008; Hammiche and Maiza, 2006). Roots of *Desmostachya bipinnata* (L.) Stapf and *Cyperus routundus* L. are applied externally on carbuncles and those of *C. rotundus* are also used to treat acne (Qureshi and Bhatti, 2008).

RESPIRATORY CONDITIONS

Respiratory tract infections are treated by about 14 per cent of the species in this study (Figure 21.1). *Solanum surattense* is commonly used for chronic coughs, asthma, bronchitis and sore throat (Qureshi and Bhatti, 2008; Savithramma *et al.*, 2007; Goodman and Ghafoor, 1992), roots of *Zaleya pentandra* (L.) Jeffrey have anti-phlegmatic activity and are also used against influenza (Qureshi and Bhatti, 2008). *Acacia nilotica* is used as a demulcent (Ilahi, 2008) and *Aerva javanica* and *Capparis decidua* are frequently used to relieve cough and chest complaints (Hammiche and Maiza, 2006; Ilahi, 2008). Decoctions of ground *Achyranthes aspera* L. roots are used to treat pneumonia (Qureshi and Bhatti, 2008; Savithramma *et al.*, 2007). Whole plants of *Leucas urticifolia* (Vahl) R. Br. and leaves of *Salvadora persica* L. are used for cough and cold (Shafi *et al.*, 2001; Savithramma *et al.*, 2007).

Asthma is a common respiratory track disease in rural areas which is successfully cured using traditional medicine since generations. For instance *Tribulus terrestris* L. fruits (Husain *et al.*, 2008) are commonly used in parts of Pakistan whereas Savithramma *et al.* (2007) recorded about 80 indigenous species that are used to cure asthma in Natuvaidyulu India including the halophytes *Achyranthes aspera*, *Citrullus colocynthis*, *Salvadora persica*, *Trianthema portulacastrum* L. and *Tribulus terrestris*.

PAIN

Among the halophytes listed in this paper about 13 per cent of the species are used for pain alleviation (Figure 21.1). *Achyranthes aspera*, *Cleome brachycarpa*, *Cymbopogon jwarancusa* (Jones) Schult. and *Solanum surattense* are used for treating joint pains (Ilahi, 2008; Qureshi and Bhatti, 2008). Whole plant decoctions of *Aerva javanica* are used for headaches, while *Solanum surattense* fruits for the relief of chest pain (Hammiche and Maiza, 2006; Qureshi and Bhatti, 2008). Whole plant extracts of *Peganum harmala* are used to relieve back pain (Hammiche and Maiza, 2006) and those of *Ipomoea pes-caprae* for pain with inflammation (Agoramoorthy *et al.*, 2008). Sap obtained from *Cleome viscosa* plants is ingested for treating ear infections, pain and deafness (Qureshi and Bhatti, 2008). Similarly, *Heliotropium europaeum* var. *lasiocarpum* (F. and M.) Kazmi leaves are a good remedy for earaches (Shafi *et al.*, 2001). In their monograph Goodman and Ghafoor (1992) reported that few drops of fresh leaves extract of *Alhagi maurorum* can cure infections, pain and redness of eyes.

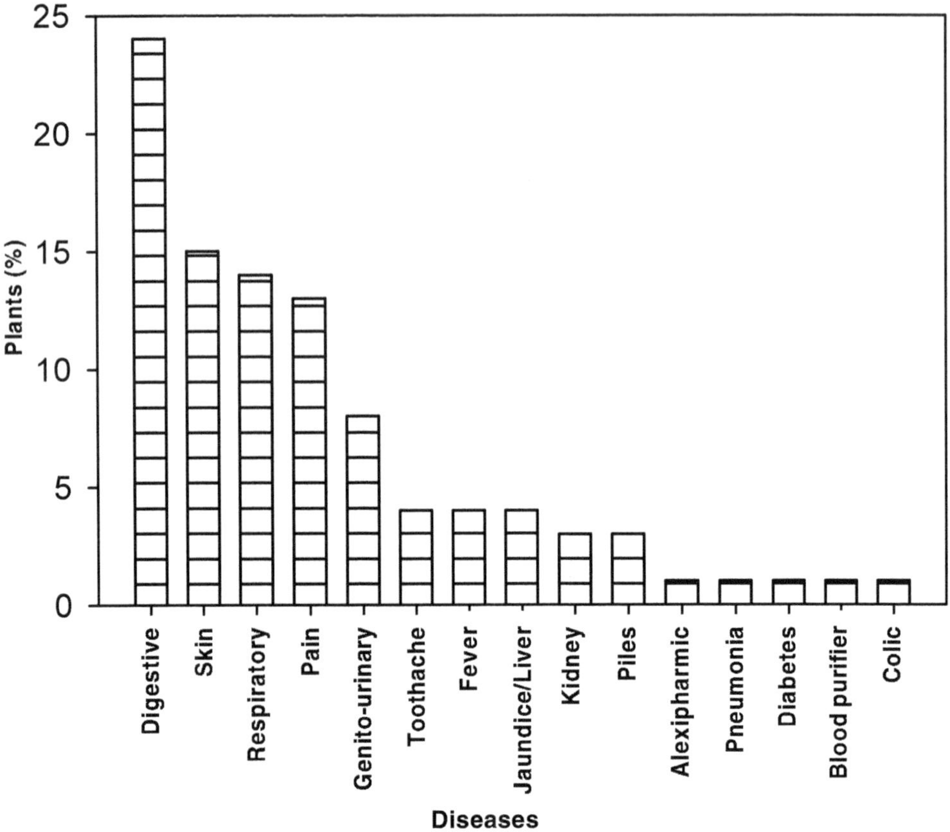

Figure 21.1: Major Ailments Treated by Local Healers Using Coastal Medicinal Plants

FEVER AND TOOTHACHE

About 8 per cent species are used to treat fevers and toothaches (Figure 21.1). Leaves of *Atriplex stocksii* when crushed to make decoction or infusion in water are said to relieve any kind of fever (Qureshi *et al.*, 2009). Tea extracted from leaves of *Cymbopogon jwarancusa* can treat prolonged fever (Qureshi and Bhatti, 2008; Goodman and Ghafoor, 1992). *Leucas urticifolia* is also used to treat fever, cold as well as gastric problems (Shafi *et al.*, 2001). Fresh plants of *Fagonia indica*, are soaked overnight and the infusion is used to cure malarial fever (Goodman and Ghafoor, 1992). *Calotropis procera* (Ait.) Ait. f. latex generally considered to be poisonous, when applied with care by experienced practitioners relieves pain in teeth and gums (Ilahi, 2008). Fruits of *Citrullus colocynthis* and decoctions of *Haloxylon stocksii* are also reported for tooth and stomachaches (Ilahi, 2008; Goodman and Ghafoor, 1992).

GENITO-URINARY CONDITIONS

About 8 per cent of halophyte species are useful in the treatment of genito-urinary and kidney related problems (Figure 21.1) *e.g.*, decoctions of *Citrullus colocynthis* and

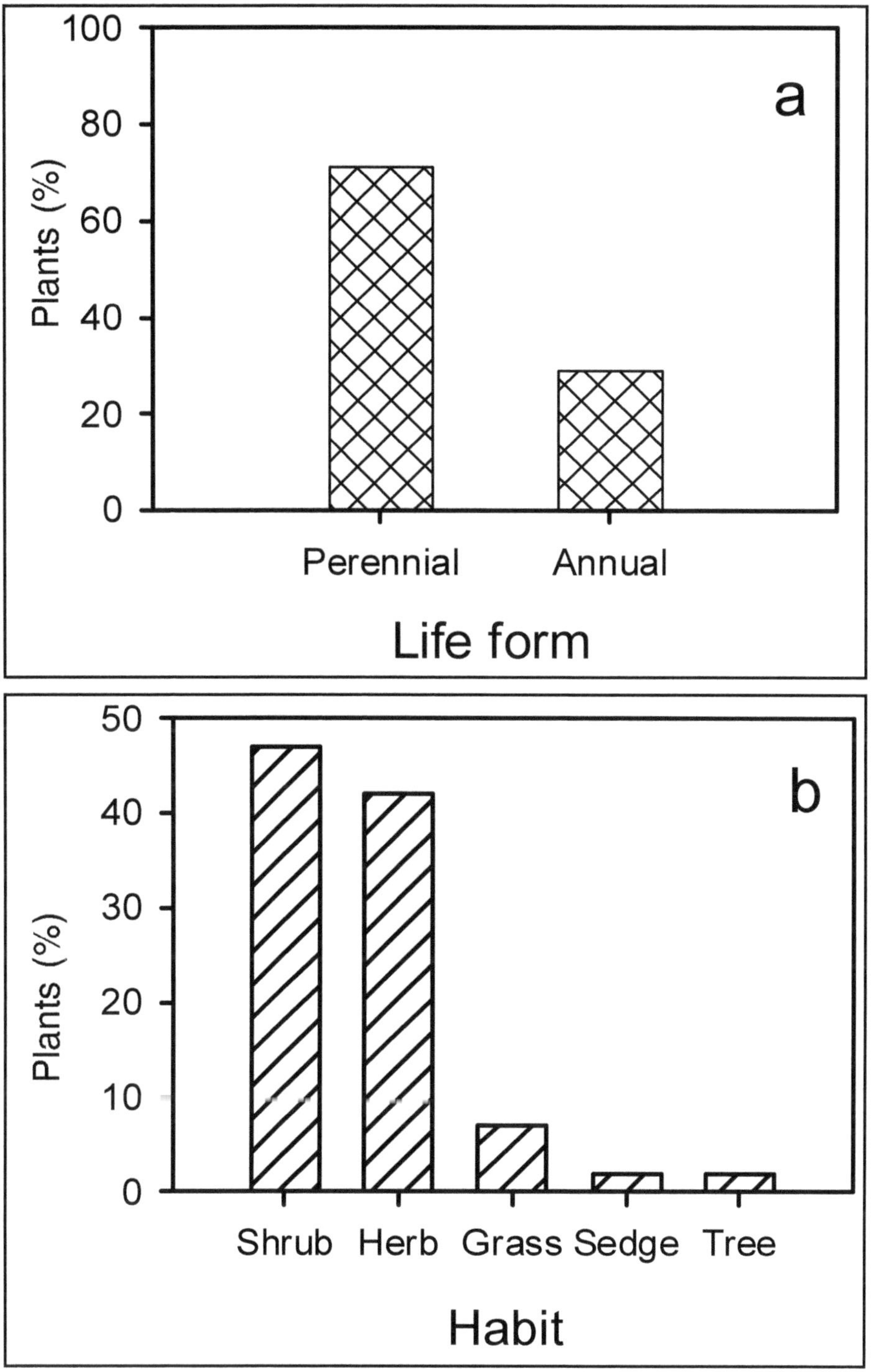

Figure 21.2: Distribution of Medicinal Plants with Respect to their (a) Life Form and (b) Habit

Withania somnifera are useful for treating leucorrhoea (Qureshi and Bhatti, 2008; Savithramma *et al.*, 2007; Hammiche and Maiza, 2006; Husain *et al.*, 2008). Fruit decoctions of *Capparis decidua* (Ilahi, 2008) and *Salvadora oleoides* (Shafi *et al.*, 2001) are used as aphrodisiac.

The whole plant decoctions from *Tribulus terrestris* are used as a potent medicine to treat Spermatorrhoea (Husain *et al.*, 2008). A small quantity of *Euphorbia caducifolia* Haines latex taken orally by males before sexual intercourse is reputed to increase prowess (Goodman and Ghafoor, 1992). Ground roots of *Achyranthes aspera* are used traditionally to relieve labor discomfort and allow ease of delivery (Qureshi and Bhatti, 2008).

Aerial parts of *Amaranthus virdis* L. are frequently utilized in the treatment of urinary tract infections especially for kidney stones and painful urination (Qureshi and Bhatti, 2008). The plant is boiled in a pot until soft and eaten for dysuria. Oral decoctions of *Alhagi maurorum* are considered to have diuretic properties (Ilahi, 2008). Infusion of *Corchorus depressus* (L.) Stocks is used to relieve irritation and pain during urination and provide relief from heat stroke in summer (Goodman and Ghafoor, 1992).

OTHER MEDICAL CONDITIONS

Leaf infusions of *Atriplex stocksii* are used to treat jaundice and other liver disorders (Qureshi *et al.*, 2009). Leaf decoctions of *Aerva javanica* are also used against jaundice while oral administration also cures yellow fever (Qureshi and Bhatti, 2008; Hammiche and Maiza, 2006). Powdered plant material of *Glinus lotoides* L. is useful as a blood purifier (Qureshi and Bhatti, 2008) and a major cure for pimples, boils, skin discoloration, jaundice, headaches, drowsiness, wrinkles and premature aging. *Citrullus colocynthis* is widely used to treat a variety of ailments using several recipes even for the same ailment such as the bitter fruits against diabetes (Qureshi and Bhatti, 2008). Initially, patients do not feel the bitterness of the fruit and the treatment is continued until the bitterness is sensed. The roots of *Achyranthes aspera* are used in the treatment of pneumonia (Qureshi and Bhatti, 2008) and *Arthrocnemum indicum* (Willd.) Moq. as an antidote against venom from bite of scorpions, snakes or poisonous insects. The plant is burnt to ash and applied directly to the affected area (Agoramoorthy *et al.*, 2008) as an alexipharmic. *Peganum harmala* infusions are used to treat colic in 2-5 week old infants (Ilahi, 2008; Hammiche and Maiza, 2006). *Citrullus colocynthis* and *Ipomoea pes-caprae* are reported for treatment of piles (Qureshi and Bhatti, 2008; Agoramoorthy *et al.*, 2008).

CONCLUSIONS

Plants continue to be almost exclusive sources of drugs for the majority of the world's population (Hamburger and Hostettmann, 1991). A variety of plant chemicals have therapeutic importance such as low molecular weight proteins and peptides (Linthorst, 1991; Dahot, 1999) or glycosides. The primary metabolites exhibit antifungal and/or antimicrobial (Robert and Seletrennikoff, 1986; Terras *et al.*, 1992) activity probably due to their ability to bind to and degrade microbial surfaces. Secondary metabolites such as phytoalexins, phenolic acids, alkaloids, saponins, flavonoids, tannins, essential oils could also be part of the plant defense against microbes (Van Etten *et al.*, 1989; Maher *et al.*, 1994).

Plant drug discovery and development are processes of collaborative effort between field practitioners and laboratory scientists. A continuous flow of raw material from the field, cooperation between practitioners and research institutes and a long-term assurance of sharing benefits of the discovery between the parties are essential at this stage. Ethno-medicinal studies typically involve thorough surveys of indigenous species, their local uses, mode of preparation, dosage and routes of administration. While chemical analyses include determination of biological (anti-bacterial, anti-fungal, anti-viral) activity, pharmacological investigations, identification of active ingredients using GC, GC-MS, LC-MS, NMR and clinical trials. However, the transformation of an active ingredient from a raw plant material to a commercial product is a long and tedious process. Flora of Pakistan accounts for 85 per cent of terrestrial plants with more than 5700 species throughout the country while the remaining 15 per cent are being documented from the remote northern areas of Pakistan (Ali and Qaiser, 1995-2005). Around 400 species of halophytes have so far been reported from Pakistan and 25 per cent of these are from the Arabian Sea coast. It is speculated that many more new reports of halophytes will be forthcoming from pristine mountainous habitats (Ali, 2008).

Domestication of medicinal halophytes is no longer a myth but is still not being pursued on war footing in salt prone areas due to two possible reasons: firstly, since salinity has not so far threatened the availability of food for humans and secondly due to a lack of awareness, acceptance and confidence in the benefits of non-conventional resources by the masses. However, global climate changes and population increase are occurring at an alarming rate are predicted to decrease the per capita water availability within the next quarter of century leading to increasing realization of need to tap new avenues in providing food and health security.

REFERENCES

Agoramoorthy, G., F.A. Chen, V. Venkatesalu, D.H. Kuo and P.C. Shea. 2008. Evaluation of antioxidant polyphenols from selected mangrove plants of India. *Asian Journal of Chemistry*. 20: 1311-1322.

Ali, S.I. 2008. Significance of Flora with special reference to Pakistan. *Pakistan Journal of Botany*. 40: 967-971.

Ali, S.I. and M. Qaiser. 1995-2005. *Flora of Pakistan*. Department of Botany, University of Karachi, Karachi.

Anonymous. 1983. Medicinal Plants, Hamdard Foundation Press, Hamdard Centre, Nazimabad Karachi-18, Pakistan. pp.1-55.

Coelho-Ferreira, M. 2009. Medicinal knowledge and plant utilization in an Amazonian coastal community of Marudá, Pará State (Brazil). *Journal of Ethnopharmacology*. 126: 159-175.

Dagar, J.C. 1995. Characteristics of halophytic vegetation in India. In: M.A. Khan and I.A. Ungar (eds.). *Biology of Salt Tolerant Plants*. University of Karachi, Karachi. pp. 255-276.

Dahot, M.U. 1999. Antibacterial and antifungal activity of small protein of *Indigofera oblongifolia* leaves. *Journal of Ethnopharmacology*. 64: 277–282.

Goodman, S.M. and A. Ghafoor. 1992. The Ethnobotany of southern Balochistan, Pakistan, with particular reference to medicinal plants. *Fieldiana: Botany New Series*, 31: 1.V-I1.84.

Gulzar, S. and M.A. Khan. 1998. Diurnal water relations of inland and coastal halophytic populations from Pakistan. *Journal of Arid Environment*. 40: 295-305.

Hamburger, M. and K. Hostettmann. 1991. Bioactivity in plants: The link between phytochemistry and medicine. *Phytochemistry*. 30: 3864–3874.

Hammiche, V. and K. Maiza. 2006. Traditional medicine in Central Sahara: Pharmacopoeia of Tassili N'ajjer. *Journal of Ethnopharmacology*. 105: 358–367.

Hassan, M., M.A. Siddique, M. Sagheer and M. Aleem. 2005. Comparative efficacy of ethanol leaf extracts of *Amaranthus viridis* l. and *Salsola baryosma* Schult. ex Roem. and Schult.) Dandy (schultes) and Cypermethrin against *Trogoderma granarium* Eeverts. *Pakistan Journal of Agriculture.* 42: 3-4.

Husain, S.Z., R.N. Malik, M. Javaid and S. Bibi. 2008. Ethnobotanical properties and uses of medicinal plants of Morgah Biodiversity park, Rawalpindi. *Pakistan Journal of Botany*. 40: 1897-1911.

Ilahi, I. 2008. Ethnobotanical studies and problems associated with regeneration of herbals in Kohat region. *Pakistan Journal of Botany*. 40: 1743-1753.

Jonathan E.M.B., C.H. Taylor and S.N. Stuart (eds.). 2004. A Global Species Assessment. *IUCN Red List of Threatened Species.* Cambridge, UK.

Khan, M.A. and M. Qaiser. 2006. Halophytes of Pakistan: Characteristics, Distribution and Potential Economic usages. In: M.A. Khan, H. Barth. G.C. Kust and B. Boer. Sabkha Ecosystems: Volume II: The South and Central Asian Countries. pp. 129–153. Springer, Netherlands.

Khan, M.A., R. Ansari, H. Ali, B. Gul and B.L. Nielsen. 2009. *Panicum turgidum*, a potentially sustainable cattle feed alternative to maize for saline areas. *Agriculture, Ecosystems and Environment*. 129: 542–546.

Linthorst, H.J.M. 1991. Pathogenesis related proteins of plants. *Critical Reviews in Plant Sciences*. 10: 123-150.

Maher, E.A., N.J. Bate, W. Ni, Y. Elkin, R.A. Dixon and C.J. Lamb. 1994. Increased disease susceptibility of transgenic tobacco plants with suppressed level of preformed phenyl propanoid products. *Proceedings of National Academy of Science.* 91: 7802–7806.

Qureshi, R. and G.R. Bhatti. 2008. Ethnobotany of plants used by the Thari people of Nara desert, Pakistan. *Fitoterapia*. 79: 468–473.

Qureshi, R., A. Waheed, M. Arshad and T. Umbreen. 2009. Medico-ethnobotanical inventory of Tehsil Chakwal, Pakistan. *Pakistan Journal of Botany*. 41: 529-538.

Rashid, S., Q. Iftikhar, M. Arshad and J. Iqbal. 2000. Chemical composition and antibacterial activity of *Suaeda fruticosa* Forssk. from Cholistan, Pakistan. *Pakistan Journal of Biological Science*. 3: 348-349.

Robert, W.K. and C.P. Seletrennikoff. 1986. Isolation and partial characterization of two antifungal proteins from barley. *Biochemica et Biophysica Acta.* 880: 161–170.

Rozema, J. and T.J. Flowers. 2008. Crops for a Salinized World. *Science*. 322: 1478-1480.

Savithramma, N., C. Sulochana and K.N. Rao. 2007. Ethnobotanical survey of plants used to treat asthma in Andhra Pradesh, India. *Journal of Ethnopharmacology* 113: 54–61.

Shafi, M.S., M.Y. Ashraf and G. Sarwar. 2001. Wild medicinal plants of Cholistan area of Pakistan. *Pakistan Journal of Biological Science*. 4: 112-116.

Song, J., H. Fana, Y. Zhaoa, Y. Jiaa, X. Dua and B. Wang. 2008. Effect of salinity on germination, seedling emergence, seedling growth and ion accumulation of a euhalophyte *Suaeda salsa* in an intertidal zone and on saline inland. *Aquatic Botany*. 88: 331-337.

Terras, F.R.G., H.M.E. Schoofs and M.F.C. De Bolle. 1992. Analysis of two novel classes of antifungal proteins from radish (*Raphanus sativus L.*) seeds. *Journal of Biological Chemistry*. 267: 15301-15309.

Van Etten, H.D., D.E. Matthews and P.S. Matthews. 1989. Phytoalexin detoxification importance for pathogenicity and practical implications. *Annual Review of Phytopathology*. 27: 143-164.

Weber, D.J., B. Gul, M.A. Khan, T. Williams, P. Wayman and S. Warner. 2001. Composition of vegetable oil from seeds of native halophytic shrubs. In: McArthur, E. Durant; Fairbanks, Daniel J., comps. 2000. Proceedings: *Shrubland Ecosystem Genetics and Biodiversity*. Proceedings RMRS-P-000. Ogden, Ut: U.S. Department of Agriculture, Forest Service Rocky Mountain Research Station.

Chapter 22

Biochemical Indicators of Salinity Tolerance in the Halophyte *Prosopis strombulifera* are Differentially Affected by NaCl and Na_2SO_4

***Analía Llanes*[1], *Mariana Reginato*[1] *Gabriela Palacio*[2] *and Virginia Luna*[1]**

[1]*Laboratorio de Fisiología Vegetal, Departamento de Ciencias Naturales, Fac. Cs. Exactas, Universidad Nacional de Río Cuarto, 5800, Río Cuarto, Argentina*

[2]*Departamento de Matématicas, Fac. Cs. Exactas, Universidad Nacional de Río Cuarto, 5800, Río Cuarto, Argentina*

ABSTRACT

P. strombulifera is an interesting model plant to study mechanisms involved in plant tolerance to extreme salinity conditions. Total aminoacids showed no quantitative but qualitative differences within treatments; proline accumulation was independent of the medium composition, showing a direct correlation between proline content and stress intensity. There was no increase in glycinbethaine levels in response to any salt. Total protein content was no modified in salt treated plants at Ψo–1.0 MPa, but increased at Ψo–1.9 MPa in all treatments; at Ψo–2.6 MPa only NaCl caused extra total protein synthesis; $SO_4^=$ treated plants had reduced protein content while bisaline treated plants had and intermediate value. Total soluble carbohydrates increased proportionally to salinity increase, with changes in the root/shoot partitioning; glucose and fructose accumulated mainly in $SO_4^=$ treated plants roots with a noticeable decrease in sucrose content. Leaves of NaCl-treated plants showed a RWC higher than controls, with a decrease in SD (lower number of larger stomata). These characteristics maybe an adaptive mechanism of injury alleviation in response to excessive Na^+ accumulation in tissues, diminishing the transpiration rate and favouring osmorregulation and higher water use efficiency. Instead, leaves of Na_2SO_4-treated plants showed lower values of Ψw and Ψo and the highest transpiration rate, in correlation with a decrease in total and specific leaf area and an increase in stomatal density. All these particular modifications (specific for each salt) in the epidermal morphology of *P. strombulifera* maybe adaptive modifications related to water use efficiency and photosynthesis control. General metabolism was also differentially affected by both salts. These results strengthen the need to consider realistic ion composition of soils to be able to extrapolate lab experiments to field conditions.

Keywords: *Prosopis strombulifera, Na_2SO_4 and NaCl, Ionic effect, Biochemical markers.*

INTRODUCTION

Soil salinity is one among the several environmental stresses causing drastic changes in growth, physiology and metabolism of plants and threatening plant productivity around the globe. Salt stress induces various biochemical and physiological responses and affects almost all plant processes; then, tolerance to abiotic stresses is very complex at the whole plant and cellular levels [1, 2]. This is in part due to the complexity of interactions between stress factors and various molecular, biochemical and physiological phenomena affecting plant growth and development [3]. Currently, there are no economically viable technological means to facilitate crop production under stress conditions. However, development of crop plants tolerant to environmental stresses is considered a promising approach, which may help satisfy growing food demands of developing and under-developed countries. Development of crop plants with stress tolerance, however, requires, among others, knowledge of the physiological mechanisms and genetic controls of the contributing traits at different plant developmental stages. In the past two decades, biotechnology research has provided considerable insights into the mechanism of abiotic stress tolerance in plants at the molecular level [4]. For example, though stress tolerance mechanisms may vary from species to species and at different developmental stages [5], basic cellular responses to abiotic stresses are conserved among most plant species. Furthermore, different abiotic stress factors may provoke osmotic stress, oxidative stress and protein denaturation in plants, which lead to similar cellular adaptive responses such as accumulation of compatible solutes, induction of stress proteins, and acceleration of reactive oxygen species scavenging systems [3]. One of the most common stress responses in plants is overproduction of different types of compatible organic solutes. Compatible solutes are low molecular weight, highly soluble compounds that are usually nontoxic at high cellular concentrations. Generally, they protect plants from stress through different mechanisms including contribution to cellular osmotic adjustment, detoxification of reactive oxygen species, protection of membrane integrity and stabilization of enzymes/proteins; but the primary function of compatible solutes is to maintain cell turgor and thus the driving gradient for water uptake [2]. Compatible solutes fall into three major groups: amino acids, quaternary amines and sugars. Although proline is known to confer osmotic tolerance during stress conditions, its specific role during plant growth is not completely clear.

The biosynthetic pathway of proline diverged between bacteria and plant systems [6]. Genes encoding most of the enzymes associated with the synthesis and degradation of proline were cloned and partially characterized, but the factors regulating the expression of these enzymes are largely unidentified. In the last decade, several attempts were made to increase the level of proline accumulation in plants by transferring genes associated with its biosynthetic pathway. Tolerance to abiotic stress, especially to salt, and improved plant growth, was observed in a variety of transgenics that were engineered for overproduction of proline [7]. Proline seems to have diverse roles under osmotic stress conditions; in addition to its role as an osmolyte, proline contributes to stabilizing sub-cellular structures (*e.g.* membranes and proteins), scavenging free radicals [6], and buffering cellular redox potential under stress conditions. It may also function as a protein compatible hydrotrope, alleviating cytoplasmic acidosis, and maintaining appropriate $NADP^+$/NADPH ratios compatible with metabolism [8]. Also, rapid breakdown of proline upon relief of stress may act as a sink for carbon and nitrogen for stress recovery and provide sufficient reducing agents that support mitochondrial oxidative phosphorylation and generation of ATP for repairing of stress-induced damages. Furthermore, proline is known to induce expression of salt stress responsive genes which possess proline responsive elements (*e.g. PRE*, ACTCAT) in their promoters [9].

In plants, glycine betaine is also a representative member of this group of osmolytes. Genetic engineering for glycine betaine biosynthesis in non-accumulating plants has been extensively reported, and transgenic plants expressing bacterial choline-oxidizing enzymes displayed increased tolerance to various stresses such as high salt concentrations and extreme temperatures [10]. A number of sugars have also been targeted for engineering of compatible-solute overproduction.

On the other hand, plants cope with stress through changes in morphology and physiology, increasing the ability to capture resources, or increasing tolerance to the stressor. An example is the change in shape and rate of leaf growth and stomatal characteristics [11], and regulation of water content and leaf water potential during stress. Halophytic plants are able to adjust their water potential to a more negative level than the soil water potential through ion and compatible solutes accumulation to maintain turgor and osmotic adjustment [12].

Prosopis (Leguminosae subfamily Mimosoideae) is a genus with many important arboreal and shrub-like species present in saline zones in America. One such species is the spiny shrub *Prosopis strombulifera* (Lam.) Benth [13], frequently found in the salinized areas of central Argentina [14]. In previous studies we demonstrated that 70 per cent of hydroponically cultured plants of *P. strombulifera* showed a halophytic response to NaCl. Treated plants were more vigorous, had a higher leaf number with smaller leaflets and harder and darker spines than the controls. Growth was slowly inhibited when NaCl concentrations exceeded 500 mM NaCl [15]. These results indicate that *P. strombulifera* is within the maximum tolerance limits reported for halophytic plants being much more tolerant than most *Prosopis* species. In relation to ion accumulation, our results showed that *P. strombulifera* accumulated great quantities of Na^+ in leaves at the expense of K^+ when cultured in high salinity, and the Na/K ratio changed depending on the salinizing agent used. Na^+ and K^+ content was equal (16 g/100gDW) when optimal growth was observed in NaCl treated plants (Ψo= -1.88 MPa, 500 mM) being Cl^- content only 5 g/100gDW. Na_2SO_4-treated plants accumulated SO_4^{2-} from the beginning of salinization with visible toxicity symptoms such as chlorosis, necrosis, and finally, leaf abscission [16]. A possible explanation for the adaptive success of this species when growing in high NaCl concentrations could involve a delicate balance between an effective Na^+ and K^+ compartmentation in vacuoles, Cl^- exclusion and osmotic adjustment through production of high endogenous levels of specific compatible solutes for cytoplasm protection under high internal Na^+ concentrations thus contributing to osmotic balance between compartments as well as to the overall turgor maintenance.

The aim of this work was to determine the endogenous levels of total amino acids, proline, gycine-betaine, total proteins and sugars in roots and leaves of *P. strombulifera* seedlings cultured in iso-osmotic solutions of NaCl, Na_2SO_4 and their mixture, in correlation with possible changes in leaf area, stomatal density, stomatal area, leaf osmotic potential and relative water content in controls and treated seedlings.

MATERIALS AND METHODS

Seeds of *P. strombulifera* were collected from south-western San Luis, Argentina. Geographically, the location is 33 43'S, 66 37'W at 400-500m of altitude, with a temperate thermal regime (*i.e.*, an annual average temperature of 15-20 °C). This area belongs to the carob tree forest located in a saline depression between the annual 300 to 400 mm isohyets in the El Monte Phytogeographic Region. The soil has a franc sandy texture, with abundant calcareous material and moderate salinity (8000 mho/cm2 electrical conductivity at the surface). Pods were collected at random from 100 plants within the same population, and pealed. Seeds were selected visually for uniform size and healthy aspect. They were scarified with sulphuric acid for 10 min and then washed overnight under flowing water. Before sowing they were rinsed in distilled water and placed in Petri dishes with two layers of water saturated filter papers at 37 °C for 24 h [17]. Germinated seeds with 20 mm long radicles were cultured in hydroponic condition in two black trays per treatment per experiment (200 seedlings per tray) with 10 per cent of full strength Hoagland's solution. The seedlings were grown in a growth chamber under a 16 h light (200 μmol $m^{-2}.s^{-1}$) (28 °C): 8 h dark (20 °C) cycle and 70 per cent relative humidity. After one week, seedlings were transferred to a 25 per cent full strength Hoagland's solution (osmotic potential -0.11 MPa). The pH of the medium was 6 in all cases, and an aquarium tubing system with a peristaltic pump was used to provide aeration. The complete experiment was performed three times. A simple aleatory design [18] with four treatments was used. After 21 d salt treatments were started by adding NaCl and Na_2SO_4 pulses of 50m mol L^{-1} and 38 mmol L^{-1} respectively for monosaline treatments and the iso-osmotic mixture for bisaline treatments, every 48 h until final osmotic potentials (Ψo) = -1, -1.88 and -2.6 MPa were reached, corresponding to plants aged 35, 57, and 72 d respectively. Bisaline solutions were obtained by mixing equal volumes of their respective monosaline solutions at each corresponding osmotic potential.

AMINO ACIDS, PROTEINS AND SUGAR ANALYSIS

For total amino acids determination the ninhydrine method was used. The proline content was evaluated according to the method of Magné and Larher [19].

Quaternary ammonium compounds (QACs) were extracted and measured as glycine betaine (GB) equivalents according to Grieve and Grattan [20]. Total proteins were analyzed by Bradford [21]. Total soluble carbohydrates (TSC) concentration was measured by phenol-sulphuric method being the absorption determined by spectrophotometry at 490 nm.

Total Leaf Area, Stomatal Density and Hydric Relations

The total leaf area was determined with the Image-Pro Plus 2.0.Program [22]. The stomatal density (SD) and stomatal area (SA) were determined by the printing technique and extraction of epidermis. The preparations were photographed with a Zeiss Axiophot microscope, and Image-Pro Plus 2.0 was used for measuring length, width and area of stomata. The relative water content (RWC) was determined based on dry weight, turgid weight and fresh weight of the samples. Osmotic potential was determined by using a cryo-osmometer (Semi Micro K-7400, Knauer).

Statistical Analysis

Data were analyzed using STATA TM version 4.0 (StataCorp, 1995). For all variables two factors were considered: osmotic potential and salt treatment (control, NaCl, Na_2SO_4 and salt mixture). Normality and homogeneity of variance were verified, comparisons were performed with t-Student test. When significant differences were observed the Tukey test a posteriori was used. P values below 0.05 were considered statistically significant.

RESULTS AND DISCUSSION

Compatible solutes accumulation helps to maintain water balance in a functioning cell and a growing plant under salt stress conditions. Several studies have reported an increase in the total aminoacid pool during salt treatments in several species [23, 24], while in others such as the halophytes *Aegiceras corniculatum* a decrease was observed [25]. In our work, *P. strombulifera* seedlings had no differences in total amino acid content as a result of salinity (data not shown). However, a remarkable proline accumulation occurred under increasing salinity but irrespective of the chemical composition of salt (Figure 22.2). Indeed, this amino acid was increased by 200 per cent since the beginning of all salt treatments in coincidence with other works which argumented that proline content in plants increases in a linear relationship with conductivity and salinity in the medium [6], but contradicts Meloni *et al.*, [31] finding that proline accumulation in *Prosopis alba* was not affected by salinity (0, 300, 600 mm NaCl); they suggested that this amino acid would not play an important role in the salinity tolerance mechanism of *Prosopis* sp. In *Medicago truncatula*, proline accumulation in response to NaCl stress was higher in aerial parts (13 fold) than in roots (8 fold) compared to controls [26]. A similar effect was observed in *P. strombulifera* where proline accumulation in leaves was much more important than in roots. This fact could be related to the greater ion accumulation in leaf vacuoles

($Na^+ + K^+$ content was 32 g/100gDW when optimal growth was observed in NaCl treated plants at Ψo= -1.88 MPa, 500 mM) raising the need for organic molecules production to counteract osmotic balance and/or ion secuestration, scavenging free radicals, and buffering cellular redox potential under stress conditions [6]. Some authors argue that proline accumulation is one of the main plant adaptations to salinity which increases reflect its important role in osmotic adjustment [27]. However, our results indicate that proline would be a stress intensity signal but not a tolerance indicator.

According to the literature, there is also a linear relationship between leaf osmotic potential and glycine betaine content in several species under salt stress conditions [29]. However, *P. strombulifera* plants did not accumulate glycine betaine in our experiments (Figure 22.2). Total soluble protein content in roots and leaves of *P. strombulifera* treated seedlings only changed at moderate salinity (-1.8 MPa), without detectable differences between the different salts (Figures 22.1 A and B). However, at high salinity higher protein accumulation was measured in NaCl treated seedlings, while in those treated with sulphate protein accumulation was lower (Figures 22.1C and D). This result could be related to the deleterious effect of sulphate salt on plant growth, which was alleviated with the salt mixture. Like other cellular constituents, sugar levels are affected by stress (Prado *et al.*, 2000). In *P. strombulifera* the total soluble sugar level increased with salinity (Figures 22.1 A and B) in coincidence with Kerepesi and Galiba [33] who reported that glucose, fructose and sucrose increased by NaCl exposure in wheat plants. However, Alamgir and Ali [34] reported increased soluble sugar content in some rice genotypes, but a decrease in others. In our experiments, total soluble carbohydrates increased concomitantly with the increase in salt concentration, being more noticeable in roots than in leaves, which suggests an effect on stem/root carbon partitioning. These results correlate with the maintenance and/or induction of root elongation at low osmotic potential as adaptive response to salinity, as well as a significant increase in fibers and lignified tissues observed in our previous work [17, 35]. Studies on the halophyte *Atriplex nummularia* growing at concentrations from 0 to 600 mM NaCl showed that the RWC in leaves of treated plants was higher than in controls. For this species succulence was an efficient mechanism to adjust osmotically [36]. Instead, *P. strombulifera* has the ability to maintain a higher RWC in leaves of seedlings growing in high NaCl concentrations without succulence (Figure 22.4). Thus, it appears that the extreme salt tolerance of this species would be due to its capability of osmotic adjustment. Also, the high water content in its leaves help to ease the impact of the excessive Na^+ accumulation in the tissues which is favored by leaf anatomical adaptations such as the particular distribution of chlorenchyma around the leaf veins [17] and the modifications of stomatal area (Figure 22.3) that enable *P. strombulifera* a more efficient use of water under these experimental conditions. Meanwhile, plants treated with high sulfate concentrations (-2.6 MPa) failed on making an efficient osmotic adjustment showing an extremely negative water potential (Figure 22.4) causing water imbalance which, in comparison with NaCl-treated plants, correlates directly with the marked decrease in specific and total leaf area and an increase in the DE and DCE with smaller stomata, probably as additional efforts of Na_2SO_4-treated plants to survive.

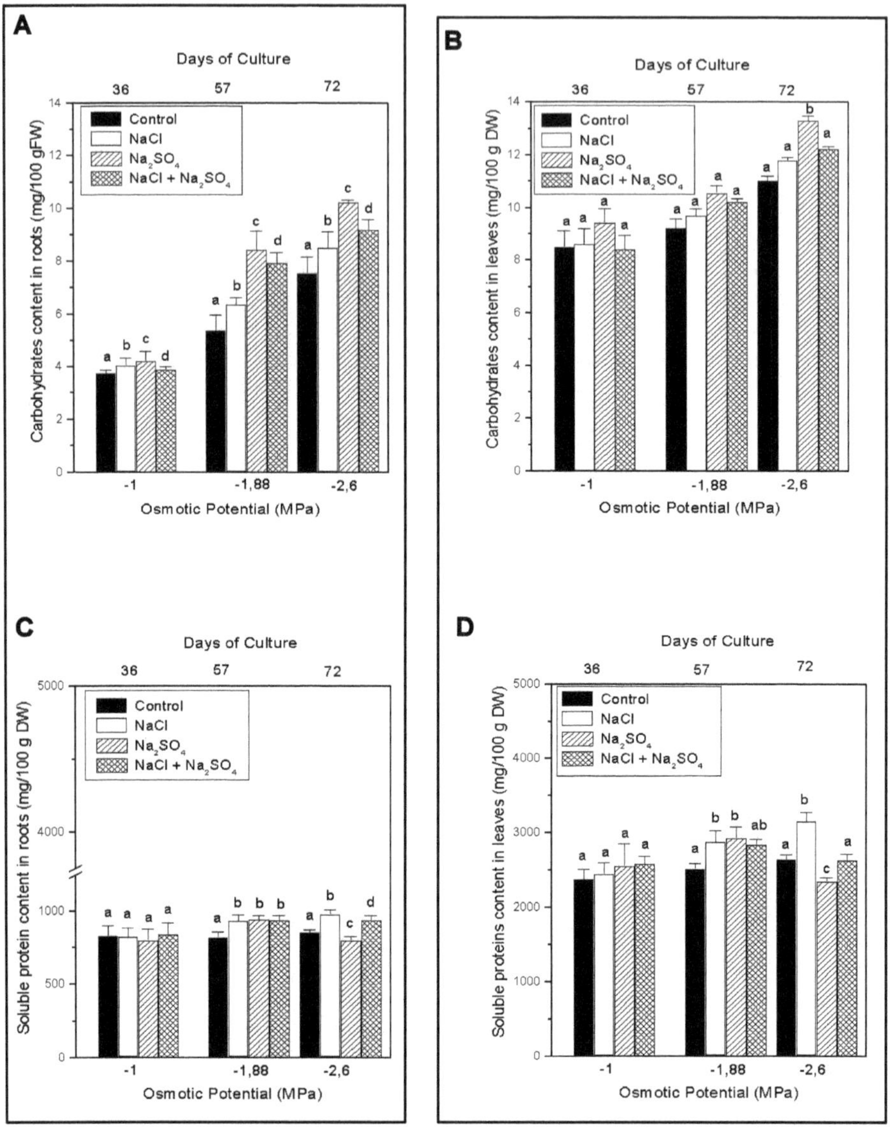

Figure 22.1: Effects of NaCl, Na_2SO_4 and their Iso-Osmotic Mixture on Total Compound Levels in Leaves and Roots of *Prosopis strombulifera* Plants.

A and B: Carbohydrates; C and D: Soluble proteins. Bars represent mean ± S.E. Different letters above bars represent significant differences ($p < 0.05$) among treatments.

Figure 22.2: Effects of NaCl, Na_2SO_4 and their Iso-Osmotic Mixture on Proline and Glycine-Betaine in Roots (A) and Leaves (B) of *Prosopis strombulifera* Plants.

Bars represent mean ± S.E. Different letters above bars represent significant differences ($p < 0.05$) among treatments.

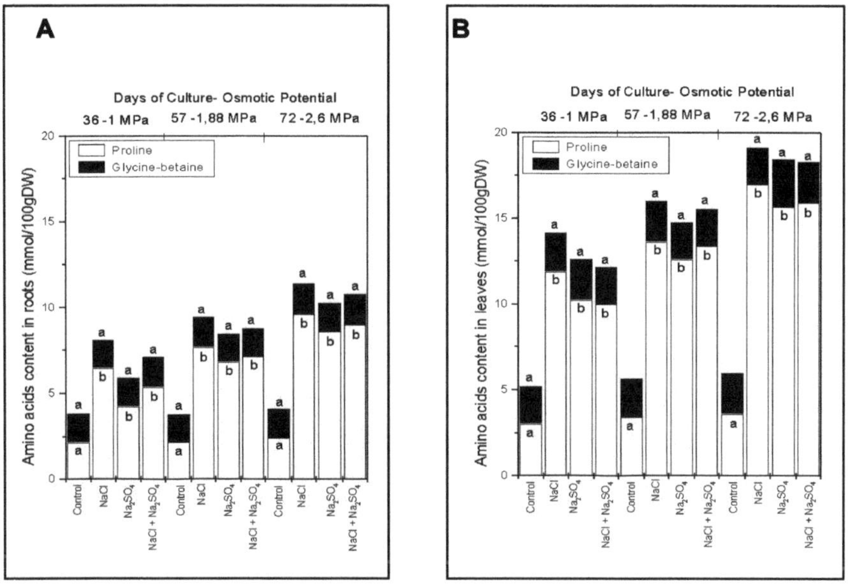

Figure 22.3: Effects of NaCl, Na_2SO_4 and their Iso-Osmotic Mixture on Stomatal Area, Stomatal Density and Total Leaf Area in *Prosopis strombulifera* Plants

Bars represent mean ± S.E. Different letters above bars represent significant differences ($p < 0.05$) among treatments.

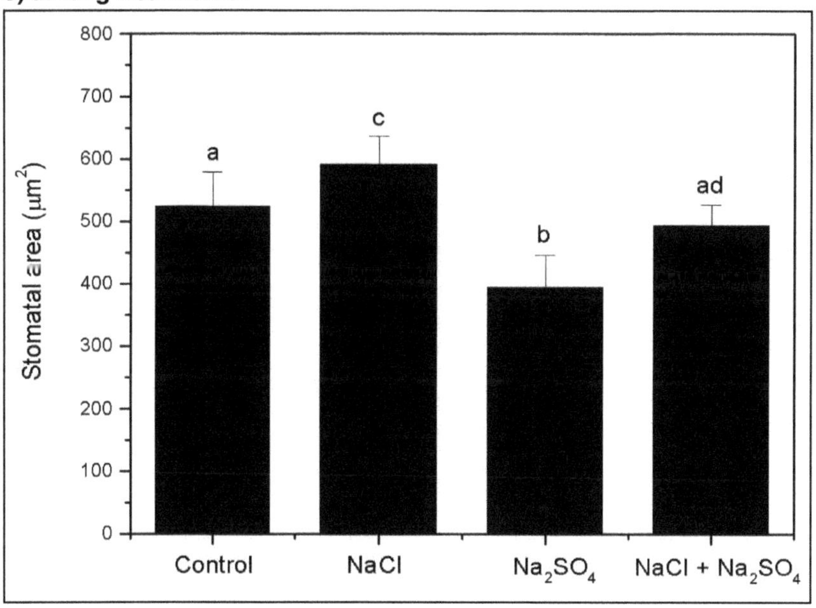

Contd...

Figure 22.3–Contd...

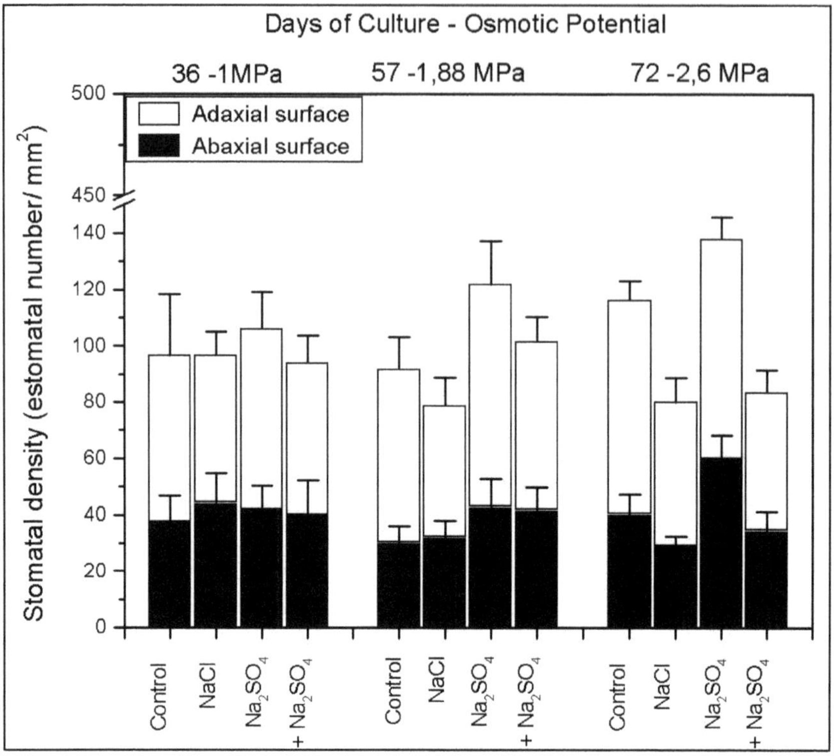

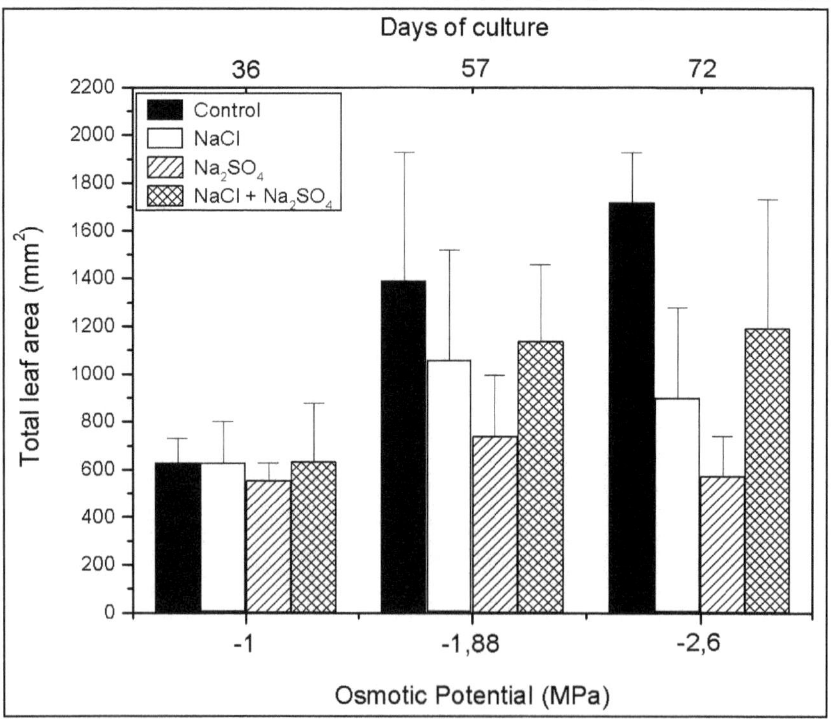

Figure 22.4: Effects of NaCl, Na_2SO_4 and their Iso-Osmotic Mixture on Relative Water Content and Osmotic Potential in Roots and Leaves of *Prosopis strombulifera* Plants

Bars represent mean ± S.E. Different letters above bars represent significant differences ($p < 0.05$) among treatments

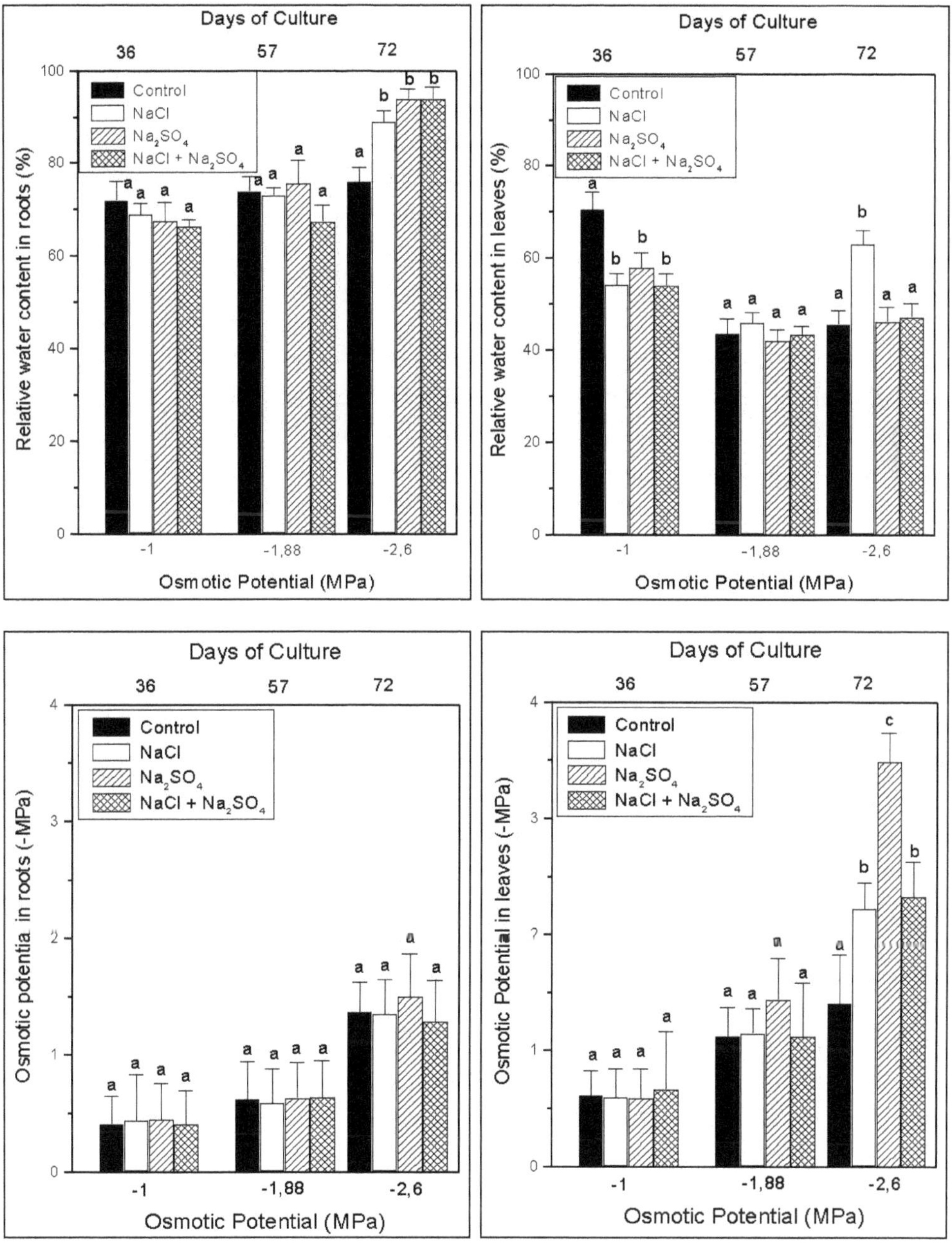

CONCLUSIONS

P. strombulifera showed a halophytic response to NaCl but strong general growth inhibition with iso-osmotic solutions of Na_2SO_4. In bisaline-treated plants, almost all morphological, physiological and biochemical parameters showed intermediate values within those measured when monosaline solutions were used, showing a partial alleviation of SO_4^{2-} toxicity and a direct influence of this anion on the overall metabolism. Our results provide evidence on the fact that some osmolytes considered as biochemical markers for salt stress tolerance such as proline, would instead play a role as stress-intensity signals. Other putative candidates as biochemical markers for salt stress tolerance in halophytes such as specific phenolic compounds (condensed tannins) and polyalcohols are currently under study in our lab.

ACKNOWLEDGEMENTS

This work was supported by CONICET, SECYT-UNRC and ANCYPT-Argentina grants to V. Luna.

REFERENCES

1. M. Foolad, L. Zhang, P. Subbiah. Genome 46: 536–545 (2003).
2. M. Ashraf, P. Harris. Plant Science 166: 3-16 (2004).
3. J. Zhu. Annu. Rev. Plant Biol. 53: 247-273 (2002).
4. V. Prabhavathi, J. Yadav, P. Kumar, M. Rajam. Mol. Breeding 9: 137–147 (2002).
5. M. Foolad, G. Lin. Journal. Am. Soc. Hort. Sci. 126. 216–220 (2001).
6. P. Kavi Kishor, S. Sangam, R. Amrutha, P. Sri Laxmi, K. Naidu, K. Rao, K. Sreenath Rao, K. Redd, P. Theriappan, N. Sreenivasulu. Current Science 88: 424-438 (2005).
7. N. Roosens, F. Bitar, K. Loenders, G. Angenon, M. Jacobs. Mol. Breed. 9: 73–80 (2002).
8. P. Hare, W. Cress. Plant Cell and Environment 21: 535-553 (1997).
9. V. Chinnusamy, A. Jagendorf, J. Zhu. Crop Sci. 45: 437–448 (2005).
10. M. Sakamoto and N. Murata. Plant Cell Environ. 25: 163-171 (2002).
11. A. Benavides-Mendoza, A. Salazar-Torres, F. Ramírez-Godina, V. Robledo-Torres., H. Ramírez-Rodríguez, R. Maiti. Terra Latinoamericana 22: 41-47 (2004).
12. M. Khan, I. Ungar, A. Showalter. Ann. Bot. 85: 225-232 (2000).
13. A. Burkart, Arbo. 57: 450-525 (1976).
14. J. Cantero, A. Cantero, J. Cisneros. Editorial Fundación Universidad Nacional de Río Cuarto, Córdoba, Argentina (1996).
15. L. Sosa. D Thesis, Universidad Nacional de Río Cuarto, Córdoba, Argentina (2005).
16. M. Reginato. D Thesis, Universidad Nacional de Río Cuarto, Córdoba, Argentina (2009).
17. H. Reinoso, L.. Sosa, L. Ramírez, V. Luna. Can. J. Bot. 82: 618-628 (2004).

18. R. Steel, J. Torrie. Segunda Edición. McGraw-Hill, México. 622 (1995).
19. M. Magné and F. Larher. Anal. Biochem. 200 115–118 (1992).
20. S. Grieve and S. Grattan. 1983. Plant Soil 70: 303-307 (1983).
21. M. Bradford.Anal. Biochem. 72:248-254 (1976).
22. D. García Vargas. D. Thesis. Universidad Autónoma de Barcelona (2006).
23. M. Muthukumarasamy, S. Gupta, R. Panneerselvam. Biol Plant 43: 67–72 (2000)
24. Y. Wang and N. Nil. Journal Hort. Sci. Biotech. 75: 623-627 (2000).
25. Parida, A. Das, B. Mittra, P. Mohanty.. Z. Naturforsch. 59: 408-414 (2004).
26. P. Armengaud, L. Thiery, N. Buhot, G. Grenier-De March, A. Savoure. Physiol. Plant. 120: 442-450 (2004).
27. S. Kumar, A. Reddy, C. Sudhakar.Plant Science 165: 1245-1251 (2003).
28. M. Khedr, A. Abbas, A. Wahid, W. Quick, G. Abogadallah. J. Exp Bot 54: 2553-2562 (2003).
29. M. Park, Z. Jekni., A. Sakamoto, R. Yuwansiri, N. Murata, T. Chen. J. Plant 40: 474-487 (2004).
30. D. Meloni, M. Gulotta, C. Martínez, M. Ovila. J. Physiol. 16: 39-46 (2004).
31. H. Reinoso, L. Sosa, M. Reginato, V. Luna. World J. of Agricul. Sci. 1: 109-119 (2005).
32. C. Prado, M. Boero,J. Gallardo, J. Gonzalez.. Bot Bull Acad Sin 41: 27–34 (2000).
33. G. Kerepesi and G. Galiba. Crop Sci. 40: 482-487 (2000).
34. A. Alamgir and M. Ali. Bangladesh J. Botany 28: 145–149 (1999).
35. H. Reinoso, L. Sosa, M. Reginato, Luna V. World J. of Agricul. Sci. 1: 109-119 (2005).
36. M. Araujo, J. Silveira, T. Almeida, I. Rocha, D. Morais, R. Viégas. Rev. Bras. Eng. Agríc. Amb. 10: 848–854 (2006).

Chapter 23

An Overview of Halophyte Plant Diversity from Salt Lake Area, Konya, Turkey

G.N. *Tug*[1], A.E. *Yaprak*[1] and M. *Vural*[2]

[1]*Ankara University Science Faculty, Department of Biology, Ankara, Turkey*

[2]*Gazi University Arts and Science Faculty, Department of Biology, Ankara, Turkey*

ABSTRACT

Tuz Lake area is one of the important natural saline habitats of Turkey. It is situated within the borders of Irano-Turanian phytogeographical region experiencing a continental climate with Mediterranean influence. It was declared as "Specially Protected Area in 2000" for being one of the hotspots of Turkey with a high ratio endemism. The area has great importance floristically, ecologically and economically. However, anthropogenic activities like intensive agricultural practices, livestock breeding and climate change are threatening this unique habitat. A part of the rich plant diversity is the plants inhabiting saline habitats. Attempts have been thus made here to evalute the saline habitat floristic diversity (halophytes) and biotic pressures faced by these.

Keywords: *Tuz Lake, Konya, Halophyte diversity, Turkey.*

INTRODUCTION

Saline and saline-alkaline areas cover over 1.5 million ha of land in Turkey (Sönmez *et al.*, 1996). These areas show great structural and formation diversity. There are lagoons alongside the Mediterranean and Aegean coasts (Erikli, İztuzu and Yumurtalık lagoons etc.); saline areas at the estuaries of the rivers Seyhan, Göksu, Kızılırmak and Gediz; saline lakes, salt marshes and saline steppes in the closed basins on terrestrial parts like Tuz Lake and saline steppes surrounding the Aras valley. Out of these Tuz Lake is drawing great attention because of its ecological and economical importance. The reason being that even though there are many coastal halophytic areas in the country, halophytic endemics are concentrated mainly around Tuz Lake (Yaprak and Tuğ, 2006). It is a closed basin which isolates it and increases the ratio of endemism in the area making it a hotspot with high number of endemic and rare taxa. The lake is also important because it provides breeding and nesting sites for huge number of migrating birds.

Many floristic, ecological and phytosociological studies have been undertaken on the saline areas of Turkey.Out of these some deal directly or indirectly with the Tuz Lake area. Notable among these are; Birand (1961), Yurdakulol (1974), Kürschner (1983) Sorger(1985), Uslu (1988), Yurdakulol and Ercoskun (199), Guvensen (1994), Yurdakulol *et al.* (1996), Karaer *et al.* (1997), Aydoðdu *et al.* (2001), Kurt *et al.* (2004), Yildirimli and Dogru-Koca (2005,2006), Guvensen *et al.* (2006), Aksoy and Hamzaoglu (2006), Karaveliogullari *et al.* (2006), Vural *et al.* (2006), Ozturk *et al.* (2008), Yaprak (2008), and Hamzaoðlu and Aksoy (2009). This paper presents an overview of the studies conducted in the Tuz Lake area, its climatic, edaphic and phytosociological characteristics as well as evaluation of the halophytic plant taxa distributed around the lake together with their life forms, phytogeographical regions, status of endemism and IUCN categories.

BRIEF DESCRIPTION OF THE STUDY AREA

Tuz Lake is the second largest lake of Turkey and second most saline lake after Dead Sea in the world; with a salinity value of 32.9 per cent (Dogdu and Sagnak, 2008).It is a shallow saline tectonic lake located in the of Central Anatolian georaphical and Irano-Turanian phytogeographical region of Turkey. It is situated at 38° 25′–39° 10′ N latitudes and 35° 5′–33° 48′ E longitudes, at an altitude of approximately 900 m above the sea level, with a depth of about 1.5 m. The lake was declared as a specially protected area by Turkish government on September 14, 2000. The plant diversity in the area changes and increases with a decrease in the salinity with an inncrease in the distance from the lake shore, but due to harsh climatic conditions and aridity, only plants adapted to the arid conditions survive. During the dry summer months when water evaporates salts in water form a thick crust over the soil surface. There are three saltworks in the area which provide the most of the salt consumed in Turkey (Kilic and Uyanik, 2001).

CLIMATE OF THE STUDY AREA

Four meteorological stations close to the study area were chosen to interpret the climatic characteristics of the region. The harsh and long arid period lasts for about 4-5 months in the area (Figure 23.1). In general it experiences semi-arid very cold and arid very cold Mediterranean bioclimates. The climatic data of the study area is presented in Table 23.1, according to the Emberger (1932).

A comparison of this data with that recorded during 1970-1992 reveals, there has been an increase of 1°C in the mean temperature of July during 1993-2005. The total precipitation has decreased with a value ranging between 1.2-11.6 mm. All these mean that the aridity of the region has become more intense during the last two decades (Ekercin and Ormeci, 2009).

Table 1: Bioclimatological synthesis of Tuz Lake area.

Station	*Altitude (m)*	*P (mm)*	*M (°C)*	*M (°C)*	*PE*	Q_2	*S*	*Precipitation Regime*	*Bioclimate*
Kulu	1010	387.0	29.8	-6.0	48.1	37.92	1.61	Sp.W.F.Su.	Semi arid very cold Mediterranean
Cihanbeyli	969	326.5	30.7	-4.9	40.1	32.06	1.3	Sp.W.F.Su.	Arid very cold Mediterranean climate
Aksaray	965	348.6	30.7	-4.0	34.5	35.05	1.12	Sp.W.F.Su.	Semi arid very cold Mediterranean
Serefli kochisar	975	377.2	30.1	-3.8	31	38.86	1.02	Sp.W.F.Su.	Semi arid very cold Mediterranean

P: Mean annual precipitation (mm), M: Maximum mean temperature of the hottest month (°C), m: Minimum mean temperature of the coldest month (°C), PE: Summer precipitation (mm), S: aridity index S=PE/M, Q_2: Pluviothermic quotient $Q_2=2000xP/M^2-m^2$, Sp: Spring, W: Winter, F: Fall, Su: Summer.

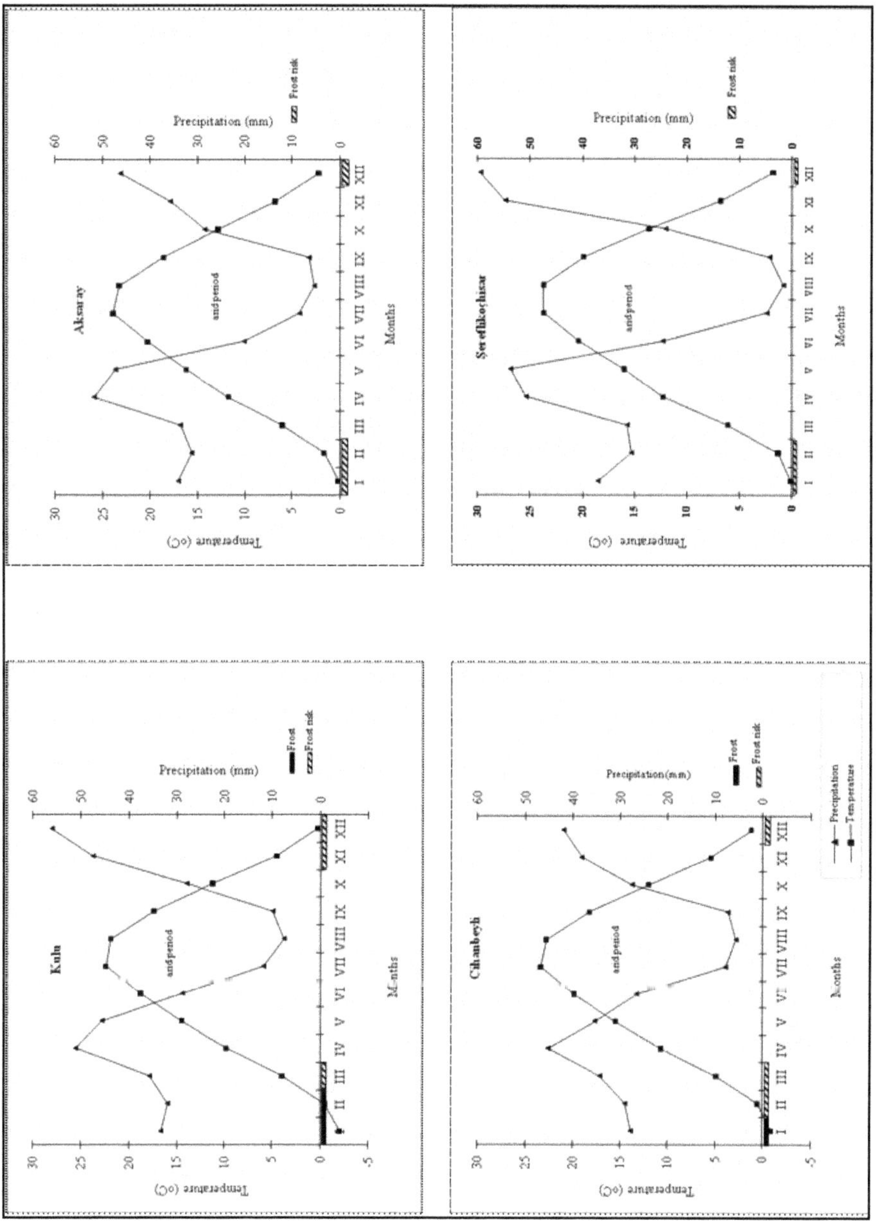

Figure 23.1: Ombrothermic Diagrams from Tuz Lake

EDAPHIC FEATURES AND THE PLANT ZONATION

Soil samples were taken from surface monthly throughout the vegetation period and prepared for the analysis. EC and the pH were determined from saturation paste of the soil using EC meter and pH meter (Ozturk *et al.*, 1997). Soils around Tuz Lake show saline and saline-alkaline characteristics. EC is greater than 4 m mohs/ds which is the indicator of salinity and pH varies between 7.5–8.7.

Three plant zones were determined around the lake area according to the changes in soil salinity.

The seasonal EC and pH values of these zones are given in Table 23.2. The number of plant taxa increases with decreasing salinity. First zone is flooded most of the year and has the highest salinity, it thus has the lowest number of plant taxa but the number increases as we move away from the lake. Only *Halocnemum strobilaceum, Salicornia perennans* and *S. freitagii* tolerate these harsh conditions and dominate this zone. *H. strobilaceum* forms patchs where salinity increases in other zones. Second zone is characterized by a decrease in salinity and increase in plant diversity, and is dominated by halophytes. The third and the outermost zone also referred to as saline steppe has xero-halophytes and non-halophytic salt resistant taxa. Although these zones have exact borders, the edaphic parameters change gradually which can be explained by the presence of biotic factors like competition (Tuğ 2006).

VEGETATION OF THE STUDY AREA

Vegetation studies around Tuz Lake started with the study of Birand (1961). Birand (1961) and Yurdakulol (1974) defined some associations from the area but they did not place them under any upper syntaxa. These studies were followed by those of Yurdakulol and Ercoskun (1990) who carried out the first study in the area following ICPN (Weber *et al.*, 2000). During last two decades the interest in the area has increased and syntaxa from the area are summarized as follows;

Class: Salicornietea Br-Bl. et Tx 1931

Order: Halostachyetalia (Grossheim) E. Topa 1938

Ass: *Halocnemtum strobilacei* (Keller) E. Topa 1938

Alliance: *Lepidio caespitosi-Limonion iconici* Aydogdu, Hamzaoglu and Kurt 2002

Ass: *Lepidio caespitosi-Limonietum iconici* Aydoðdu, Hamzaoglu and Kurt 2002

Ass: *Limonio tamaricoidis-Puccinellietum* Aydogdu, Hamzaoglu and Kurt 2002

Ass: *Sphenopodo divaricati- Halaocnemetum strobilacei* Aydogdu, Hamzaoglu and Kurt 2002

Ass: *Suaedo anatolicae-Salsoletum nitrariae* Aydoðdu, Hamzaoglu and Kurt 2002

Ass: Bupleuro gracili-Limonietum gmelini Yurdakulol ex Hamzaoglu 2009

Order: *Juncetalia maritimi* Br-Bl. 1931

Alliance: *Inulo aucheranae-Elymion salsi* Aydogdu, Hamzaoglu and Kurt 2002

Ass: *Inulo aucheranae-Elymetum* sals Aydogdu, Hamzaoglu and Kurt 2002

Table 23.2: Mean Seasonal EC and pH Values from Tuz Lake

	Spring						Summer						Fall					
	EC (aS/m)			pH			EC (aS/m)			pH			EC (aS/m)			pH		
	1	2	3	1	2	3	1	2	3	1	2	3	1	2	3	1	2	3
Kulu	79.90	14.69	32.31	7.8	8.4	8.7	84.40	5.12	21.84	8.0	8.7	8.6	76.68	19.58	38.18	7.5	8.5	8.8
Cihanbeyli	132.49	36.55	1.66	7.6	8.1	8.4	123.63	27.29	8.54	7.9	8.4	8.4	121.40	37.33	13.83	7.6	8.3	8.4
Şereflikoçhisar	131.51	37.53	5.32	7.8	8.1	8.8	106.04	30.65	6.33	7.6	8.3	8.5	153.37	94.64	11.52	7.5	8.1	8.4
Aksaray	137.79	87.11	12.27	7.6	7.8	8.6	150.11	59.32	15.61	7.5	8.3	8.4	150.11	65.27	21.87	7.3	7.7	8.0

Ass: *Eragrostio collinae-Puccinellietum anatolicae* Aydogdu, Hamzaoglu and Kurt 2002

Class: Astragalo-Brometea Quezl 1973,

Order: *Onobrychido armenae-Thymetalia leucostomi* Akman, Ketenoglu, Quezel 1985

Alliance: *Achilleo wilhelmsii-Artemision santonici* Aydogdu, Kurt, Hamzaoglu et Cansaran 2004

Ass: *Achilleo wilhelmsii-Artemisietum santonici* Aydogdu, Kurt, Hamzaoglu et Cansaran 2004

Ass. *Artemisio scopariae-Peganetum harmalae* Aydogdu, Kurt, Hamzaoglu et Cansaran 2004

THREATS IN THE TUZ LAKE AREA

Tuz Lake is a closed basin and the water sources of the lake are mainly undergroundwaters, others being rainfall, small temporary water flows drying out during summer, drainage waters from agricultural areas together with sewages. The Lake is getting smaller as a result of changes in the precipitation regimes together with increasing evaporation and overuse of undergroundwaters. The main uses being irrigation. The area is surrounded by agricultural areas which are spreading throughout the lake and for irrigation, undergroundwaters are used without any control. Irrigation is done by deep wells, so the groundwater level is getting deeper each year. It is assumed that there are 10.000 wells in the basin and most of them are not licensed and there is not any control mechanisms over them (Dengiz and Baskan, 2009; Dengiz *et al.*, 2009). The plant species, which are capable of living under these extreme conditions are very sensitive for changes in environmental conditions, so the habitat lost eventually leads to loss in biodiversity. Since it is a closed basin, it acts like a sink for all pollutants. The waters accumulating in Konya basin and causing salinization of agricultural areas are drained into Tuz, together with this drainage water domestic wastewater and the wastewater from industry also enter the lake (Ormeci and Ekercin 2005; Kilic, 2003). The pollution sources are mainly municipal drainage systems of Konya, Aksaray and Sereflikochisar which has a population of 2.5 million and the fertilizers and pesticides from agricultural areas are widely used here. An increase in the stress over the species is increaing due to agricultural activities which cause a decrease in the biodiversity and increase in water and soil pollution.

There are some other problems like a dump at Eskil. Although this area is very rich in biodiversity and has high endemism ratio, it is used as an open dump and the uncontrolled burning place for dump. Overgrazing is another problem that threatens the lake. Especially in summer, the livestock breeding activities result in the destruction of vegetation. Stubble burning in agricultural areas cause fires at natural saline habitats and because of the ignorance of the importance of these areas by public, the biodiversity loss becomes one of the major problem. The plant list of the area is given in Table 23.3 with the comments of life forms, phytogeographical regions, endemism status and IUCN categories of each species.

According to the data published by other workers and our investigations nearly 100 taxa from 62 genera and 25 families were recorded from the area. The families with highest number of genera and taxa are; *Poaceae* 6-10, *Asteraceae* 8-13, *Chenopodiaceae* 14-27, *Plumbaginaceae* 2-9, and *Caryophyllaceae* 5-5. The distribution of these taxa in chorology are; Irano-Turanian 63, Euro-Siberian 2, Mediterranean 3, Biregional 14, Polyregional 14, Common 2 and Unknown 2. The study area belongs to the Irano-Turanian phytogeographical region which explains the dominance of the species from this area. Hemicryptophytes comprise the dominant group according to the life forms and the second dominant group is therophytes. Phanerophytes

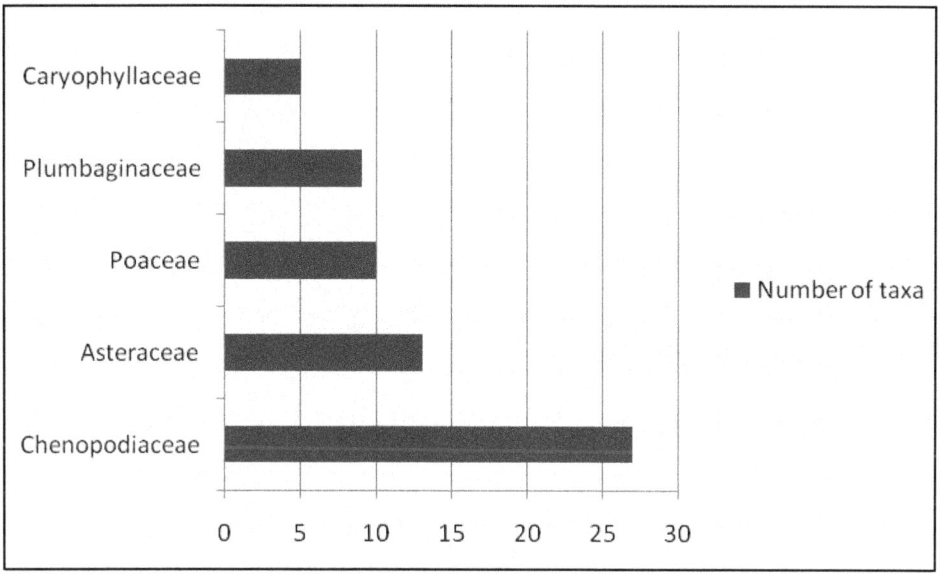

Figure 23.2: Distribution of Number of Taxa in Families

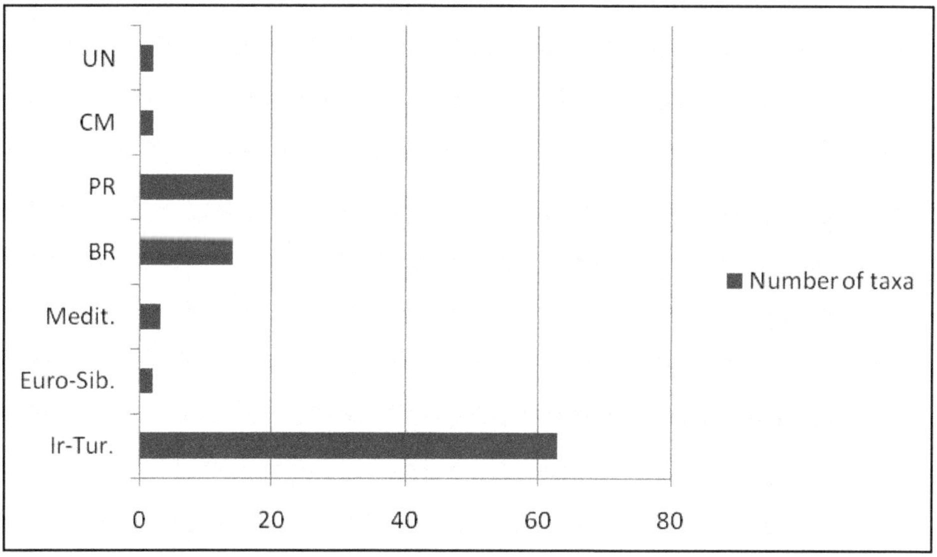

Figure 23.3: Distribution of Number of Taxa in Phytogeographical Regions

represented by only one taxon can be accepted as an indication of stress caused by both salinity and aridity. 47 of the taxa in the list are endemic to Turkey and the endemism ratio of the area is 47 per cent. The distribution of halophytic taxa in IUCN categories is as follows; VU 18, LC 5, EN 7, CR 14, NT 5 and DD 2. Halophytic taxa around Tuz Lake are mostly used as animal feed and they have also some other traditional usages as sources of drugs, nutrients or fuel for heating which are not very common.

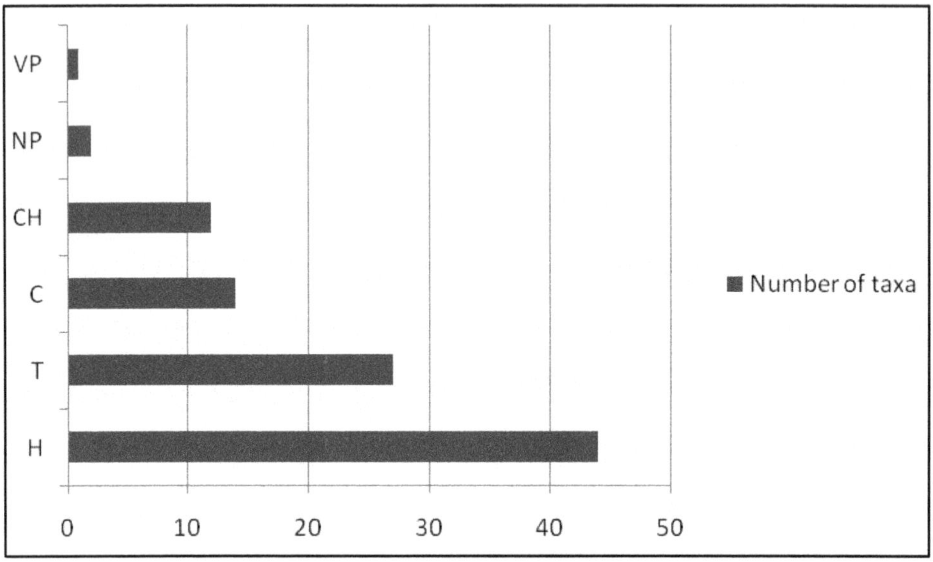

Figure 23.4: Distribution of Number of Taxa in Life Forms

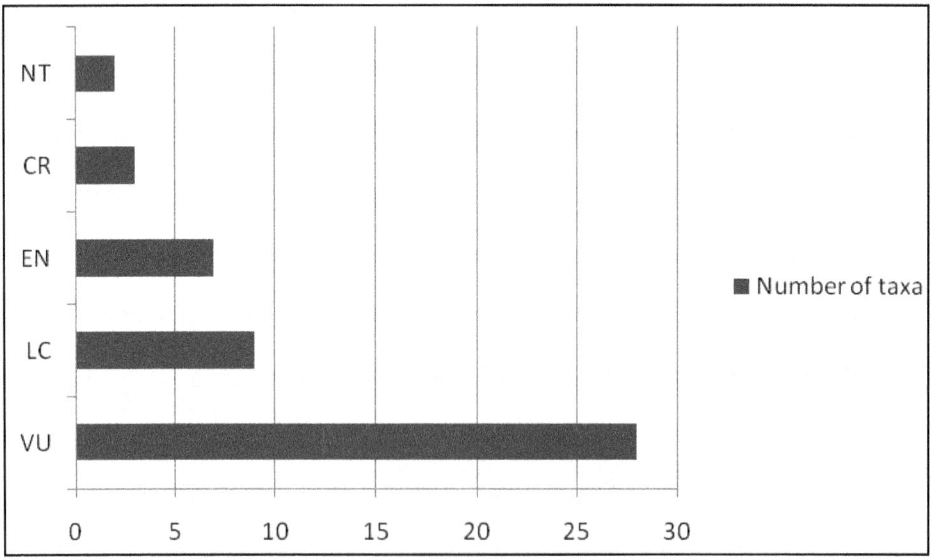

Figure 23.5: Distribution of Number of Taxa in IUCN Threatened Categories

Table 23.3: Halophytic Plants of the Tuz Lake

Family	*Taxa*	*L.F*	*C*	*E*	*IUCN*
Apiaceae	*Falcaria falcarioides* (Bornm. and Wolff) Wolff	H	IR-TUR	X	
	Ferula caspica Bieb.	H	BR		
	F. halophila Peşmen	H	IR-TUR	X	EN
Asteraceae	*Achillea sieheana* Stapf	H	IR-TUR	X	VU
	Artemisia santonicum L.	CH	EURO-SIB		
	Centaurea halophila Hub.- Mor.	T	IR-TUR	X	DD
	C. tuzgoluensis Aytac and H. Duman	H	IR-TUR	X	CR
	Cousinia birandiana Hub.-Mor.	H	IR-TUR	X	LC
	Inula aucherana DC.	C	IR-TUR		
	Onopordum davisii Rech. f.	H	IR-TUR	X	NT
	Senecio salsuginea H. Duman and Vural	H	IR-TUR	X	CR
	Taraxacum bessarabicum (Hornem.) Hand.-Mazz. subsp. *bessarabicum*	H	PR		
	T. farinosum Hausskn. and Bornm.	H	IR-TUR	X	LC
	T. mirabile Wagenitz	H	IR-TUR	X	VU
	T. tuzgoluense Yıld. and Dogru-Koca var. *tuzgoluense*	H	IR-TUR	X	VU
	T. tuzgoluense Yıld. and Dogru-Koca var. *eskilense* Yıld. and Dogru-Koca	H	IR-TUR	X	CR
Boraginaceae	*Onosma halophilum* Boiss. et Heldr.	H	IR-TUR	X	VU
Brassicaceae	*Lepidium caespitosum* Desv.	H	IR-TUR	X	NT
	L. cartilegineum (J. May) Thell. subsp. *crassifolium* (Waldst. and Kit) Thell.	H	BR		
Caryophylla-ceae	*Dianthus aydogduii Y. Menemen* and *E. Hamzaoglu*	H	IR-TUR	X	CR
	Gypsophila oblanceolata Bark.	C	IR-TUR	X	VU
	Saponaria halophila Hedge et Hub.-Mor.	H	IR-TUR	X	CR
	Silene salsuginea Hub.-Mor.	H	IR-TUR	X	EN
	Spergularia media (L.) C.Presl.		H	CM	
Chenopo-diaceae	*Atriplex laevis* C.A.Meyer	T	UN		VU
	A. tatarica L. var. *tatarica*	T	UN		
	Camphorosma monspeliaca L. subsp. *lessingii* (Litw.) Aellen		CH	BR	
	Chenopodium chenopodioides (L.) Aellen	T	PR		
	Cyathobasis fruticulosa (Bunge) Aellen	CH	IR-TUR	X	VU
	Halanthium kulpianum (Koch) Bunge	T	IR-TUR		
	Halimione verrucifera (Bieb.) Aellen		CH	BR	

Contd...

Table 23.3–Contd...

Family	*Taxa*	*L.F*	*C*	*E*	*IUCN*
	Halocnemum strobilaceum (Pall.) Bieb.	CH	PR		
	Kalidiopsis wagenitzii Aellen	CH	IR-TUR	X	CR
	Microcnemum coralloides (Loscos and Wagenitz Pardo) Font-Quer subsp. *anatolicum*	T	BR	X	VU
	Panderia pilosa Fisch. and Mey.	T	IR-TUR		
	Petrosimonia brachiata (Pall.) Bunge	T	IR-TUR		
	P. nigdeensis Aellen	T	IR-TUR	X	VU
	Salicornia freitagii Yaprak et Yurdakulol	T	IR-TUR	X	CR
	S. perennans Willd.	T	BR		
	Salsola anatolica Aellen	T	IR-TUR	X	EN
	S. crassa M. Bieb.	T	BR		
	S. inermis Forssk.	T	IR-TUR		VU
	S. laricina Pall.	CH	IR-TUR		
	S. macera Litw.	T	IR-TUR		
	S. nitraria Pall.	T	IR-TUR		
	S. stenoptera Wagenitz	T	IR-TUR	X	LC
	Suaeda altissima (L.) Pall.	T	BR		
	S. carnosissima Post	T	IR-TUR		
	S. confusa Iljin	T	BR		
	S. eltonica Iljin	T	BR		VU
	S. prostrata Pall. subsp. *anatolica* Aellen	T	IR-TUR	X	VU
Cyperaceae	*Cladium mariscus* (L.) Pohl	C	PR		
	Cyperus noeanus *Boiss.*	H	IR-TUR	X	DD
	Schoenoplectus litoralis (Schrader) Palla	C	CM		
Fabaceae	*Astragalus ovalis* Boiss. et Ball	H	IR-TUR	X	EN
	Sphaerophysa kotschyana Boiss.	H	IR-TUR	X	NT
	Tetragonolobus maritimus (L.) Roth.	H	PR		
Frankeniaceae	*Frankenia hirsuta* L.	CH	PR		
	F. pulverulenta L.	T	PR		
Guttiferae	*Hypericum salsugineum* Rabson and Hub.-Mor.	H	IR-TUR	X	EN
Iridaceae	*Gladiolus halophilus* Boiss. and Heldr.	C	IR-TUR	X	VU
Juncaceae	*Juncus heldreichianus* Marsson ex Parl. subsp. *heldreichianus*	C	MEDIT		
	J. heldreichianus Marsson ex Parl. subsp. *orientalis* Snog.	H	IR-TUR		
	J. maritimus Lam.	C	PR		

Contd...

Table 23.3–Contd...

Family	*Taxa*	*L.F*	*C*	*E*	*IUCN*
Juncaginaceae	*Triglochin palustris* L.	C	PR		
Lamiaceae	*Salvia halophila* Hedge	H	IR-TUR	X	VU
Liliaceae	*Allium sieheanum* [Hausskn. ex] Kollmann	C	IR-TUR	X	LC
	A. vuralii Kit Tan	C	IR-TUR	X	CR
	Asparagus lycaonicus P. H. Davis	H	IR-TUR	X	CR
	A. persicus Baker	H	IR-TUR		
Linaceae	*Linum seljukorum* Davis subsp. *seljukorum*	T	IR-TUR	X	EN
Orobancha-ceae	*Cistanche salsa*	VP	IR-TUR		
Plantaginaceae	*Plantago crassifolia* Forsskal	H	MEDIT		VU
Plumbagina-ceae	*Acantholimon halophilum* Bokhari *halophilum*	CH	IR-TUR	X	CR
	Limonium adilguneri Yıld. and A. Dogru-Koca	CH	IR-TUR	X	CR
	L. anatolicum Hedge	CH	IR-TUR	X	CR
	L. bellidifolium (Gouan) Dumort.	H	BR		
	L. globuliferum (Boiss. and Heldr.) O. Kuntze	H	IR-TUR		
	L. gmelinii (Willd.) O. Kuntze	H	EURO-SIB		
	L. iconicum (Boiss. and Heldr.) O. Kuntze	H	IR-TUR	X	NT
	L. lilacinum (Boiss. and Bal.) Wagenitz	H	IR-TUR	X	NT
	L. tamaricoides Bokhari	H	IR-TUR	X	EN
Poaceae	*Aeluropus littoralis* (Gouan) Parl.	C	PR		
	Crypsis aculeata (L.) Aiton	T	PR		
	Leymus cappadocicus (Boiss. Et Bal.) Melderıs	H	IR-TUR		VU
	Puccinellia anisoclada (V. Krecz.) V. Krecz. subsp. *melderisiana* Kit Tan	H	IR-TUR	X	CR
	P. bulbosa (Grossh.) Grossh.) subsp. caesarea Kit Tan	C	IR-TUR	X	CR
	P. convoluta (Hornem.) P. Fourr.	H	BR		
	P. intermedia (Schur) Janch.	H	BR		
	P. koeieana Melderıs subsp. *anatolica* Kit Tan	H	BR	X	LC
	Sphenopus divaricatus (Gouan) Reicheb.	T	PR		
	Sporobolus virginicus (L.) Kunth.	C	IR-TUR		
Primulaceae	*Glaux maritima* L.	C	PR		
Santalaceae	*Thesium compressum* Boiss. and Heldr.	T	IR-TUR		
Scrophularia-ceae	*Verbascum helianthemoides* Hub.-Mor.	H	IR-TUR	X	VU
	V. pyroliforme (Boiss. and Heldr.) O. Kuntze subsp. *pyroliforme*	H	IR-TUR	X	VU
	V. sinuatum L.	H	MEDIT		

Contd...

Table 23.3–Contd...

Family	*Taxa*	*L.F*	*C*	*E*	*IUCN*
Tamaricaceae	*Reaumuria alternifolia* (Lab.) Britten	CH	IR-TUR		
	Tamarix symrnensis Bunge	NP	PR		
Zygophyllaceae	*Nitraria schoberi* L.	NP	IR-TUR		

C: Cryptophyte, CH: Chamaephyte, H: Hemicryptohpyte, T: Therophyte, NP: Nano-Phanerophyte, VP: Vascular Parazite, IR-TUR: Irano-Turanian, EUR-SIB: Euro-Siberian, MEDIT: Mediterarnean, BR: Bi-Regional, PR: Poly-Regional, CM: Common, UN: Unknown, VU: Vulnarable, EN: Endangered, LC: Least Concern, NT: Near Threatened, CR: Critically Endangered, DD: Data Deficient.

CONCLUSIONS

Salinization of soils all over the globe is leading towards an increasing attention on the halophytic plant species flourishing on these habitats.These plants have some adaptations to live under these harsh conditions which allow them to serve as a gene source for salt resistant plant taxa which could serve as a source for potential agricultural crops Flowers (1975), Flowers *et al.* (1986) and Jennings (1968, 1976). According to Chapman (1974) species capable of tolerating 0.5 per cent or more NaCl are regarded as halophytes. Salt marshes comprise almost 10 per cent of the total terrestrial area of the world (O'Leary and Glenn 1994). The area of unnatural saline habitats is increasing especially around the agricultural areas as a result of inadequate drainage, wrong irrigation practices and changing climatic conditions, whereas the area of natural saline areas is decreasing because of some adverse biotic activities (Ozturk *et al.*, 2007). The country is thus facing a great threat of loosing its great halophytic diversity. Tuz Lake is an important area economically as being one of the most important salt sources of Turkey. Although it is a "Specially Protected Area", it is under the risk of increasing disturbance due to changing precipitation regimes, global warming, wrong irrigation practices, pollution and overgrazing. There is a great need for taking some precautionary measures for its sustainable management. Halophytic species have great importance in today's life because of the soil salinization problems. The agricultural areas loose their productivity and, their remediation or usage can be maintained by using halophytes or the genes from the halophytes. They can be a solution to the hunger or phytoremediation of agricultural areas.

REFERENCES

Akman, Y., Ketenoğlu, O. and Quezel, P., 1985. A new syntaxon from Central Anatolia. Ecologia Mediterranea. 10:111-221.

Akman, Y., Quezel, P., Barbero, M., Ketenoglu, O. and Aydogdu, M., 1991. La vegetation des steppes, pelouses ecorchees st a xerophytes epineux de l'Antitaurus dans la partie sud-ouest de l'Anatolie. Phytocoenologia, 19(3):391-428.

Aksoy, A., Hamzaoğlu, E., 2006. Vegetation zones in the salty marshes of Cenral Anatolia and natural borders of agricultural usage (Turkey). In. M. Ozturk, Y.

Waisel, M.A.Khan and G. Gork (eds). Biosaline Agriculture and Salinity Tolerance in Plants. Birkhauser Verlag/Switzerland. pp: 109-116.

Aydogdu, M., Hamzaoglu E. and Kurt, L. 2002. New halophytic syntaxa from Central Anatolia (Turkey). Israel Journal of Plant Sciences, 50: 313-323.

Aydogdu, M., Kurt, L., Hamzaoglu E., Ketenoglu, O., Cansaran, A. 2004. Phytosociological studies on salty steppe communities of the Central Anatolia, Turkey. Israel J. Pl. Sci., 52(1): 71-79.

Birand, H. 1961. Orta Anadolu bozkirinda vejetasyon incelemelerinin ilk sonuclari: I. Tuz Golu corakcil bitki birlikleri, Tarim Bakanligi Ilmi Rapor ve Arastirma Serisi, Toprak Su Umum Mudurlugu Nesriyatý Sayý: 103, Ankara.

Braun-Blanquet, J., 1931. Aperçu des groupements vegetaux du Bas Languedoc. S.I.G.M.A. Communication no: 9.

Chapman, V.J., 1974. Salt marshes and salt deserts of the world. Ecology of Halophytes Ed. By R.J. Reimold and W.H. Queen. Academic Press. Inc. New York and London. A Subsidiary of Harcourt Brace Jovanovich Publishers. pp: 3-19.

Dengiz, O., Ozcan, H., Koksal, E.S., Baskan, O., Kosker, Y., 2009. Sustainable natural resource management and environmental assessment in the Salt Lake (Tuz Golu) Specially Protected Area. Environ. Monit. Assess. (online first) doi: 10.1007/s10661-009-0749-4

Dengiz, O., Baskan O., 2009. Land quality assessment and sustainable land use in Salt Lake (Tuz Golu) specially protected area. Environ. Monit. Assess. 148:233-243.

Dogu, M.S., Sagnak, C., 2008. Climate change, drought and over pumping impacts on grounwaters: two examples from Turkey. Synthesis of Proceedings, Balwois 2008-Conference on water observation and information system for decision support. Ohrid, Republic of Macedonia. 27-31 May.

Ekercin, S., Ormeci, C., 2009. Evaluating climate change effects on water and salt resources in Salt Lake, Turkey using multitemporal SPOT imagery. Environ. Monit. Assess. Doi: 10.1007/s10661-099-0840-x

Ekim, T., Koyuncu, M., Vural, M., Duman, H., Aytac, Z. and Adiguzel, N. 2000. Red Data book of Turkish Plants (*Pteridophyta* and *Spermatophyta*). Turkish Association for the Conservation of Nature, Van Centennial Univ. Press, Ankara (in Turkish).

Flowers, T.J. 1986. Halophytes. *Quarterly Review of Biology*, 61, 313–337.

Flowers, T.J. 1975. Halophytes.–In: Baker, D.A. and Hall, J.L. (eds), Ion transport in plant cells and tissues. Pp. 309-334. North-Holland Publishing Co., Amsterdam.

Flowers, T.J.,Hajibagheri, M.A., Clipson, N.J.W. 1986. Halophytes.-Quart. Rev. Biol., 61: 313-337.

Freitag, H., Vural, M., Adiguzel, N. 1999. A remarkable new *Salsola* and some new records of *Chenopodiaceae* from Central Anatolia, Turkey. Willdenowia, 29(1/2): 123-139.

Freitag, H. and Duman, H. 2000. An unexpected new taxon of *Salsola* (*Chenopodiaceae*) from Turkey. Edinburgh J. Bot., 57(3): 339-348.

Guvensen, A., Gork, G., Ozturk, M., 2006. An overview of the halophytes in Turkey. In Sabkha Ecosystems. Volume II: West and Central Asia. Khan, M.A. (Ed). Springer. pp:9-30.

Hamzaoglu, E., Aksoy, A., 2009. Phyto-sociological studies on the halophytic communities of Cenral Anatolia. Ekoloji, 71:1-14.

Hamzaoglu, E., Duran, A. and Akhani, H. 2005. A New Genus Record for the Flora of Turkey: *Tetradiclis* Stev. ex M. Bieb. (*Zygophyllaceae*). Turk. J. Bot., 29: 403-407.

Jennings, D.H. 1968. Halophytes, succulence and sodium in plants-a unified theory. New Phytol., 67(4): 889-911.

Jennings, D.H. 1976. The effects of sodium chloride on higher plants. Biol. Rev. Cambridge Philos. Soc., 51: 453-486.

Karaveliogullari, F.A., Vural, M., Polat, H., 2006. Two new taxa from Central Anatolia, Turkey. Israel Journal of Plant Sciences, Volume 54(2): 105–111.

Kilic, A.M., 2003. Tuz Golu'nde Ortaya Cikan Kirlenme ve Kimyasal Acidan Gol Suyunun Incelenmesi. *V. Ulusal Cevre Muhendisligi Kongresi, Bildiriler Kitabý,* 01-04 Ekim, 584-592, Ankara.

Kilic, A.M., Uyanık, E., 2001. Tuz Golu'nde olusan kirlenmenin gol uzerindeki etkilerinin arastirilmasi. 4. Endustriyel Hammaddeler Sempozyumu Bildiriler Kitabı, Izmir, pp: 135-149.

Kurt, L., Hamzaoglu, E., Tug, G.N., Evren, H. 2004. A study on the relationships of salinity and endemism ratios in the surrounding of Salt Lake (Inner Anatolia, Turkey), F.U. Fen ve Muhendislik Bilimleri Dergisi, 16(4):745-756.

Kurschner,H. 1983. Vegetationsanalytische Untersuchungen an Halophytenfluren Zentralanatoliens (Türkei). Tübinger Atlas des Vorderen Orients: Beih. Reihe A 11 (TAVO). Wiesbaden: Reichert,. 89 S.

O'Leary, J.W. and E.P. Glenn. 1994. Global distribution and potential for halophytes, pp. 7-17 in V.R. Squires and A. Ayoub (eds.), Halophytes as a resource for livestock and rehabilitation of degraded lands. Kluwer.

Ormeci, C., Ekercin, S., 2005. Uzaktan algilama teknigi ile Tuz Golunde su kalitesi degisim analizi. TMMOB Harita ve Kadastro Muhendisleri Odasý 10. Turkiye Harita Bilimsel ve Teknik Kurultayý. Ankara.

Ozturk, M., Pirdal, M., Ozdemir, F., 1997. Plant Ecology Applications. Ege Univ., Science Fac.157, 129 pp, Univ. Press, Bornova-Izmir.

Ozturk, M., Guvensen, A., Sakcalı, S., Gork, G., 2008. Halophyte plant diversity in Irano-Turanian phytogeographical region of Turkey. In Biosaline Agriculture and High Salinity Tolerance Chedly A., Ozturk, M., Ashraf, M. and Grignon, C (Eds). Birfhauser-Switzerland. pp:141-155.

Ozturk,M., Sakcali,S., Bahadir,H., Gork,G., Beyazgul,M. 2007. Salinity-Alkalinity and Pollution Problems in the Gediz Basin-Turkiye. International Conference on Environment: Survival and Sustainability 19-24 February, Near East University, Nicosia-Northern Cyprus Volume 7: MT-11: Environmental Science and Technology, pp: 2935-2941.

Quezel, P., 1973. Contribution a l'etude phyto-sociologique du massif du Taurus. Phytocoenologia, 1:131-222.

Sonmez, B., Agar, A., Bahceci, I., Mavi, A., Yarpuzlu, A., 1996. Turkiye Coraklýk Islah Rehberi, T.C. Basbakanlik Koy Hizmetleri Genel Mudurlugu A.P.K. Dairesi Baskanligi, Toprak ve Su Kaynaklarini Arastirma Sube Mudurlugu, Yayýn no: 93, Rehber No:12.

Topa, E., 1939. La vegetation des halophytes du nord de la Romanie en connexion avec calle du retse du pays. S.I.G.M.A. Communication no:7.

Tug, G. N., 2006. Factors effecting the zonation of halophytic vegetation around Tuz Lake and its environs. Ankara University Graduate School of Natural and Applied Science (PhD. Thesis).

Vural, M., Duman, H., Aytac, Z., Adigüzel, N., 2006. *Saponaria karapinarensis, Senecio salsuginea* and *Centaurea tuzgoluensis,* three new species from central Anatolia, Turkey. Belgian Journal of Botany, Volume 139, Number 2, 252-260.

Wagenitz, G. 1959. 14. Neue und bemerkenswerte Chenopodiaceen Inneranatoliens.– Ber. Deutsch. Bot. Ges., 72(3): 151-158.

Weber, H.E., Moravec, J. and Theurillat, J.-P., 2000. International Code of Phytosociological Nomenclature. 3rd edition. Journal of Vegetation Science 1: 739-768.

Yaprak, A.E., 2008. A taxonomical revision of *Salicornia*, *Sarcocornia* and *Arthrocnemum* genera (Chenopodiaceae) in Turkey. Ankara University Graduate School of Natural and Applied Science (PhD. Thesis).

Yaprak, A.E. and Yurdakulol, E., 2008. *Salicornia freitagii* (Chenopodiaceae), a new species from Turkey. Annales Botanici Fennici. 45:207-211.

Yildirimli, S. and Dogru-Koca, A. 2005. A new species and variety of *Taraxacum, T.tuzgoluensis* and *T.tuzgoluensis* var. *eskilensis* (Asteraceae) from Tuzgolu, Turkey. Ot Sistematik Botanik Dergisi. 12(2): 1-12

Yildirimli, S. and Dogru-Koca, A. 2006. A new species *Limonium adilgunerii* Yildirimli and Dogru-Koca (Plumbaginaceae) from edge of Tuzgolu, Inner Anatolia Turkey. Ot Sistematik Botanik Dergisi. 13(1): 11-16.

Yurdakulol, E. and Ercoskun, T. 1990. Ecological and Syntaxonomical studies on the arid land vegetation in Central Anatolia. Doga- TR. J. of Botany, 14:109-123.

Yurdakulol, E.,1974. Investigations on the plant sociology of arid marsh vegetation in the Konya plain. Bitki, 1(2):257-277.

Chapter 24

Biodiversity and Threats of Natural Ecosystems in Iran: A Case Study from Golestan National Park, NE Iran*

H. Akhani and R. Khoshravesh

Department of Plant Sciences, School of Biology, College of Science, University of Tehran, P.O. Box 14155-6455, Tehran, Iran

ABSTRACT

Iran provides a diversity center for evolving unique biological diversity and a corridor for migration and an exchange route of many plant and animal species of various biogeographical origins. Golestan National Park (GNP) is the most rich area of biodiversity in northeastern parts of Iran, the first Iranian National Park and a UNESCO designated reserve, which was established as Almeh and Ishaki Protected area in 1957. The mesic parts of the Park support deciduous lowland and montane forests with a rich phanerophytic vegetation and provide habitats for the growth of several Arcto-Tertiary relict plants. The mammals found in the area are red deer, roe deer, wild boar, brown bear and leopard. The steep rocky outcrops of the South-facing scrubs provide a savanna-like vegetation dominated by a dense C_4 grassland. These habitats are ideal living habitats for the Persian ibex. The undulating hills and mountains of the eastern, northern and central parts of the area are covered by *Juniperus* woodlands and *Artemisia Stipa* steppe rich in many Irano-Turanian elements. These habitats prefer the populations of Urial sheep. The flat plains of the Dashte Mirzabaylu with a semi-desert

---

* This paper is dedicated to the former Director of Golestan National Park Mr. Yahya Shahkuhmahali, a keen nature conservationist who devoted his life for the protection of this Park in Golestan province and surrounding areas. He was killed during a confrontation with poachers when he was doing his conservation duties in Saluk Protected area on 10.7.2005.

vegetation of *Artemisia* steppe and halophytic vegetation provide a habitat for the goitered gazelle. The vascular plants of the area have been well documented. So far, a total of 1364 vascular plants belonging to 570 genera and 116 families have been recorded from the area which corresponds to ca. 19 per cent of all known species in Iran. Several endemic and subendemic species are known from the area. The Park suffers from a lack of management programme and several destructive activities like overgrazing, poaching and fire endangering unique ecosystems of the area. As a consequence of severe habitat degradation and deforestation around and inside the Park, several floods took place since 2001 resulting in the devastation of large parts of the riverine and lowland forests around the Madrasu river. The underway reconstruction and widening of the main so called "Asian road" which passes through the Central parts of the park, is a critical threat to the Park. These developments stress the fact that national and international organizations and NGOs should join hands to rescue this unique biosphere reserve in SW Asia.

Keywords: *Golestan National Park, Biodiversity, Endemism, Degradation, Conservation.*

INTRODUCTION

Iran is the country of extremes (Zohary, 1973). The great diversity of natural resources are certainly among the main reasons for the flourishing of civilizations in Iran (Firuz, 1974). Iran covers an area of 1.6 million km^2 with altitude ranging from 20 meters below sea level along the Caspian Sea to 5776 meter at the peak of Demavand. It lies between 25° to 39° 45′northern latitude and from 44° to 63°18′eastern longitude experiencing extremely different climatic conditions. We find the driest deserts on earth such as Lut desert as well as areas getting 2000 mm annual rainfall like SW Caspian forests.

The country is a center of diversity for evolving unique biological diversity and a corridor for migration and exchange route for many plant and animal species of various biogeographical origins. The highlands of the country serve as a corridor for montane floral exchange and migration route between Iran, Caucasus and Anatolia, Central Asia, Afghanistan and Himalayas (Noroozi *et al.*, 2008). The interior and southern low plains provide a route for floral exchange and migration path for arid flora between Central Asian deserts towards the north, and Arabian/African and Indian deserts towards the South and Southeast. The intricate geological and climatic history is not only reflected by active florogeny in many plant groups, but also by geographical position between North and South. In the North flora of glacial fluctuation is protected and in the South we come across extreme aridization. The occurrence of many relict species and interesting plant communities are evidences of such an antiquity. In the South Caspian forests many Arcto-Tertiary relict species survive as endemics and sub-endemics (Leestmans,2005; Frey *et al.*, 1999; Tralau, 1963).

Out of the total surface of Iran ca. 7 per cent is designated as protected area including 19 National Parks, 92 Protected Areas, 33 Wild Life Refugees and 16 National Natural Monuments. Golestan National Park is a UNSECO designated biosphere reserve. The protection history of the area goes 53 year back when this area was designated by Iran Game Council as "Almeh and Ishaki Protected Area" in 1957. In 1962 it was named as "Almeh Protected Area", in 1963 "Mohammad Reza Shah Park", in 1964 "Mohammad Reza Shah Wildlife Park" and in 1974 "Mohammad Reza Shah National Park", as the first National Park in Iran. In 1978 the Qorkhod Protected Area -located in the east of the Park with 34,000 hectares- was included into the Park, so the area of the Park increased to 126,000 hectares. After the Iranian Revolution in 1979 its name was changed to Golestan National Park with an exclusion of Qorkhod Protected Area from the Park and a reduction of the area to 87,242 hectares (Akhani, 1998). Golestan, the land of flowers, was the local name of the area for many years, clearly because of extremely diverse flowering plants growing in the area.

Golestan National Park mirrors the biodiversity and landscape of Iran. It is situated in NE Iran, E Golestan, NW of Northern Khorassan "Khorassane Shomali"and N of Semnan provinces, a 35 km part of the extremely busy, so-called "Asian road"; connecting north and central Iran with Northeast Iran; passes through the forest zone and along the southern border of the Park. The area is located between two completely different climates, namely the temperate sub-humid climate in the West and the cold-arid climate in the east and south. Often the climate changes drastically within a few hundred meters moving from W to E, N to S or vice versa, along a small topographic gradient. Mean annual precipitation ranges between 140 mm in Dasht-e Kalpush, located in the south of the Park, and 870 mm in Minudasht. Based on vegetation evidences the northern slopes of Alu-Baq and area above the Qorqon rocks receives more than 1000 mm/year of rainfall. The Park has a high topographic and complicated geomorphology with altitude varying between 400 in the west of the Park to 2411 m at the mountain peak of Divar Kaji. A detailed description of the physical structure, climate and geology of the area has been given by Akhani (1998). This paper aims to review and update the existing knowledge of the biodiversity of the area with highlighting the critical conservation status and threats to the ecosystems of the Park.

VEGETATION

Vegetation of the Park is extremely diverse and complicated. The western and central parts are composed of closed deciduous forests; shown in green in the satellite picture of the Park. We may still find virgin forests of the South Caspian zone in some parts of the Park located in Qorghon forests and around Sulukli Lake. The forests are dominated by broad-leaved deciduous trees and shrubs such as *Quercus castaneifolia, Carpinus betulus, Parrotia persica, Zelkova carpinifolia, Tilia caucasica, Fraxinus excelsior, Ulmus glabra, Diopyros lotus, Acer velutinum, Populus caspica, Mespilus germanica, Crataegus microphylla* and *Prunus divaricata*. The closed forests are interrupted by rocky outcrops and deep cliffs whose vegetation differs from the lowlands to the highlands. In most parts *Carpinus orientalis* is present as sparse or very open bushland. The lowland rocky outcrops are covered by a savanna-like vegetation with rich C_4 grasslands, unique for a temperate forest (Akhani and Ziegler, 2002). The occurrence of this relict C_4 savanna-like grassland is due to high summer temperatures and frequent summer rains in the area. Towards the highlands the montane forestis replaced by open *Quercus macranthera* forest, *Juniperus excelsa* woodland and in the subalpine peaks by carpet-like patches of *Juniperus sabina* and *J. communis*. In the eastward side the forest zone in the Center of the Park (Tunnel area), there is a transition zone of c. 3 km where *Paliurus spina-christi* is dominant. It is intermixed with a number of drought-resistance shrubs which eventually change into *Stipa* and *Artemisia* steppe in the Southern border of the Park. In the valleys of the eastern and North-Central parts of the Park, still the accumulation of clouds and effect of microclimatic conditions result in the formation of gallery forests dominated by *Acer monspessulanum* ssp. *turcomanicum*. In altitudes between 1300-1600 m in NW and SW parts of the Park the forest zone is followed by a belt of open scrub, with many shrubs and small trees. These thickets are very rich in woody and herbal taxa. The common shrubs include: *Crataegus pentagyna*, *C. azarolus* ssp. *pontica*, *Cerasus microcarpa*, *Cerasus pseudoprostrata*, *Cotoneaster ovatus*, *C. nummularioides*, *Rosa* spp., *Malus orientalis*, *Pyrus boissieriana*, *Prunus divaricata*, *Rhamnus pallasii*, *Colutea buhsei*, *Berberis* spp., *Lonicera floribunda*, *L. bracteolaris*, *L. iberica* and *Cornus australis*. The herbal vegetation often covers up to 100 per cent of the ground which consists of many grasses belonging to the genera *Stipa*, *Hordeum*, *Festuca* and *Dactylis*.

Southern, Eastern and Northern flat plains bordering the Park are driest parts, covered by extensive *Artemisia* steppe. The Mirzabaylu plain is situated at an altitude of 1200-1250 m. This is the habitat of a small population of Gazelle. The occurrence of some salt springs supply salt water to the lowlands of this plain which thus abounds in halophytic communities. Major halophytic plant communities have species like *Anabasis aphylla*, *Caroxylon dendroides* (=*Salsola dendroides*), *Kaviria tomentosa* (=*S. tomentosa*), *Suaeda physophora*, *S. microsperma* and on moist parts *Phragmites australis*. There is a great diversity among the Chenopod species growing in this part with diverse photosynthetic behaviour (Akhani and Ghasemkhani, 2007). Almeh valley lies in the northern sector of this plain, where an isolated Saxaul community (*Haloxylon ammodendron*) is found. This is a biological rarity where a true desert shrub accompanies trees like *Juniperus excelsa* and *Acer monspessulanum*. This valley leads to Almeh area (1750-1850 m), a well known locality for many Middle Eastern botanists, and a hot spot of biodiversity in the Park. The area with several shallow valleys is covered by *Acer monspessulanum* and slopes by *Juniperus excelsa*. The flat plains with deeper soils are covered by mixed grasslands and some thorn-cushions like *Onobrychis cornuta*.

On the subalpine summits of Divar-Kaji, Soltan Hoopi and Gerieh Sar (belonging to the Aq-Mazar range), with altitudes from 2000 2400 m, *Juniperus sabina* and *J. communis* form 'carpet-like' prostrate formations. The coverage is up to 100 percnet. *Acer monspessulanum* ssp. *turcomanicum*, *Ephedra major* ssp. *procera*, *Rhamnus cathartica*, *Lonicera iberica* and *Ribes melananthum* belong to the woody species associated with a number of alpine and subalpine chamaephytes and hemicryptophytes. The subalpine meadows around snow bed depressions and near mountain springs are dominated by *Elymus elongatiforme*.

A considerable part of the former cultivated areas, demolished and/or evacuated villages before the establishment of the Park in the forest zone are covered by dense communities of the invasive fern *Pteridium aquilinum*.

In 1994 some local people and park rangers informed that there is a lake in the park. It was much surprising because by its permanent water and extensive aquatic vegetation was neither shown in any map or documented anywhere. The area inhabited several isolated species in the park which were new records for Iran (Akhani and Scholz, 1998, Akhani, 1998-1999, Joharchi and Akhani, 2006). The isolation and protection of this water body is very important because the native flora of most of the freshwater wetlands of the South Caspian forests are threatened due to expansion of invasive fern *Azolla filiculoides*, once introduced after the revolution for enhancing nitrogen fixation in rice fields (Khoshravesh *et al.*, 2009).

FAUNA

The main criterion for considering the Golestan National Park as a protected area in the earlier times of its establishment was the very rich and unique wild life of particular attraction for hunters. The presence of both lowland and montane virgin forests, Juniper woodlands, montane steppes and *Artemisia* steppes in flat plains associated with a complicated topography and rich water resources provide diverse habitats for many large vertebrates. According to the current information (Kiabi *et al.*, 1994; Madjnoonian *et al.*, 1999) a total of 69 species of mammals, belonging to 21 families and 50 genera as well as 150 species of birds belonging to 42 families and 89 genera have been reported from the Park. The last individual of the extinct Caspian Tiger (*Panthera tigris*) was shot in the Golestan National Park in 1959 (Vuosalo, 1976). There are reports of living three individuals of the endangered Cheetah (*Acinonyx jubatus*) in the northeast of the area back ca. three decades ago (Kiabi, 1978), but these have not been seen anymore in the area. The large mammals of the Park include Urial sheep (*Ovis ammon*), maral or red deer (*Cervus elaphus*), roe deer (*Capreolus capreolus*), goitered gazelle (*Gazella subgutturosa*), Persian ibex (*Capra aegagrus*), wild boar (*Sus scrofa*), brown bear (*Ursus arctos*), leopard (*Panthera pardus*), fox (*Vulpes vulpes*), hare (*Lepus capensis)*, jackals (*Canis aureus*), wild cats (*Felis catus*), and wolf (*Canis lupus*). Among the smaller mammals following species are worth mentioning Hedgehog (*Erinaceus europaeus*), shrew (*Sorex minutus*), and bat (*Myotis blythii*). The wild sheep, Persian ibex, gazelle and maral are the most attractive animals for hunters. Unfortunately the populations of these animals are drastically decreasing since 1979. Today we find populations of Persian ibex on steep rocky slopes from Adam Chaqran to Tunnel, Chonde Abbas, Soolegerd and Qorqon (Kiabi, 1975). Urial sheep has the largest population size of the large mammals in the area. This animal prefers gentle or steep slopes dominated by *Juniperus excelsa,* and can be found in the Almeh valley, Soolegerd and the northern parts of the Park. The small population of gazelles in the *Artemisia* steppe on Mirza-Baylu plain and some individuals in the northeast of the Park (around Soolegerd) are of great interest, because they can easily be observed by the tourists.

Among the most interesting bird species we find pheasant (*Phasianus colchicus*) and chukar (*Alectoris chukar*) in the Park with good populations; former living in forest openings and scrubs of the Central and West and later on the undulating slopes of the steppe zone of the Park in the South, North and Central parts. So far five globally threatened bird species including *Aegypius monachus*, *Aguila heliaca*, *Falcon aumanni* and *Crex crex*, six regionally threatened or declining species including *Pernis apivorus*, *Gypaetus barbatus*, *Neophron percanopterus*, *Gyps fulvus*, *Aquila pomarina* and *Falco cherug*, nine species with a limited distribution in the Middle East and 16 rare species of the European bird fauna have been recorded from the Park (Madjnoonian *et al.*,1999). One of the rare species of the Park is *Lanius schach* which is found in Soolegerd area in the NE of the Park. Eight fishes have been reported from the rivers of the Park (Abdoli, 1993), however, no information is available about these at present after the heavy floods which devastated their habitats since 2001.

Little is known about the invertebrates and in particular the insects of the Park. According to the collections made during zoological excursions between 1997-2001 by the Austrian zoologists several groups of Park fauna were investigated (Gutleb and Wieser, 2002). Table 24.1 presents a statistical analysis of the results summarized from different workers. The best represented families of butterflies (Lepidoptera) are Nuctuideae with 141 species followed by Gelechiidae, Tortricidae, Pyralidae and Geometridae. Other groups of invertebrates are reported by much fewer number of collected species including 54 species of Heteroptera, 28 Arachnids, 21 species of Coleoptera, 10 species of Trichoptera, nine species of Neuroptera and *Potamon ibericum* (Potanidae) the only species of Crustacea.

Table 24.1: List of Invertebrates Reported from Golestan National Park Based on the Austrian Zoological Excursion in 1997 and 2001 (Gutleb and Wieser, 2002). Collection sites: (1) Tange Gol, (2) Dasht, (3) Sharleq, (4) Almeh, (5) Almeh vally, (6) Mirzabaylu and (7) Ghez Ghaleh

Name of Group	*No. Species*	*No. of Genera*	*No of Families*	*Collection Sites*	*Authors and Pages*
Lepidoptera	502	264	37	1, 2, 4, 5, 6	Wieser, Huemer and Stangelmair, pp. 52-81
Trichoptera	10	5	-	1	Malicky, p. 81
Coleoptera	21	18	4	1, 3, 5	Plössl, pp 81-83
Heteroptera	54	49	16	1, 2, 4	Heiss, pp. 87-97
Neuroptera	9	8	5	1, 2, 3, 4, 5, 6	Hölzel and Rauschm pp. 97-98.
Arachnida	28	26	17	1, 2, 4, 6	Komposch, pp. 98-103
Crustacea	3	1	1	2, 4, 7	Petutschnig pp. 104-107

BRYOPHYTES AND FUNGI

Bryophyte flora of the Park was studied by Frey and Kürschner (1977) and Kürschner *et al.* (2000). There reports show that 121 species belonging to 64 genera and 35 families are distributed in the Park area,which corresponds with 28 per cent of all reported species from Iran out of a total number of 437 species (Akhani and Kürschner, 2004). Many species (31 taxa) have been reported as a first record from Iran from the territory of the Park. The Pottiaceae with 21 species and Brachytheciaceae with 16 species are the most diverse bryophyte groups in the Park. The first family favors more dry habitats and the second occurs as epiphyte, or grows on rotten wood and wet rocky places. The epiphytes are well represented in the forest zone of the Park including pleurocarp mosses like *Amblystegium serpens, Anomodon attenuatus, Homalothecium sericeum, Leucodon immersus, Neckera besseri, N. complanata, Palamocladium euchloron* and *Pseudoleskeella laxiramea*, and the leafy liverworts (Jungermanniopsida) like *Porella platyphylla, Radulla complanata, Frullania dilatata* and *Metzgeria furcata*. The soil living species are also very rich in shady forests such as *Eurhynchium flotowianum, Eurhynchium pulchellum, Plagiomnium rostratum, Fissidens bryoides, Plagiothecium platyphyllum, Eurhynchium praelongum* in the mesic part and *Bryum capillare, Tortula mucronifolia* and *Syntrichia caninervis* var. *pseudodesertorum* in the xeric part of the Park. The rocky habitats both in the mesic and xeric zones of the GNP are very rich in bryophytes such as *Frullania dilatata, Grimmia pulvinata, Hypnum cupressiforme, Schistidium apocarpum, Tortula muralis, Porella platyphylla, Thamnium alopecurum, Bryum argenteum, Trichostomum crispulum,* in forest zone and *Grimmia orbicularis, Orthotrichum cupulatum* and *Syntrichia caninervis* var. *pseudodesertorum* in xeric parts of the Park. Two aquatic species including *Ricciocarpos natans* and *Drepanocladus aduncus* were found from the isolated lake Sulukli located in the montane forests and the northwestern parts of the Park (Frey and Kürschner,1977; Kürschner *et al.*, 2000).

From bryogeographical point of view nearly 95 per cent of the taxa belong to the (sub) cosmopolitan and Northern elements. Examples of such species are *Marchantia paleacea, Bryum argenteum, B. capillare* and *Tortula muralis*. Three species, *Leucodon immersus, Palamocladium euchloron*, and *Pseudoleskeella laxiramea* are endemic for the Near and Middle East. The last is an exclusively South Caspian endemic (Figures 1-3 in Kürschner *et al.*, 2000).

A list of the known fungi of the Park has been reported by Abbasi (2005) based on his own studies on rust fungi (Uredinales) and other fungal groups reported in other published works. The rust fungi with 79 taxa from 7 genera and 4 (-5) families are highlighted by the enormous diversity of genus *Puccinia* which has 54 taxa followed by *Uromyces* with 13 taxa. Among other fungal groups the Homobasidiomycetes with 205 species are the most represented group in the area (Hallenberg, 1981). Nine species of Ustilaginomycetes (Basidiomycota), 20 species of the Erysiphales, nine species of Pyrenomycetes, one species of Discomycetes and one species of Loculoascomycetes (Ascomycota) and three anamorphic fungi are known from the Park. Within these groups, the following fungal genera are represented by five or more species: *Erysiphe* (14), *Agaricus* (5), *Peniophora* (6), *Phellinus* (8 species), *Trametes* (7).

The Park is very rich in Lichen diversity. A first list of 49 species belonging to 30 genera was reported by Sohrabi (in Akhani, 2005).

FLORA, PHYTOGEOGRAPHY AND ENDEMIC SPECIES

An updated list of known families and the respective number of species and genera are given in Table 24.2. In most cases the families are accepted according to new classification of Angioserpm Phylogeny Group (APG II, 2003). The circumscription and number of genera in Chenopodiaceae is increased based on recent molecular phylogeny studies in Salsoleae (Akhani *et al.*, 2007). The number of vascular species known in the Park is 1364 belonging to 570 genera and 116 families. This corresponds to ca. 1/5 of the total flora of the country, considering the surface area of the park which is only 0.06 per cent of Iran. The well represented genera in the Park are *Astragalus* with 45 species followed by *Allium* (10), *Veronica* (17), *Silene* (16), *Vicia* (16), *Cousinia* (16), *Centaurea* (15) and *Carex* (15). In spite of extensive floristic studies in the area, still the botanical collections keep on additing to the list of the flora. Many rare species of Iran are only known from the Park's boundaries. Many floristically interesting species have been found in the Park, *e.g.* a wild population of *Narcissus tazetta* and a population of *Allium vavilovii* as the closest relative to the cultivated onion (Akhani, 1999; Fritsch *et al.*, 2001). Some species show highly interesting disjunction such as *Elatine hydropiper*, a very rare aquatic plant (Joharchi and Akhani, 2006), *Linosyris vulgaris*, *Carex pseudocyperus* (Akhani, 1999), *Hordelymus europaeus* and *Heteropogon contortus* (Akhani and Scholz, 1998).

Table 24.2: List of Plant Families, Number of Species and Genera in Golestan National Park (Updated after Akhani 1998). New families added since 1998 are in bold and those with an asterisk show that the number of species or their status and name has been changed. The circumscription of families with a few exceptions is according to Angiosperm Phylogeny Group classification (APG II 2003).

Major Groups and Families	*Genera*	*Species*
Pteridophytes		
Aspleniaceae	3	5
Dryopteridaceae *	5	7
Equisetaceae*	1	3
Dennstaedtiaceae	1	1
Marsileaceae *	1	1
Ophioglossaceae	1	1
Polypodiaceae	1	2
Pteridaceae	2	2
Gymnosperms		
Cupressaceae	1	3
Ephedraceae	1	3
Taxaceae	1	1
Angiosperms		
Basal and *inserti sedis* families		
Ceratophyllaceae	1	1
Cynomoriaceae	1	1
Eudicots		
Aceraceae	1	6
Amaranthaceae	1	6
Anacardiaceae	1	1
Apiaceae (Umbelliferae)*	40	58

Contd...

Table 24.2–Contd...

Major Groups and Families	*Genera*	*Species*
Apocynaceae (incl. Asclepiadaceae) 5		7
Aquifoliaceae	1	1
Asteraceae (Compositae)*	75	171
Berberidaceae * (incl. Podophyllaceae) 3		5
Betulaceae	1	2
Boraginaceae*	18	43
Brassicaceae (Cruciferae)*	41	73
Campanulaceae	2	5
Capparidaceae	2	2
Caprifoliaceae	2	5
Caryophyllaceae*	21	57
Celasteraceae	1	2
Chenopodiaceae*	27	56
Cistaceae	2	4
Convolvulaceae*	3	8
Cornaceae	1	1
Corylaceae	1	2
Crassulaceae	4	9
Cucurbitaceae	1	1
Cuscutaceae	1	4
Dipsacaceae	4	7
Ebenaceae	1	1
Elaeagnaceae	1	1
Elatinaceae	1	1
Ericaceae*	1	1
Euphorbiaceae*	3	14
Fabaceae (Papilionaceae)	20	122
Fagaceae	1	2
Frankeniaceae	1	1
Fumariaceae*	2	6
Gentianaceae	1	1
Geraniaceae	3	12
Grossulariaceae	1	1
Hamamelidaceae	1	1
Hypericaceae (Guttiferae)	1	7
Juglandaceae	1	1

Contd...

Table 24.2–Contd...

Major Groups and Families	*Genera*	*Species*
Lamiaceae (Labiatae)	28	66
Lentibulariaceae	1	1
Linaceae	1	4
Loranthaceae	1	1
Lythraceae	1	1
Malvaceae*	4	6
Moraceae	2	2
Oleaceae	2	4
Onagraceae	2	5
Orobanchaceae*	1	10
Oxalidaceae	1	1
Paeoniaceae	1	1
Papaveraceae*	5	9
Plantaginaceae	1	4
Platanaceae	1	1
Plumbaginaceae	3	8
Polygalaceae	1	1
Polygonaceae*	4	25
Portulacaceae	1	1
Primulaceae	5	5
Punicaceae	1	1
Ranunculaceae*	11	28
Resedaceae	1	2
Rhamnaceae	2	6
Rosaceae*	20	45
Rubiaceae	6	20
Rutaceae	2	3
Salicaceae	2	6
Santalaceae	1	2
Saxifragaceae	1	1
Scrophulariaceae*	12	41
Solanaceae	5	13
Tamaricaceae	2	4
Thymelaeaceae	3	3
Tiliaceae	1	1
Ulmaceae	3	5

Contd...

Table 24.2–Contd...

Major Groups and Families	*Genera*	*Species*
Urticaceae	2	3
Valerianaceae	2	9
Verbenaceae	1	1
Violaceae	1	6
Vitaceae	1	1
Zygophyllaceae	5	6
Monocots		
Alismataceae	1	1
Alliaceae*	1	16
Amaryllidaceae*	3	4
Araceae	2	2
Asparagaceae*	1	3
Asphodelaceae*	1	4
Colchicaceae	1	5
Cyperaceae*	5	28
Dioscoreaceae	1	1
Iridaceae*	3	11
Juncaceae	2	8
Hyacinthaceae*	4	6
Lemnaceae	2	4
Liliaceae s. str. *	3	19
Orchidaceae	10	22
Ruscaceae*	2	3
Poaceae (Gramineae)*	55	119
Potamogetonaceae*	1	2
Sparganiaceae	1	1
Typhaceae	1	2
Zannichelliaceae	1	1

Total: 116 families, 570 genera and 1364 species

Flora of Golestan National Park belongs to two main phytogeographical areas: The Hyrcanian Province of Euro-Siberian area in the forest zone and Khorassan-Kopetdagh Province of Irano-Turanian area in steppe and Juniper woodlands. However, in S-facing rocky outcrops the presence of several Mediterranean elements is worth mentioning. Among the (sub) Mediterranean species two showy species *Sternbergia fischeriana* and *S. lutea* are frequent in scrubs and rocky outcrops. Other Mediterranean species include *Caucalis platycarpa, Convolvulus cantabrica, Cynoglossum creticum, Medicago rigidula, Asteriscus spinosus, Petrorhagia prolifera, Teucrium polium var. tensum, Trifolium angustifolium, T. scabrum, Vicia amphicarpa, V. lutea, Jasminum fruticans, Stipa bromoides, Carex hallerana, Daucus guttatus, Laser trilobum, Orlaya daucoides, Fumana arabica, Onosma pusilla* (Akhani and Ziegler, 2002; Rechinger, 1989).

Among the several endemic species of the Hyrcanian province in the Park following species are outstanding: *Parrotia persica, Acer velutinum, Quercus castaneifolia, Alyssopsis mollis* (an endemic Cruciferae), *Fritillaria kotschyana, Allium chelotum, Epipactis rechingeri, Centaurea hyrcanica, Alnus subcordata, Ilex spinigera, Saponaria bodeana, Primula heterochroma, Rhynchocorys maxima, Echium amoenum Asyneuma amplexicaule, Colutea buhsei, Cydonia oblonga, Cynoglossum kandavanensis, Euonymus velutina, Froriepia subpinnata, Hesperis hyrcana, Orchis adenocheila, Populus caspica, Pyrus boissieriana,* and *Veronica siaretensis.*

There is a strong floristic link between the flora of Golestan and the forest around the Black Sea known as Euxinian flora. Examples of common Hyrcanian and Euxinian species are; *Acer cappadocicum, Allium paradoxum, Bupleurum marschallianum, Colchicum speciosum, Corydalis angustifolia, Crataegus microphylla, Crocus speciosus, Danaë racemosa, Digitalis nervosa, Dipsacus strigosus, Eleutherospermum cicutarium, Iranecio othonnae, Linum nervosum, Mespilus germanica, Poa masenderana, Quercus macranthera, Rhamnus spathulifolia, Serratula quinquefolia, Stachys byzantina, Steveniella satyrioides, Trinia leiogona,* and *Zelkova carpinifolia.*

Irano-Turanian species with ca. 41 per cent consist of large parts of the flora of the area. A considerable part of the Irano-Turanian species are endemic in the Khorassan-Kopetdag floristic province (Akhani, 1998; Kamelin, 1970). Examples of these are *Acantholimon pterostegium, Acer monspessulanum* ssp. *turcomanicum, Allium cristophii, A. helicophyllum, A. vavillovii, Alyssum lanceolatum, Astragalus basineri, A. brevidens, A. cartilaginous, A. khoshyailensis, A. pendulinus, A. pseudoindurascens, A. ufraensis, Buffonia sintenisii, Centaurea iljinii, C. sintenisiana, Corydalis chionophila, Cousinia arctotidifolia, C. freynii, C. leucantha, C. meshhedensis, C. smirnowii, C. turcomanica, Crucianella sintenisii, Erysimum ischnostylum, E. kerbabaevii, Ferula badrakema, Fritillaria raddeana, Iris fosterana, I. kopetdaghensis, Onobrychis sintenisii, Oxytropis kopetdaghensis, Phlomis cancellata, Scutellaria litwinowii, Silene indeprensa, Stachys subaphylla, Sterigmostemum ramosissimum, Tulipa micheliana, Veronica khorassanica* and *Vicia venulosa.*

The Park is putative for having several local endemic and subendemic species. Several species have their type locality from the Park. Here is the list of species with high conservation value:

Astragalus memoriosus Pakravan, Nasseh and Maassoumi

A rare species was known from a locality in the South of the Park near Cheshmeh Khan.

Centaurea golestanica Akhani and Wagenitz

An isolated species confined to the grassy vegetation with scattered shrubs in the Northwestern border of the Park near Koilar.

Cousinia qarehbilensis **Rech. f.**

The species was described from the type locality near Robate-Qarebil in the eastern border of the Park. It was not recollected since its discovery in 1975.

Crocus almehensis **C. D. Brickell and B. Mathew**

An interesting yellow-flowered species with a limited range in Golestan. The species is known locally as Zaferane Almeh (Almeh Saffron).

Crucianella platyphylla **Ehrend. and Schönb.-Tem.**

An interesting relict species described from ca. 100 SE of the Park. The species is a chasmophytic plant in subalpine rocky areas in the forest zone of the Park. It was also found in similar ecological conditions in Jahan Nama Protected area (Jafari and Akhani, 2008).

Diaphanoptera stenocalycina **Rech. f. and Schiman-Czeika**

A local endemic and endangered species, known only in a small area, in the easternmost parts of the Park. The habitat of the species has been degraded due to ploughing of the Mirza-Baylu plain and digging of a canal around the eastern and southern border of the Park.

Eriocycla ghafforiana **Akhani**

An interesting chasmophytic species described from the cliffs at the northwestern border of the Park and found also near Gorgan (Akhani and Ziegler 2002).

Festuca akhanii **Tzvelev**

The species was described newly from grasslands close to scrub vegetation in South-Central parts of the Park (Tzvelev 1997).

Johrenia golestanica **Rech. f.**

A very late flowering species growing in the open scrubs mostly in association with S-facing rocky outcrop communities.

Leutea laseroides **Akhani**

A chasmophytic and very interesting species which was first named as *Laser rechingeri*. After having more studies it was proved that the correct generic position of this species is *Leutea* because of a published name *Leutea rechingeri*, a new name was given (Akhani, 2005). A recent molecular phylogentic study confirms the position of this species within *Leutea* (Valiejo-Roman *et al.*, 2006).

Plantago podlechii **Akhani**

A rare species of the mixed open scrub and grassland communities in Yelagh is located in Southern parts of the Park.

Poa golestanensis **H. Scholz and Akhani**

This species dominates the wet meadows around isolated Sulukli Lake and subalpine meadows near Divar Kaji.

Salvia **spec. nov. (inedit.).**

Seseli kiabii tortuosum **L. ssp. *kiabii* Akhani**

An autumn flowering species which its subspecific epithet commemorated to the Iranian ecologist Dr. B. Kiabi and an investigator of the wild life of the Golestan National Park. The taxon has a wider distribution in the eastern parts of the Caspian lowlands in association with rocky outcrop communities.

Silene sojakii **Melzh.**

A very rare and an endangered species known only from type locality ca. 3 km from Dasht. This enigmatic plant has not been recollected in spite of several attempts. Some speculations about the rarity or probable extinction of the species have been provided in Akhani (1998).

Stachys subaphylla **Rech. f.**

A distinct endemic species inhabiting some rocky areas of the northeastern parts of the Park. The type locality of the species is located ca. 40 km E of the area.

CONSERVATION AND THREATS

The most critical problem in the conservation of nature in Iran in general, and Golestan National Park in particular, is the poor management of the Department of Environment (DOE). In the last three decades due to long-term Iran-Iraq war and post-war reconstruction activities, there has been little interest in the conservation issues, therefore DOE has not improved much in management, directorship and hiring professional staff. Most protected areas and national parks are managed by the outdated rules existing from the time when this organization was initially established as "Iran Game Council" in 1957. Since then population of Iran has increased drastically and the need for food has eventually lead to an expansion of agricultural land, over grazing and overexploitation of natural resources. An increase in urban areas, industrialization and intensive road construction provide easy access to many of the formerly very little disturbed natural habitats. Both the cheap oil and gasoline price in Iran (one of the cheapest rate in the world), and increase in the car production have increased the destructive activities due to biotic influences in many ways, such as transportation of livestock to remote areas, deforestation, pouching, littering, cultivation and transfer of land to industrial and urban areas. Therefore, most protected areas and national parks in the country have been severely damaged by the mismanagement during the last three decades. The critical status of Golestan National Park; keeping in view its importance, demands an acceptable conservation level. The facts about the park are summarised below.

FLOODING AND ASIAN ROAD

The largest impact is the passage of the so-called "Asian road" through the Park, which passes along the river valley of Madrasu, dividing it into the Northern and Southern parts. This is a very busy road in Iran connecting Central and North of Iran with Northeast. Hundreds of wild animals get killed annually; *e.g.*, two leopards in 2009; due to vehicle accidents. The road was completely destroyed by several floods, biggest ones in summer 2001 and 2002. The flood on 11 July, 2001 was the most catastrophic in the history of the Park. The dam near Dasht village at the southern border of the Park got demolished increasing the flood level which washed out several meter deep soils and completely destroyed the road, the riverine forests and nearly 700 people who were on the road or camping in the Park lost their lives. Several other floods followed this causing more damages to the park every year due to soil erosion and loss of vegetation. Many environmental NGO'S and DOE experts strived hard before these floods took place to detour the road outside the Park. A warning in this connection was given by the first author (Akhani 1998). It was clearly mentioned then that every year hundred thousands of visitors coming from the west to visit the holy city of Mashhad or coming from the east to visit North and Central Iran pass through this road. Many of these travellers spend at least a few hours in the Park. These visitors cause many problems such as; making fire, cutting branches, using detergents alongside the water courses (particularly along Madrasu River), disturbing wildlife, fishing along the river, littering etc. There are no teaching programmes or educated personnel to guide the visitors or distribute information sheets in order to increase public awareness on environmental protection in the Park. Strong arguments were raised after the floods to detour the road in order to prevent further damage to the travellers or camping people in the Park. An open letter was sent to the former president of Iran Mr. Khatami on April 2003, signed by 243 university professors and environment experts. The detouring of the road was approved by the Iranian cabinet as a cresult of this letter, but the project never realized. Ministry of Road and Transport insisted to reconstruct the present road by starting a very destructive project which has completely devastated the river ecosystem and large parts of still undamaged forests. Recently very deep concrete walls have been constructed with a depth of up to seven meters and the road was widened from 11 m to 50 m, thus changing it into a highway. A second open letter to the President of Iran was mailed in October 2006, singed by 152 university professors, but the activities in the Park did not cease. The new expanded road provides strong barrier for wild life passage in the area, increasing the vehicle accidents, and preventing development, succession and regeneration of natural forests.

Overgrazing

Most parts of the marginal areas of the Park are suffering from intrusion of the herds of surrounding villages, most critical part being Dashte Shad (formerly Dashte Shah), which has been occupied by large herds entering deep into the Park. Overgrazing has destroyed large parts of the *Crataegus* scrub which belongs to a unique transitional ecosystem in the Park. Less grazing problems are present in nearly all bordering parts of the area. Examples are Dasht, Yakhti-kalan, Soolegerd, Koilar, Lohondor, Khan-Doushan etc.

Pouching

Perhaps one of the most serious conservation problems in the Park is pouching. Despite the efforts to prevent this, it seems that the present conservation methods, rules and personnel are not efficient enough to overcome this problem. The degradation of habitats associated with intensive pouching cause drastic reduction in the wild life in the area. According to Kiabi *et al.* (2004), the number of Maral (*Cervus elaphus maral*) populations was 1897–2096 during 1976-1978 (Kiabi 1978). Since then the population size has declined to 500 in 2003. These authors insisted on the pouching, road accidents and livestock grazing as the major factors affecting the Maral population adversely. In order to control hunting it is clear that the traditional conservation methods are not efficient. The population of Iran has doubled since seventies, when this area was designated as a National Park. The economic conditions have worsened and the salary of the Park personnel is very low. Presently there are ca. 50 park rangers, one-third of these are usually off. It is imperative to increase training and education for the Park personnel, pay appropriate salaries and hire more and younger staff. Furthermore there are cultural and economic reasons for hunting such as high number of poor and unemployed people living in the area

Fire

Climatic transition causes an instability in the weather conditions of the Park and presence of strong winds adds to this. The dense ground vegetation of the shrublands which becomes dry during dry summer months help in the spread of even a small fire into an uncontrolable disaster. The potential sources for man-made fires are presence of high number of uncontrolled visitors and people illegally entering into the Park. Natural fires seem to be unlikely in the area. According to information provided by the park rangers and local people the oldest and the largest fire; which damaged large parts of the central parts of the Park; took place between 1950–1953, before the Park became a protected area. According to Ghaemi (cited in Safaian *et al.*, 2005) 67 fires have taken place in the Park during 1957-1997. In 1995 alone ca. 6 fires burnt the area badly. One of these occurred in Yelagh plain on 29–31 August. It was a major catastrophe in the history of the Park like the flood of 2001. In 2006 many parts of the grassland and scrub vegetation in the Central parts of the Park were damaged by fire. In fact deliberate fires, fires caused by poachers, shepherds, by the people entering the Park illegally, and people coming to the Park to harvest the forage or collect mushrooms and medicinal plants are all responsible for the fire accidents. Moreover, many tourists passing on the road stay for shorter times but thay too make fires. Safaian *et al.* (2005) investigated the influence of fire in Yelagh plain for five years after its occurrence in 1995. They did not find significant changes in the diversity. The coverage of both annual and perennial grasses increased in comparison with non-burnt area. However, the coverage of woody species decreased considerably in burnt area. They concluded that fire is a preventive factor in the succession of ecosystem towards its climax stage, but dominance of grasses can be regarded as condition for further fire occurrence.

Invasive and Introduced Species

Several invasive and weedy plants have occupied some parts of the Park after the destruction of natural vegetation by floods. Several invasive and weedy species belonging to genera *Conyza*, *Chenopodium*, *Setaria*, and *Amaranthus* and *Rapistrum rugosum* are distributed widely in degraded and flooded river sides. Some species like *Atriplex canescens* have been cultivated in the Southern parts of the Park, their dense canopies cause damaging effects on the native flora.

Ploughing and Construction of a Canal

The provincial authorities of office of environmental protection have decided to allow ploughing and cultivation of introduced species into the Mirza-Baylu plain, in order to increase forage for the small gazelle population living there; but this is a wrong management plan. The whole area were ploughed and prepared for cultivation. After strong opposition from the first author, the cultivation was stopped, but ploughing has largely damaged the plain. Another wrong management was making a barrier to prevent passing of gazelle from the Park into surrounding areas. DOE constructed a deep canal along the southern and eastern border of the Park. However, this canal has increased the drainage and soil erosion and desertification in the area. Both activities show that adverse results in improving the gazelle population have caused sever damage to the natural vegetation of the area.

There are many other impacts including harvesting of grasslands in many parts of the area, collection of forest fruits, mushrooms and medicinal plants, as well as rare plants by tourists and even students and scientists which adds to the above mentioned threats.

In conclusion we can say that the Park is suffering from critical management. The poor conservation level in most parts of the Park is far beyond the national and international rules which need to be applied in the national park. Therefore a number of national and international efforts and NGO activities are required in order to improve the protection status of the Park. Present protection level cannot guarantee conserving the unique biodiversity of this first and most important Iranian National Park in the holarctic.

ACKNOWLEDGMENTS

This paper presents the results of the research project "Geobotanical studies in Different Parts of Iran IV" supported by the Research Council, University of Tehran and the joint project between Office of Department of Environment of Golestan Province and the University of Tehran under the act 102. Present research studies have been supported by two grants from German Academic Exchange Service (DAAD) during 1993-1997 and 2002. The help provided by the "Institute of Systematic Botany of the Munich University", in particular Prof. (Dr.) D. Podlech is highly appreciated. Our special thanks go to the rangers, managers and directors of the Park for their help, friendship during field trips and their generous efforts to protect the area.

REFERENCES

Abbasi M. 2005: The fungal flora of Golestan National Park. In: Akhani, H. (ed.) The Illustrated Flora of Golestan National Park 74-75 and 342-354. *University of Tehran Press*. Tehran.

Abdoli A. 1993: Fishes of Golestan National Park's river. *Abzeeyan* 4(3):32-33 + XIII. (In Farsi).

Akhani H. 1998: Plant biodiversity of Golestan National Park, Iran. *Stapfia* 53: 1-411.

Akhani H. 1999: Studies on the flora and vegetation of the Golestan National Park, NE Iran III. Three new species, one new subspecies and fifteen new records for Iran. *Edinburgh J. Bot.* 56 (1): 1-31.

Akhani H. 2005: The Illustrated Flora of Golestan National Park, Iran. Vol. 1. *Tehran University Press*. Tehran.

Akhani H., Kürschner H. 2004: An annotated and updated checklist of the Iraninan Bryoflora. *Crypt. Bryol.* 25: 315-347.

Akhani H. and Scholz H. 1998. Studies on the flora and vegetation of the Golestan National Park, NE Iran II: A new *Poa* and some new and noteworthy grass records for Iran. *Edinburgh J. Bot.* 55 (3): 443-453.

Akhani H. and Ghasemkhani M. 2007: Diversity of photosynthetic organs in Chenopodiaceae from Golestan National Park (NE Iran) based on carbon isotope composition and anatomy of leaves and cotyledons. *Beih. Nov. Hedw.* 131: 265-277.

Akhani H., Edwards G. E., Roalson, E. 2007: Diversification of the Old World Salsoleae s.l. (Chenopodiaceae): molecular phylogenetic analysis of nuclear and chloroplast datasets and a revised classification. *Intern. J. Plant Scienc.* 168 (6): 931–956

Akhani H., Ziegler, H. 2002: Photosynthetic pathways and habitats of grasses in Golestan National Park (NE Iran), with an emphasis on the C_4 grass dominated rock communities. *Phytocoenologia* 32 (3): 455-501.

Angiosperm Phylogeny Group 2003 An update of the Angiosperm Phylogeny Group classification for the orders and families of flowering plants: APG II. *Bot. J. Linn. Soc.* 141:399-435.

Gutleb B., Wieser, C. (eds.) 2002: Ergebnisse einer zoologischen Exkursion in den Nordiran, 2001. *Carinthia* II. 192./112. 33-140.

Firuz E. 1974: Environment Iran. National Society for the Conservation of Natural Resources and Human Environment. Tehran.

Frey W., Kürschner, H. 1977: Studies on the bryophyte flora and vegetation of the Mohammad Reza Shah National Park, N. Iran. *Iranian. J. Bot.* 1 (2): 137-153.

Frey W., Kürschner H., Probst W. 1999: Flora and vegetation, including plant species and larger vegetation complexes in Persia. In: Yarshater, E (ed.), *Encyclopaedia Iranica* 10 (1): 46-63.

Fritsch R. M., Matin, F., Klass, M. 2001: *Allium vavilovii* M. Popovet Vved. and a new Iranian species are the closest among the known relatives of the common onion *A. cepa* L. (Alliaceae). *Gen. Res. Crop Evol.* 48: 401 408.

Hallenberg N. 1981: Synopsis of wood-inhabiting *Aphyllophorales* (*Basidiomycetes*) and *Heterobasidiomycetes* from. N. Iran. *Mycotaxon* 12 (2): 473-502.

Jafari, S. M., Akhani, H. 2008: Plants of Jahan Nama Protected Area, Golestan Province, N. Iran. Pakistan Journal of Botany 40(4): 1533-1554.

Joharchi M. R., Akhani H. 2006: Notes on the flora of Iran 6: Eight new plant records from Iran collected from Khorassan and Golestan Provinces (NE Iran). *Rostaniha* 7 (Suppl. 2): 131-140.

Leestmans R. 2005: Le refuge Caspian et son importance en biogeography. *Linn. Belg.* 20 (3): 97-102.

Kamelin R. V. 1970: Specific phyto-geographical features in the flora of the Soviet part of the Kopet-Dag Range. *Bot. Zhurn.* 55 (10): 1451-1462. (In Russian).

Khoshravesh, R., Akhani, H., Eskandari, M., Greuter, W. 2009: Fern and fern allies of Iran. Rostaniha 10 (Suppl. 1): 1-129.

Kiabi H. B. 1975: Ecological studies of the Persian Ibex (*Capra hircus aegagrus*) in Mohammad Reza Shah National Park, Iran. Unpublished M. Sc. Thesis. Michigan State University.

Kiabi H. B. 1978: Ecology and management of Maral (*Cervus elaphus maral*) in northeastern Iran, 1976-1978. Ph. D. Dissertation. Michigan State University.

Kiabi B. H., Ghaemi, R. A. Jahanshahi, M., Sasani, A. 2004. Population status, biology and ecology of the Maral, *Cervus elaphus maral*, in Golestan National Park, Iran. *Zool. Mid. East* 33: 125-138.

Kiabi H. B., Zehzad B., Farhang Darreh-Shoori B., Majnounian H., Goshtasb Meigouni, H. 1994: Golestan National Park. Department of the Environment. Tehran. (In Persian).

Kürschner H., Papp B., Akhani, H. 2000: New records to the bryophyte flora of Iran. Studies on the flora and vegetation of the Golestan National Park IV. *Nova Hedw.* 71 (3-4): 509-518.

Madjnoonian, M. Zehzad B., Kiabi B., Darreh-Shoori B. F., Meigouni H. G. 1999: Golestan National Park (Biosphere Reserve). Department of Environment. Tehran. (In Persian).

Noroozi, J., Akhani, H., Breckle, S. W. 2008: Biodiversity and phytogeography of the alpine flora of Iran. Biodiversity and Conservation 17(3): 493-521.

Rechinger K. H. 1989: Fifty years of botanical research in the Flora Iranica Area (1937-1987). - In: Tan K. (ed), Plant taxonomy, phytogeography and related subjects: The Davis and Hedge Festschrift. pp. 301-349. Edinburgh.

Safaian M., Shokri, M., Ahmadi M. Z., Atrakchali A., Tavili A. 2005. Fire influence on the grassland vegetation in Golestan National Park (Alborz Mts. Iran). *Pol. J. Ecol.* 53 (1): 435-443.

Tralau H. 1963: Asiatic Dicotyledonous affinities in the Cainozoic Flora of Europe. *Kungl. Svenska Vetenskapsakad.* Handl. 4, Ser. 9, 3: 1-87 + 5 plates.

Tzvelev, N. N. 1997: A new species of the genus *Festuca* (Poaceae) from Iran. Botaniènyj Žurnal 82(4): 118-119.

Valiejo-Roman C. M., Terentieva E. I., Samigullin T. H., Pimenov M. G., Ghahremani-Nejad F., Mozaffarian V. 2006: Molecular data (nrITS-sequencing) reveal relationships among Iranian endemic taxa of the Umbelliferae. *Fedd. Repert.* 117: 367-388.

Vuosalo E. 1976: Once there was a tiger. *Wildlife Mar* 18 (3): 1267-129.

Zohary M. 1973: Geobotanical foundations of the Middle East. 2 vols. Gustav Fischer Verlag.

Chapter 25

Characterizing and Screening Oil Degrading Microbes for Land and Beach Reclamation in Indonesia

Dwi Susilaningsih[1], Theresia Umi Harwati[1], Yuki Kasai[2], Kazuya Watanabe[2], Fumiyoshi Okazaki[3], Shigeaki Harayama[3], Yantyati Widyastuti[1] and Bambang Prasetya[1]

[1]*Research Center for Biotechnology, Indonesian Institute of Sciences, Jl Raya Bogor Km 46 Cibinong Bogor Indonesia 16911*

[2]*Marine Biotechnology Institute, 3-75-1 Heita, Kamaishi, Iwate 026-0001*

[3]*National Institute of Technology and Evaluation, 2-5-8 Kazusakamatari, Kizarasu, Chiba*

ABSTRACT

Biological treatment has become increasingly popular as a remediation method for soils and groundwater contaminated with petroleum hydrocarbon, chlorinated solvents, and pesticides. Studies to date suggest that indigenous microbes suitable for bioremediation exist in soils and water columns. *In situ* bioremediation strategies may involve the stimulation of native bacterial strains or the introduction of nonnative microorganisms. These microbes then transform contaminants into non- or less-hazardous chemicals. Major problem efficacy of microbe-driven bioremediation is limited by at least three factors:

(*i*) Bioavailability of the toxic substrate to be degraded,

(*ii*) Ease with which the toxic substrate can diffuse or be taken up by the degrading microorganisms, and

(*iii*) Capability of the degrading microorganisms to survive in the polluted environment.

In this regard, we have isolated the oil degrading bacteria from soil and marine sources and screened those isolates for surfactant synthesizer. A total of

103 out of 463 isolates showed positive surfactant-degrading properties. By means of partial 16 S DNA analyses, it has been found that the majority of taxa are related to *Alcanivorax, Pseudomonas, Bacillus, Bortetela, Brucella, Acenitobacter, Staphia, Lysobacter, Talasosophira* and *Rhodobacteraceae*. In particular the soil strains were found proposed as new genus and species. Further studies of 20 selected from marine strains on biosurfactant properties assay showed that they were capable of lowering the surface- and interfacial water tension from 74 mN/m to 40-65 mN/m and from 24 mN/m to 6–10 mN/m, respectively. In addition, most of the surfactants were capable of emulsifying hydrocarbon (crude oil) of 0.01 to 0.15 units, comparable to 0.08 units of synthetic surfactant (20 per cent Tween). Further observation showed that the majority of the surfactants were able to degrade a long chain of alkane, but not branched alkane, with a recovering rate of 20-80 per cent. Preliminary application test showed 4 strains capable for enhance the porosity of soil-contaminated with oil (soil-sludge).

Keywords: *Bacteria, Biosurfactant, Bioremediation, Marine-Soil reclamation, Petroleum hydrocarbon.*

INTRODUCTION

Degradation of chemical and petroleum pollutants in terrestrial and marine areas by microbes is friendly approval method. Although physical and chemical treatments are much faster than bioremediation, but the treatment is hazardous and generates next pollutant in the environment. Bioremediation of petroleum in the environment is tent to utilization of native microbes in the niche and improves the product of synthesized surfactant which is known as emulsifier, mediator and digester of pollutant (Desai and Banat, 1997).

Biosurfactants are produced by a variety of microbes, secreted either extracellularly or attached to parts of cells, predominantly during growth on water-immiscible substrates ranging from low to high-molecular-weights. In general, low-molecular-weight biosurfactants are effective in lowering the surface and interfacial tensions while high-molecular-weight biosurfactans are efficient in emulsifying oil droplets. The chemical structures and properties of biosurfactants are thus diverse, and therefore it is reasonable to assume that different groups of biosurfactants have different biological roles. One of the possible applications of biosurfactants is the remediation of oil-polluted environments because product purity is not important and the products themselves are generally non-toxic. A number of studies have shown that biosurfactants are effective for the abatement of soil and marine oil pollution providing better biodegradability and lower toxicity than synthetic surfactants (Finnerty, 1994; Garcia-Junko *et al.*, 2001; Makkar and Cameotra, 2002). Unlike synthetic surfactants, microbially produced compounds are biodegradable and particularly well suited for environmental applications, such as bioremediation and dispersion of oil spills. Compared with synthetic compounds, biosurfactants offer the advantages of little, or no, environmental impact and the possibility of their *in situ* production.

It is well known that petroleum hydrocarbons are major pollutants in environments as a result of domestic waste, oil production, transportation activities, refuse from coastal oil refineries and accidental spillage of fuels. Petroleum is a viscous liquid mixture that contains thousands of compounds mainly consisting of carbon and hydrogen. Evaporation and photo-oxidation are two major weathering processes, while ultimate and complete degradation is accomplished mainly by micro flora, bacteria being dominant in this function. However, natural microbial degradation occurs relatively slowly in environments due to unfavorable growth conditions such as low temperatures and lack of nitrogen and phosphorous nutrients, and a low availability of hydrophobic hydrocarbons to oil-degrading microorganisms. To overcome the latter problem, certain bacteria may produce biosurfactants to stimulate the biodegradation of hydrocarbons.

The objectives of the present study, therefore, include; screening of surfactant producing bacteria, characterization of the surfactants, and application of the biosurfactant.

METHODOLOGY

Screening for Biosurfactant-producing Bacteria

Soils samples were collected from oil-contaminated area at Cepu and Semarang, Central Java. Water samples were collected from Jakarta bay, its surroundings and Seribu Island (Jakarta), which often suffered environmental damages by oil-spills from ships and domestic industries. Microorganisms capable of degrading petroleum hydrocarbons were enriched in agar medium containing crude oil. Growing bacteria were indicated as the bacteria capable for oil-degradation. Particularly, surfactant-producing strains were identified by the formation of halo around single colonies, and then collected. The methods of Alley and Brown (2000) were used.

16S rRNA Gene Sequence Determination

Total genomic DNA was isolated from late-exponential-phase cells in agar medium containing 1000 ppm (w/w) Arabian crude oil using a kit of quick-qiagen. PCR amplifcation of 16S rRNA genes was obtained using the forward primer 16F27 (5« AGAGTTTGATCMTGGCTCAG3 «) and the reverse primer 16R1492 (5« TACGGYTACCTTGTTACGACTT 3«). Direct sequence determination of PCR-amplifed DNA was carried out using an automated DNA sequencer and Taq cycle sequencing reactions according to the protocols of the manufacturer (Perkin-Elmer, Applied Biosystems, 2005). Sequence data were initially aligned using the Clustal X program (www.ebi.ac.uk/clustalw) with 16S rRNA gene sequences compiled in the Ribosomal Database Project (RDP). Phylogenetic relationships were estimated using NJ plot of the Phylogeny Inference Package (phylip version 3.4, 2005). Homologies were searched by gene-bank database of NCBI (www.ncbi.nlm.nih.gov).

Characterization and Determination of Surfactant Properties

The supernatant of cultures were evaluated for their characters and compositions. The surface active tension and interfacial tension were analyzed using the drop shape analysis (DSA) equipment (Kruss DSA100, 2006, Germany). The emulsification activity was done following the Navon-Vanezia methods (1995). The sugar content was determined by total sugar analysis using phenol-sulfuric assay (Taylor, 1995), lipid/fat was analyzed by Bligh-Dryer method (1956), and protein was determined by Bradford method (1976).

Assay for the Biodegradation of Hydrocarbon and Crude Oil

Analysis of hydrocarbon and crude oil recovery was performed by Gas Chromatography/mass spectrophotometer (GC/MS) with a 6890/5973 instrument using PETRSCA1 methods as described by Harayama (1997). Cell cultures (2 ml) were extracted by 3 ml dichloromethane (Merck, USA), shaking reciprocally for 2 minutes and stand for few minutes for separation into two phases, the lower phase was transferred into a new tube, and the extraction was repeated until yellow interface disappeared. A tea-spoonful of sodium sulfate (Merck, USA) was applied to remove water followed by concentration under nitrogen gas stream to a final volume of about 250 µL and then transferred into GC-vial, the samples were ready for measurement in the GC/MS. HP19091B-102 with silica column was used with the following conditions: initial oven temperature 250 °C, injection temperature 320 °C, flow rate of the mobile phase 54.1 mL/min and running time 50 min.

Preliminary Test for *ex situ* Application

Soil-sludge containing oil from contaminated site in Kalimantan Area was used for trial. Selected bacteria (Tanah 33-cp, Tanah 34-cp, Tanah 35-cp and Tanah 38-cp) were introduced into the soil with ratio 1 ml culture at 10^8 bacterial numbers for 50 gram soil. 30 days incubation was done in triplicate series. Weight, porosity and texture of soils were observed. The experimental procedures and monitoring were following Radijevic and Bashkin methods (1999) with little modification.

RESULTS AND DISCUSSION

Microbial remediation or bioremediation of hydrocarbon and crude oil-contaminated soils and marine environments are an emerging technology involving the application of biosurfactant. The surfactants produced or synthesized by microbes (biosurfactants) have met the expectation because they are environmental friendly and do not generate pollutants. To understand the role of biosurfactant function in the community, isolation, characterization and determination the strains that produce surfactant were used to evaluate surfactant properties. Therefore, in order to know response in the actual contaminant site, preliminary study on application of bacteria and their product (surfactant) in the soil containing oil contamination was done.

SCREENING OF BIOSURFACTANT-PRODUCING AND HYDROCARBON-DEGRADING BACTERIA

Using the quick screening by plate assay technique a total of 103 isolates out of 463 tested had been obtained. Out of these 103 isolates, 66 showed double capability for hydrocarbon removal and surfactant production. Based on the speed and size of halo formation 20 strains were chosen as model research for further study (Table 25.1). This indicates the strength of surfactant properties such as the surface active or interfacial tensions activity. They should have correlation with the surfactant function towards the hydrocarbon compounds, for instance in breaking the hydrocarbon chain, lowering surface and interfacial tensions, enlarge the movement area of the bacteria or to make soluble the compounds into water phase. There is an interaction of biosurfactant with the bacterial cells. In the biodegradation works, there is a tendency to follow two mechanisms: first, it can solubilize hydrophobic compounds within micelle structures, effectively increasing the apparent aqueous solubility of the organic compound and its availability for uptake by cells. Second, it can cause the cell surface to become hydrophobic, thereby increasing the association of the cell with slightly soluble substrate (Pembrey *et al.*, 1999; Al-Tahhan *et al.*, 2000).

Table 25.1: Candidates of Surfactant-Producing-Bacteria Screened from Indonesian Marine Environments

No	*Code of Selected Isolates*	*BLAST Search for Partial 16S rDNA*	*Homology (per cent)*
1	ID05-P-006Bt	AY633972.1 *Nocardioides alkalitolerans* strain KSL-12	96
2	ID05-P-009Bt	AY683531.1 *Alcanivorax dieselolei* strain NO1A clone 1	99
3	ID05-P-018Bt	AY683537.1 *Alcanivorax dieselolei* strain B-5	99
4	ID05-P-020Bt	AY683537.1 *Alcanivorax dieselolei* strain B-5	99
5	ID05-P-031Bt	AY683531.1 *Alcanivorax dieselolei* strain NO1A	99
6	ID05-P-032Bt	AY683531.1 *Alcanivorax dieselolei* strain NO1A	99
7	ID05-P-045Bt	AY683531.1 *Alcanivorax dieselolei* strain NO1A clone 1	99
8	ID05-P-049Bt	AY683531.1 *Alcanivorax dieselolei* strain NO1A	99
9	ID06-KWF.4Bt	AY918481.1 *Klebsiella pneumoniae* strain 3.4T	99
10	ID06-KWF.13Bt	DQ666628.1 *Pseudomonas aeruginosa* strain RsB-29	99
11	ID06-KWF.14Bt	AY445076.1 *Muricauda aquimarina* strain SW-72	96
12	ID06-KWD.13Bt	AM111012.1 *Lysobacter* sp.	96
13	RCO-CO-74	AF137358 *Pseudomonas aeruginosa*	100
14	RCO-CO-76	AF137358 *Pseudomonas aeruginosa*	100
15	RCO-CO-83	AJ494734 *Bacillus pumilus* El-44-7	99
16	RCO-CO-91	AY394865 *Alcanivorax* sp. EPR 6	100
17	RCO-CO-93	AY331355 *Pseudomonas* sp. MFY72	99
18	RCO-CO-95	AF137358 *Pseudomonas aeruginosa*	100
19	RCO-CO-97	EF198251 *Thalassospira* sp. MACL12B	99
20	RCO-Phe-107	AY331355 *Pseudomonas* sp. MFY72	99

Because the enrichment samples were conditioned for isolating degrading hydrocarbon bacteria, the majority of isolates of surfactant-producer obtained play an important role for the *Alcanivorax, Pseudomonas, Bacillus* and *Acenitobacter* (Table 25.1). Some of the isolates for hydrocarbon degradation were capable for eliminating crude oil and remove several PAH compounds. *Alcanivorax* is a ubiquitous marine petroleum oil-degrading bacterium with an unusual physiology specialized for alkane metabolism. This "hydrocarbonoclastic" bacterium degrades an exceptionally broad range of alkane hydrocarbons but few other substrates. *Pseudomonas, Bacillus* and *Acenitobacter* are famous for the ability to synthesize the extracellular compounds around the cells, possibly polysaccharides, lipid and protein which have surfactant-like properties depending on the substrate sources and the environment they are living in (Noordman *et al.*, 2002; Bodour *et al.*, 2003).

Since the oil degrading microbes were extensively observed in temperate areas, rare references from tropic areas therefore had high probability and new genus and species were found (Harwati *et al.* (a-c), 2008-2009). Family of Rhodobacteraceae from class Alphaproteobacteria dominated the oil-contaminated soils. However, bacteria showed less growth in the habitat, maybe due to lack of nutrient and air-flow in the soils. Many reports were distinguishing lack of nitrogen and phosphorus soluble in the soils and water columns as major factor of limitation in the bacterial growth (Harayama, 1997; Venkanteswaran *et al.*, 1999; Kasai *et al.*, 2002). It is suggested that the petroleum bioremediation process must involve the substances with solublizing, emulsifying and detergent like compounds excreted from the bacteria or others.

Surfactant Properties of Selected Isolates

Surfactant is defined by their surface active molecules which exhibit the emulsifying activity, reducing surface and interfacial tension, and their critical micelle concentration. Surface active tension values of selected isolates were variable from 40 mN/m to 65 mN/m, and water tension of 74 mN/m (Table 25.2). The highest activity was shown by group *Pseudomonas* since strains possess the ability to synthesize various surfactants such as rhamnolipid, in their late-exponential and stationary phase of growth under the limiting of nitrogen and iron concentrations (Oschner and Reiser, 1995). These values were lower when compared to the activity of synthetic surfactant (Tween). In concentration of 20 per cent v/v Tween could reduce the water tension up to 40 mN/m. However, bacterial surfactants possibly were less toxic than the Tween and have selective functions.

The interfacial tension observations of selected strains were relatively comparable to the synthetic surfactant of 20 per cent v/v Tween (Table 25.2). Surfactants that lower interfacial tension are particularly effective in mobilizing bound hydrophobic molecules and making them available for biodegradation. It is suggested that surfactant synthesized by *Pseudomonas* and *Bacillus* groups have low-molecular-weight, low critical micelle concentration for increasing the apparent solubility of hydrocarbon in hydrophobic cavities of micelles (Ron and Rosenberg, 2001; Bodour *et al.*, 2003).

The hydrocarbons emulsifying activities were comparable to the chemical surfactant of 20 per cent v/v Tween, *i.e.* around 0.01-0.15 units as compared to 0.08 units. The highest activities were shown in groups of *Bacillus, Acenitobacter* and *Alcanivorax* (Table 25.2). These results suggest that these groups of bacteria may synthesize the high-molecular-weight biosurfactants. The exocellular polymeric surfactants composed of polysaccharides, proteins, lipopolysaccharides, peptides, lipoproteins and their mixtures. These complexes of biopolymers were possibly produced by bacteria (Lin *et al.*, 1994; Oschner and Reiser, 1995; Al-Tahhan *et al.*, 2002).

Table 25.2: Surfactant Properties of 20 Selected Isolates

No	*Code of Isolates*	*Plate Assay*	*Lowering Surface Tension Activity (mN/m)*	*Lowering Interfacial Tension Activity (mN/m)*	*Emulsification Activity (Relatives unit)*
1	ID05-P-006Bt	+	68.23	6.68	0.042
2	ID05-P-009Bt	++	66.34	7.22	0.037
3	ID05-P-018Bt	++	62.59	5.11	0.055
4	ID05-P-020Bt	+	60.76	7.88	0.023
5	ID05-P-031Bt	+	61.27	9.56	0.046
6	ID05-P-032Bt	+	61.16	7.54	0.041
7	ID05-P-045Bt	+++	44.67	11.53	0.044
8	ID05-P-049Bt	++	63.35	8.66	0.059
9	ID06-KWF.4Bt	+++	64.34	7.23	0.066
10	ID06-KWF.13Bt	+++	61.39	4.98	0.075
11	ID06-KWF.14Bt	+++	62.34	8.43	0.083
12	ID06-KWD.13Bt	+++	59.87	10.68	0.134
13	RCO-CO-74	+++	56.87	9.87	0.063
14	RCO-CO-76	+++	55.49	10.22	0.054
15	RCO-CO-83	++	44.31	8.96	0.044
16	RCO-CO-91	++	57.99	8.77	0.023
17	RCO-CO-93	+++	56.86	7.45	0.038
18	RCO-CO-95	+++	40.58	8.34	0.047
19	RCO-CO-97	+++	60.22	9.44	0.056
20	RCO-Phe-107	+++	59.99	10.22	0.048
21	Tween 20 (10% v/v)	-	43.20	6.98	0.001
22	Water	-	78.8	25.5	0.00

Notes: +: formed small halo (d ≤1 cm), ++: formed medium halo (1-2 cm), +++: formed halo 2 cm d ≤

Averages nominal from triplicate calculations and triplicate treatments.

Preliminary Characterization of the Surfactant Products

Early steps of characterization of the biosurfactants showed that some of the surfactant compounds positively contain lipids, proteins and sugars or their mixture (Table 25.3). These indicate that many types of surfactants were obtained. Further studies on their specificity such as their chemical structures and function in nature, particularly in the biodegradation of hydrocarbon in marine environment should be done.

Table 25.3: Partial Characterization Type of Surfactants from Several isolates Note: Not determined

Codes of Isolates	Assays		
	Carbohydrate Content (ppm)	Protein Content (ppm)	Lipid content (g/100ml)
ID05-P-009Bt	22,1875	226,8182	0,955
ID05-P-018Bt	16,1458	140,4545	nd_
ID05-P-031Bt	16,5938	340,4545	0,71
ID05-P-032Bt	21,3958	122,2727	0,39
ID05-P-045Bt	33,4688	167,7273	nd_
ID05-P-049Bt	16,2813	213,1818	0,275
ID06-KWF.14Bt	13,9688	99,5455	nd_
ID06-KWD.13Bt	7,4479	135,9091	0,7

Monitoring of Hydrocarbon Removal by Cultures

Most of the selected strains could remove crude oil with a recovery rate of 60-90 per cent of the base concentration. Some of them can remove long chain alkanes, several PAH in a rate of 20-80 per cent and branch alkanes with a rate of 2-20 per cent, respectively (Figure 25.1). These showed the specific function of each strain, some isolates seem to have function for emulsifying but do not degrade others but tend to degrade a particular long chain alkanes, PAH or branched alkanes, while others have no emulsifying nor degrading hydrocarbon activity, just producing biosurfactants. These phenomena are very interesting in order to know their functions in marine bioremediation, especially hydrocarbon degradation. Major physiological role of biosurfactants is to permit microorganisms to grow in water immiscible substrates by reducing the surface tension at the phase boundary, thus making the substrate more readily available for uptake and metabolized. In addition, biosurfactant may also be involved in the adhesion of the microbial cells into the hydrocarbon substrates in the waters (Fietcher, 1992; Ron and Rosenberg, 2001).

Preliminary Application the Microbes and their Product to the Soil-oil-Contaminated

We found the oil-contaminated soils are forming sludge which is not permeable (the water can not pass through the soils). After treatment with mixture of bacteria (4 strains) during 1 month, the soil structure changed from sludge to the soil particles; oily-surface was disappearing and porosity changed to absorb the water. The dry weight of soils decreased (Table 25.4). Soil fertility is defined as soil with good soil texture, structure, friability and type. Soil structure refers to the way soil particles clump together, while friability is the ability of a soil to crumble. Good quality soils tend to form clumps that crumble easily, whereas poor soils (*e.g.* sandy soils) do not clump therefore lack structure. On the other hand, sandy soils are friable while clays are not (Rodojevic and Bashkin, 1999).

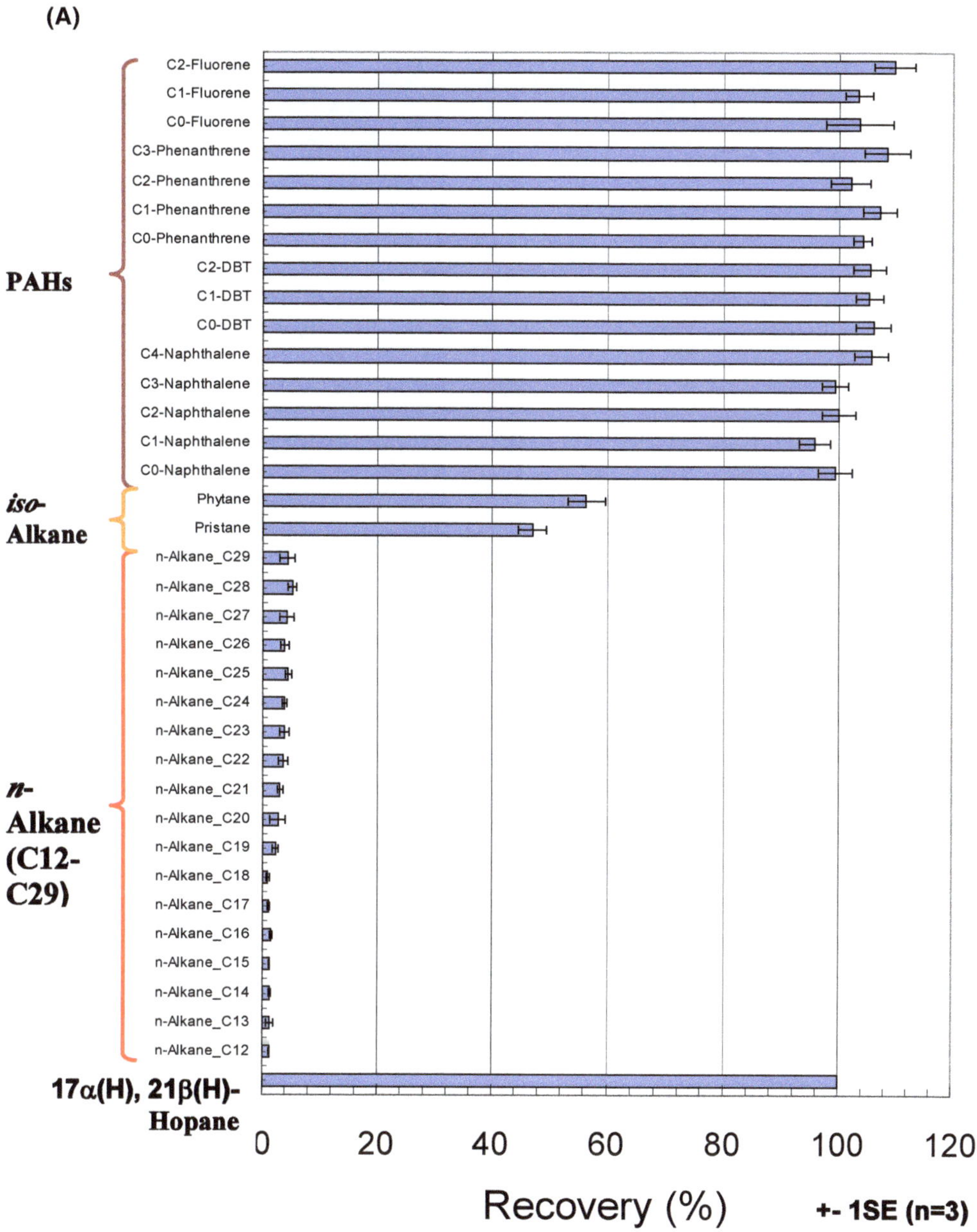

Figure 25.1: Remaining Concentration of Hydrocarbon After Treated with ID05-P-032Bt (A)

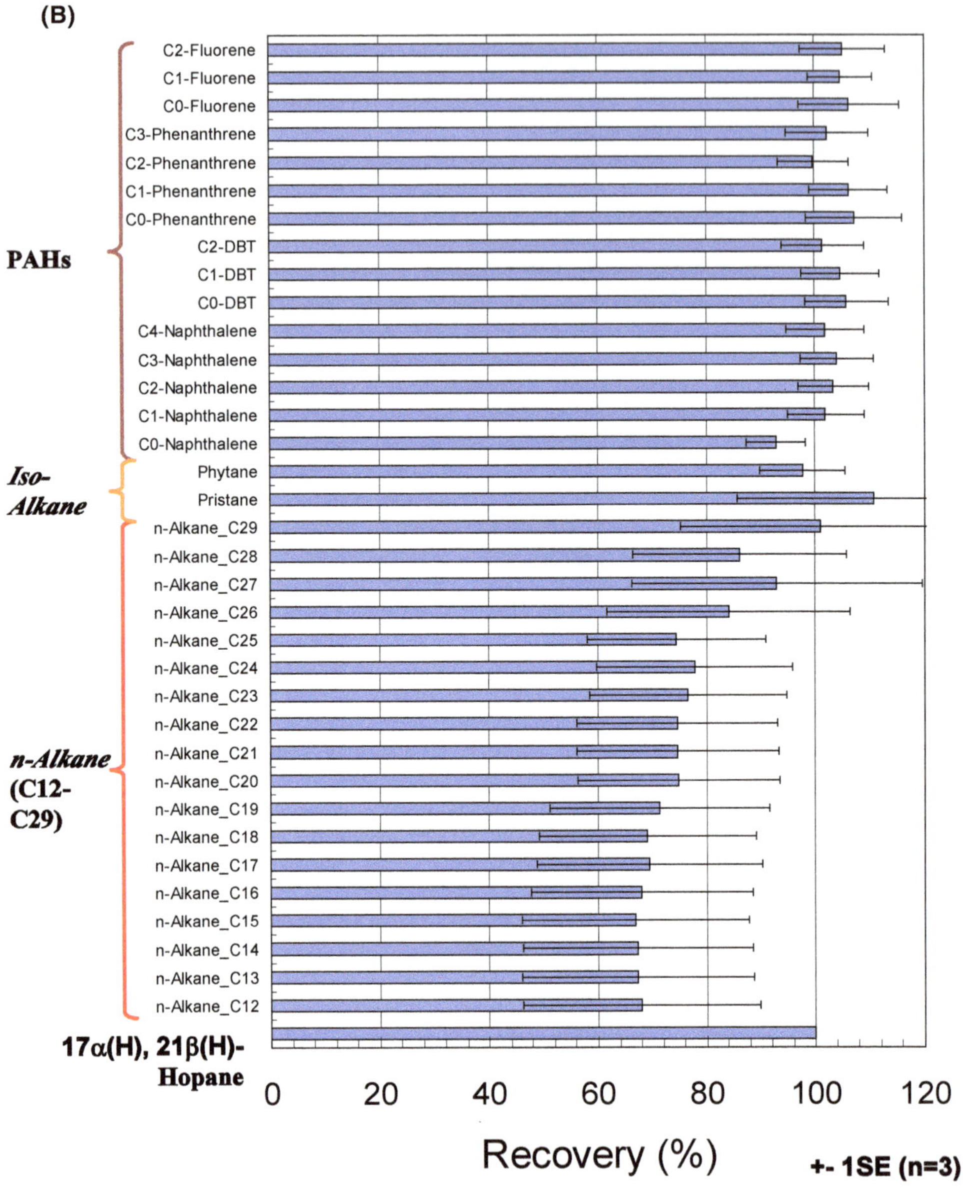

Figure 25.2: Remaining Concentration of Hydrocarbon After Treated with ID06-KWD.13Bt (B).

The treatment was done in triplicate with 30 days incubation period.

Table 25.4: Observation of Oil-Contaminated Oil Treated Using Selected Bacteria (Tanah 33-cp, Tanah 34-cp, Tanah 35-cp and Tanah 38-cp)

Soil Material	*Bulk Density (g cm^{-3})*		*Particle Size (µm)*		*Visual Appearance*	
	Before	*After*	*Before*	*After*	*Before*	*After*
Kalimantan sludge	1.2–1.5	0.9–1.1	1–1.7	0.03–0.62	Oily, sludge, smell of petroleum	Not oily, sandy, no smell of petroleum

CONCLUSIONS

A number of surfactant-producing bacteria were isolated from Indonesian soil and marine environments, and characterized by their capability of degrading hydrocarbon and PAH compounds. Due to the lack references from tropic area, we found several numbers, proposed to be new genus and species. Surfactant properties analysis showed that various types of surfactants were isolated, for instance lipid-, protein-, carbohydrate-containing or their mixtures. Analysis showed that some of these surfactants were able to eliminate hydrocarbon, PAH or both. It is suggested that each strain and its surfactant production potential has specific functions in marine environment. Preliminary application of the bacteria cultures introduced in to the soils resulted in the bacteria which improved the texture and structure of the soil. However, further elucidation and physico-chemical analyses of each surfactant are needed to find out their functions in oil bioremediation in nature.

ACKNOWLEDGEMENT

This research was supported by NEDO-NITE Japan, JICA Japan and Indonesia Institute of Sciences (LIPI) Indonesia.

REFERENCES

Alley JF and Brown LR. 2000. Use of sublimation to prepare solid microbial media with water-insoluble substrate. Appl. And Environ. Microbial. 66 (1); 439-442.

Al-Tahhan RA, Sandrin TR, Bodour AA and Maier RM. 2000. Rhamnolipid-induced removal of lipopolysaccharide from Pseudomonas Aeruginosa: Effect Cell surface properties and interaction with hydrophobic substrates. Appl. And Environ. Microbial. 66 (8); 3262-68.

Bligh, E.G. and Dryer, W.J. 1995. A rapid method of total lipid extraction and Purification. Canada Journal of Biochemistry and Physiology, 37, 911-917.

Bradford, M.M. 1976. A rapid and sensitive method for the quantities of microgram quantities of protein utilizing the principle of protein dye binding. Analytical chemistry, 72, 248-254.

Bodour AA, Drees KP and Maier RM. 2003. Distribution of biosurfactant-producing bacteria in undistributed and contaminated arid southwestern soils. Appl. And Environ. Microbial. 69 (6); 3280-87.

Desai, J.D., and Banat, I.M. 1997. Microbial production of surfactant and their commercial potential. Microbial. And Mol. Biol. Rev., 61, 47-64.

Fietcher A. 1992. Biosurfactants: moving torards industrial application. Trends in Food science and Technology November, 3; 286-292.

Finnerty WR. 1991. Biosurfactant in environmental biotechnology. Curr. Opinion in Biotechnol., 5;291-295.

Garcia-Junco M, Olmedo ED and Ortega-calvo JJ. 2001. Bioavailability of solid and non-aqueous phase liquid (NAPL)-dissolved phenanthrene to the biosurfactant-producing bacterium *Pseudomonas aeruginosa* 19SJ. Environ.Microb 3(9); 561-569.

Harayama S. 1997. Polycyclic aromatic hydrocarbon bioremediation design. Curr. Opinion in Biotechnol. 8:268-273.

Harwati TU (a), Kasai Y, Kodama Y, Susilaningsih D and Watanabe K. 2008. *Tranguilimonas alkanivorans* gen.no., sp.nov., an alkane-degrading bacterium isolated from Semarang Port in Indonesia. Int. Journal of Systematic and evolutionary Microbiology, 58; 2118-2121

Harwati TU (b), Kasai Y, Kodama Y, Susilaningsih D and Watanabe K. 2009. *Tropicimonas isoalkanivorans* gen.no., sp.nov., an branched-alkane-degrading bacterium isolated from Semarang Port in Indonesia. Int. Journal of Systematic and evolutionary Microbiology, 59; 388-391.

Harwati TU (c), Kasai Y, Kodama Y, Susilaningsih D and Watanabe K. 2009. *Tropicibacter napthalenivorans* gen.no., sp.nov., and branched-alkane-degrading bacterium isolated from Semarang Port in Indonesia. Int. Journal of Systematic and evolutionary Microbiology, 59; 393-396.

Kasai Y, Kishira K and Harayama S. 2002. Bacteria belonging to the Genus Cycloclasticus play a primary role in the degradation of aromatic hydrocarbons released in a marine environment. Appl. And Environ. Microbiol. 68(11); 5625-5633.

Lin SC, Minton MA, Sharma MM and Georgiou G. 1994. Structural and immunological characterization of a biosurfactant produced by *Bacillus licheniformis* JF-2. Appl. And Environ. Microbiol. 60 (1); 31-38.

Makkar RS and Cameotra SS. 2002. An update on the use of unconventional substrates for biosurfactant production and their new applications. Appl Microbiol Biotechnol, 58;428-434.

Navon-Vanezia S, Zosim Z, Gottlieb A, Legmann R, Carmelli S. Ron EZ and Rosenberg E. 1995. Alasan, a new bioemulsifier from *Acenitobacter radioresistens*. Applied and Environmental Microbiology, 61(9); 3240-3244.

Noordman WH, Wachter JHJ, de Boer GJ and Janssen DB. 2002. The enhancement by surfactants of hexadecane degradation by *Pseudomonas aeruginosa* varies with substrate availability. Journal of Biotechnology, 94; 195-212.

Oschner UA and Reiser. 1995. Autoinducer-mediated regulation of rhamnolipid biosurfactant synthesis in *Pseudomonas aeruginosa*. Proc. Natl. acad. Sci, USA, 92; 6424-6428.

Pembrey RS, Marshall KC and Schneider PR. 1999. Cell surface analysis techniques; what do cell preparation protocols do to cell surface properties? Applied and Environmental Microbiology, 65(7); 2877-2894.

Radojevic M and Bashkin VN. 1999. Practical environmental Analysis. The Royal Society of Chemistry, Cambridge, UK.

Ron EZ and Rosenberg E. 2001. Natural role of Biosurfactants. Environ. Microbiol. 3(4); 229-236.

Taylor KACC. 1995. A modification of the phenol sulfuric acid method of total sugar determination. Appl. Biochem and Biotechnol, 53; 207-214.

Venkanteswaran K, Kanai S, Tanaka H and Miyachi S. 1999. Vertical distribution and biodegradation activity of oil-degrading bacteria in the Pacific Ocean. J Mar Biotechnol 1; 33-39.

Chapter 26

Resources Use Optimization in Main Food Legume Crops Production in River Nile State (RNS) of Sudan

***Elgilany A. Ahmed*[1], *Hamid H.M. Faki*[1] *and Hashim A. Elobeid*[2]**
[1]*Ministry of Science and Technology, Agricultural Economics Research and Policy Center, Sudan Agricultural Economics and Policy Research Centre (AEPRC), Shambat, Khartoum North, Khartoum, Sudan*
[2]*University of Khartoum, Faculty of Agriclture, Department of Agric.Economics, Shambat, Khartoum, Sudan*

ABSTRACT

In Sudan, the tenants are using numerous crops in order to intensify production in an attempt to improve home food security and income. Seasonal legume crops are regarded as essential crops within the crop combination in RNS namely, faba beans, kidney beans and chick peas. They are a major part of the daily diet for the Sudanese. Moreover, they play an important role in sustaining the productivity of the farming systems through fixation of atmospheric nitrogen. The RNS is considered as one of the main supplier of legume crops to the country. The crops are commonly produced under pump irrigation from the River Nile. The production of seasonal legume crops in the State are facing numerous constraints namely inefficiency in resource use, low productivity level and high production costs. The paper aims to optimize the available resources use in winter seasonal crops vis-à-vis production of food legume crops. A paper was prepared out in RNS to establish resource combination levels that maximize gross margins from food legume crops that commonly grow within the combination. Primary data was collected by using structured questionnaires for (70) randomly selected respondents from Elzeidab scheme public irrigated scheme of RNS as a case study. A linear programming technique through the General Algebraic Modeling

System (GAMS) program was used to assess the optimal combining resources in seasonal legume crops. The model results revealed that tenants can get higher returns by allocating more resources namely land, water, labor and capital to the food legume crop production. Higher net benefits would be from food legume crops production and least from their exclusion. The RNS tenants should therefore, be guided on how to optimally and efficiently utilize their resources and be encouraged to grow food legume crops that give production and yield advantages, earn high returns and contribute significantly to farm sustainability and alleviate malnutrition in RNS.

Keywords: *Resources use optimization, Environmental conservation and improvement.*

INTRODUCTION

Natural resources (*e.g.* water and land) constitute the most important factors of agricultural production. In Sudan, their scarcity and importance rose from population pressure on land. The potential of irrigation for raising both food production and living standards of the rurals has since long been recognized. Generally, a third of the current crop production comes from the one-sixth of irrigated arable land. This reflects the improved reliability of crop production and the greater intensity of land-use follow in the removal of seasonal water supply constants. The problem therefore is how to balance demand and supply of resources under these conditions and the solution lies in resources use optimization.

The trend of farmers in the country towards diverse crop combinations is a dominant practice as means of increasing resource efficiency. This is earned by experience to avoid agricultural risk, such as pests and unfavorable climatic conditions like high temperature, low moisture and others. Many studies report that RNS has a comparative advantage with seasonal cash and food crop production namely, wheat, faba bean, chick pea, dry bean, onion, vegetables, spices, sorghum, maize, potato and fodder beside some perennial crops. This paper focusses on faba beans, kidney beans and chick pea as most important three legume crops in the RNS, because they have a significant role in the diets of the Sudanese people and contribute substantially to the economy of the country. Salih *et al.* (1995) reported that the food legume crops are gaining further importance as a source of protein for the majority of the population who cannot afford animal products because of escalating prices.

All the mentioned seasonal crops are grown in both private and public irrigated schemes of the state, but the public ones are regarded as the main suppliers for these crops on the basis of area and high number of tenants. The production in the public schemes is based on payment of fixed water charges. This paper covers Elzeidab public irrigated scheme as a case study to implement the scheme. The farm management is fully under the tenants' control, while the government is considered as a water seller besides preparing for agricultural policies. Although, the research work is important to assess and recommend for the ideal crop combination, but in RNS a limited research has been conducted to establish resource combination level that will maximize the tenants net returns. In the last three decades lot of research work has been done at RNS, through many organization programs such as International Center for Agricultural Research in the Dry Area (ICARDA) and International Fund for Agricultural Development (IFAD)in collaboration with Agricultural Research and Technology Corporation (ARTC), but the majority of these studies stressed on cereals and legume crops as pure stands and they provided information on production, productivity, breeding improvement, agronomy, microbiology, crop protection, technology transfer and others, but not on the cereal-legume combination particularly for winter season which is the main season in RNS of Sudan. The optimum resource levels in the crop combinations that maximize the tenant' profits have not been adequately assessed. This study therefore aimed to determine the optimal levels of farm resource combination namely, water, land, labor and capital, that might guide the tenants of the RNS in general and tenants of the public irrigated schemes of the State in particular, to make rational economic decision regarding efficient reallocation of the limited resources.

METHODOLOGY

The study was carried in Elzeidab public irrigated scheme of RNS. The crops here are commonly produced under pump irrigation from the River Nile to some extent as well as from undergroundwater. The farming system of the RNS is characterized mainly as not fully mechanized system, the winter season is considered the main season for producing cereal and legume crops. Recently, the State enlarged animal production activities and oil crops. The study depends mainly on primary data which was collected by using structured questionnaires for (70) randomly selected respondents through probability proportional method from Elzeidab scheme public irrigated scheme of RNS as a case study. Integrated techniques and tools were used to assess resource use in Elzeidab scheme of RNS, GAMS, Crop Wat.4 and Excel and SPSS software programs were employed to achieve the objectives of this study. The data collected focussed on the dominant crop combination and the available resource allocation. There are a number of tools that can be used to optimize resource use, one of these being linear programming technique (LP). A General Algebraic Modeling System (GAMS) program was used to assess the optimal combining resources in seasonal legume crops.

LP is known as a mathematical technique for finding the best uses of a firm's limited resources. The adjective "Linear" is used to describe a relationship, which is directly and precisely proportional, "Programming" refers to the use of certain mathematical techniques so as to get the best solution to a problem involving limited resources. In this study, LP technique was used to achieve the optimal solution for legume crops in the crop combination of Elzeidab farming system. From the collected data, the average farm resources, yield and gross margins by feddan were computed and entered in the General Algebraic modeling System (GAMS) software program for optimization analysis. The model was specified with gross margins maximization as the objective function as:

$$Max\ Z = \sum_{j=1}^{n} Cj\ Xj \qquad \text{............(1)}$$

Such that

$$\sum_{j=1}^{n} aijXj \; d'' \; bi, \; all \; j = 1 \; to \; m \qquad \text{............(2)}$$

and

$$Xj \; e\geq 0, \; all \; j = 1 \; to \; n \qquad \text{............(3)}$$

where,

Z: Objective function value.

Xj: Is the level of the jth the farm activity, such as the acreage of wheat grown. Let n denote the number of possible activities; the *j*= 1 to n.

Cj: The objective value, in this case the forecasted gross margin of a unit of the jth activity (SD per feddan).

aij: Quantity of the ith resource available (*.i.e.* cubic meter of water, feddan of land, days of labour or other required quantities of inputs) required to produce one unit of the jth activity.

m: Denote the number of resources; then *i = 1 to m*.

bi: Amount of the ith resource available (*e.g.* cubic meter of water, feddan of land, days of labour or other required quantities of inputs).

The aim was to find the cropping system (defined as a set of activities levels X_j, $j = 1$ *to n*) that has the highest possible total gross margin, *Z*, but doesn't violate any of the fixed resource constraints or involve any negative activity levels.

Equation (1) is the objective function, which maximizes the gross margins from one feddan of legumes crops and other seasonal crops.

Equation (2) shows the limits on the levels of the available resources (*.i.e.* cubic meter of water, feddan of land, days of labour or other required quantities of inputs) that tenant can apply to produce the mentioned crops.

Equation (3) which is a non-negativity condition, states that all resources used in the production process and output must be equal to or greater than zero, meaning that negative use of resources and negative of production is impossible. The coefficients represent the average requirement of the ith activity (enterprise), calculated on per feddan basis.

A scenario analysis was undertaken based on model solution to know the importance of food legume crops in the crop combination through the change in value of the objective function (gross margin) and the allocation of the resources under study over the dominant crop combination.

RESULTS AND DISCUSSION

Legume crops are the most important food source in Sudan. These are major part of the daily diet for the population, providing major portion of the protein requirement in the country. They also play an important role in sustaining the productivity of the farming systems in the country, because of their beneficial affect on the soil through the fixation of atmospheric nitrogen (Ageeb, 1995). The three crops were grown at the same time as other seasonal crops and some perennial crops. The majority of the crops of crop combination are regarded as high profitable crops and that might be one the competition causes particularly under the resource scarcity. The observed increase in world production of cool season food legumes during the 1980s was up to 90 per cent, caused by a tripling of pea production in the developed countries, while in the developing countries the increase in chickpea and lentil contributed only 10 per cent and faba bean remained stagnant.

Figure 26.1 gives the average land allocations per farm for Elzeidab tenants among the adopted crop combination in season 2005–06, and it is obvious that there are numerous choices for Elzeidab tenants to determine the best crop combination. Furthermore, the competition exceeds the traditional seasonal crops to the competition of the legumes with each other. The distribution of legume crop areas within the farm is usually determined by several factors of which market-induced ones are the main. Figure 26.2 shows the distribution of legume crop areas among the tenants' farms.

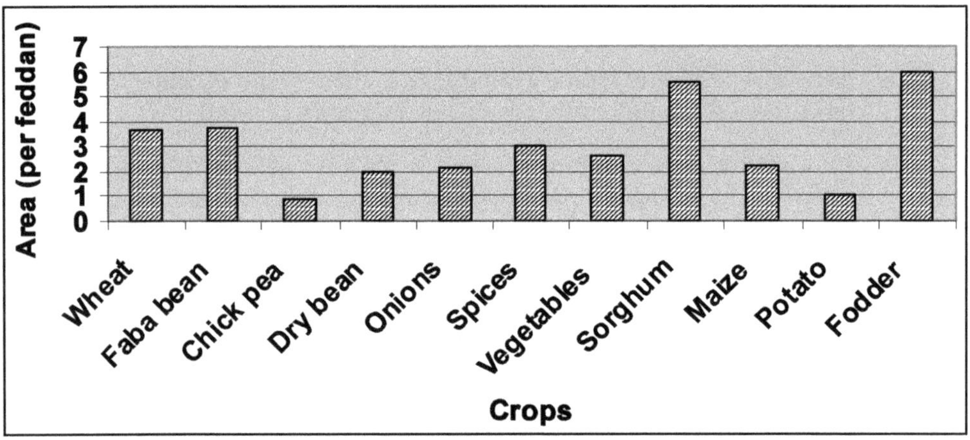

Figure 26.1: Land Allocations per Farm for Elzeidab Tenants 2005–06

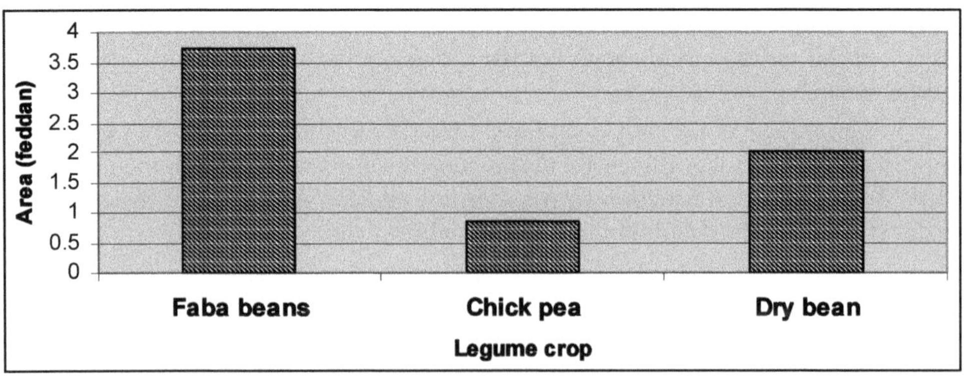

Figure 23.2: Distribution of Legume Crop Areas Among the Tenants' Farms of Elzeidab Scheme

Figure 26.2 shows the variation between the food legume crops of Elzeidab scheme. The tenants of the scheme devoted the greatest portion of their farm to faba bean crop, ranked by dry beans crop while chick peas occupied the smallest area within tenants' farm. This result is not strange, because faba bean is the most important food legume all over the country, and it constitutes the main dish on the breakfast and dinner tables for a large sector of Sudanese people. Although, chick peas is an important cash crop in Sudan and could achieve high profit for the State tenants but historically this crop faces a marketing problem. The production and consumption of this crop fluctuates in the country and during all the year except in the Holy month of Ramadan which witnesses the peak of its consumption by the fasting Muslim population.The dry beans were grown in small areas at the River Nile banks or islands as favourable options of land or/and high terrace land of the scheme, but the last decade witnessed considerable production expansion. The greatest portion of the farm land among the legume crops was devoted for faba bean crop, followed by dry beans crop, while chick pea covered the lowest area as given in Table 26.1.

Table 26.1: Distribution of Elzeidab Surveyed Farmers According to the Area, Yield and Production of the Legume Crops with Compared with ARTC Yield

Crops	*Area (Fed)*	*Yield (kg/fed.)*	*Production (kg)*	*ARTC Yield (kg/fed.)*	*Gab Per cent*
Faba bean	3.73	488.70	1822.85	1500.00	67
Chick pea	0.85	414.00	351.90	1250.00	66
Dry bean	2.00	540.00	1080.00	1200.00	55

Table 26.1 also shows the impacts of adopting the recommendation of Agricultural Research and Technology Corporation (ARTC) and the new technologies on the level of crop productivity. The achieved yields of legume crops under study were generally low when compared to the yields of ARTC. It is obvious, that the actual yield level of legume crops grown by Elzeidab tenants is below the one of the ARTC, that the minimum and maximum yield gap percentage is 67, 66 and 55 per cent for faba beans, chick peas and dry bean crops respectively, indicating a much potential yield gap for the legume crop yield.

OPTIMAL CROP PRODUCTION

The most important season in the Northern Region is winter; hence the tenants pursue the best crop combination to achieve satisfactory returns. According to the importance of the winter season, the resource use and availability might be described as fully utilized during this season. Many study mention that introducing seasonality in the model would further restrict the model solution and will likely lead to lower value of the objective function.

The information received from the model running implies the objective function value (returns), the optimal crop combination, utilized resource accompanied by their respective marginal productivities. The analysis also provides some other relevant results as shown in Table 26.2. It also represents the actual and optimal cultivated area for the crop combination. The model results devoted the optimal land only for chick pea and dry bean as 8.62 and 1.38 feddan respectively, while the rest of the crops did not appear in the optimal plan. The results also enlighted the fact that the actual returns from crop production are SD 399487.28, while the optimal returns are SD 891596.73; which is more than the actual returns by SD 492109.45 and equal to 123 per cent of the actual one.

Table 26.2: Optimal Resource Use and Cropping Pattern Plan for Eleidab Scheme Tenancy

Item	*Actual*	*Optimal*	*Units*
Resource use			
Total land	10	10	Feddan
Total irrigation water	28573	15384.42	Cubic meter (m^3)
Total labour	191	124	Man-day
Total capital	179532	122237.61	SD
Returns: objfn value (Z)	399487.28	891597.73	SD
Cropping pattern			
Wheat	1.1	–	Feddan
Faba bean	1.1	–	Feddan
Chick pea	0.3	8.62	Feddan
Dry bean	0.6	1.38	Feddan
Onions	0.6	–	Feddan
Spices	0.9	–	Feddan
Vegetables	0.8	–	Feddan
Sorghum	1.7	–	Feddan
Maize	0.7	–	Feddan
Potato	0.4	–	Feddan
Fodder	1.8	–	Feddan

RESOURCE USE AND CONSTRAINTS

It was found that on an average tenants had up to 10 feddan, 28573 cubic meters (m^3), 191man-day of labour and SD 179532 (about US$ 870) as capital available to implement the mentioned crop combination in the season. The levels at which these resources were used in the food legume crops were achieved (Table 26.1).

The most important and serious season for RNS of Sudan tenants is the winter season, which is regarded as the successful one, affecting the selection of the best crop combination due to comparative advantage among the agricultural systems. According to the characteristics of the winter season, the resource use and availability might be described as fully utilized during this season. The actual and optimal levels of the resources are presented in Table 26.2 per season. The total optimal area in the plan solution is 10 feddan, equal to 100 per cent of the total available land used in the actual solution. It is clear that under the optimal plan, all available land used and devoted for chick pea as 8.62 feddan and the rest 1.38 feddan for dry bean due to their high returns when compared to other seasonal crops in the study area.

From Table 26.2 the optimal and actual water used are 15384.42 m^3 and 28573 m^3 respectively. The table shows the optimal quantities of water used for the different seasonal crops during this study. Moreover, it represents that increase in water began from October till January and the required amount decreased towards harvest. The mentioned period from October to January known as the most active period during the year in the RNS coincides with the growing of winter cash crops.

It was also obvserved that the optimal plan distributed the available land only for two legume crops namely chick pea and dry bean as 8.62 and 1.38 feddan respectively, but did not include faba bean with other two legumes.

The research designed a scenario based on model solution to confirm the importance of food legume crops in the crop combination. Although, chick pea and dry beans achieved the highest area portions among the dominant crop combination, they are regarded as the important cash crops due to their high returns for the scheme tenants, but unfortunately, they suffer from some marketing problems. The production and consumption of these crops fluctuates all over the country and during long period of the year particularly post-harvest, except the short period for chick pea in the Holy month of Ramadan which witnesses the peak of its consumption by the fasting people. The model assumed the decline of both chick pea and dry bean prices as a dominant phenomenon in the RNS markets, but left faba bean to assess the impacts of low returns on one hand, and see existence impacts of faba bean crop as an important cash crop within the crop combination under study.

Seasonal price variation during the year is a normal feature of agricultural products in RNS. Price of food crops in general follows a common seasonal pattern. Prices decrease to a low level immediately after harvest and rise there-after until the next season as village merchants and assemblers store some supplies to meet the continuous demand (Ijami, 1994). It was detected that the decline of chick pea and dry bean production prices would expectedly lead to a fall in gross margins, but the margin would remain positive.

Table 26.3 shows the results of the new model solution. The model assumed the decline of chick pea and dry bean prices by 65 and 55 per cent of their average prices, respectively. The scenario analysis found that the effect of their price change was influenced by the gross margin, the tenants' returns and all the resources under the study.

The optimal return was found to be SD 254528.7457 which is less than the basic solution by 37 per cent. The optimal levels of the resources used were found to be 8.254 feddan, 22399.143 m^3, 182 Man-day and SD 163490.002 of the total land, irrigation water, labour and capital, respectively. It is clear that the results obtained confirm the importance of food legume crops within seasonal crop combination and their absence will affect the farm sustainability by reducing the tenants' returns by 37 per cent as compared to their actual returns, decreasing the cultivated area by 18 per cent. The results are appointed to unemployed for the other resources as given in Table 26.3.

Table 3. Impact of low prices of chick pea and dry beans.

Item	Actual	Optimal	Units
Resource use:			
Total land	10	8.254	Feddan
Total irrigation water	28573	22399.143	Cubic meter (m^3)
Total labour	191	182	Man-day
Total capital	179532	163490.002	SD
Returns: objfn value (Z)	399487.28	254528.746	SD
Cropping pattern:			
Wheat	1.1	–	Feddan
Faba bean	1.1	3.110	Feddan
Chick pea	0.3	–	Feddan
Dry bean	0.6	–	Feddan
Onions	0.6	1	Feddan
Spices	0.9	0.044	Feddan
Vegetables	0.8	1.048	Feddan
Sorghum	1.7	1.750	Feddan
Maize	0.7	–	Feddan
Potato	0.4	1.302	
Fodder	1.8	–	

The distribution of the cultivated area was found to be 3.110 feddan for faba bean and 1.0, 0.044, 1.048, 1.750, 1.302 feddan for onions, spices, vegetables, sorghum and potato, respectively as shown in Table 26.3. This result indicates faba bean as the most profitable crop as mentioned in the above applied scenario, and it could be one of the best option for the RNS tenants due to its high value. Moreover the food legume crops fix atmospheric nitrogen, which is made available to the non-legume crops resulting into higher yields of the non-legume crops if grown rotationally than if grown as pure stand. Consequently they assist in farm sustainability. It was also oberved that ignoring food legume crops will be irrational decision and it might affect the optimality of resources allocation and hence the farm returns and sustainability.

CONCLUSION AND POLICY IMPLICATIONS

Appropriate combination of land, water, labour and capital resources for producing main food legume crops combined with the other seasonal crops in RNS of Sudan is very important and should be well designed and applied. The policy maker of Elzeidab scheme should consider the economics use of the scarce resources to increase farm gross margins, conserve soil fertility and farm sustainability. This study revealed that the main food legume crops in Elzeidab of RNS are regarded as the most valuable crops when compared to the other seasonal crops. The tenants of the scheme are not optimally allocating the available resources and devoted small portion of their farm land for main food legume crops and hence need to be guided. It was found that chickpea crop is most profitable crop all over Elzeidab crop combination, followed by dry beans and faba beans respectively. The tenants could increase their returns if they allocated their resources optimally and might earn much higher if they devoted more land for main food legume crops. The research work by ARTC confirms the possibility of obtaining high yields. The scenario analysis revealed that the decline of chick pea and dry bean prices by 65 and 55 per cent of their average prices respectively, led to decrease of the tenants' returns and affected all the resources under the study. Extension work is needed to transfer improved technologies to scheme tenants; in addition, the scheme tenants also need to be encouraged to dedicate more on food legume crops as high value crops when designing crop combination.

REFERENCES

Ageeb, O. A. (1995). Production and Improvement of Cool–Season Food Legume in Sudan. In Salih, H. Ageeb, O.A. Saxena, M.C. and Solh, M.B. (eds), ICARDA.

Salih, H.S. and Farouk, A. (1995). Faba bean Improvement Breeding and Improvement. In: Production and Improvement of Cool-Season Food Legume in Sudan. Salih, H. Ageeb, O. A. Saxena, M.C. and Solh, M.B. (eds), ICARDA.

Ijami, A.A. (1994). Efficiency and equity Effects of Market Access on Agricultural productivity in Sudan. A Case study Small Holder along the R.

Chapter 27

An Investigation on the Intestinal Parasites and their Larval Stages on Vegetables and Fruits of Mosul City

Zohair I.F.Rahemo and Mohammad H. Mahmood

Department of Biology, College of Science, University of Mosul, Mosul, Iraq

ABSTRACT

A survey was carried out from September 2003 to August 2004 to detect the types and prevalence of parasitic larvae, eggs and protozoal stages on some fresh vegetables and fruits. A total of 780 samples (500 vegetables and 280 fruits) were collected from five local markets and four vegetable gardens in Mosul city. Two techniques (floatation and sedimentation) were used for examination of samples. The vegetables included: celery (*Apium graveolens*), lettuce (*Lacteal sativa*), leek (*Allium porrum*), onion (*Allium cepa*), radish (*Raphanus sativus*) and carrot (*Daucus carota)* and fruits included: mulberry (*Orus alba*), fig (*Ficus carica),* grape (*Vitis vinifera*), strawberry (*Fragaria grandiflora*) and date palm (*Phoenix dactylifera*).The survey revealed that samples of vegetables and fruits were contaminated with fifteen parasitic stages, including cysts of *Entamoeba histolytica* (2 per cent), *Entamoeba coli*(11.66 per cent), *Iodamoeba butschlii*(3.66 per cent), *Endolimax nana* (2.16 per cent), and *Giardia lamblia* (13 per cent), oocysts of coccidian (22 per cent), trophozoites of soil amoebae(4.16 per cent), eggs of *Taenia* spp.(6.16 per cent), *Hymenolepis nana* (3.83 per cent), *Ascaris lumbricoides* (8.5 per cent), and *Trichuris trichiura* (3.33 per cent), rhabditiform larvae of hookworm (2.5 per cent), and *Strongyloides stercoralis* (3.83 per cent).

***Keywords**: Vegetables, Fruits, Parasites, Mosul-Iraq.*

INTRODUCTION

Intestinal parasites cause severe health problems to human beings and animals thus threatening the social and economic status (Markel *et al.*, 1999).Chang (1994) believed that 1/4 of human population are infected by intestinal worms. Bundy and De-Silva (1998) published an article entitled; "can we deworm this wormy world". In Turkey Ulukanlgil and Seyrek (2003) wrote an article on the "battle against worms". They refer to the ways about deworming activities and difficulties facing this problem in developing societies. Furthermore, climate and other environmental factors such as humidity, temperature, topography all play important role in the distribution of parasites (Roberts and Janovy, 2005). In addition to this inadequate sanitation and using human faeces as fertilizers also play an important role in the parasites distribution. Beuchal (1998) conluded from his study that vegetables and fruits are vehicles for transmission of diseases. Present investigation aims to put forward the contamination with parasites or their larval stages in different vegetables and fruits consumed by peoples in Mosul city (Iraq).

METHODOLOGY

A total of 780 specimens of vegetables and fruits were screened for the parasites or their larval stages. Out of these 500 specimens were vegetables belonging to 8 species namely, celery, parsley, carrot, radish, garden cress, leek, onion and lettuce. The fruit samples were 280 belonging to 5 species namely, mulberry, fig, grape, strawberry, and date palm. Smears were prepared using normal saline (NS) solution, some examined directly or stained with Lugols iodine. Different solutions were prepared and examined by flotation technique using saturated sugar solution, and sedimentation technique (Brown and Neva, 1983). The parasites or their larval stages were estimated and identified by microscopical examination. Duncan multiple tests were used to analyze the number recovered (Al-Rawy, 1984).

RESULTS AND DISCUSSION

As shown in Table 27.1 a total of 15 parasite stages were recovered from the vegetables and fruits available in Mosul region. Among these 5 types were protozoal cysts, coccidian oocysts, soil ameobae, 4 types helminth eggs, in addition to two types of nematode larvae. It was concluded that floatation (FL) method using sugar solution was best as compared to the sedimentation (SD) technique for the detection of occurrence of parasite stages on vegetables and fruits. Similar results have been reported by Brown and Neva (1983), and Felix *et al.* (1996) working on strawberries of Mexico.

Entamoeba histolytica (Protozoa Cysts)

It was less abundant in vegetable and fruits examined in the present study, 2 per cent FL and 1.33 per cent SE, but in other study it was less than that found on leafy vegetables by Opara and Udoidung (2002). Similar percentage was found on strawberry investigated by Flisser (1994). While cysts of *Entamoeba coli, Iodomoeba butschlii, and Endolimax nana* were 11.6, 3.66, 2.16 per cent by FL, it was 7.5, 2.5, 1.5 per cent respectively by SD. Higher percentage was reported on lettuce by Al-Mukhtar and Al-Dabagh (1991).These three intestinal parasites are of less medical importance, as they are commensals (Roberts and Janovy,2005).

Table 27.1: Percentages of the Occurrence of Parasites and Parasitic Stages on Vegetables and Fruits

Name of Parasite Protozoal Cysts	*Method of Diagnosis Sedimentation SD%*	*Method of Diagnosis Flotation FL%*
Entamoeba histolytica	1.33	2
Entamoeba coli	7.5	1.66
Iodomoeba butschlii	2.5	3.66
Endolimax nana	1.5	2.16
Giardia lamblia	9.0	13
Coccidian oocysts	18	22
Soilamoebae	2.5	4.16
Eggs of Helminthes		
Taenia spp.	6.16	5.51
Hymenolepis nana	3.83	3.5
Ascaris lumbricoides	8.5	7.66
Trichuris trichiura	3.33	3.33
Hook worms(rabditiform L)	2.33	2.5
Strongyloides stercoralis(rabditiform L)	3.16	3.5
Hook worms (rabditiform L)	4.0	4.5
Strongyloides stercoralis(filariform L)	5.16	5.83
Total average	6.44	5.09

Cysts of *Giardia lamblia*

These were found in higher rate among protozoan cysts, 13 per cent in FL and 9 per cent in SD. This parasite is more common in children and even in infected animals in Southern Africa (Appleton *et al.*, 1995). Furthermore Roberts and Gjerd(2001) in Norway, found a high percentage of the cysts of this parasite (20 per cent) on lettuce which is more than that reported in the present study.

Soil Amoebae

Free living amoebae were reported as 4.16 per cent in the FL, 2.5 per cent in the SD. These amoebae have power to infect human causing meningitis (Roberts and Janovy, 2005). Atkins and Saygi (2003) isolated these soil amoebae from many types of soils and water in Turkey specially *Acanthamoeba* and *Naegleria* spp. It is known that these amoebae belong to three genera: *Naegleria, Acanthoamoeba, and Hartmanella* which are very distinct from ordinary free living amoebae (Roberts and Janovy, 2005).

HELMINTHS EGGS

Eggs of *Ascaris lumbricoides*

These eggs were more abundant compared to other worms. The percentage was 8.5 in FL, 7.66 in SD. This is a high percentage possibly because human faeces are used as organic fertilizer in farms which make these eggs abundant in soil and in turn on vegetables. In other studies higher percentages have been reported by Opara and Udoidung (2002).

Eggs of *Taenia* or *Echinococcus* spp.

Percentage was 5.51 for FL, 6.16 for SD. The eggs of *Taenia* and *Echinococcus* are very similar and very difficult to be distinguished. Such high percentages are possibly due to the high rate of egg production of *Echinococcus* which are present in the intestine of dogs and they are in close contact with the soil where vegetables and fruits are grown. The rate is less than that found on lettuce examined by Al-Mukhtar and Al-Dabagh (1991).

Eggs of *Hymenolepis nana*

The percentages found were 3.83 in FL, 3.5 in SD. This high rate of occurrence is possibly due to abundance of mice and rats and other rodents on farms in addition to human contamination. The percentage is more or less close to the percentage found on parsley by Shalal (2005).

Eggs of *Trichuris trichura*

The percentage were 3.33 in FL and SD, which is low because these eggs are known to be sensitive to environmental condition such as dryness, direct sun light (Klass, 1987).

Larvae of *Strongyloides stercoralis*

The percentage of rhabditiform larvae was 3.5 in FL, 3.16 in SD, while the filariform larvae were 5.83 in FL, 5.16 in SD.

CONCLUSIONS

The survey revealed that the samples of vegetables and fruits were contaminated with fifteen parasitic stages. In general pollution of vegetables (-FL52 per cent, SD 46.5 per cent) was higher than fruits (FL 17 per cent, SD 15.5 per cent). Fruits are less polluted by parasites and parasitic stages because they are not in close contact with soil or with irrigation water or animal faeces.

REFERENCES

Abdulla, I. A. (2002). The role of some vegetables in transmission of intestinal parasites in Mosul city, Iraq. Iraqi J. Agricult.3 (3): 132-138.

Al-Mukhtar and Al-Dabagh 1991. The occurrence of parasites of public health significance on lettuce and celery sold in local markets in Mosul_Iraq. J. comm.. med 4(2); 167-175.

Al-Rawy, K.M. 1984. Introduction to Statistics. Mosul University Press, Iraq.

Akin, Z. and Saygi, G.2003. Studies on *Acanthamobae* and *Naegleria* spp. Isolated from various water and soil specimens. Acta Parasitol. Tur. 27(2): 117-121.

Appleton, C.C.; Sharp, B.L. and Sauer, D.L.1995. Wetlands and water related parasitic diseases of man in Southern Africa. In: Cowan G.L. (ed) Wetlands of South Africa, Department of Environmental affairs and Tourism, Pretoria (Species Publication): 227-246.

Beuchat,L.R.1998. Surface decontamination of fruits and vegetables eaten raw: a review. Food safety unit, World Health organization WHO/FSF/FOS/98.2.

Brown, H.W. and Neva, F.A.1983. Basic Clinical Parasitology. 3rd ed. Appleton-Century-Crofts, New York.

Bundy, D.A.P. and De-Silva,N.R.1998. Can we deworm this wormy world. Brit. Med.Bull.54 (2): 421-432.

Chan, I.; Bundy, D.A. and Kan, S.P.1994. A small-scale survey of intestinal helminthes infections among the residents near Paske, Laos. Kor. J. Parasitol. 36(1); 55-58.

Felix, N.S.: Wastavino, G.R.; Arteaga, I.H.: Bravo, M.C. and Schettion,P.M.S.1996. Parasite search in Strawberries from Irapuato, Guanajuato and Zamoira, Michoacan(Mexicop). Archives Med.Res.27(2): 229-231.

Flisser,A.1994. Taeniasis and cysticercosis due to *Taenia solium*. In Tsieh Sur. Ed. Prog. In clinical Parasitolol.vol 4, Boca Raton,FL: CRC press; 97

Klass, J.1987. Lumen dwelling helminthes. In JB Horad, Clinical and pathogenic Microbiology.The CV Mosby Comp.Missouri.

Markell, E.K.; John, D.T. and Krotoski, W.A.1999. Medical Parasitology. 8ted. W.B. Saunders Company, Philadelphia.

Opara, K.N. and Udoidung, N.I.2002. Parasitic contamination on leafy vegetables: A function of the leaf are index (LAI). Glob.J. Pure and Applied Sci.9 (1); abstract.

Roberts, L. and Janovy,J.R.2005.Gerald D. Schmidt and L.Karry S.Robersts Foundations of Parasitology.McGraw Hill Seven Edition. McGraw Hill Higher Education.

Robertson, L.J. and Gjerd, B. 2001.Occurrence of parasites of fruits and vegetables in Norway. J. Food Port.64 (11) 1793-1798.

Saygi, G.; Ozcelik, S.; and Poyraz, O.1995. A survey of intestinal parasites in students of adult education center in Sivas, Turkey. J. Egypt.Soc. Parasitol.25 (2); 303-310.

Shalal, M.H. A.2005. Evaluation of Parasitic Pollution in Vegetables and Pasteur herbs in Mosul. M.Sc. thesis, College of Veterany Medicine, Mosul University, 1-65.

Ulukanlgil, M. and Seyrek, A.2003. Battle against worms: difficulties in a developing society. Acta Parasitol. Turc.27 (1) L49-55.

Chapter 28

Effects of Continuous Thermophilic and Mesophilic Treatments on Maturity of Corncob Compost and Reducing of Composting Time

M. *Kianirad*[1], M. *Muazardalan*[2], *Gh. Savaghebi*[2], M. *Farahbakhsh*[2] and S. *Mirdamadi*[1]

[1]Department of Biotechnology,
Iranian Research Organization for Science and Technology (IROST), Tehran, Iran
[2]Department of Soil Science Engineering,
University College of Agriculture and Natural Resources,
University of Tehran, Karaj, Iran

ABSTRACT

This study focuses on the effect of temperature on the compost quality and the processing time. Two temperature levels of 46°C (thermophilic) and 30°C (mesophilic) have been considered. A mixture of 60 per cent ground corn cob, 30 per cent dried sports grass and 10 per cent bio-activator was used as composting material. The composting process lasted for 120 days. The maturity indices at the end of composting indicate that, for thermophilic treatment, the compost shows an improved quality as compared to the mesophilic treatment. The maturity indices, namely carbon to nitrogen ratio, water extractable organic carbon, ratio of cation exchange capacity to total organic carbon and germination index were determined for the high temperature treatment. On the basis of the compost characteristics, the processing time for thermophilic and mesophilic treatments was 75 days and 105 days, respectively. It can be concluded that treatment at the high-temperature level considered in this study presents a more suitable option for improvement of composting.

***Keywords**: Composting, Thermophilic, Mesophilic, Maturity indices, Composting time, Corncob.*

INTRODUCTION

Corn is one of the most widely planted crops in the world. According to FAO (2008), the annual production worldwide is about 695 × 109 kg. The major production regions are American continent (53 per cent), Asia (29 per cent), and Europe (11 per cent). Corn cob is an important by-product of corn production: for every 100 kg of corn grain, approximately 18 kg of corn cobs are produced (Torre *et al.*, 2008, Chiellini *et al.*, 2009). A large quantity of corn cob remains unused as lignocellulosic waste, or is used as animal feed. Owing to their chemical composition, corn cobs show great potential as suitable raw material for producing compost (Arvanitoyannis and Tserkezou, 2008). Corn cob compost, used instead of peat, is useful in amendment of soil properties, horticulture and hydroponic systems (Butler *et al.*, 2008). On the other hand, various environmental factors for improving quality and economy of composting processes, including agroresidual composting, have not been well developed. Moreover, the cumulative volume of agroresidual and lengthy composting time induce many problems such as low quality of compost, high cost of production, etc. Compost maturity is a critical factor affecting compost quality for specific agronomic objectives. Maturity has been a confusing and often subjective issue for producers and growers. This is mainly due to lack of objective and quantitative methods to classify these products; therefore, the maturity index appears to be a useful tool in assessing compost quality (Sanchez and Royse, 2009). Among environmental parameters, temperature is a key factor in composting, which has frequently been used to judge the efficiency and degree of stabilization and maturity (Bueno *et al.*, 2008).

Active composting is typically characterized by thermophilic phase that sanitizes the product and allows a high rate of decomposition followed by a lower temperature phase that allows the product to stabilize while still decomposing at a slower rate. Optimum composting temperatures, based on maximizing decomposition and degradable carbohydrates such as xylan, were found in the range of thermophilic condition (Arvanitoyannis and Kassaveti, 2007; Butler *et al.*, 2008). Some researchers have found that temperatures of this magnitude are not required to produce a high-quality product. Other studies have indicated that lower temperatures might allow more microbial activity, and storage could affect compost maturity (Butler *et al.*, 2001). According to a recent study (Sanchez and Royse, 2009), at high-temperature treatment the chemical fractions of straw mineralized more rapidly as compared to low temperature treatment. Therefore, the fact that issues of compost operating temperature in active and curing stages should be controlled within the low- or high-temperature ranges is a question that remains unsettled. Many tests have been proposed to assess the effect of different ranges of temperature on maturity and rate of composting time. The ratio of carbon to nitrogen (C/N) is the most accepted criterion related to the compost stability, yet it is not suitable for maturity index. The cation exchange capacity (CEC) and the ratio of CEC to total organic carbon (CEC/TOC) have been used to evaluate the degree of maturity and humification of compost (Zmora-Nahum *et al.*, 2005). Also, germination index (GI), water extractable organic carbon (WEOC), and ratio of light emission at 465 nm to that at 665 nm (E4/E6) are the main factors applied for estimating the degree of composting maturity (Bueno *et al.*, 2008).

The main purpose of this investigation was to study the effect of temperature on corn cob composting, the maturity parameters, and the composting time. Two temperature levels of 46°C (thermophilic) and 30°C (mesophilic) have been considered. Results of this study are expected to provide a sound foundation and give new information towards a better control of composting process and recycling of agroresiduals.

MATERIALS AND METHODS

Preparation of Composting Material

The corn cob was dried at 40°C in an oven for 12 h, cut into small pieces (1–3 cm) and ground to pass through a 0.4 mm screen. The composting raw material used in this study was a mixture of 60 per cent ground corn-cob, 30 per cent dried sports grass and 10 per cent bio-activator (processed sheep manure). Weighing 15 kg, the mixture was placed on a tray at a height of 20 cm to provide proper aeration. The moisture range was maintained at 55–60 per cent by constant addition of water during the process. Two temperature levels of 46°C (thermophilic) and 30°C (mesophilic) were used during the process. Aeration was carried out at a rate of 0.4 L min^{-1} and maintained by turning over the composting heap automatically for 4 h day^{-1}. The materials were manually mixed and homogenized every 4 days. The temperature of materials was measured based on a 1-day interval during the active stage process, and a 7-day interval for the matured stage. The materials were allowed to undergo self-heating at two treatment temperatures. Triplicate experimental samplings were also performed during different intervals for physico-chemical analyses. Samples were dried at 60°C in an oven for 24 h and ground to 0.4 mm for the analyses.

Composting Apparatus

The composting process was conducted in a static solid state bioreactor. The system consisted of three trays with two separate sections for mesophilic and thermophilic treatments. Each tray within the reactor had the same volume constructed for schedule pipe and aeration frames. The reactor was equipped with controlling instruments such as heat generator, air blower, humidity meter and three thermometer sensors placed in different locations of the trays.

PHYSICO-CHEMICAL ANALYSES

Chemical Analyses

The pH and electrical conductivity (EC) were determined in 1:10 (w/v) extracts. The TOC was measured using an organic carbon analyzer (Model ELTRA CS 2000). Total nitrogen (T-KN) was measured by the Kjeldahl method (Gerhard Automatic system). CEC was determined with BaC12-triethanolamine following the method of Lax *et al.* (1986). To measure WEOC, 20 g of compost was weighed and extracted with 200 mL of de-ionized water by shaking for 24 h. The extracts were centrifuged at 7000 rpm for 25 min and filtered through 0.45 µm Millipore membranes. The water extracts were immediately analyzed for organic carbon.

UV Spectrophotometry

Following procedure was tentatively adopted according to the method proposed by Zbytniewski and Buszewski (2005). The E4/E6 ratio was determined by dissolving humic substances. A mixture of suitable amounts (1 g) of air dried compost was weighed and added to a small volume (50 mL) of 0.5 M NaOH, shaken for 2 h and centrifuged at 6000 rpm for 25 min. The optical density absorbance of the solution at 465 nm (E4) and 665 nm (E6) was measured using a UV-spectrophotometer (Model Unicam 8620).

Germination Index (GI)

Phytotoxicity was evaluated according to the method proposed by Mari *et al.* (2003). The GI was determined as follows: 20 g of dry weight compost were extracted with 200 mL of distilled water, stirred for 2 h and then centrifuged at 9000 rpm. Ten *Lepidium sativum* seeds were evenly distributed on filter paper in petri dishes (10-cm diameter) and moistened with 5 mL of the compost extract. Three replicate dishes for each sample were incubated at 25°C for 3 days. The numbers of germinating seeds was measured. As a control, 5 mL of distilled water was replaced with the extract at every treatment.

Statistical Analysis

Statistical analysis of the data was performed using SPSS v.13.0 for Windows Statistical Software Package.

RESULTS AND DISCUSSION

Temperature Profiles

Figures 28.1 and 28.2 indicate the effects of temperature profile under continuous thermophilic and mesophilic treatments during the preliminary self-heating and the constant temperature treatment. As expected, the temperature of the composting process began to rise soon after the establishment of composting conditions (Sanchez and Royse, 2009). The temperature in active stage at thermophilic treatment gradually increased to a maximum value of about 51°C on day 35, and remained at the same level till day 39. In the matured stage, the temperature slowly declined to a nearly constant value of 47°C until the end of composting (Figire 28.1). Temperature in the active stage with mesophilic treatment quickly reached a maximum value of about 53°C on day 12, and remained at the same level until day 15.

In the cooling stage, the temperature dropped to about 30°C on days 26–32 (Figure 28.2). Temperature in the matured stage reached 40°C and remained at 30°C until day 120. Differences between the two treatments are presumably due to a higher proportion of readily degradable carbon, and intensive microbial activity at mesophilic compared to thermophilic treatments (Arvanitoyannis and Tserkezou, 2008). The peak temperature in mesophilic treatment during the active stage was 48–53°C, and, at the end of the process, it was concluded that higher organic carbon molecules could not get degraded. In contrast, in thermophilic treatment, microbial thermophiles adapted to the new conditions and hence organic carbon degradation and

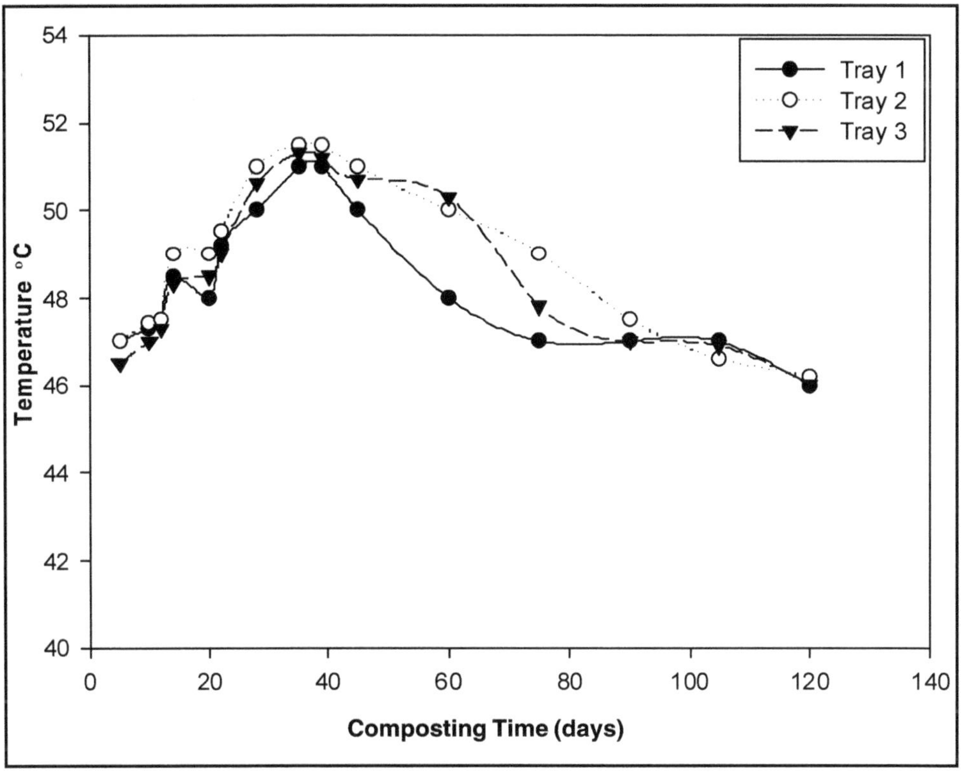

Figure 28.1: Temperature Profiles During Compost Processing at Thermophilic (46°C) Treatment

transformation to humic substances continued (Hachicha *et al.*, 2006; Sanchez and Royse, 2009).

EVALUATION OF COMPOST PROCESSING

The physicochemical properties of both treatments are shown in Table 28.1. EC values increased to 19.74 mS cm^{-1} and 17.86 mS cm–1 for thermophilic and mesophilic treatments, respectively. During the active stage, the ammonification of proteins, amino acids and peptides lead to the release of ammonium ion in the solution phase and contributed to the increase in EC. Moreover, in the active stage, intense mineralization takes place, which is manifested by a considerable increase of EC. The percentage of TOC estimated in both treatments was significantly different as compared to that of the initial stage, while no significant differences were observed in either of the treatments (Table 28.1). However, in thermophilic treatment, TOC decreased more than that in mesophilic case. As can be seen in Table 28.1, the C/N ratio in thermophilic and mesophilic treatments showed a significant ($P < 0.01$) decrease by the end of the composting process 13.9 for thermophilic and 17.69 for mesophilic treatments. Moreover, the final results showed that in thermophilic treatment the C/N ratio reached 15.73 on day 75, but did not stabilize because of continuous decomposition of organic matter.

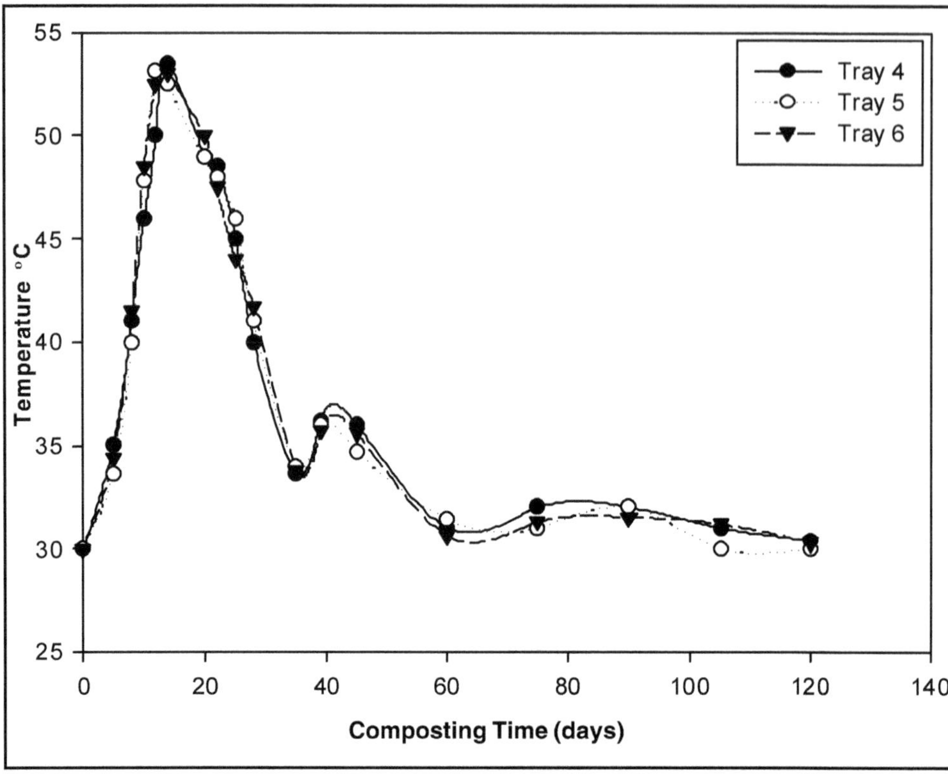

Figure 28.2: Temperature Profiles During Compost Processing at Mesophilic (30°C) Treatment

However, in the case of mesophilic treatment, the C/N ratio stabilized to 17.7 on day 105. Arvanitoyannis and Kassaveti (2007) suggest that a C/N ratio of 20 would reflect a satisfactory degree of composting maturity. In addition, Iglesias-Jimnez and Perez-Garcia (1992) reported that composts are usually considered mature with a C/N ratio of 10–15. Some authors have suggested that changes in the C/N ratio reflect organic matter decomposition and stabilization during composting; therefore, C/N ratio cannot be used as an absolute indicator of compost maturity (Tiquia, 2003; Domeizel *et al.*, 2004; Zbytniewski and Buszewski, 2005, Benito *et al.*, 2006). The water extractable organic carbon (WEOC) consists of phenolic compounds, sugars such as xylan, hemicellulose, organic and amino acids, peptides and degradable organic matter (Zdena *et al.*, 1999). The WEOC in both treatments showed that, by the end of the process, the amount of easily biodegradable organic compounds such as carbohydrates, produced at the initial stage, decreased to 1.69 and 1.48 per cent organic carbon for mesophilic and thermophilic treatments respectively (Table 28.1).

The results show that WEOC reached 1.64 per cent organic carbon on day 75 which was not significant at the end of the process ($P < 0.01$), whereas in thermophilic treatment, WEOC decreased to 1.75 per cent organic carbon on day 105. In matured

Table 28.1

Treatment	*Stage*	*pH*	*EC*	*TOC*	*TK-N*	*C/N*	*CEC*	*WEOC*
	Initial	6.4±0.05	8.3±0.3	45.5±0.7	1.2±0.1	38.03±0.5	9.56±2.1	4.21±0.4
	Active stage	8.3±0.1	16.4±0.9	38.5±0.35	1.44±0.11	26.79±0.75	30.06±2.5	2.87±0.1
	End active stage	8±0.12	19.16±1.1	37.27±0.8	2±0.11	18.65±0.5	59.26±2.8	1.87±0.05
Thermophile	Matured	7.53±0.21	20.32±0.87	36.02±0.56	2.29±0.09	15.73±0.76	77.08±2.7	1.64±0.07
	End process	7.3±026	22.72±1.4d	34±0.45	2.45±0.13	13.9±0.83	75.14±2.5	1.48±1
	Active stage	8.85±0.15	15.39±0.67	39.04±0.72	1.05±0.07	37.2±0.64	15.62±1.5	2.56±0.15
Mesophile	End active stage	7.63±0.25	16.72±0.8	37.2±0.9	1.74±006	21.4±0.58	35±2.2	2.29±0.08
	Matured	6.41±0.31	17.65±1.2	35.93±1.1	2.03±0.12	17.7±1	62.15±2.8	1.75±0.09
	End process	6.53±0.34	17.83±1.15	35.9±1	2.08±0.1	17.3±1.1	64.62±3	1.69±0.05

compost, soluble organic carbon is present as humic substances, which are resistant to further degradation, thus explaining the increased stability of the compost. A decrease in water-soluble carbon is often used as an indicator of compost maturity (Garcia *et al.*, 1991). Nevertheless, a limit of WEOC < 1.7 per cent can be used as a good criterion to define the degree of maturity (Bernal *et al.*, 1998). Cation exchange capacity (CEC) of organic matter is an indicator of maturity (Stevenson, 1994). The CEC values for the two treatments increased significantly during composting (Table 28.1).

Towards the end of the process, very high CEC values of 75.14 cmole kg^{-1} and 64.62 cmole kg^{-1} were reached at high-and low-temperature treatments respectively. Although the value of CEC at low-temperature treatment was more than that of the high-temperature treatment during active composting, the value in the high-temperature treatment composting reached much earlier during the matured stage. In processed composts, the remaining carbon compounds in the compost have more functional groups (*e.g.* carboxyl groups) that can adsorb cations. The maturity process is influenced by the increased oxidation of organic matter that leads to a rise in CEC. The proposed CEC value used as an index of maturity is more than 60 cmolc kg–1 (Butler *et al.*, 2001). Therefore, our results underscore the importance of temperature in changing the CEC and consequently predicting the rate of humification or maturity. In order to compare the maturity of different kinds of composts, an index such as CEC/TOC, which eliminates the errors incurred due to the different total organic carbon can be used. Therefore, CEC/TOC has been used in this work to study the maturity of the compost. In thermophilic treatment, CEC/TOC was found to be 2.21 at the end of process, whereas in mesophilic treatment this ratio was 1.8 (Figure 28.3). The established minimum value of 1.8 for the CEC/TOC ratio could be considered as a base for maturity index (Bernal *et al.*, 1998). Changes in the E4/E6 ratio for both treatments are presented in Figure 28.3.

For both treatments, E4/E6 ratio decreased and reached its minimum on day 15, reaching its maximum value at the active stage. Finally, the lower ratio of E4/E6 was achieved in thermophilic and mesophilic treatments, stabilizing on days 75 and 90, respectively. As a matter of fact, in mesophilic treatment, the active stage is the phase of intensive microbial attack, characterized by rapid loss of readily decompostable organic substances, which in the matured stage is characterized by gradual decomposition of more resistant compounds such as lignin and aromatic compounds (Stevenson, 1994). Our results for the thermophilic treatment show large values of the E4/E6 ratio associated with the active stage.

However, a lower ratio of E4/E6 was noticed in the matured stage. Based on the E4/E6 ratio in the humic substance, it was concluded that this ratio could be used to evaluate the degree of humification and maturity. According to Wang *et al.* (2004) a low ratio of E4/E6 is indicative of a high degree of condensation of aromatic compounds, whereas a high E4/E6 ratio displays a low degree of aromatic condensation, indicating the presence of a large proportion of aliphatic structures. The germination index (GI) is one of the most important parameters for evaluating toxicity and the degree of compost maturity. As shown in Figure 28.4, in active stage both composts showed low GI values which increased during the composting process

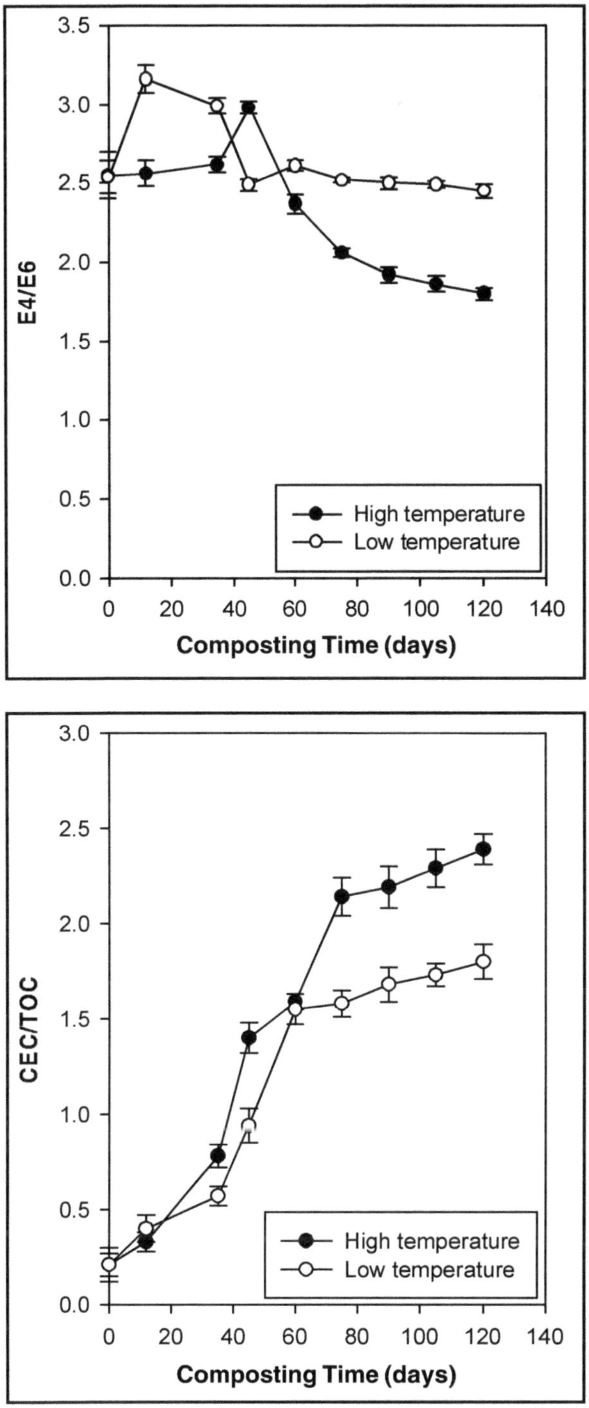

Figure 28.3: Evaluation of E4/E6, CEC/TOC and in Thermophilic and Mesophilic Treatments at Different Stages of Composting

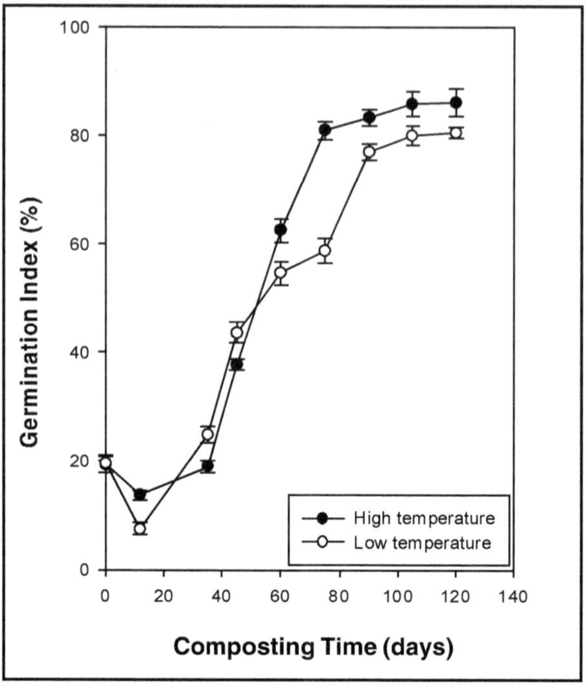

Figure 28.4: Evaluation of Germination Index (per cent) in Thermophilic and Mesophilic Treatments at Different Stages of Composting

due to decomposition and metabolism of the phytotoxic organic compounds. It has been reported that a GI > 80 per cent indicates the disappearance of phytotoxins in compost (Tiquia, 2003). Our findings indicate that GI increased during matured stage of composting to values of 80.5 and 86 per cent in mesophilic and thermophilic treatments on day 120 respectively. Therefore, the percentage GI in high-temperature treatment is significant ($P < 0.05$) compared to the low-temperature treatment.

COMPOSTING TIME

The composting time could be changed while the compost was incubated in different temperature levels, according to the changes in physiochemical characteristics. Most compost should be actively processed for 90-120 days to be considered stable. Our study showed that the best results of maturity indices such as C/N, WEOC, CEC/TOC, E4/E6 and GI have been obtained in thermophilic treatment on the 75th day, whereas that of mesophilic treatment stabilized on the 105th day ($P < 0.01$). The additional processing time, which results in compost production is referred to as 'compost maturity' (Butler *et al.*, 2001). The C/N ratio usually decreases during the composting process and, consequently, is sometimes used to indicate compost stability. The correlation between compost characteristics and composting time for the two temperatures were examined further through stepwise linear regression analysis. The R^2 values of compost characteristics versus composting time are presented in Table 28.2. Results revealed that further compost characteristics are correlated to mesophilic and thermophilic treatments. Except the E4/E6 ratio, other useful compost characteristics in mesophilic treatment significantly decreased ($P < 0.05$), whereas in thermophilic treatment the above values were more significant ($P < 0.01$). Therefore, the duration of composting process in thermophilic treatment should be predicted.

Table 2: R 2 Values of Compost Characteristics Versus Composting Time

Compost Characteristic	*Thermophilic*	*Mesophilic*
pH	0.023	0.358
EC	0.88	0.52
TOC	0.918	0.612
N	0.928	0.799
C/N	0.917	0.789
CEC	0.894	0.881
CEC/TOC	0.936	0.9
WEOC	0.886	0.693
E4/E6	0.672	0.351
GI	0.961	0.935

R^2 >0.5 were significant at p<0.05 and R^2 >0.65 were significant at p<0.01

EC: electrical conductivity (mS cm^{-1}); TOC: total organic carbon (per cent); CEC; cation exchange capacity (cmole kg^{-1}); WEOC: water-extractable organic carbon (per cent organic carbon) All determinations were carried out in triplicate and Dunkan test at P <0.05 were used to determine the signicant dierences between treatment measures.

CONCLUSIONS

Composting time and quality of compost are important factors influencing the economy of the composting process, and are strongly affected by the treatment temperature. Two temperatures of high and low level have been considered in the treatment of corn cob composting. The thermophilic treatment was the best choice for improvement of the process, due to increased activities of thermophilic microbes at the matured stage. Under this condition, the lignocellulose compounds show higher degradation, resulting in reduced C/N compared to the mesophilic treatment. Additionally, greater amounts of water extractable carbon is utilized by micro-organisms. The phytotoxic compounds are greatly reduced during the thermophilic treatment, resulting in higher germination index. The transformation of the organic carbon under thermophilic conditions produces more functional groups, which cause higher CEC/TOC ratio. Moreover, the thermophilic treatment shortens the composting time due to the faster onset of the matured stage compared to the mesophilic treatment. Based on the findings of this work, corn cob could be considered as a promising raw material for compost manufacture. More investigations should be carried out to determine the economical benefits of the results of this study.

ACKNOWLEDGEMENTS

Mehran Kianirad gratefully acknowledges financial support from the Department of Soil Science Engineering of University College of Agriculture and Natural Resources, University of Tehran. The composting process was conducted in the Department of Biotechnology at IROST.

REFERENCES

Arvanitoyannis, I.S. and Kassaveti, A., 2007. Current and potential uses of composted olive oil waste. International Journal of Food Science and Technology, 42, pp. 281–295.

Arvanitoyannis, I.S. and Tserkezou, P., 2008. Corn and rice waste: a comparative and critical presentation of methods and current and potential uses of treated waste. International Journal of Food Science and Technology, 43, pp. 958–988.

Bernal, M.P., Paredes, C., Sanchez-Monedro, M.A. and Cegarra, J. 1998. Maturity and stability parameters of composts prepared with a wide range of organic wastes. Bioresource Technology, 63, pp. 91–99.

Butler, T.A, Sikora, L.J., Steinhilber, P.M. and Douglass, L.W., 2001. Compost age and sample storage effects on maturity indicators of biosolids compost. Journal of Environment Quality, 30, pp. 2141–2148.

Butler, T.J., Han, K.J., Muir, J.P., Weindorf, D.C. and Lastly, L., 2008. Dairy manure compost effects on corn silage production and soil properties. Agronomy Journal, 100, pp. 1541–1545.

Chiellini, E., Cinelli, P., Ilieva, V.I., Imam, S.H. and Lawton, J.W.,2009. Environmentally compatible foamed articles based on potato starch, corn fiber, and poly(vinyl alcohol). Journal of Cellular Plastics, 45, pp. 17–32.

Domeizel, M., Khalil, A. and Prudent, P., 2004. UV spectroscopy: a tool for monitoring humification and for proposing an index of the maturity of compost. Bioresource Technology, 94, pp. 177–184.

FAO., 2008 FAOSTAT statistics database http://faostat.fao.org/site/567/DesktopDefault.aspx?PageID=567.

Hachicha, S., Chtourou, M., Medhioub, K. and Ammar, E. 2006.Compost of poultry manure and olive mill wastes as an alternative fertilizer. Agronomy Sustainable Development, 26, pp. 135–142.

Iglesias-Jiménez, E. and Perez-Garcia, V., 1992. Determination of maturity indices for city refuse composts. Agronomy Ecosystem Environment, 38, pp. 331–343.

Lax, A., Roig A. and Costa, F., 1986. A method for determining the cation exchange capacity of organic materials. Plant and Soil, 94, pp. 349–355.

Mari, I., Ehaliotis, C., Kostsou, M., Balis, C. and Georgakakis, D., 2003. Respiration profiles in monitoring the composting of by-products from the olive oil agro-industry. Bioresource Technology, 87, pp. 331– 336.

Sanchez, J.E. and Royse, D.J., 2009. Scytalidium thermophilum-colonized grain, corncobs and chopped wheat straw substrates for the production of *Agaricus bisporus*. Bioresource Technology, 100, pp. 1670–1674.

Stevenson, F.J.,1994. Humus chemistry. Genesis, composition, reactions.New York: Wiley, pp. 496.

Tiquia, S.M., 2003. Evaluation of organic matter and nutrient composition of partially decomposed and composted spent pig litter. Environment Technology, 24, pp. 97–107.

Wang, P., Changa, C.M., Watson, M.E., Dick, W.A., Chen, Y. and Hoitink, H.A.J., 2004. Maturity indices for composted dairy and pig manures. Soil Biology and Biochemistry, 36, pp. 767–776.

Zbytniewski, R. and Buszewski, B., 2005. Characterization of natural organic matter (NOM) derived from sewage sludge compost Part 1: chemical and spectroscopic properties. Bioresource Technology, 96, pp. 471–478.

Zmora-Nahuma S., Markovitcha, O., Tarchitzkyb, J. and Chena, Y., 2005. Dissolved organic carbon (DOC) as a parameter of compost maturity. Soil Biology and Biochemistry, 37, pp. 209–211.

Chapter 29

Green House Gas Emmisions from Cement Manufacturing Industry and Mitigation Measures: Tanzania Country Report

Dugushilu Mafunda

Principal Research Officer, Tanzania Commission for Science and Technology, P.O. Box 4302, Dar es salaam, Tanzania

PREAMBLE

The United Republic of Tanzania is located on the East coast of Africa between parallels 1° S and 12° S and meridians 30° E and 40° E. It extends from Lake Tanganyika in the West, to the Indian Ocean in the East; Lake Victoria in the North, to Lake Nyasa and River Ruvuma in the South. It borders Kenya and Uganda in the North, Rwanda and Burundi in the North-west, Democratic Republic of Congo to the West, Zambia to the South-west and Malawi and Mozambique to the South. Mainland Tanzania has an area of about 939,702 square kilometers and the Islands of Zanzibar and Pemba, in the Indian Ocean, occupy an area of 2,643 square kilometers. The average temperature is between 20° C and 32° C. Average annual rainfall approximately ranges from 600 mm to 1,800 mm per year under normal conditions and depending on the altitude of a place from sea level. The average duration of the dry season is 5 to 6 months. Tanzania is one of the largest countries in Africa with a total population of 34.6 million. The population is growing at an average rate of 2.1 per cent per annum (Tanzania census, 2004).

BACKGROUND INFORMATION

Green house gases (GHG) are gases in an atmosphere that absorb and emit radiation within the thermal infrared range. This process is the fundamental cause of the greenhouse effect. The main greenhouse gases in the Earth's atmosphere are, carbon dioxide, methane, nitrous oxide, water vapour and ozone. Human activities since the start of the industrial era around 1750 have increased the levels of greenhouse gases in the atmosphere. The 2007 assessment report compiled by the IPCC points out that increase in anthropogenic greenhouse gas concentrations is very likely to have caused most of the increases in global average temperatures since the mid-20th century. Measured atmospheric concentrations of carbon dioxide are currently 100 ppmv higher than pre-industrial levels. Carbon dioxide, methane, nitrous oxide and three groups of fluorinated gases (sulfur hexafluoride, HFCs, and PFCs) are the major greenhouse gases and the subject of the Kyoto Protocol, which came into force in 2005. Of these CO_2 takes the major share.

Main sources of greenhouse gases from human activities are:

- ✰ Burning of fossil fuels.
- ✰ Land use change (mainly deforestation in the tropics) account for up to one third of total anthropogenic CO_2 emissions.
- ✰ Livestock enteric fermentation and manure management.
- ✰ Use of chlorofluorocarbons (CFCs) in refrigeration systems, and use of CFCs and halons in fire suppression in systems and manufacturing processes.
- ✰ Agricultural activities, including the use of fertilizers, that lead to higher nitrous oxide (N_2O concentrations) and paddy rice farming (higher methane concentration-CH_4).

There are seven sources of CO_2 from fossil fuel combustion.

1. Solid fuels (*e.g.*, coal): 35 per cent.
2. Liquid fuels (*e.g.*, gasoline, fuel oil): 36 per cent.
3. Gaseous fuels (*e.g.*, natural gas): 20 per cent.
4. Flaring gas industrially and at wells: <1 per cent.
5. Non-fuel hydrocarbons: < 1 per cent.
6. Cement production: 3 per cent.
7. Shipping and air transport: 4 per cent.

The TIRDO (Tanzania Industrial Research and Development Organization) Report concentrates on CO_2 emission from cement production.

Cement Production as a Source of GHG's

The raw materials for Portland cement production are a mixture (as fine powder in the 'Dry process' or in the form of a slurry in the 'Wet process') of minerals containing Calcium oxide, Silicon oxide, Aluminium oxide, Ferric oxide, and Magnesium oxide.

In the clinker, the raw mixture is heated in a cement kiln, a slowly rotating and sloped cylinder, with temperatures increasing over the length of the cylinder up to a peak temperature of 1400-1450 °C. A complex succession of chemical reactions take place as the temperature rises. The peak temperature is regulated so that the product contains sintered but not fused lumps. Sintering consists of the melting of

25-30 per cent of the mass of the material. The resulting liquid draws the remaining solid particles together by surface tension, and acts as a solvent for the final chemical reaction in which alite is formed. Too low a temperature causes insufficient sintering and incomplete reaction, but too high a temperature results in a molten mass or glass, destruction of the kiln lining, and waste of fuel. The resulting material is clinker. On cooling, it is conveyed to storage. The enthalpy of formation of clinker from calcium carbonate and clay minerals is ~1700 kJ/kg. However, because of heat loss during production, actual values can be much higher. The high energy requirements and the release of significant amounts of carbon dioxide make cement production a great concern for global warming.

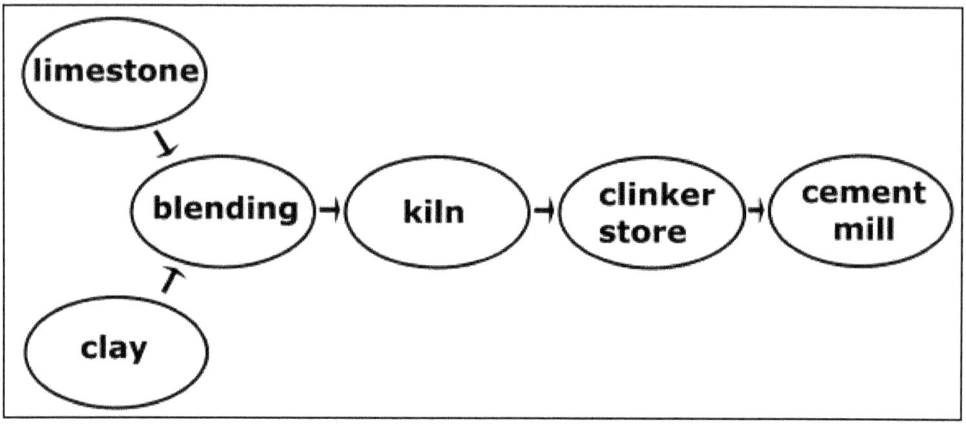

Figure 29.1: Cement Production Process

Cement manufacturing can cause environmental impacts at all stages of the process. These include emissions of airborne pollution in the form of dust, gases, noise and vibration when operating machinery and during blasting in quarries, consumption of large quantities of fuel during manufacture, release of CO_2 from the raw materials during manufacture. Equipment to reduce dust emissions during quarrying and manufacture of cement is widely used, and equipment to trap and separate exhaust gases are coming into increased use.

Tanzania Initiatives on Climate Changes

Tanzania ratified the UN Framework Convention on Climate Change (UNFCCC) and subsequently acceded to the Kyoto Protocol in April, 1996 and February, 2003, respectively. In order to meet its obligations under the climate change convention, Tanzania has implemented several climate change related project activities, including two National Communications to the UNFCCC and the Technology Needs Assessment for Climate Change. The country has also put in place a number of environmental and sectoral policies and plans which are intended to increase its ability to cope with current environmental problems as well as with additional risks posed by climate change. The National Environmental Action Plan (NEAP) of 1994 was a first step towards incorporating environmental concern into national planning and development. The National Environmental Policy (NEP) was promulgated in

1997 to provide a framework for mainstreaming environmental considerations into the decision making process in the country. In particular, the NEP highlights the importance of integrating environmental management in several Sectoral programmes and policies.

In view of Tanzania's vulnerability to climate variability and change, an in depth assessment of impacts of climate change and climate variability need to be undertaken. The results will ensure that, strategies are evolved to ensure that options for adaptation and mitigation which are pursued do not unduly sacrifice national development endeavours. This Country Report (CEEST Report) is part of these initiatives. Centre for Energy, Environment, Science and Technology (CEEST) is undertaking the on-going Tanzania National Project on climate change.

Modern cement use cannot be divorced from population, urbanization and constructions. Expansion of urban areas in many towns and cities influences the increase of cement production for construction. This is exacerbated by the increase of population of many towns and cities which increases the demand for cement for construction and hence CO_2 emissions. However cement production is one of the emitters of CO_2. As an effort to balance between human development in towns and cities and environment, the main objective of this Country Report (CEEST Report) is to establish the levels of CO_2 emission from cement production and analyze potential mitigation measures and identify appropriate technological options in the cement manufacturing industry.

PREVIOUS COUNTRY REPORTS: AN OVERVIEW

Energy audit carried out by TIRDO in over 150 industries/companies in the country noted that energy use in industry is inefficient and is characterised by low awareness on energy management; existence of poor housekeeping to the energy utilizing equipment; poor power factors; and existence of inefficient equipment and systems.

The Report noted that good housekeeping measures will ensure proper operation, upkeep and maintenance of energy consuming equipments. Typical energy saving from good housekeeping will have a maximum of 10 per cent of the total energy consumed by the equipment.

The house keeping measures as identified by TIRDO involve:

- ☆ Switching off motors, lights, air conditioners, when they are not in use.
- ☆ Cleaning heat transfer surfaces of air conditioners. Dust deposit impairs its heat transfer efficiency up to 20 per cent.
- ☆ Dusting off lighting fixtures and replacement of burnt ones. Burnt fluorescent tube lights will continue to consume over 10 per cent of its rated power.

The Report also noted that power factor to electrical equipment is poor. High power factor is a way of utilizing energy efficiently and it is beneficial to both the Tanzania Electrical Supply Company (TANESCO) and to their consumer. TANESCO requires consumers to restrict their reactive power factor above 0.9. Low power factor causes a large drop of volts in generators and ultimately high generation cost.

Power factor correction is defined as:

1. $(kVA)^2 = (kW)^2 + (kVAr)^2$ *(Eqn.1)*
2. $\text{Cos}\,(\varphi) = kW/KVA$ *(Eqn.2)*

where,

kW: Active power

kVAr: Reactive power

Cos (φ): Existing power factor

kVA: Maximum demand

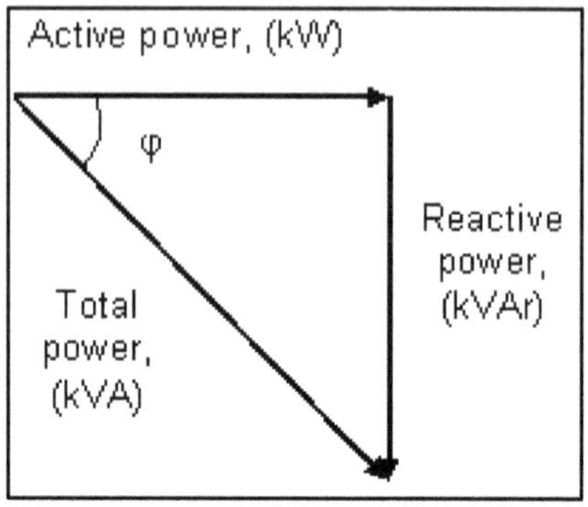

Inefficient Electric Equipment and Systems

The TIRDO Report noted that there exists inefficient equipments and systems in both TANESCO as well as electricity consumers. This increases transmission and distribution losses. High rating of transformers and cables are therefore required. It further noted that during the electricity consumption low voltage results in inefficient operation of equipment such as motors and lighting to consumers as well as high monthly demand (kVA) charges. Main loss types from motors are core loss; stator and rotor resistance, friction and windage loss and stray load loss. This causes a large voltage drop across cables. The Report points out for a need for extra voltage regulating/stabilizing equipment to be installed.

PRODUCTION TRENDS IN INDUSTRY SECTOR

Traditional Tanzania manufacturing industry sector constitutes mainly of food processing factories (coffee, tea, sugar, maize, cigarettes, beverages, etc.) and export crops (tobacco, pyrethrum, sisal twine, textile, etc.). Other industries are mining, wood and wood products, leather, metal engineering, cement etc. TIRDO Report shows that the beer and textile had the highest production (1996-2004). Generally, industry production trend in the country stagnated from 1986-2004 (Figure 29.1) for most industries especially Aluminum, corrugated iron, iron sheets, paints, pyrethrum, wood products, fish nets, sisal, cigarette, and breweries.

However, the production for beer and textile industries remarkably increased from 1995 to 2004 (Figure 29.2). On the other hand the industrial production for gold and gemstone stagnated between 1975-1986 (Figure 29.3) and started to increase between 1986-2002 (Figure 29.3).

ENERGY CONSUMPTION TRENDS IN INDUSTRY SECTOR

The on-going CEEST Project has established that industry consumes about 46.04 per cent; domestic sector consumes 34.81 per cent while transmission losses takes

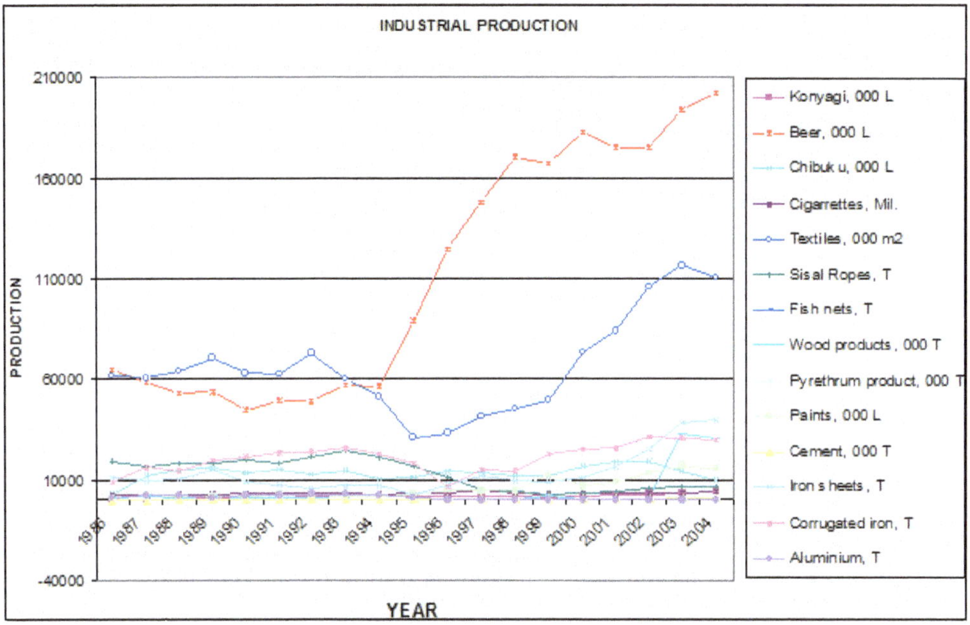

Figure 29.2: Industrial Production for Various Factories, 1996–20004 [CEEST, 2009]

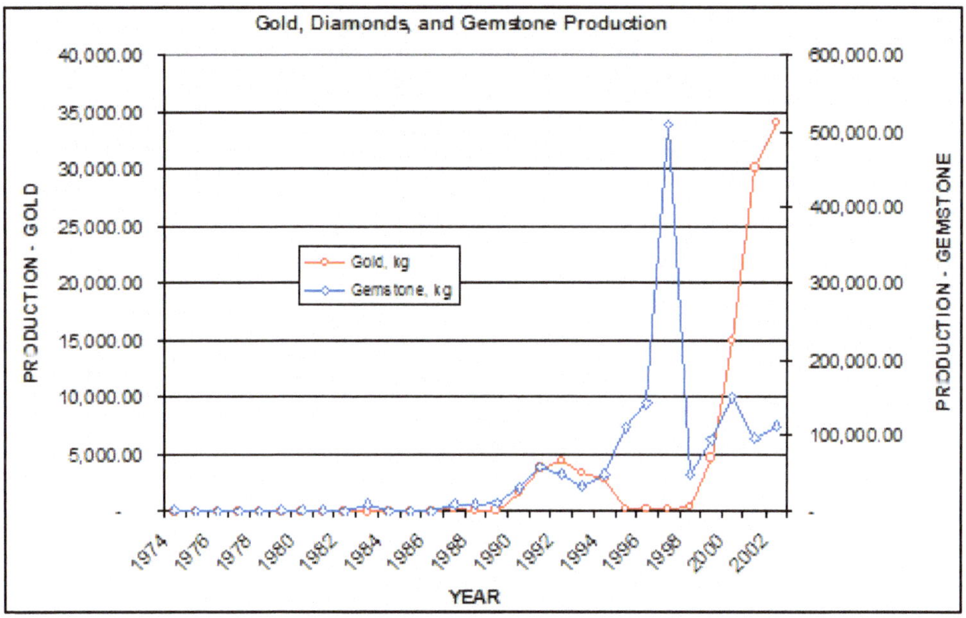

Figure 29.3: Industrial Production for Gold and Gemstone, 1975-2002 [CEEST, 2009]

19.15 per cent of the electricity generated. The project noted that industry consumption has increased from 395.00 to 861.7 Mil. KWh between 1975–2002 (Table 29.1, Figure 29.1) while domestic consumption increased from 91-1, 2000.29 Mil.KWh between 1975–2002 (Table 29.1, Figure 29.1), a remarkable transmission loss between 72-714 Mil. KWh between 1975-2002 (Table 29.1). Overall consumption has increased from 558 -2,796 Mil. KWh

Table 29.1: Electricity Consumption, Mil. KWh [Source: CEEST, 2009]

Year	*Domestic*	*Industrial*	*Transmission Loss*	*Total*
1975	91.00	395.00	72	558.00
1980	163.00	575.00	59	797.00
1985	230.65	592.35	229	1,052.00
1990	391.86	912.14	325	1,629.00
1995	730.55	900.45	246	1,877.00
2000	1,268.44	644.56	571	2,484.00
2001	1,142.03	872.97	727	2,742.00
2002	1,220.29	861.71	714	2,796.00
Per cent Share	34.81(per cent)	46.04 (per cent)	19.15 (per cent)	?

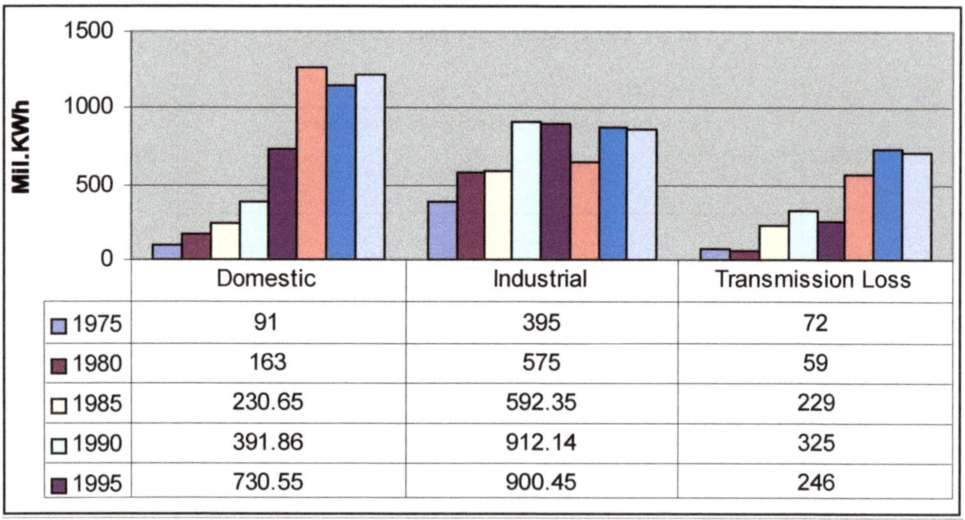

Figure 29.4: Electricity Consumption, Mil KWh

Petroleum Consumption

The CEEST Report noted that LPG consumption has been decreasing from 3.85 02.7 Gg for year 1995–2002 and IDO has also been decreasing from 38.63–11.82 for year 1995–2002. However FO has remained almost steady (116.97–100.77Gg) during the same period. Overall, fuel consumption has decreased from 159.46 to 118.29 between 1995 to 2005 (Table 29..2). The Project further noted that petroleum consumption of industry is 78.8 per cent from Fuel Oil (FO) and 19.24 per cent from Industrial Diesel Oil (IDO) and 1.87 per cent for LPG (Table 29.2). Generally, the Project established that industry consumes 55 per cent of all LPG imported in Tanzania are consumed by industries and almost all the imported IDO and FO.

Table 29.2: Petroleum Consumption [Source: CEEST, 2009]

Fuel Type	*Consumption, Gg*								*Total*	*Share*
	1995	*1996*	*1997*	*1998*	*1999*	*2000*	*2001*	*2002*		
LPG	3.85	4.92	1.54	1.92	2.05	1,80	2.00	2.70	20.77	1.87
IDO	38.63	33.25	23.89	18.56	37.65	20.99	25.73	14.82	213.52	19.24
FO	116.97	114.12	105.67	114.38	109.57	114.28	99.97	100.77	875.73	78.89
Total	**159.46**	**152.29**	**131.10**	**134.85**	**149.26**	**137.08**	**127.69**	**118.29**	**1,110.02**	**100.00**
Average								**138.75**		

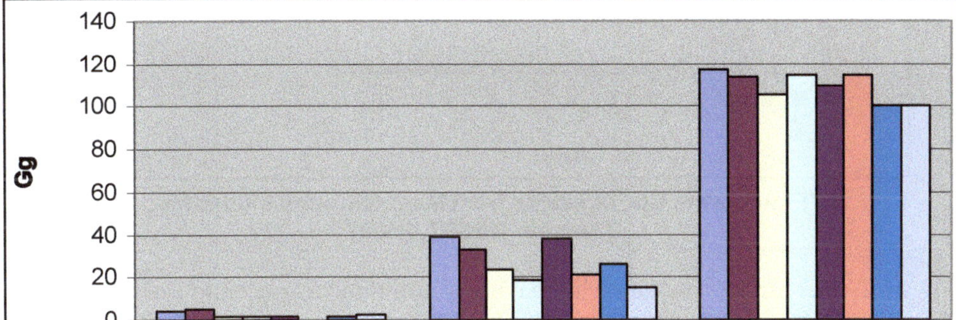

	LPG	IDO	FO
1995	3.85	38.63	116.97
1996	4.92	33.25	114.12
1997	1.54	23.89	105.67
1998	1.92	18.56	114.38
1999	2.05	37.65	109.57
2000	0	20.99	114.28

Figure 29.5: Petroleum Consumption, Gg

It was established that electricity, FO, IDO, LPG are major energy sources consumed in the manufacturing industries. The production in manufacturing industries is influenced by energy consumption and increases with increase of energy consumption.

GHG EMISSIONS IN CEMENT INDUSTRY

Emissions of CO_2 from petroleum oil consumption from various manufacturing and construction industries have generally increased between 1995–2005

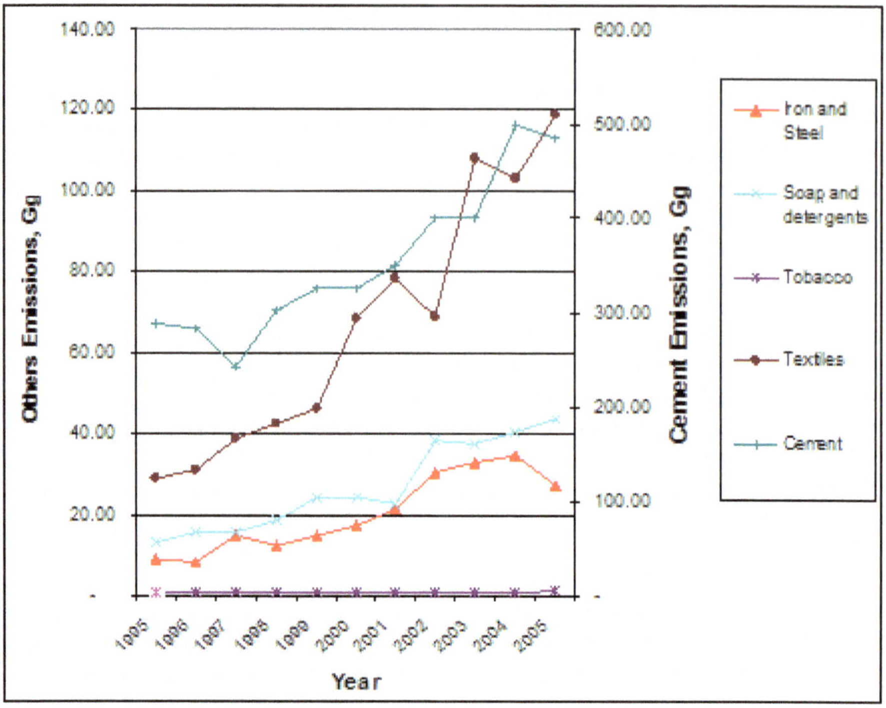

Figure 29.6: CO_2 Emission from Manufacturing Industries [Source: CEEST, 2009]

(Figure 29.6), with a remarkable increasing trend in the cement and textile factories. CO_2 emission from cement has increased from 280–490 Gg during 1995–2005 followed by textile industry which increased from 30 to 120 Gg. The emission for other manufacturing industries was generally lower than these industries.

CO_2 from Clinker Production

The CEEST Project points out that the proportion of energy use in the whole cement processing is the highest in clinker production (80 per cent), followed by finish grinding (12 per cent) and material preparation (crushing, proportioning and drying) (8 per cent). The high energy use in clinker can be reduced by using efficient technologies. However, there are various efficient technologies for rotary kilns. CO_2 emission from clinker is registered as 411.5 Gg which can be reduced to 106.99 Gg by using efficient technologies. The Project quantifies an overall emission saving by efficient technologies as 26 per cent.

Emission from Electric Consumption

The CEEST Project shows that GHG emissions increased from 197.50 to 430.86 tCO_2 between 1975-2002 (Table 29.4), with electricity consumption increasing from 395-900.45 Gg during the same period. Figure 6 shows that levels of emissions increase with increase of electricity consumption. This therefore, shows that GHG emission is highly influenced by energy utilization efficiency.

Table 29.3: CO_2 Emission from Clinker Production [Source: CEEST, 2009]

Measure	*Value*
Cement production in year 2000, MT	833,000.00
Clinker fraction in cement	0.95
Clinker production in year 2000, MT	791,350.00
CO_2 emission factor for the clinker in the Portland cement, Gg CO_2/T clinker	0.000520
CO_2 emission from the clinker production, Gg	411.50
CO_2 reduction by efficient technologies	106.99

Table 29.4: Electricity Consumption [Source: CEEST, 2009]

Year	*Electricity Consumption (kwH)*	*Emission tCO_2*	*Emission Gg CO_2*
1975	395.00	197,500.00	197.50
1980	575.00	287,500.00	287.50
1985	592.35	296,176.14	296.18
1990	912.14	456,068.19	456.07
1995	900.45	450,225.53	450.23
2000	644.56	322,281.91	322.28
2001	872.97	436,483.84	436.48
2002	861.71	430,857.09	430.86

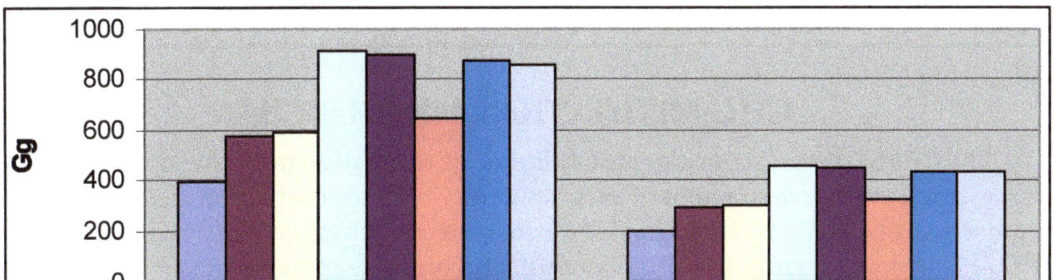

	Electricity Consumption	Emission $GgCO_2$
1975	395	197.5
1980	575	287.5
1985	592.35	296.18
1990	912.14	456.07
1995	900.45	450.23

Figure 29.7: GHG Emission from Electricity Consumption (Gg)

Emissions from the Use of FO Vs Natural Gases

The project shows that there is reduction in the GHG emission potential when natural gas is used. It is around 38.66Gg for CO_2, 0.00363 Gg for CH_4 and 0.000181 Gg for N_2O (Table 29.5). So far only Twiga Cement Industry uses natural gas while Mbeya and Tanga cement Industries use FO/IDO. Retrofitting Mbeya and Tanga cement by natural gas fuel would reduce the CO_2 emission accordingly since natural gas has low CO_2 emission factor compared to IDO/FO

Table 29.5: GHG Emission Reduction Potential from Switching from FO/IDO Use of Natural Gas [Source: CEEST, 2009]

Measure Description	*Value*
Emission level from use of fuel oil, Gg	
Emission from fuel oil use, Kg CO_2 (at 77, 400 kgCO_2 per TJ fuel oil)	140.46779
Emission from fuel oil use, Kg CH_4 (at 3 CH_4 per TJ fuel oil)	0.005444
Emission from fuel oil use, Kg N_2O (at 0.2 kg N_2O per TJ fuel oil)	0.000363
Emissions level from use of natural gas, Gg	
Expected emission from natural gas use, Kg CO_2 (at 56, 100 kgCO_2 per TJ natural gas)	101.811926
Expected emission from natural gas use, Kg CH_4 (at 56, 100 kg CH_4 per TJ natural gas)	0.001815
Expected emission from natural gas use, Kg N_2O (at 56, 100 kg N_2O per TJ natural gas)	0.000181
Emission reduction potential from use of natural gas, Gg	
CO_2	38.655865
CH_4	0.00363
N_2O	0.000181

GHG MITIGATION MEASURES

The CEEST Project considers the followings mitigation measures for industrial sector; good energy management practices, power factor correction, use of efficient motors, use of efficient cement manufacturing processes and energy switching from IDO/FO to natural gases in cement manufacturing.

Good Energy Management Practices

Good energy management practices in industrial sector include: switching off equipment that are not in use, proper upkeep of the equipment, curbing leaks of compressed air systems and steam, proper turning of burners, and managing load management. Majority of these measures require minimal or no investment for their implementation. Good housekeeping practices have instant energy saving of up to 10 per cent of the total energy used.

Efficient Motor Application

Motors account for 75 per cent of all the electrical energy consumed in industry. Typical motor losses from premium efficient motors are given in Table 29.4. GHG emissions reduction by using efficient motors at a combined margin grid emission factor of 0.5 tCO_2/MWh) is 6.91 Gg CO_2.

Power Factor Correction

The CEEST Project shows that new power factor of 0.98 will reduce CO_2 emissions by 108.52 tCO_2 with a monthly reduction of 35,395,792.98 kVA (Equivalent to 217,047 MWh) (Table 29.6).

Table 29.6: GHG Emission Reduction [Source: CEEST, 2009]

Measure	*Value*	
After Improvement		
New power factor	0.98	
New maximum demand	69,718,986.18	kVA
New reactive load	13,873,903.08	kVAr
Size of capacitor bank	66,006,413.34	kVAr
Cost of capacitor bank	6,864,666,987,703.58	Tsh/kVAr
Monthly kVA reduction	35,395,792.98	kVA
Monthly cost saving	271,131,774,253.53	Tsh
Yearly cost saving	3,253,581,291,042.32	Tsh
Simple pay back period, years	2.11	years
Green House Emission Reduction		
Maximum demand reduction	35,395,792.98	kVA
Equivalent MWh	217,047.00	MWh
Grid emission factor	0.5	tCO_2/MWh
Emission reduction	108,523.50	tCO_2

Use of Efficient Cement Manufacturing Process

Though the annual capacity of Twiga factory is 1,200 MT; Mbeya factory (300,000 MT) and Tanga factory (750,000 MT), only a total of 924,000 MT/year of Portland cement are produced annually. This shows that the production of cement factories is below their capacities. All the three factories use conversional efficient technologies of dry process. It is only Twiga factory which uses natural gas. The other two cement factories use Industrial Diesel Oil and/Fuel Oil (Table 29.7).

GHGs Mitigation by Fuel Switching in Cement Production

The Project shows that the mitigation measures reduce CO_2 from 140.46779 to 101.811926, CH_4 from 0.005444 to 0.001815 and N_2O from 0.000363 to 0.001815. Generally, the identified mitigation measures show that by applying the proposed mitigation measures there will be a significant reduction of CO_2 by 42.45 per cent, followed by CH_4 (29.47 per cent) and N_2O (14.87 per cent) (Table 29.8).

Table 29.7: Typical Efficient Technologies for Rotary Kilns in Cement Factories [Source: CEEST, 2009]

Measure	Fuel Savings (GJ/T)	Electric Saving (KWh/T)
Convert pre-heater kiln to pre-calciner kiln	0.16-0.7	-
Upgrade long dry kiln upgrade to pre-heater/pre-calciner kiln	1.4	-
Upgrade older dry kiln upgrade to multi-stage pre-heater kiln	0.9	-
Convert to reciprocating grate cooler	0.27	3
Improve combustion system	0.1–0.5	-
Indirect firing	0.015–0.022	-
Optimize heat recovery/upgrade clinker cooler	0.05 -0.16	2
Seal replacement	0.011	-
Low temperature heat recovery for power	-	20-35
High temperature heat recovery for power	-	22
Low pressure drop cyclones	-	0.7–4.4
Efficient Kiln Drive	-	0.55 -3.9

Table 29.8: Fuel Switching

Fuel	Fuel Oil	Natural Gas	Reduction Potential	Per cent
CO_2	140.46779	101.811926	38.655865	42.45
CH_4	0.005444	0.001815	0.00363	29.47
N_2O	0.000363	0.001815	0.000282	14.87

CONCLUDING REMARKS

The CEEST Project establishes that, cement manufacturing sector is the major emitter of GHGs. The Project identifies GHG mitigation measures as follows:

- ☆ Energy management;
- ☆ Power factor correction;
- ☆ Use of efficient motors;
- ☆ Use of efficient technologies in cement production; and
- ☆ Fuel switching from FO/IDO to natural gases in cement producing factories.

The Project further shows that on implementing the switching from fuel oil to natural gases in cement factories, it is possible to cut down CO_2, CH_4, and N_2O emissions by 42.45, 29.4, and 14.88 per cent respectively. The overall effectiveness of cutting down the CO_2 emissions from the industry and manufacturing sector lies in increasing efficiency of the cement production industry.

RECOMMENDATIONS

Maximization of the effectiveness in GHG's mitigation measures in Tanzania requires:

- ✰ More emphasis on improving efficiency in the existing three cement factories.
- ✰ Retrofitting the Tanga Cement and Mbeya Cement Companies for using natural gas.
- ✰ Adherence to good energy management practices.

REFERENCES

Luhanga, M.L., Mwandosya, M.J. and Nyenzi B.S. (1997), Vulnerability and Adaptation to Climate Change Impacts in Tanzania. *CEEST Research Report Series.*

United Republic of Tanzania (2003), *Initial National Communication under the United Nations Framework Convention on Climate Change (UNFCCC).* Vice President's Office, Dares Salaam, Tanzania

United Republic of Tanzania (2005): *National Strategy for Growth and Reduction of Poverty (NSGRP).* Vice President's Office, Dares Salaam, Tanzania.

United Republic of Tanzania, (2003): *Vulnerability Assessment Report,* Prime Minister's Office, Dares Salaam, Tanzania.

URT, 2004: Tanzania Census.

www.ingramcontent.com/pod-product-compliance
Ingram Content Group UK Ltd.
Pitfield, Milton Keynes, MK11 3LW, UK
UKHW022216051026
15624UKWH00001B/192